22nd International Laser Physics Workshop

(LPHYS'13)

Journal of Physics: Conference Series Volume 497

Prague, Czech Republic
15 – 19 July 2013

ISBN: 978-1-63266-722-9
ISSN: 1742-6588

Printed from e-media with permission by:

Curran Associates, Inc.
57 Morehouse Lane
Red Hook, NY 12571

Some format issues inherent in the e-media version may also appear in this print version.

Printed by Curran Associates, Inc. (2014)

For permission requests, please contact the Institute of Physics at the address below.

Institute of Physics
Dirac House, Temple Back
Bristol BS1 6BE UK

Phone: 44 1 17 929 7481
Fax: 44 1 17 920 0979

techtracking@iop.org

Additional copies of this publication are available from:

Curran Associates, Inc.
57 Morehouse Lane
Red Hook, NY 12571 USA
Phone: 845-758-0400
Fax: 845-758-2634
Email: curran@proceedings.com
Web: www.proceedings.com

TABLE OF CONTENTS

Journal of Physics: Conference Series **497** (2014) 011001 doi:10.1088/1742-6596/497/1/011001

Preface

Dear Readers,

The 22nd annual International Laser Physics Workshop, LPHYS'13, took place in the City of Prague, the Czech Republic. The conference gathered 387 participants from 34 countries. It was hosted by the Institute of Physics at the Academy of Sciences of the Czech Republic and the Czech Technical University in Prague.

At this occasion, the LPHYS'13 Steering and the Advisory & Program Committees of the conference would like extend its sincere gratitude to Professor Miroslav Jelínek, Co-Chair and Head of the Local Organizing Committee and his team for the outstanding job performed on organizing, arranging, managing and putting in order the conference that lead to its successful resulting conclusion. As a result of scientific seminars at the Workshop, we would like to present to you the following conference proceedings published in this volume of the IOP Journal of Physics: Conference Series.

Please be advised that the 23rd annual International Laser Physics Workshop (LPHYS'14) will take place from 14–18 July, 2014 in Sofia, Bulgaria, hosted by Institute of Electronics at the Bulgarian Academy of Sciences.

LPHYS'13 Steering and Advisory & Program Committees

Peer review statement

All papers published in this volume of *Journal of Physics: Conference Series* have been peer reviewed through processes administered by the proceedings Editors. Reviews were conducted by expert referees to the professional and scientific standards expected of a proceedings journal published by IOP Publishing.

Recent developments in femtosecond filamentation

A Houard, Y Liu and A Mysyrowicz

Laboratoire d'Optique Appliquée, ENSTA ParisTech, Ecole polytechnique, CNRS, Palaiseau, F-91762, France

E-mail: andre.mysyrowicz@ensta-paristech.fr

Abstract. We review recent developments in the field of femtosecond laser filamentation.

1. Introduction

We review recent developments in the field of femtosecond filamentation. In a first part, we describe the physical effects contributing to induce filamentation during the propagation of intense short laser pulses through transparent media. The importance of group velocity dispersion is illustrated in the case of fused silica. In the normal dispersion regime, the pulse evolution during propagation is complex and depends critically on input parameters such as the beam convergence, pulse duration, laser intensity. By contrast, a simple reproducible behavior is obtained in the anomalous dispersion region. The filamentary pulse is then closely parented to a lossy spatio-temporal soliton.

In a second part, we discuss the interaction between two air filaments in the normal dispersion regime. Interference between the fields of the crossing pulses leads to the formation of a plasma grating. We show how one can characterize the plasma density, its decay and the mechanism for this decay. Both ambipolar diffusion and plasma recombination play a role in molecules, whereas the decay is dominated by ambipolar diffusion in atoms. We show how energy can be exchanged between two crossing filaments due to scattering by a moving plasma grating. We also discuss the interaction between two counter-propagating filaments. The two filaments act as a retro-reflecting mirror: a third probe pulse of same frequency, incident at an arbitrary angle on this mirror, is retro-reflected with a spectacular improvement in beam profile quality and contrast ratio (ratio between precursor light and the pulse peak intensity).

In a third part, we discuss two applications making use of filaments in air. The first concerns the generation of THz pulses. A laser pulse undergoing filamentation emits a radially polarized THz pulse along a forward oriented cone. We discuss the origin of this effect and describe methods to enhance the THz emission by several orders of magnitude. This includes the application of an electric field, either static or optic, or the constructive THz emission from several filaments. The second application is the triggering of guided electric discharges by filaments and their use as RF antennas.

2. Filamentation in the normal/anomalous dispersion regime

Ultrashort, intense laser pulses are now commonly available, thanks to the development of laser-amplifiers systems based on Ti: Sapphire as the active medium and chirped pulsed amplification (CPA) technique. Pulses at 800 nm emitted by such lasers have typically a duration between 10 and 100 fs. A strongly non linear propagation regime sets in when such pulses, even with a modest energy of a few millijoules, are launched through atmosphere. This nonlinear propagation is complex and involves a subtle interplay between many effects. No analytical or semi-analytical treatment can be used to describe it. One must resort to a numerical treatment. This is usually done by solving an extended non linear Schrödinger equation for the evolution of the pulse envelope treated as a scalar quantity. It allows a proper description in the paraxial approximation for rapid temporal changes, down to the single optical cycle limit, but fails to describe fast spatial changes, on a scale comparable to the wavelength. We refer to references [1, 2] for a description of the numerical code and its limits.

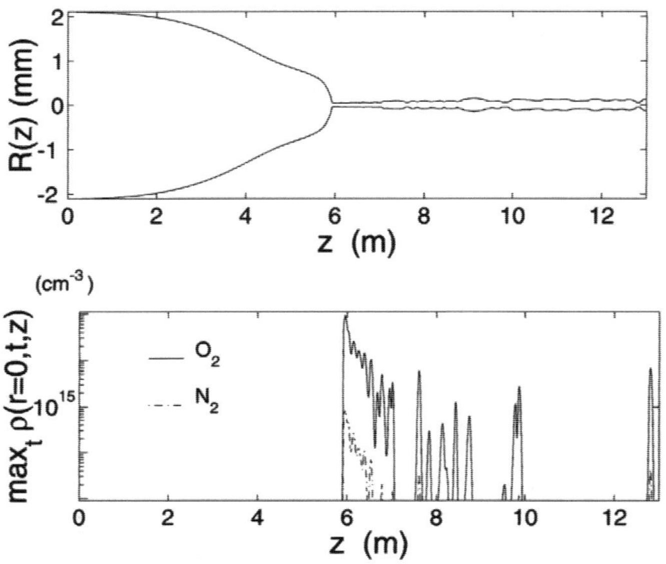

Figure 1. (a) Beam width of a collimated femtosecond laser pulses during filamentation propagation in air, obtained from numerical simulation. (b) The corresponding plasma density as a function of propagation distance.

Figure 1 shows the result of a simulation for the propagation in air of an initially collimated pulse of Gaussian profile (pulse duration: 100 fs, pulse energy: 5 mJ, beam diameter: 40 mm, wavelength: 800 nm). The first effect coming into play is the optical Kerr effect. It induces a change of the medium refractive index according to the law:

$$n \cong n_0 - n_2 \, I(r,t). \tag{1}$$

This leads to beam self focusing, because the index is larger around the beam center, where the intensity is highest, than on the wings and therefore bends the pulse wavefront. The effect is cumulative upon propagation because the focusing reinforces the change of n which in turn increases the intensity. This self focusing overcomes the defocusing effect of diffraction and leads to a collapse of the beam upon itself, if the initial pulse peak power exceeds a critical value P_{cr} [3-5] where

$$P_{cr} \; = \; 3.72\lambda_0/8\pi n_0 n_2. \tag{2}$$

In the example given, the initial region dominated by the optical Kerr effect extend over a distance of 6 meters at which point the beam diameter decreases abruptly and the pulse intensity increases correspondingly.

When the collapsing pulse intensity reaches a value above 10^{13} W/cm², multi-photon ionization of air molecules sets in. The creation of free electrons leads to a defocusing effect. Indeed, the refractive index decreases according to the law:

$$n \cong n_0 - \frac{\rho(r,t)}{2\rho_c}, \tag{3}$$

where $\rho(r,t)$ is the density of free electrons and $\rho_c = \varepsilon_0 m e \omega_0^2 / e^2$ the critical plasma density above which the plasma becomes opaque (m_e and e denote the electron mass and charge). The plasma of highest density is created at the center of the beam and acts therefore as a defocusing lens. The combined dynamic interplay between beam self focusing and defocusing leads to the formation of a contracted pulse with an intense ionizing core that maintains a radius of ~ 50 μm over large distances. The filament core is surrounded by a reservoir of laser energy that sustains the core by supplying an inflow of energy. Figure 1(b) illustrates the action of the plasma. It surges and acts to prevent collapse when the pulse intensity reaches ~ 10^{15} W/cm². It first gives rise to a continuous meter long plasma string, followed by a quasi-periodic cycles of focusing and defocusing. At larger distances, plasma surges only intermittently. In the regions between and beyond these occasional surges of ionization, the nonlinear propagation is dominated by a dynamic competition between self focusing and diffraction (bright channels formation). It should be stressed that the behavior shown in figure 1 is not generic but depends on initial beam conditions (beam size and convergence, pulse energy, pulse duration, etc). With multi TW lasers, continuous plasma strings have been observed at a distance of 1 km with the plasma string extending over more than 50 m [6]. Bright channels persist over several km. [7-8]

There is a profound restructuring of the pulse time profile during filamentation. Figure 2 shows the calculated evolution of an initial converging pulse with initial Gaussian profile upon propagation in Argon. A complex pulse splitting pattern takes place, which evolves during propagation and leads to the fortuitous emergence of isolated compressed pulses at certain distances. Pulses of initial duration 35 fs have been shown to compress down to 5 fs during filamentation in Argon [11]. It should be stressed again that the location and duration of such shortest isolated pulses depends on initial conditions.

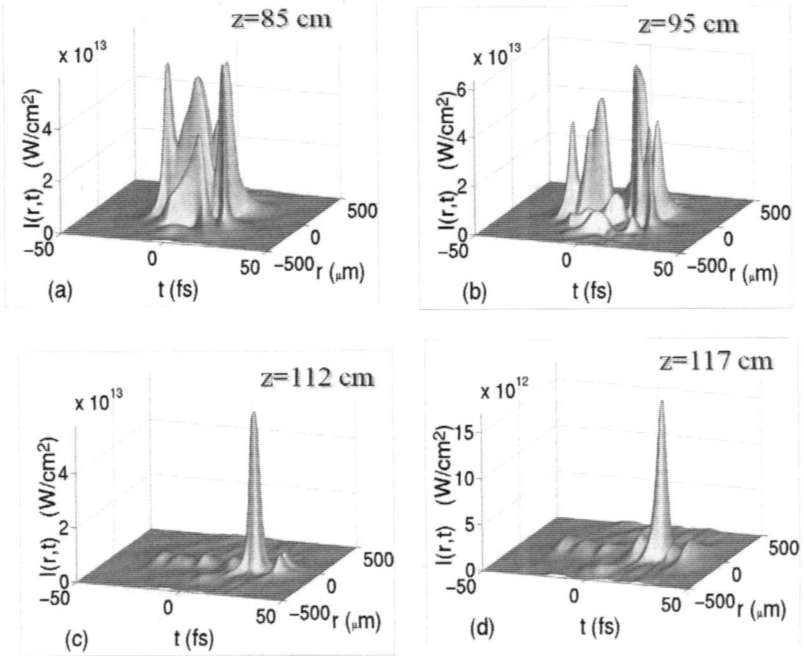

Figure 2. Calculated evolution of an initial converging femtosecond pulse with initial Gaussian profile upon propagation in Argon.

A spectacular effect of filamentation is beam profile self-cleaning. It occurs in the region dominated by the Kerr effect, during the catastrophic collapse. The filament core acquires a Townes mode profile that is retained in the bright channels emerging after the end of the ionization region [12-13].

We now address the influence of group velocity dispersion on the filamentation process. One can readily understand why it plays an important role. Consider the evolution of the pulse spectrum in the self-focusing region. Because of self-phase modulation [14], instantaneous frequencies are created that shift to the red (blue) the ascending (descending) part of the pulse, according to the following formula:

$$\omega(t) = -\frac{\partial \Phi}{\partial t} \approx \omega_0 - \frac{n_2 \omega_0}{c} z \frac{\partial I(r,t)}{\partial t}. \tag{3}$$

In the normal dispersion region, red frequencies move faster than blue frequencies and therefore there is a detrimental pulse stretching during propagation. In the anomalous dispersion region, by contrast, red frequencies have a slower velocity than blue components. The generated frequencies are swept back to the main pulse. As a consequence, one expects lower threshold for filamentation and more stable filamentary pulses in the region of negative group velocity dispersion (GVD).

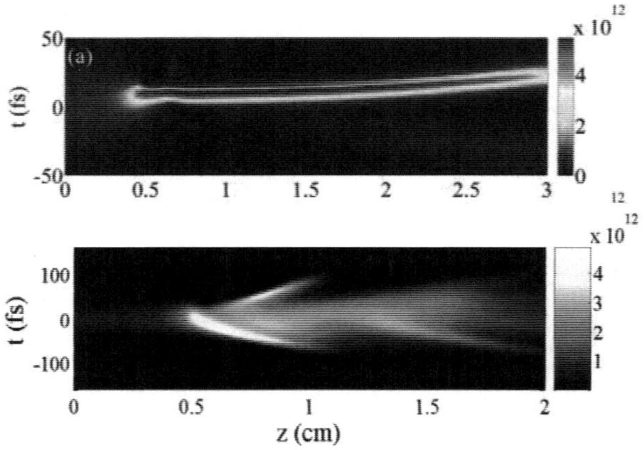

Figure 3. Numerical calculation of the pulse evolution during filamentation in silica for a femtosecond pulse with a central wavelength at 1.6 μm (a) and at 800 nm (b).

Experimental verification of the role of the dispersion regime would be difficult to achieve in gases, because of the lack of extended regions of negative group velocity. Transparent solids such as fused silica offer a compact system where it can be experimentally verified [15]. Indeed, there is a wide frequency range of anomalous dispersion starting around 0.97 μm and extending beyond 3.85 μm in SiO_2. The critical power in SiO_2 is a few MW instead of GW for gases. Figure 3 shows a comparison between the calculated filamentation in both dispersion regions. With positive GVD, the behavior is complex, showing complex pulse breaking, analog to what happens in air. With negative GVD, a single compressed pulse is formed after 0.5 cm of propagation. The slightly subluminal pulse then conserves its diameter and duration over the remaining length of the 3 cm thick sample.

Figure 4. Side photography of the blue fluorescence of SiO2 induced by a self-guided laser pulse at 1.9 μm with P = 30 Pcr. The laser pulse is focused on a 70 μm spot (FWHM) at the left surface of the 3 cm thick sample.

Experiments confirm these simulations. Figure 4 shows a side image of the blue luminescence emitted by the sample at high intensity I >10^{13} W/cm². The 20 µm wide filament is formed after 0.5 cm and is maintained over the rest of the sample. Figure 5 compares the beam profile emerging from the sample to the profile measured at an equivalent distance in the absence of sample. Pulse self cleaning, a signature of filamentation is manifest, as well as beam contraction. Measurement of pulse duration of the pulse exiting the sample is shown in figure 6. In agreement with the numerical simulations, the pulse contracts in time and keeps the contracted value over the sample length. From the numerical calculations, one deduces that the filament core has a duration of 8 fs. Together with its surrounding reservoir, the pulse duration becomes 25 fs, in agreement with the measurement where the central core and its reservoir are not resolved. Thus, the filamentation process in the anomalous region leads to the formation of a pulse that is almost invariant with propagation and is therefore closely parented to a spatiotemporal soliton. It must be underlined however that the pulse is lossy since the pulse consumes energy to ionize the medium. Also, part of the light energy is dissipated in the form of a conical emission. [16]

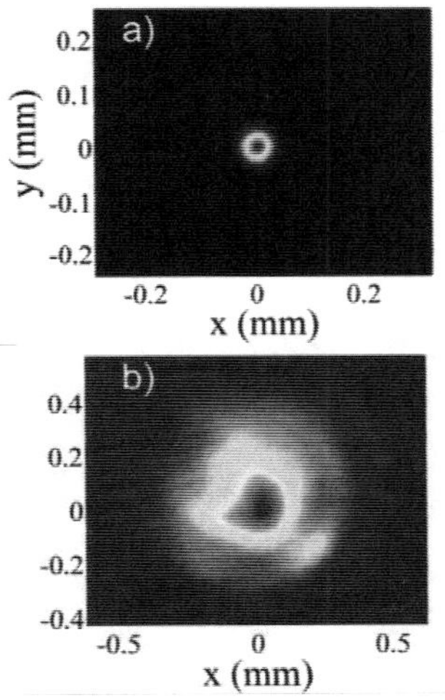

Figure 5. Beam profile of the laser pulse at 1.9 µm (a) with filamentation and (b) without filamentation in the fused silica sample. Intensity scale is normalized.

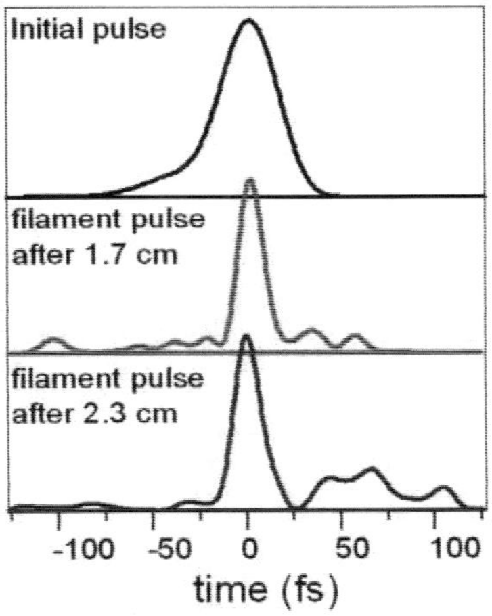

Figure 6. Measurement of the 1.9 µm pulse temporal shape after propagation in the sample performed with a WIZZLER.

3. Interaction between two filaments

Interaction of filaments has been observed to give rise to many interesting phenomenon such as energy exchange between filaments, enhanced third harmonic [17-19], plasma grating formation [20-22], molecular lensing effect [23], coherent control of filament fusion or repulsion [24-25], etc. In the following we discuss several aspects of the interaction between two filaments that enable to extract properties of the plasma created during filamentation, to exchange energy between crossing filaments and to improve spectacularly a third pulse probe pulse of same wavelength.

3.1. Plasma grating

3.1.1. Temporal dynamic of a plasma grating. Field interference of two crossing filamentary laser pulses give rise to laser intensity spatial modulation in the overlapping area. As a result, a strong modulation of the plasma density is created due to the highly nonlinear nature of optical ionization. This modulated plasma structure, i.e. plasma grating, is formed instantaneously, on the time scale of the pulse duration τ_p. On the other hand, its subsequent decay is long and can last for several hundreds of picoseconds in atomic gases. It is important to analyze this decay in view of possible applications. Recently, we measured the decay process of the plasma grating in different molecular and atomic gases. In the experiment, we employed a probe pulse at 400 nm co-propagating with one of the pump pulse. This 400 nm probe pulse is expected to be diffracted into the propagation direction of the other pump pulse in the presence of the non-sinusoidal plasma grating. Therefore, the measured 400 nm signal as a function of the temporal delay between the probe pulse and the two synchronized pump pulses allows characterizing the temporal evolution of the plasma grating. More details of the experimental setup can be found in ref [26]

It was found that in molecular gases like N_2 and O_2 the plasma grating decays rapidly, with a lifetime about 10-20 ps (Figures 7(a) and 7(b)). However, in atomic gases such as Ar and Xe the plasma grating persists for more than 300 ps (Figures 7(c) and 7(d)). The decay of a plasma grating process is expected to be due to free electron recombination and ambipolar diffusion. To obtain insight on the respective importance of the two effects, we varied the crossing angle of the two pump pulses. The resulting variation of grating period leads to a change of the ambipolar diffusion while the free electron recombination remains the same. From such experiments, we conclude that the grating evolution is ruled by ambipolar diffusion in atomic gases and by a combination of ambipolar diffusion and collision-assisted free electron recombination in molecular gases. From the systematic measurements, electron diffusion and recombination coefficients are extracted for Ne, Ar, Kr, Xe, N_2, O_2, CO_2 and air at 1 bar [27].

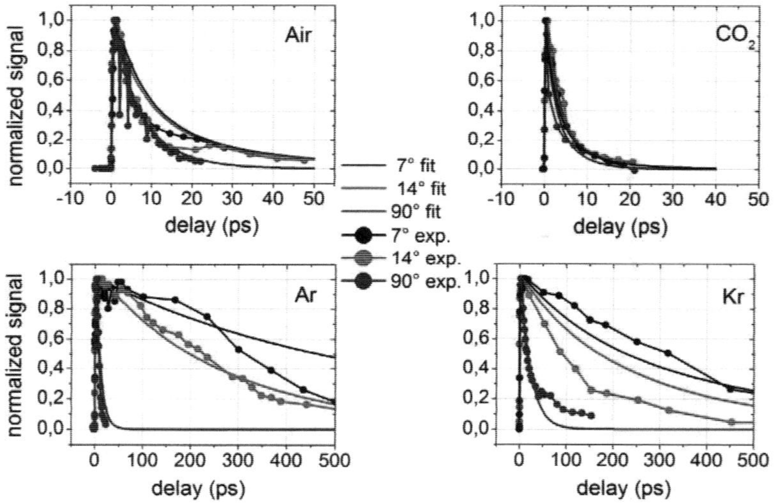

Figure 7. Normalized intensity of diffracted probe signal as a function of the delay τ (ps) for crossing angles φ of 7°, 14° and 90° in different gases (Air, CO_2, Ar and Kr). Measurements are represented by dots. Calculations are represented by solid lines.

3.1.2. Energy exchange between filaments in the presence of travelling plasma grating. The first experimental observation energy exchange between two filaments was reported in 2009 by Bernstein *et al.* [28]. They demonstrated an energy exchange ratio of 7% between two filament-forming pulses with slight frequency difference. In their work, a travelling refractive index grating is formed due to the Raman effect of air molecules. As a result, the classical two-beam coupling (TBC) scheme applies [28-29] and laser energy transfers from the high frequency pulse to the low frequency pulse. This provides a control over filaments propagation by modifying their characteristics in flight.

Recently, we observed a new regime of energy exchange where laser energy transfers from the low frequency pulse to the high frequency one. This new type of energy exchange was also found in monatomic noble gases such as Ar and Xe, excluding a traditional TBC mechanism based on a retarded optical Kerr nonlinearity. We attribute this new energy exchange to the effect of a moving plasma grating formed at the crossing point of the two filamentary pulses. Due to its plasma nature, this refractive grating can sustain very high laser intensity. With tightly focused laser pulses, we have demonstrated energy exchange efficiency of 50% in air. The result is presented in figure 8 [22].

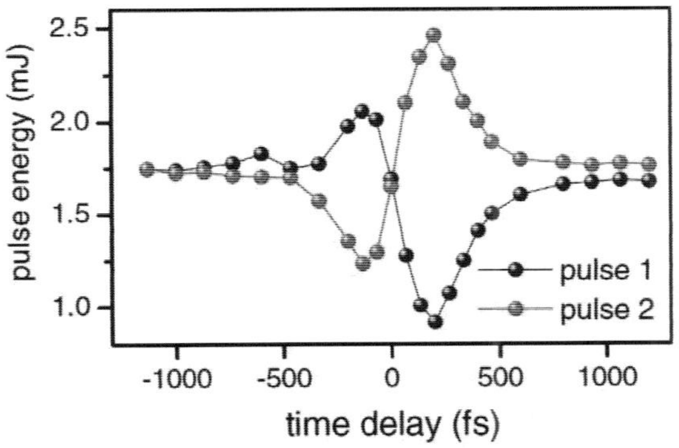

Figure 8. Energy exchange between two filaments crossed with an angle of 20°.

3.1.3. Plasma grating in motion. To detect the movement of a plasma grating, we employed an experimental setup similar to that in *3.1.1.* except that there was a difference in the wavelength of the two pump pulses [30]. We spectrally analysed the diffracted 400 nm pulse with a spectrometer. The movement of the plasma grating imparts a frequency shift on the probe pulse owing to the Doppler effect. Figure 9 presents the spectrum gravity of the reflected 400 nm signal pulse as a function of the time delay (frequency difference) between the two pump pulses with linear chirp. The central frequency of the signal pulse shifts proportional to the frequency difference between the pump pulses, i.e., to the velocity imparted to the grating. A calculation of the measured frequency shift based on the classical Doppler effect agrees well with this result [30].

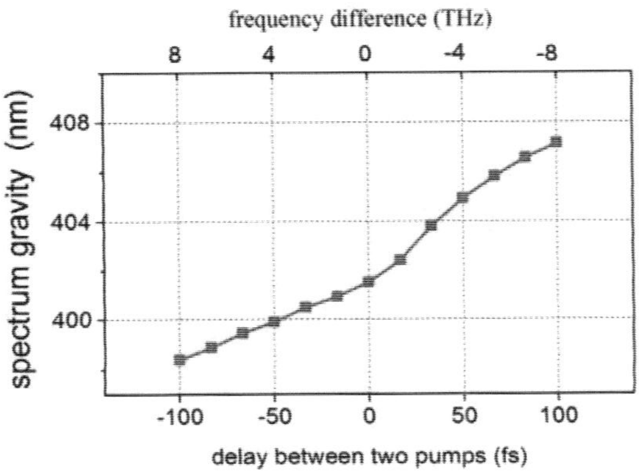

Figure 9. Spectrum center of gravity of the signal pulse as a function of the time delay between the two pump pulses.

Journal of Physics: Conference Series **497** (2014) 012001 doi:10.1088/1742-6596/497/1/012001

3.2. Filaments mirror

A special geometry is obtained when two filaments meet head on [31]. The encountering filaments form a mirror to a third probe pulse of same frequency impinging at an arbitrary angle. The mirror has special properties because the reflection occurs in the retro-direction. In addition, the reflected beam profile and pulse contrast ratio is dramatically improved. Figure 10 shows the beam fluence profile of the probe pulse before and after reflection. One can understand this spectacular property by considering the plasma grating formed by either pump with the probe pulse. As shown in section *3.1.1*, the other pump will diffract into the direction of the incoming probe. Therefore, the reflected pulse is replaced by a fraction of the pump pulses with self-cleaned beam profile. The profile fits indeed a Townes mode profile, as expected if the retro-signal corresponds to a diffraction from the filaments. In addition, the small size (4500 μm^3) acts as a pulse cleaner removing distortions in the phase profile of the incoming probe pulse.

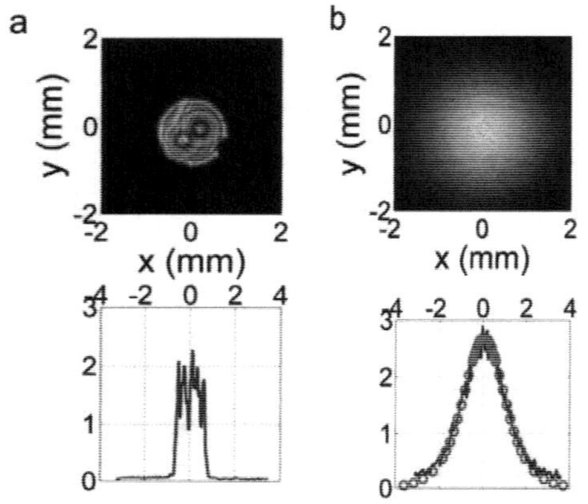

Figure 10. Fluence profile of a femtosecond pulse before (a) and after (b) reflection on the filament mirror, as recorded in a single shot with a CCD camera.

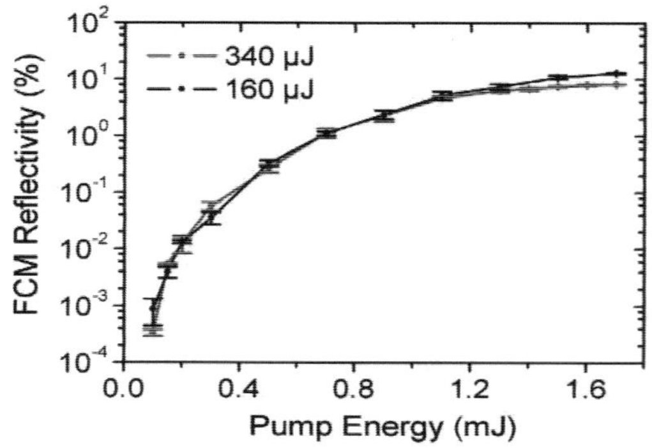

Figure 11. Reflectivity as a function of the pump pulses energy. Probe pulse has energy of 160 and 340 μJ and is incident at an angle of 90°. The three pulses are coincident in time.

Figure 11 shows the reflection of the filament mirror as a function of pump laser intensity for a probe incident at 90° from the filaments. Before the onset of ionization, the reflection is very small, because no plasma grating is present. There is a very steep increase of the reflection once the threshold for pump beam filamentation is reached, due to the high-order dependence of the ionization rate with intensity, followed by a near saturation, with a reflection efficiency of 20-30%, reflecting the intensity clamping inside a filament [1]. At an incident angle of 30°, the reflection increases to 40%.

This leads to the possibility of temporal pulse cleaning by removing light precursors. Indeed, if the chronology of the pump and probe pulses is adjusted such that they all meet at the same time, any precursors present in the pump and probe pulses will not be reflected because no plasma mirror is formed at the time of their encounter, due to the corresponding lower intensity. The time profiles of the incident and retro-reflected pulse have been measured with a third order ω-2ω cross-correlator. No precursor was detected in the reflected pulse within the dynamic range of the detector, as shown in figure 12.

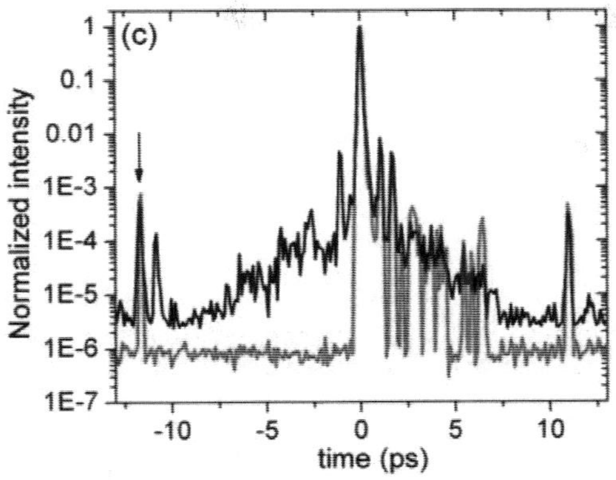

Figure 12. Temporal contrast of the probe (in black) and of the reflected beam (red) measured with a third order cross correlator.

4. Applications of femtosecond laser filaments

Intrigued by the spectacular effects accompanying filamentation process, researchers have demonstrated plenty of scientific applications of the filaments such as triggering and guiding of lightning [32], Light Detection and Ranging Technique (LIDAR) [33], few-cycle light pulse generation [34], virtual antenna [35], micromachining in transparent solids [36], Terahertz (THz) pulses generation [37], control of aerodynamic flows [38], water condensation [39] and snow formation [40]. Some applications such as few-cycle pulse generation are already quite matured, while others like THz generation have just been discovered and require further investigation. In the following two sub-sections, we will detail two applications: THz wave generation by filaments and control of electric discharges with filaments.

4.1. THz generation by one filament, two filaments

In 2007, D'Amico *et al.* reported a forward THz pulsed radiation stemming from a femtosecond laser filament in air [37]. They observed that the THz is radiated in a hollow-cone structure in the laser propagation direction, with a radial polarization (see figure 13). The authors developed a Transition-Cherenkov model which reproduced the above observed features [41]. In this model, the heavily damped longitudinal plasma oscillation produced in the wake of the propagating pulse is at the origin of the THz pulsed emission. Terahertz (THz) radiation produced by ultrashort laser pulses in air attracted much attention in recent years, since it has some unique properties. The THz source can be positioned at the proximity of a remote target by displacing the onset of filamentation. This solves the long-standing problem of the poor propagation of THz through atmosphere due to strong absorption by water vapor. In contrast to other THz generation techniques with short laser pulses, there is no obvious limit to the laser power that can be used, since there is no material in the path of the pulse that can be damaged.

However, the energy conversion efficiency from laser to THz radiation by a single filament is low, on the order of 10^{-9} [42]. Therefore, it is desirable to improve the conversion efficiency and control the physical properties of this Terahertz radiation. Several techniques have been developed along this line.

In a first scheme, an external electric field is applied on the filament either longitudinal or transverse. A THz enhancement by 3 orders of magnitude is obtained with an applied field E_{ext} ~ 10 kV/cm (figure 13). At the same time, the radiation pattern and polarization property of this amplified emission keep the same as that of the pure transition-Cherenkov THz radiation [43]. This amplified THz pulse is of particular significance since a radially polarized THz wave has been demonstrated to be compatible with the propagating mode of a metal wire THz waveguide [44]. Although, this method of enhancement suffers from the fact that the electric field must be applied on the plasma region of the filament, not an easy task for stand-off THz applications.

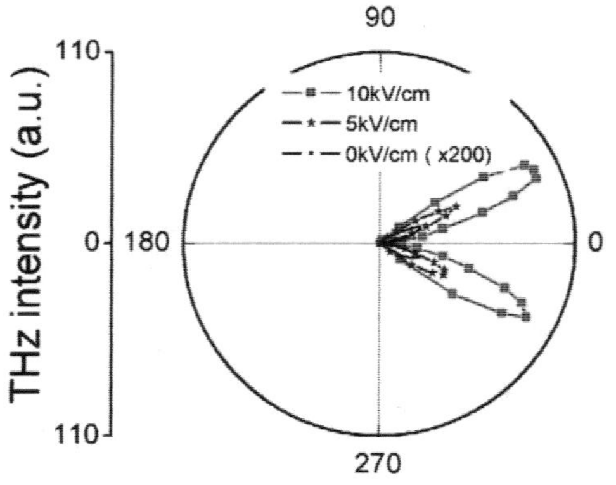

Figure 13. Amplification of the transition-Cherenkov THz emission. The pure transition-Cherenkov THz (E_{ext} = 0 kV/cm) is multiplied by a factor of 200 for visibility.

For most THz applications, a THz beam with maximum intensity on the axis is preferable. This requires the presence of a transverse current in the filament and an orthogonal electric field should be able to fulfill this task. In the experiment, we applied an external electric field perpendicularly to the filament axis. With the increase of the external electric field strength, the asymmetrical radiation pattern becomes more and more symmetrical (figure 14). As a result, a THz beam with above desired feature was obtained when E_{ext} exceeds 3 kV/cm. More important, the THz intensity was enhanced by 3 orders of magnitude with respect to the transition-Cherenkov THz with E_{ext} = 9 kV/cm. Concerning the THz polarization, we found it is linear polarized and the polarization plane is dictated simply by the direction of the external field. In addition, we demonstrated that the polarity of the THz can be easily controlled by the direction of the electric field [45].

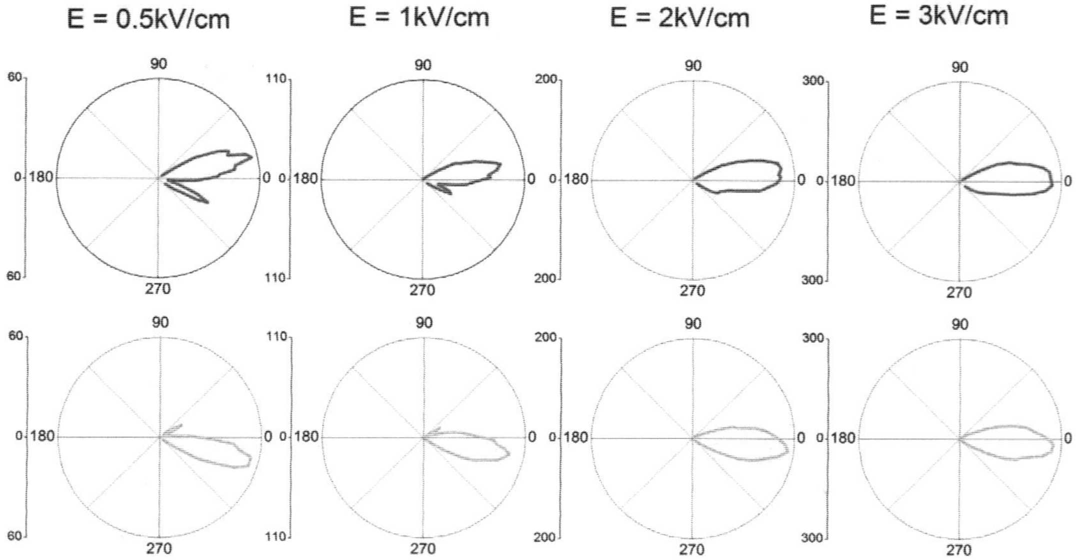

Figure 14. THz Radiation pattern from a laser filament in the presence of increasing transverse electric field. In the top (bottom) row, the external electric field is directed downward (upward).

Another approach is to use an oscillating transverse field, by combining a short IR pulse at 800 nm with its harmonic generated in a thin BBO crystal. After the first demonstration of this two-color method [46], it has been extensively studied with a millimeter scale long filament (L ~ λTHz) [47-52]. It was found that the optical biased at second harmonic frequency can lead to a THz amplitude enhancement by two orders of magnitude compared to the air plasma driven by just 800 nm femtosecond laser pulses [47, 49]. The relative phase between the fundamental laser field and its second harmonic was found to be critical for the yield and polarity of the THz radiation [47, 53]. Recently, this method was also extended to long air plasma (L >> λ THz) obtained in the filamentation regime [54-55]. For such long filaments, it was confirmed that the two-color scheme is more efficient than just a pulses at 800 nm. But, the temporal separation of the 800 nm and 400 nm pulses during their propagation inside the long filaments due to group delay dispersion of air sets a limit of interaction length (~ 2 m) and results in a strong THz yield decrease for standoff THz generation [55].

One advantage of the filaments-based THz source lies in the fact that it is laser damage-free due to its plasma nature, and one can expect high energy THz radiation with powerful incident femtosecond laser pulses. However, it is well known that multiple filamentation forms when the incident laser power well exceeds the critical power for self-focusing in air P_{cr} (~ 5 GW [1]). The underlying mechanism of multiple filamentation is the imperfection of the laser beam spatial profile as well as the turbulence in air during propagation. As a result, the process of multiple filamentation changes from shot-to-shot, which hinders the usage of multiple filaments as a stable THz source. Is there possible solution?

Inspired by the phased antenna array in the RF domain, we recently proposed and demonstrated experimentally the coherent synthesis of THz radiations of an array of organized filaments [56]. We find that the THz intensity scales up with N^2 provided that proper filament separation and laser pulse time delays are chosen. Moreover, the THz radiation pattern can be controlled, which is a useful property for many applications. For example, in figure 15 we presented the THz radiation patterns obtained from two filaments separated by a distance of 2.4 mm for increasing time delay between the two pulses. Upon increase of the delay τ_p, the radiation pattern becomes asymmetric. With τ_p = 3.7 ps, the radiation is totally directed along one lobe. This is particular interesting for applications where irradiation of targets with intense laser pulse should be avoided. Our simulations based on the Cherenkov-Transition model reproduced well our experimental observations.

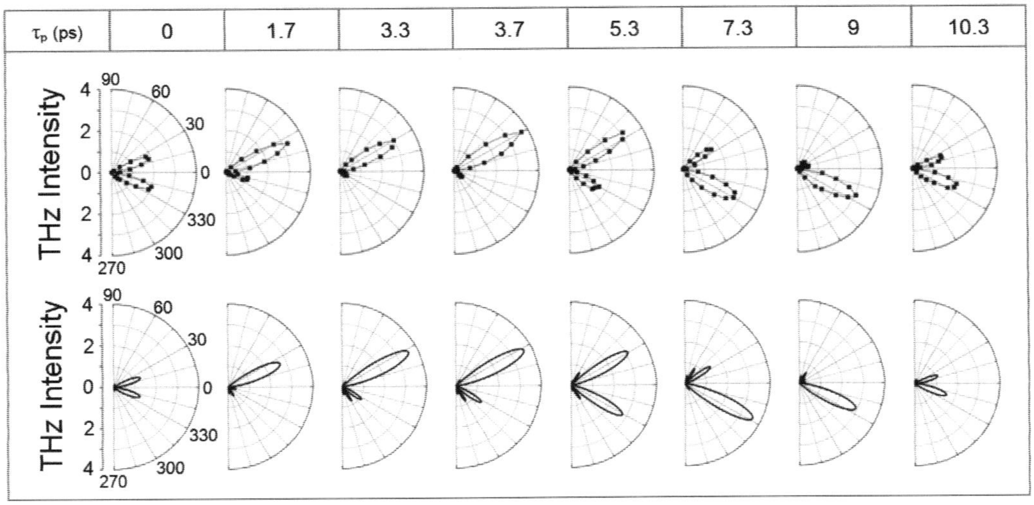

Figure 15. Evolution of emission diagrams of the THz radiation from two neighbouring filaments as a function of the temporal delay between the two laser pulses. The two filaments are separated by d = 2.4 mm. Top row: experimental results. Bottom row: calculations.

We further extrapolated the simulation to a higher number of filaments. We have calculated the THz radiation distribution for N = 4, 6, 8, 9, 12, 16 filaments organized in a square grid. With a properly optimized delay, the total THz energy can be channeled towards a preferential direction. With two filaments and zero delay, a butterfly wing pattern appears in the XOZ plane (figure 16 (a)). A single leaf radiation pattern is obtained with τ_p = 3.5 ps (figure 16(b)). For a larger number of filaments, similar leaf-shape radiation pattern appear with properly chosen pulse delays, as presented for 16 filaments in figure 16(c), where its peak intensity is found to be ~ 250 times that of a single filament.

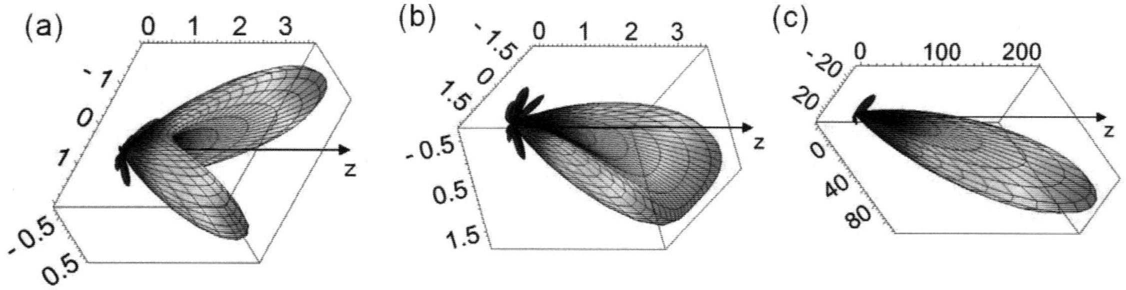

Figure 16. Calculated THz radiation diagram from two parallel filaments separated by 2.4 mm without time delay (a) and with a time delay τ_p = 3.5 ps (b). (c) Calculated THz radiation diagram from 16 filaments with proper time delays.

4.2. Control of electric discharges with laser filaments
Plasma strings produced by femtosecond laser filaments proved to be particularly useful for remote manipulation of high voltage discharges. The physical mechanism is the following [57-58]. In the presence of a static electric field, free electrons formed during filamentation release their kinetic energy in the gas by Joule heating, leading to the appearance of a depressurized channel at the center of the filament path. The resulting low density column offers a privileged path for discharge [59-60]. In parallel, a fraction of the free electrons does not recombine on the parent ions but become attached to neutral oxygen molecules. Such loosely bound electrons can be easily released by current heating, leading to a decrease of the leader inception voltage [61]. As a consequence, a filament decreases the breakdown voltage in a gas by more than 30% and guides the discharge over the perfectly straight path defined by the laser.

Femtosecond filament can trigger and guide megavolts discharges over several meters [62-64], carry high DC currents with reduced losses [65] or deviate arcs from their natural path [66]. These properties are of great interest for applications such as the laser lightning rod [32], virtual plasma antennas for radiofrequency transmission, plasma aerodynamic control or high voltage switch.

4.2.1. High current spark gap. A previous experiment had shown that plasma filaments were able to initiate high current DC discharge with reduced losses [65]. To analyse further this property we studied the triggering of atmospheric kA spark gap by laser filamentation [67]. The beam (300 mJ, 700 fs at a central wavelength λ_0 = 800 nm) was focused in air with a 5 m focal length lens leading to a quasi-homogeneous plasma column with an effective length of 2 m and a diameter of ~ 3 mm. The experimental spark gap was installed in the middle of the plasma column. Figure 17 gives a schematic overview of the experimental setup. Electrodes with a central hole optimized the contact between the laser created plasma column and the high voltage. The high voltage (up to $U_0 \leq 30$ kV) was applied to the spark gap set in parallel with a 50 Ω resistance by activating a conventional switch T. It created on the spark gap electrodes a voltage pulse with rise time (10% - 90%) ~ 50 ns and exponential decay τ =

RC ≈ 14 μs. In this case the laser triggered discharges yielded a current maximal amplitude reaching ~12.5 kA with a 30 kV charge.

An important parameter of the spark gap is the jitter, which is usually defined as the standard deviation of the switching time delay. Figure 18 presents the measured jitter for different gap lengths D as a function of U_{norm}, the breakdown voltage normalized to the self-breakdown voltage threshold in static regime. One can see that it is possible to have a jitter of 0.2-0.4 ns in the region $0.7 < U_{norm} < 1.2$ with a 10 mm gap. The measured jitter is close to the time resolution of the oscilloscope (1 GHz and 10 GSa).

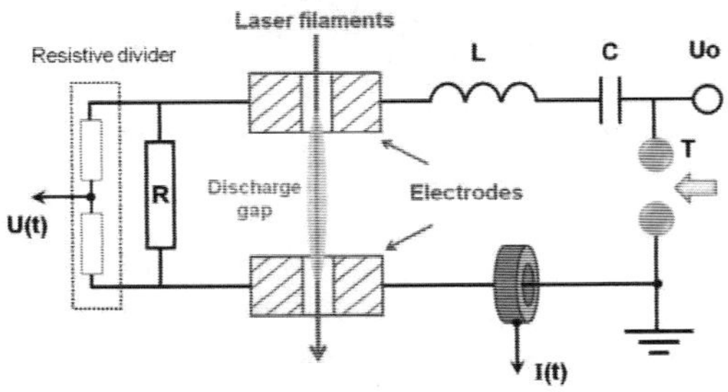

Figure 17. Schematic view of the experimental set-up with a pulsed voltage applied to the axial gap switch. The laser beam passes through the axial holes in the electrodes.

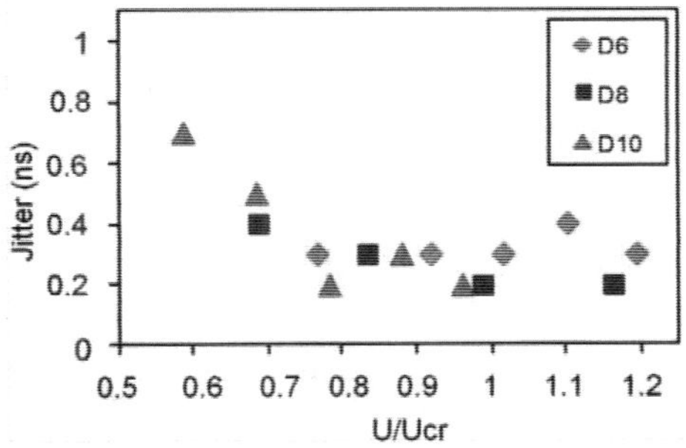

Figure 18. Jitter of the switch as a function of normalized voltage for gaps of 6, 8 and 10 mm.

Triggering via filamentation has several advantages. Since filament generated plasma column can be created at a distance of several hundreds of meters from the laser, remote triggering can be achieved. Because filaments have a near constant longitudinal electron density over several meters, they connect the two electrodes quasi-instantaneously, circumventing the complex process required in the case of an initial localized plasma, and reducing thereby the jitter. Thus, this spark gap combines the advantages of large inter-electrode air gaps capable of switching high voltages and the low jitter of narrow spark gaps or SF_6 filled cells. The long and uniform plasma channels formed by the filament also offer the possibility to trigger simultaneously several spark gaps with an excellent synchronization.

4.2.2. Tesla coil discharges. Experiment on triggering and guiding of large scale (0.5 – 3 m) discharges by use of laser-induced ionized filament have been performed so far with unipolar pulsed or DC high voltage sources, where the electric field direction remains constant, typically with MARX generators [61]. Recently demonstration of large discharge triggering and guiding by femtosecond laser filaments using a Tesla coil generator has been reported by our group and by Henrikson *et al.* [68-69].

The Tesla coil generator is basically a voltage elevator transformer, where the coupling of two resonant R L C circuits allows obtaining voltage bursts of voltage oscillations. Our system delivers voltage pulses oscillating at 100 kHz with peak amplitude ~ 350 kV. The particularity of our Tesla transformer is that it can be synchronized with an external TTL signal with a temporal jitter below 20 nanoseconds. The laser beam is focused in air producing a continuous plasma column about 2 m long, touching tangentially the two spherical electrodes connected to the Tesla output.

Figure 19. Photography of a Tesla guided discharge of 1.7 m.
The laser beam comes from the left.

The discharges can be triggered and guided with 100 % success; however successful operation requires a precise timing of the trigger with respect to the laser arrival time. Optimum repeatability is achieved when the laser initiates the discharge during a maximum of the oscillatory voltage at the output of the Tesla. No electric discharge is produced by the Tesla generator in the absence of laser when the electrode gap is larger than 32 cm. Guided discharges activated by the filament are obtained up to an inter-electrode distance of 200 cm corresponding to a decrease of 80 % of the breakdown voltage. With optimal laser parameters, reproducible guided discharges of 1 meter at a repetition rate of 10 Hz were demonstrated.

Measurement of voltage and discharge current allows an estimation of the plasma column resistance around 1 kΩ. This resistance appears to increase linearly with the gap length. The influence of the laser input energy on the triggering and guiding effect is also studied showing that the multifilamentation regime significantly increases the probability to trigger long discharges.

4.2.3. Virtual RF antenna. An application of long laser guided electric discharge is the virtual plasma antenna [35]. Plasma antennas, where plasma replaces metal as the conducting element, have long been known and used [70]. However, most designs use low-pressure plasmas confined inside solid dielectric vessels. We experimentally demonstrated a functional plasma antenna in air, which brings about many advantages like tunability in a large frequency range (100 MHz – 1 GHz), stealth when de-activated and quick reconfiguration capabilities.

In this experiment we used the filament induced high-voltage electric discharge generated by the compact Tesla coil as a plasma antenna at atmospheric pressure. The guided discharge had a length of about 100 cm. Radio-frequency (RF) power was injected in the plasma by means of an inductive coupler in the form of a hollow metallic cylindrical cavity, fed by a 35 W solid-state RF amplification chain. Radio emission was then detected using a remote patch antenna with a 100 MHz - 1 GHz bandwidth (see setup in figure 20). As shown in the example given in figure 21, when the coupler is excited at 990 MHz and no plasma is present, there is no signal observed at this frequency (black curve), whereas a clear emission peak appears when the laser guided discharge is generated (red curve). RF energy coupling in the plasma has consequently been achieved, which resulted in the plasma column behaving as an emitting antenna. The emission level from a copper rod with a length similar to that of the plasma is approximately equal to four times the plasma signal strength, demonstrating the applicability of this technology in real situations.

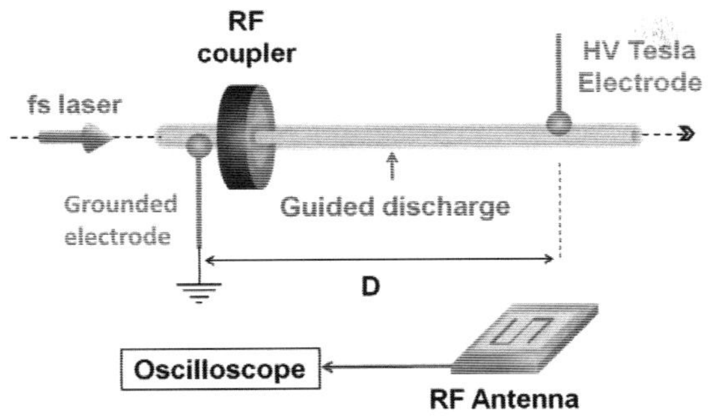

Figure 20. Experimental setup used for RF coupling in the plasma.

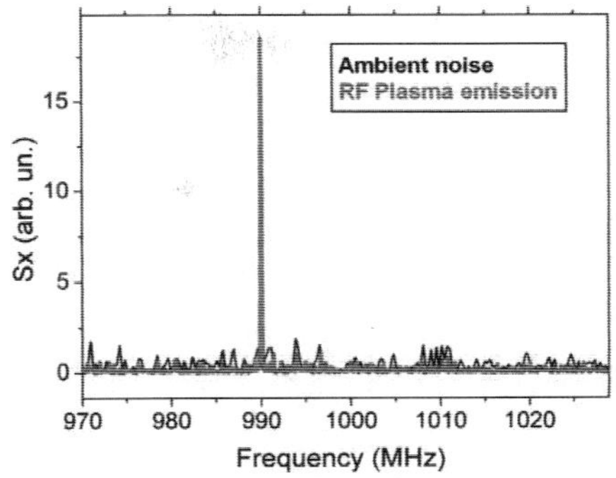

Figure 21. RF signal strength with (red) or without (black) the plasma column when the coupler is excited at 990 MHz. The insert corresponds to a zoom of the highlighted region around 990 MHz.

5. Conclusion

In summary, the physics underlying femtosecond laser filamentation has been introduced. Several examples of interaction between filaments have been described. Finally we have discussed applications of femtosecond filamentation to the generation of THz pulses in air and to the initiation and control of high voltage discharges.

Acknowledgments

Authors wish to acknowledge contributions from A. Couairon for the numerical studies and from M. Durand, Y. Brelet, B. Prade, A. Jarnac, J. Carbonnel, S. Mitryukovskiy, Y. B. André, G. Point, B. Forestier and L. Arantchouk for experimental measurements. This work has been partially supported by a grant from the French DGA (grant EPAT n° 200795091), by ANR (grant n° ANR-2010-JCJC-0401-01) and by LASERLAB-EUROPE (grant agreement n° 284464, EC's Seventh Framework Programme).

References

[1] Couairon A and Mysyrowicz A 2007 Femtosecond filamentation in transparent media *Phys. Rep.* **441** 47
[2] Couairon A, Brambilla E, Corti T, Majus D, Ramiez-Gongora O. and Kolesik M 2011 Practitioner's guide to laser pulse propagation models and simulation *The European Physical Journal Special Topics* **199** 5
[3] Marburger JH 1975 Self Focusing: Theory *Progress in Quantum Electronics* **4** 35
[4] It has been shown by Arizona group that the value of Pcr obtained for a cw laser beam from the Marburger formula is not strictly valid for short intense pulses [5]. Nevertheless expression (2) gives a useful order of magnitude.
[5] Polynkin P and Kolesik M 2013 Critical power for self-focusing in the case of ultrashort laser pulses *Phys. Rev. A* **87** 053829
[6] Durand M *et al.* 2013 Kilometer range filamentation *Opt. Express* **21** 26836
[7] Mechain G *et al.* 2004 Long range self-channeling of infrared laser pulses in air: a new propagation regime without ionisation *Appl. Phys. B* **79** 379
[8] It has been recently proposed that the main process leading to filamentation is a saturation of the refractive index followed by a change of its sign at high intensities with multiphoton ionization playing a minor role (HOKE model) [9]. Recent careful measurements of the plasma density in filaments disagree with this model and support the standard model [10].
[9] Bejot P *et al.* 2010 Higher-order Kerr terms allow ionization-free filamentation in gases *Phys.*

Rev. Lett. **104** 103903

[10] Wahlstrand J K, Cheng Y-H, Chen Y-H, and Milchberg H M 2011 Optical Nonlinearity in Ar and N2 near the Ionization Threshold *Phys. Rev. Lett.* **107** 103901

[11] Zaïr *et al.* 2007 Spatio-temporal characterization of few-cycle pulses obtained by filamentation *Opt. Express* **15** 5394

[12] Moll K, Gaeta A and Fibich G 2003 Self-Similar Optical Wave Collapse: Observation of the Townes Profile *Phys. Rev. Lett.* **90** 203902

[13] Prade B, Franco M, Mysyrowicz A, Couairon A, Buersing H, Eberle B, Krenz M, Seiffer D and Vasseur O 2006 Spatial mode cleaning by femtosecond filamentation *Optics Letters* **31** 17 2601

[14] Agrawal G P 1989 *Nonlinear fiber optics* (Boston, MA: Academic Press)

[15] Durand *et al.* 2013 Self guided propagation of ultrashort laser pulses in the anomalous dispersion region of transparent solids: a new regime of filamentation *Phys. Rev. Lett.* **110** 115003

[16] Durand *et al.* 2013 Conical emission and extreme blueshifted continuum peaks from filamentation in the regime of anomalous dispersion in fused silica *Phys. Rev. A* **87** 043820

[17] Suntsov S, Abdollahpour D, Papazoglou D G, and Tzortzakis S 2010 Filamentation produced third harmonic in air via plasma enhanced third-order susceptibility *Phys. Rev. A* **81** 033817

[18] Yang X. *et al.* 2009 Noncollinear interaction of femtosecond filaments with enhanced third harmonic generation in air *Appl. Phys. Lett.* **95** 111103

[19] Liu Y, Durand M, Houard A, Forestier B, Couairon A and Mysyrowicz A 2009 Efficient Generation of Third Harmonic Radiation in Air Filaments: a Revisit *Optics Commun.* **284** 4706

[20] Suntsov S, Abdollahpour D, Papazoglou D G, and Tzortzakis S 2009 Femtosecond laser induced plasma diffraction gratings in air as photonic devices for high intensity laser applications *Appl. Phys. Lett.* **194** 251104

[21] Yang X, Wu J, Peng Y, Tong Y, Lu P, Ding L, Xu Z, and Zeng H 2009 Plasma waveguide array induced by filament interaction *Opt. Lett.* **34** 3806

[22] Liu Y, Durand M, Chen S, Houard A, Prade B, Forestier B, and Mysyrowicz A 2010 Energy exchange between femtosecond laser filaments in air *Phys. Rev. Lett.* **105** 055003

[23] Varma S, Chen Y-H and Milchberg H M 2008 Trapping and Destruction of Long-Range High-Intensity Optical Filaments by Molecular Quantum Wakes in Air *Phys. Rev. Lett.* **101** 205001

[24] Xi T-T, Lu X, and Zhang J 2006 Interaction of Light Filaments Generated by Femtosecond Laser Pulses in Air *Phys. Rev. Lett.* **96** 025003

[25] Shim B, Schrauth S E, Hensley C J, Vuong L T, P Hui, Ishaaya A A and Gaeta A L 2009 Controlled interactions of femtosecond light filaments in air *Phys. Rev. A* **81** 061803

[26] Durand M, Jarnac A, Liu Y, Prade B, Houard A, Tikhonchuk V and Mysyrowicz A 2012 Dynamics of plasma gratings in atomic and molecular gases *Phys. Rev. E* **86** 036405

[27] Jarnac A, Durand M, Liu Y, Prade B, Houard A, Tikhonchuk V and Mysyrowicz A 2014 Study of Laser Induced Plasma Gratings Dynamics in Gases *Optics Commun.* **312** 35

[28] Bernstein A C, McCormick M, Dyer G M, Sanders J C, and Ditmire T 2009 Two-Beam Coupling between Filament-Forming Beams in Air *Phys. Rev. Lett.* **102** 123902

[29] Boyd R W 2008 *Nonlinear Optics 3rd ed.* (San Diego: Academic Press)

[30] Durand M, Liu Y, Forestier B, Houard A, and Mysyrowicz A, 2011 Experimental observation of a traveling plasma grating formed by two crossing filaments in gases *Appl. Phys. Lett.* **98** 121110

[31] Jarnac A, Durand M, Houard A, Liu Y, Prade B, Richardson M, Mysyrowicz A 2014 Spatio-temporal cleaning of a femtosecond laser pulse through interaction with contra-propagating filaments in air *to be published*

[32] Zhao X M, Diels J-C, Braun A, Liu X, Du D, Korn G, Mourou G, and Elizondo J M 1994 Use

of self-trapped filaments in air to trigger lightning in *Ultrafast Phenomena*, Springer Series in Chemical Physics. New York: Springer-Verlag **60** 233
[33] Kasparian J *et al.* 2003 White-light filaments for atmospheric analysis *Science* **301** 61
[34] Hauri C P, Kornelis W, Helbing F W, Couairon A, Mysyrowicz A, Biegert J and Keller U 2004 Generation of intense, carrier-envelope phase-locked few-cycle laser pulses through filamentation *Appl. Phys. B* **79** 673
[35] Brelet Y, Houard A, Point G, Prade B, Arantchouk L, Carbonnel J, Andre Y-B, Pellet M, Mysyrowicz A 2012 Radiofrequency plasma antenna generated by femtosecond laser filaments in air *Appl. Phys. Lett.* **101** 264106
[36] Yamada K, Watanabe W, Toma T, Itoh K, and Nishii J 2001 In situ observation of photoinduced refractive-index changes in filaments formed in glasses by femtosecond laser pulses *Opt. Lett.* **26** 19
[37] D'Amico C, Houard A, Franco M, Prade B, Couairon A, Tikhonchuk V T, Mysyrowicz A 2007 Conical forward THz emission from femtosecond laser filamentation in air *Phys. Rev. Lett.* **98** 235002
[38] Dufour G, Fornet B and Rogier F 2013 Numerical modelling of supersonic flow actuated by laser–induced plasma International *Journal of Aerodynamics* **3** 122
[39] Rohwetter P *et al.* 2010 Laser-induced water condensation in air *Nature Photonics* **4** 451
[40] Ju J *et al.* 2012 Laser-filamentation-induced condensation and snow formation in a cloud chamber *Opt. Lett.* **37** 1214
[41] D'Amico C *et al.* 2008 Forward THz radiation emission by femtosecond filamentation in gases: theory and experiment *New Journal of Physics* **10** 013015
[42] Houard A, Liu Y, Mysyrowicz A and Leriche B 2007 Calorimetric detection of conical THz radiation from femtosecond laser filaments in air *Appl. Phys. Lett.* **91** 241105
[43] Liu Y, Houard A, Prade B, Diaw A, Tikhonchuk V T and Mysyrowicz A 2008 Amplification of Transition-Cherenkov Terahertz Radiation of Femtosecond Filament in Air *Appl. Phys. Lett.* **93** 051108
[44] Wang K and Mittleman R 2004 Metallic wire for Terahertz wave guiding *Nature* **432** 376
[45] Houard A, Liu Y, Prade B, Tikhonchuk V T and Mysyrowicz A 2008 Strong enhancement of Terahertz radiation from laser filaments in air by a static electric field *Phys. Rev. Lett.* **100** 255006
[46] Cook D J and Hochstrasser R M 2000 Intense terahertz pulses by four-wave rectification in air *Opt. Lett.* **25** 1210
[47] Xie X, Dai J, and Zhang X C 2006 Coherent control of THz wave generation in ambient air *Phys. Rev. Lett.* **96** 075005
[48] Dai J, Karpowicz N, Zhang X C 2009 Coherent polarization control of Terahertz waves generated from two-color laser induced gas plasma *Phy. Rev. Lett.* **103** 023001
[49] Thomson M D, Kress M, Loffler T and Roskos H G 2007 Broadband THz emission from gas plasma induced by femtosecond optical pulses: From fundamental to applications *Laser & Photonics Rev.* **1** 349
[50] Houard A, Liu Y, Prade B, Mysyrowicz 2008 Polarization analysis of terahertz radiation generated by four-wave mixing in air *Opt. Lett.* **33** 1195
[51] Kim K Y, Glownia J H, Taylor A J, and Rodriguez G 2007 Terahertz emission from ultrafast ionizing air in symmetry-broken laser fileds *Opt. Express* **15** 4577
[52] Kim K Y, Taylor A J, Glownia J H, and Rodriguez G 2008 Coherent control of terahertz supercontinuum generation in ultrafast laser-gas interactions *Nat. Photonics* **2** 605
[53] Liu Y, Houard A, Durand M, Prade B, and Mysyrowicz A 2009 Maker fringes in the Terahertz radiation produced by a 2-color laser filed in air Opt. Express **17** 11480
[54] Wang T J, Daigle J F, Chen Y, Marceau C, Théberge F, Châteauneuf M, Dubois J, and Chin S L 2010 High energy THz generation from meter-long two-color filaments in air *Las. Phys. Lett.* **7** 517

Journal of Physics: Conference Series **497** (2014) 012001 doi:10.1088/1742-6596/497/1/012001

[55] Daigle J F, Théberge F, Henriksson M, Wang T J, Yuan S, Châteauneuf M, Dubois J, Piché M and Chin S L 2012 Remote THz generation from two-color filamentation: long distance dependence *Opt. Express* **20** 6825

[56] Mitryukovskiy S I, Liu Y, Prade B, Houard A and Mysyrowicz A 2013 Coherent synthesis of THz fields from filament antenna array *Appl. Phys. Lett.* **102** 221107

[57] Tzortzakis S, Prade B, Franco M, and Mysyrowicz A 2000 Time-evolution of the plasma channel at the trail of a self-guided IR femtosecond laser pulse in air *Opt. Commun.* **181** 123

[58] Bodrov S *et al.* 2011 Plasma filament investigation by transverse optical interferometry and terahertz scattering *Opt. Express* **19** 6829

[59] Vidal F *et al.* 2000 Modeling the triggering of streamers in Air by ultrashort laser pulses *IEEE Trans. on Plasma Science* **28** 418

[60] Tzortzakis S, Prade B, Franco M, Mysyrowicz A, Hüller S, and Mora P 2001 Femtosecond Laser-guided Electric Discharge in Air *Phys. Rev. E* **64** 57401

[61] Comtois D *et al.* 2003 Triggering and Guiding of an Upward Positive Leader From a Ground Rod With an Ultrashort Laser Pulse—II: Modeling *IEEE Trans. on Plasma Science* **31** 387

[62] Pepin H *et al.* 2001 Triggering and guiding high-voltage large-scale leader discharges with sub-joule ultrashort laser pulses *Phys. Plasmas* **8** 2532

[63] Rodriguez M *et al.* 2002 Triggering and guiding megavolt discharges by use of laser-induced ionized filaments *Opt. Lett.* **27** 772

[64] Fuji T *et al.* 2008 Leader effects on femtosecond-laser-triggered discharges *Phys. Plasma* **15** 013107

[65] Houard A *et al.* 2007 High current permanent discharges in air induced by femtosecond laser filamentation *App. Phys. Lett.* **90** 171501

[66] Forestier B *et al.* 2012 Triggering, guiding and deviation of long air spark discharges with femtosecond laser filament *AIP Advances* **2** 012151

[67] Arantchouk L, Houard A, Brelet Y, Carbonnel J, Larour J, André Y-B and Mysyrowicz A 2013 A simple high-voltage high current spark gap with subnanosecond jitter triggered by femtosecond laser filamentation *Appl. Phys. Lett.* **102** 163502

[68] Brelet Y *et al.* 2012 Tesla coil discharges guided by femtosecond laser filaments in air *Appl. Phys. Lett.* **100** 181112

[69] Henriksson M, Daigle J-F, Théberge F, Chateauneuf M and Dubois J 2012 Laser guiding of Tesla coil high voltage discharges *Opt. Express* **20** 12721

[70] Borg G G, Harris J H, Martin N M, Thorncraft D, Milliken R, Miljak D G, Kwan B, Ng T and Kircher J 2000 Plasmas as antennas: Theory, experiment and applications *Phys. Plasmas* **7** 2198

Journal of Physics: Conference Series **497** (2014) 012002 doi:10.1088/1742-6596/497/1/012002

Spin turbulence in spinor Bose-Einstein condensates

Makoto Tsubota, Kazuya Fujimoto

Department of Physics, Osaka City University, Osaka 558-8585, Japan

E-mail: tsubota@sci.osaka-cu.ac.jp

Abstract. We summarize the recent theoretical and numerical works on spin turbulence (ST) in spin-1 spinor Bose-Einstein condensates. When the system is excited from the ground state, it goes through hydrodynamic instability to ST in which the spin density vector has various disordered direction. The properties of ST depend on whether the spin-dependent interaction is ferromagnetic or antiferromagnetic. ST has some characteristics different from other kinds of turbulence in quantum fluids. Firstly, the spectrum of the spin-dependent interaction energy exhibits the characteristic power law different from the usual Kolmogorov -5/3 law. Secondly, ST can show the spin-glass-like behavior; the spin density vectors are spatially random but temporally frozen.

1. Introduction

Quantum turbulence (QT) is currently one of the most important topics in low temperature physics [1]. QT has been long studied for superfluid helium, while the realization of Bose–Einstein condensation in trapped atomic gases has proposed another important stage of QT. After the realization in 1995 lots of experimental and theoretical works have been devoted to the issues of quantized vortices in this system [2, 3, 4, 5]. Most works have addressed the system of a small number of vortices or a vortex array under rotation, but there are very few on QT. Following a few theoretical proposals [6, 7, 8], Henn *et al.* succeeded in creating and observing three-dimensional QT in trapped ^{87}Rb BECs [9, 10]. An important advantage of atomic gases is that multicomponent BECs can be created experimentally. Multicomponent BECs allow the formation of various unconventional topological defects with complex properties that arise from the internal degrees of freedom and the interactions between different components [11, 12, 13], yielding a rich variety of superfluid dynamics [14, 15]. Hence multicomponent BECs can propose a more advanced system of QT. Two-component QT has already been studied theoretically and numerically [16, 17, 18]; the most important interest would be how two kinds of QT interact.

Recently we have studied theoretically and numerically spin turbulence (ST) in spinor BECs [19, 20, 21, 22, 23, 24]. A spinor BEC is a kind of multicomponent BECs with spin degrees of freedom [12, 13]. In contrast to the two-component BECs, interatomic interactions allow for a coherent transfer of population between different hyperfine spin states (spin-exchange collisions), which yields a fascinating physics different from two-component BECs. Spinor BECs can be another important stage of turbulence in quantum fluids. When the system is highly excited from the ground state, it goes through hydrodynamic instability to ST in which the spin density vector has various disordered direction. This ST shows characteristic behaviors different from other kinds of QT. First, the spectrum of the spin-dependent interaction energy obeys a -7/3 power law [19, 20], which is different from the traditional Kolmogorov -5/3 power law

Journal of Physics: Conference Series **497** (2014) 012002 doi:10.1088/1742-6596/497/1/012002

in turbulence [25, 26]. Secondly, the spin density vectors are spatially random but temporally frozen, which reminds us of the analogy of spin glass [22]. This article reviews such recent works on ST.

2. The Gross-Pitaevskii equations and the hydrodynamic description

We consider a two-dimensional spin-1 spinor BEC at zero temperature. The macroscopic wave functions Ψ_m with the magnetic quantum number m ($m = 1, 0, -1$) obey the Gross-Pitaevskii (GP) equations [27, 28]

$$\begin{aligned} i\hbar\frac{\partial}{\partial t}\Psi_m &= (-\frac{\hbar^2}{2M}\nabla^2 + V)\Psi_m + \sum_{n=-1}^{1}[-g\mu_B(\mathbf{B}\cdot\hat{\mathbf{S}})_{mn} + q(\mathbf{B}\cdot\hat{\mathbf{S}})^2_{mn}]\Psi_n \\ &+ c_0\rho\Psi_m + c_1\sum_{n=-1}^{1}\mathbf{s}\cdot\hat{\mathbf{S}}_{mn}\Psi_n. \end{aligned} \tag{1}$$

Here, V and $\mathbf{B}$ are the trapping potential and magnetic field. The parameters M, g, μ_B, and q are the mass of a particle, the Landé g factor, the Bohr magneton, and a coefficient of the quadratic Zeeman effect, respectively. The total density ρ and the spin density vector s_i ($i = x, y, z$) are given by $\rho = \sum_{m=-1}^{1}|\Psi_m|^2$ and $s_i = \sum_{m,n=-1}^{1}\Psi^*_m(\hat{S}_i)_{mn}\Psi_n$ with the spin-1 matrices $(\hat{S}_i)_{mn}$. The averaged density is given by $\rho_0 = \int \rho d\mathbf{r}/A$ with the system area A. The parameters c_0 and c_1 are the coefficients of the spin-independent and spin-dependent interactions. We focus on the spin-dependent interaction energy $\mathcal{E}_s = \frac{c_1}{2}\int \mathbf{s}^2 d\mathbf{r}$, whose coefficient c_1 determines whether the system is ferromagnetic ($c_1 < 0$) or antiferromagnetic ($c_1 > 0$).

The GP equations (1) allows us to obtain the hydrodynamic equations of the spin density vector, which we need to understand the scaling power law in the energy spectrum. The full hydrodynamic equations are derived by Yukawa and Ueda [29], while its simper version is available for the ferromagnetic case [30, 31, 32]. We will assume two things. First, the total density ρ is time-independent because c_0 is usually much larger than c_1. Secondly, the spin density vector takes the largest value everywhere owing to the ferromagnetic interaction. The two assumptions are confirmed in our numerical simulation [19]. Then the normalized spin density vector $\hat{\mathbf{s}} = \mathbf{s}/\rho_0$ is shown to obey the hydrodynamic equations

$$\frac{\partial}{\partial t}\hat{\mathbf{s}} + (\mathbf{v}\cdot\nabla)\hat{\mathbf{s}} = \frac{\hbar}{2M}\hat{\mathbf{s}}\times[\nabla^2\hat{\mathbf{s}} + (\mathbf{a}\cdot\nabla)\hat{\mathbf{s}}], \quad \mathbf{v} = \frac{\hbar}{2iM\rho}\sum_{m=-1}^{1}(\psi^*_m\nabla\psi_m - \psi_m\nabla\psi^*_m), \tag{2}$$

where $\mathbf{a} = (\nabla\rho)/\rho$ and $\mathbf{v}$ is the superfluid velocity [30, 31, 32]. In a uniform system ρ keeps almost constant, thus vanishing $\mathbf{a}$. As a result, we obtain

$$\frac{\partial}{\partial t'}\hat{\mathbf{s}} + \Big(\frac{\mathbf{v}}{c_s}\cdot\nabla'\Big)\hat{\mathbf{s}} = \hat{\mathbf{s}}\times\nabla'^2\hat{\mathbf{s}}. \tag{3}$$

Here space and time are respectively normalized by the coherence length $\xi = \hbar/\sqrt{2Mc_0\rho_0}$ and a characteristic time $\tau = \hbar/c_0\rho_0$ ($t' = t/\tau$, $\nabla' = \xi\nabla$), and $c_s = \sqrt{c_0\rho_0/2M}$ is the sound velocity. Another important characteristic length is the spin coherence length $\xi_s = \hbar/\sqrt{2M|c_1|\rho_0}$, which is relevant to spin structures such as domain walls and polar core vortex [33].

3. Spin turbulence with the ferromagnetic interaction

We numerically created ST by three different methods: (1) the instability of counterflow between the $m = \pm 1$ components in a uniform system [19], (2) the instability of the helical spin structure in a trapped system [20], (3) the instability by an oscillating magnetic field in a uniform system

[21]. All simulations are performed in a two-dimensional system. In every case we confirmed the -7/3 power law in the spectrum of the spin-dependent interaction energy and the spin-glass-like behavior [22], though only in the case (3) how to apply the oscillating magnetic field is different between [21] and [22]. In this section we describe the case of (1) as a typical one.

3.1. Spin turbulence and the energy spectrum

The counterflow between the $m = \pm 1$ components with a relative velocity $\mathbf{V}_R = V_R \hat{\mathbf{e}}_x$ is represented by the initial wave functions

$$(\Psi_1, \Psi_0, \Psi_{-1}) = \sqrt{\frac{\rho_0}{2}}\Big(\exp\Big[i\Big(\frac{M}{2\hbar}\mathbf{V}_R\cdot\mathbf{r} - \frac{\mu_1}{\hbar}t\Big)\Big], 0, \exp\Big[-i\Big(\frac{M}{2\hbar}\mathbf{V}_R\cdot\mathbf{r} + \frac{\mu_{-1}}{\hbar}t\Big)\Big]\Big), \quad (4)$$

where $\hat{\mathbf{e}}_x$ is a unit vector along the x direction and the chemical potentials μ_1 and μ_{-1} are equal to $c_0\rho_0 + MV_R^2/8$. The analysis of the Bogoliubov-de Gennes equation shows that the system is dynamically unstable at any relative velocity V_R [19]. Hence the initial counterflow goes through the dynamical instability towards ST.

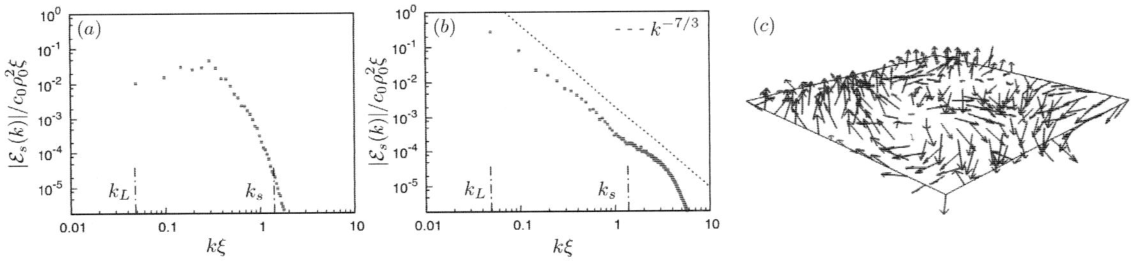

Figure 1. Time dependence of spectrum of spin-dependent interaction energy $\mathcal{E}_s(k)$ for the ferromagnetic case at $t/\tau =$ (a) 150, and (b) 3500. The dotted line in (b) refers to the $k^{-7/3}$ power law. The wave numbers k_L and k_s correspond to the system size and the spin coherence length, respectively. (c):Profile of the spin density vector corresponding to (b). This ST is obtained through the counterflow instability with $c_0/c_1 = -20$, $c_0 > 0$, $V_R/c_s = 0.78$. The coordinate is normalized by the coherence length ξ and the box size is 128×128. Space in the x and y directions is discretized into 512×512 bins.

How ST develops is shown in Fig. 1. The energy spectrum $\mathcal{E}_s(k)$ has in the early stage a peak corresponding to the most unstable mode of the dynamical instability (Fig. 1 (a)), while the energy gradually flows to the higher wave numbers. When the spin density vector is enough disordered as shown in Fig. 1 (c), the energy spectrum takes a characteristic -7/3 power law (Fig. 1 (b)).

Such a characteristic power law exhibits that this ST is not just a random disordered state but an organized self-similar state. We can find easily the similar scenario in classical turbulence (CT) [25, 26]. Three-dimensional fully developed turbulence has the inertial range, where the kinetic energy is transfered selfsimilarly from low to high wave numbers and its spectrum shows the Kolmogorov -5/3 law. This Kolmogorov law can be obtained by applying the scaling analysis due to the selfsimilarity to the hydrodynamic equations [34]. The similar derivation is available to the hydrodynamic equation (3). The present case allows us to neglect the second term in the left hand side, because $|\mathbf{v}|$ is much smaller than c_s. Let us assume that Eq. (3) is invariant for the scale transformation $\mathbf{r} \rightarrow \alpha\mathbf{r}$ and $t \rightarrow \beta t$. Then it is easy to find that the spin density vector should be transformed as $\hat{\mathbf{s}} \rightarrow \alpha^2\beta^{-1}\hat{\mathbf{s}}$. Hence we obtain $\hat{\mathbf{s}} \sim \Lambda k^{-2}t^{-1}$ with a dimensional constant Λ in the scaling region. We apply the usual dimensional analysis to the scaling region to suppose that the energy flux $\epsilon \sim \hat{\mathbf{s}}^2 t^{-1} \sim \Lambda^2 k^{-4} t^{-3}$ is independent of k. Then the energy spectrum $\mathcal{E}_s(k)$ should be represented only by ϵ, k and Λ, resulting in the spectrum $\mathcal{E}_s(k) \sim \hat{\mathbf{s}}^2 k^{-1} \sim \Lambda^{2/3} \epsilon^{2/3} k^{-7/3}$.

3.2. Spin-glass-like behavior

ST shows another characteristic behavior. In ST the spin density vectors become spatially random but temporally frozen, which reminds us of spin glass. Spin glasses are magnetic systems in which the interactions between the magnetic moments are in conflict with each other [35]. Thus, these systems have no long-range order but exhibit a freezing transition to a state with a kind of order in which the spins are aligned in random directions.

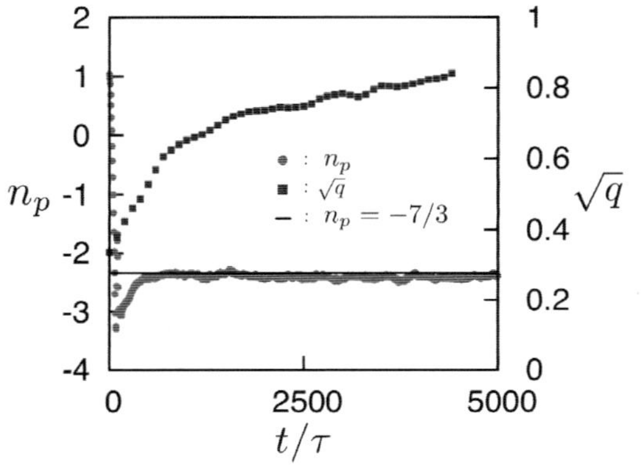

Figure 2. Time development of the exponent n_p of the spectrum $\mathcal{E}_s(k)$ and the spin glass order parameter $\sqrt{q(t)}$. The conditions and the parameters are same as those of Fig. 1. See Ref. [22] for the detail.

In order to characterize the spin-glass-like behavior in ST, we introduced an order parameter following the spin-glass theory [35, 36]. We should note that the order parameter is time-dependent. For the normalized spin density vector $\hat{\mathbf{s}}(\mathbf{r},t) = \mathbf{s}(\mathbf{r},t)/|\mathbf{s}(\mathbf{r},t)|$, its space average and the time average during the period $[t, t+T]$ could be defined as

$$[\hat{\mathbf{s}}(\mathbf{r},t)] = \frac{1}{A}\int_A \hat{\mathbf{s}}(\mathbf{r},t) d\mathbf{r}, \quad \langle \hat{\mathbf{s}}(\mathbf{r},t)\rangle_T = \frac{1}{T}\int_t^{t+T} \hat{\mathbf{s}}(\mathbf{r},t_1) dt_1. \tag{5}$$

The argument of how to take T is discussed in Ref. [22]. The time-dependent order parameter is defined as

$$q(t) = [\langle \hat{\mathbf{s}}(\mathbf{r},t)\rangle_T^2]. \tag{6}$$

If the spin density vector is completely frozen, $q(t)$ should be unity. Figure 2 shows the time dependence of $q(t)$ and the power exponent $n_p(t)$ of the spectrum of the spin-dependent interaction energy. The order parameter $q(t)$ increases obviously as the exponent $n_p(t)$ approaches $-7/3$, which means that the spin-glass-like order grows as the ST with the $-7/3$ power law develops. Of course, our system of spinor BECs differs from a magnetic system yielding spin glass. Spin glass states do not show such a power law behavior in the energy spectrum, being much different from ST. We do not know currently what causes the spin-glass-like behavior in ST [22].

4. Spin turbulence with the small spin magnitude

When the amplitude of spin density vector is small, another -1 power law appears in the low wave number region in addition to the -7/3 power in the high wave number region of the energy spectrum in ST [23]. This kind of ST is realized in two cases: (i) with antiferromagnetic interaction and (ii) with ferromagnetic interaction under a static magnetic field. In the first article [19] we found that the antiferromagnetic interaction allowed the system to develop to ST too but the power exponent of $\mathcal{E}_s(k)$ did not obey the simple scaling behavior discussed in the last section. Investigating the behavior of the ST led us to the consideration that the different behavior was attributable to the small spin magnitude. Then we noticed that ST with small spin amplitude is realized in the system with ferromagnetic interaction under a static magnetic field too, where the quadratic Zeeman term reduces the spin amplitude.

The scaling argument should start with the full hydrodynamic equation [29] described by **s** and the nematic tensor

$$n_{\mu\nu} = \frac{1}{\rho}\Psi_m^*(\hat{N}_{\mu\nu})_{mn}\Psi_n \tag{7}$$

with

$$(\hat{N}_{\mu\nu})_{mn} = \frac{1}{2}[(\hat{S}_\mu)_{ml}(\hat{S}_\nu)_{ln} + (\hat{S}_\nu)_{ml}(\hat{S}_\mu)_{ln}]. \tag{8}$$

The detail is described in Ref. [23]. Here we summarize the results.

We apply three approximations. The first is that the total density ρ is uniform. The second is that the superfluid velocity is much smaller than the sound velocity. The third is that the magnitude of the spin vector $\mathbf{s}/\rho$ is smaller than unity. Then appears a boundary wave number $k_b = 2\sqrt{|c_1|M\rho_0}/\hbar$. Applying the scaling argument to the hydrodynamic equations leads to the spectrum $\mathcal{E}_s(k) \sim \epsilon_L^{2/3}k^{-1}$ for $k < k_b$ and $\mathcal{E}_s(k) \sim \epsilon_H^{2/3}k^{-7/3}$ for $k > k_b$ with different energy flux ϵ_L and ϵ_H. We numerically obtain ST in a uniform system by the counterflow instability. The energy spectrum for developed ST is shown in Fig. 3. In the system with antiferromagnetic interaction (Fig. 3 (a)), the spectrum tends to exhibit the expected $-7/3$ power law in the high-wave-number region $k_b < k < k_s$, but it deviates from the -1 power in the low-wave-number region $k_L < k < k_b$. The system with ferromagnetic interaction under a static magnetic field exhibits more clearly both the -1 and -7/3 power laws as shown in Fig. 3 (b).

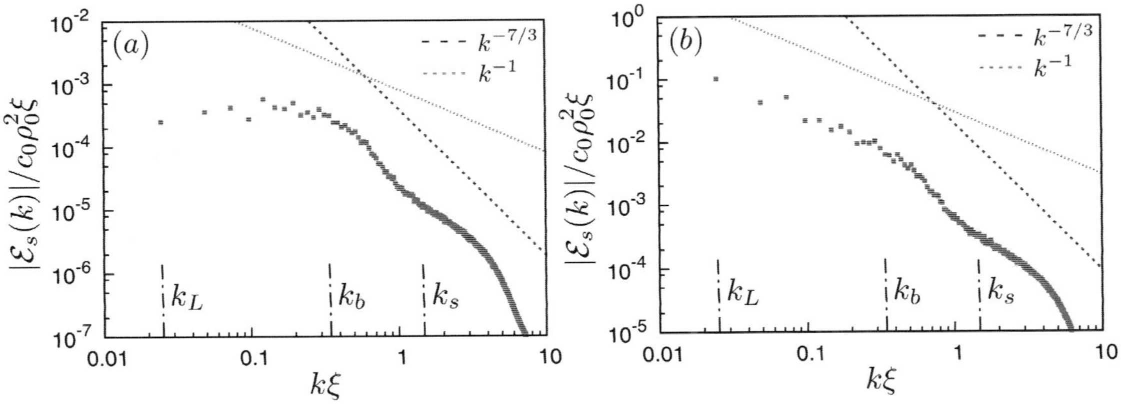

Figure 3. Spectrum of spin-dependent interaction energy $\mathcal{E}_s(k)$ for (a) antiferromagnetic interaction at $t/\tau = 3500$ and (b) ferromagnetic interaction under a static magnetic field at $t/\tau = 3000$ [23]. The wave numbers k_L and k_s correspond to the system size and the spin coherence length, respectively, and the boundary wave number k_b is given in the text.[Fujimoto *et al.*: Phys. Rev. A **88** (2013), reproduced with permission. Copyright 2013 by the American Physical Society].

We investigated the order parameter q of Eq. (6) for ST with antiferromagnetic interaction too [24]. The order parameter q does not grow, which means that spin density vectors are not frozen temporally but fluctuate. This behavior is much different from the ferromagnetic case.

5. Conclusions and discussions

Spinor Bose-Einstein condensates propose a novel system of turbulence in quantum fluids, namely spin turbulence (ST) in which the spin density vectors are disordered. When the dynamics starts from an initial state with the energy much higher than the ground state or the system is excited from the ground state by an applied field, ST appears after hydrodynamic instability. ST exhibits some characteristic statistical laws, and the behavior depends on the spin-dependent interaction. When the spin-dependent interaction is ferromagnetic, the spectrum of the interaction energy shows the -7/3 power law; this is understood through the scaling analysis for the hydrodynamic equations of spins and confirmed by the numerical simulation of the GP equations. Furthermore, the spin density vectors are spatially random but temporally frozen, which reminds us of the analogy with spin glass; the order parameter certainly grows in ST. However, the story is just more complicated when the interaction is antiferromagnetic. Then appears a boundary wave number k_b, and the scaling analysis in ST finds that the energy spectrum obeys the -1 power law for $k < k_b$ and the -7/3 law for $k > k_b$; the -7/3 power law was confirmed numerically. This energy spectrum actually comes from the small spin amplitude due to the antiferromagnetic interaction. The similar situation is expected to occur also for the ferromagnetic system under a static magnetic field that reduces the spin amplitude, in which both -1 and -7/3 power laws were confirmed clearly by the numerical simulation. The spin-glass-like behavior did not appear in the antiferromagnetic ST.

Such a spin-disordered state was experimentally created in a trapped system through the instability of the initial helical structures of spin, and actually observed [37]. Creating ST and observing the spin density vector $\mathbf{s}(\mathbf{r}, t)$ would allow us to investigate the energy spectrum $\mathcal{E}_s(k)$ and the spin-glass-like behavior.

It is very important to ask what causes the energy cascade in ST. A mechanism causing the Kolmogorov $-5/3$ power law in classical turbulences is considered to be the Richardson cascade of vortices [26, 25]. Since there are very few topological defects in our simulation of ST, however, the cascade due to them is not relevant. At present we guess the cascade of spin waves plays an important role for ST. This will be worked in reference to the wave turbulence theory [38, 39].

Usual selfsimilar power law in the inertial range of turbulence is caused by some dissipation at high wave numbers, which makes the energy flow towards high wave numbers. Our system based on Eq. (1) conserves the total energy; no dissipative mechanisms are included in our formulation. Thus it is not so trivial to understand why and how the characteristic power laws are built and sustained in $\mathcal{E}_s(k)$. Here we should note that spinor BECs have superfluid degrees of freedom (mass current) as well as spin degrees of freedom and they are coupled. Hence the spin-dependent interaction energy $\mathcal{E}_s$ is not conserved in the dynamics. While spin develops ST, superfluid may create its own turbulence. It would be quite interesting to investigate how these degrees of freedom interact with each other and develop turbulence for the whole system.

Acknowledgments

The authors thank Yusuke Aoki for the scientific collaboration and the preparation of the manuscript.

[1] Halperin W P and Tsubota M (eds) 2009 *Progress in Low Temperature Physics* vol 16.(Amsterdam: Elsevier)
[2] Fetter A L 2009 *Rev. Mod. Phys.* **81** 647
[3] Kasamatsu K. and Tsubota M 2009 *Prog. Low Temp. Phys.*, ed. Halperin W P and Tsubota M vol 16 (Amsterdam: Elsevier) p 351
[4] Fetter A L 2010 *J. Low Temp. Phys.* **161** 445
[5] Anderson B P 2010 *J. Low Temp. Phys.* **161** 574
[6] Berloff N G and Svistunov B 2002 *Phys. Rev.* A **66** 013603
[7] Parker N G and Adams C S 2005 *Phys. Rev. Lett.* **95** 145301
[8] Kobayashi M and Tsubota M 2007 *Phys. Rev.* A **76** 045603

[9] Henn E A L, Seman J A, Ramos E R F, Caracanhas M, Castilho P, Olímpio E P, Roati G, Magalhães D V, Magalhães K M F and Bagnato V S 2009 *Phys. Rev.* A **79** 043618
[10] Henn E A L, Seman J A, Roati G, Magalhães K M F and Bagnato V S 2009 *Phys. Rev. Lett.* **103** 045301
[11] Kasamatsu K, Tsubota M and Ueda M 2005 *Int. J. Mod. Phys.*B **19** 1835
[12] Kawaguchi Y and Ueda M 2013 *Phys. Rep.* **520** 253
[13] Stamper-Kurn D M and Ueda M 2013 *Rev. Mod. Phys.* **85** 1191
[14] Tsubota M, Kasamatsu K and Kobayashi M 2013 *Novel Superfluids* ed. Bennemann K H and Ketterson J B vol 1 (Oxford: Oxford University Press) p 156
[15] Tsubota M, Kobayashi M and Takeuchi H 2013 *Phys. Rep.* **522** 191
[16] Takeuchi H, Ishino S and Tsubota M 2010 *Phys. Rev. Lett.* **105** 205301
[17] Ishino S, Tsubota M and Takeuchi H 2011 *Phys. Rev.* A **83** 063602
[18] Karl M, Nozak B and Gasenzer T 2013 *Sci. Rep.* **3** 02394
[19] Fujimoto K and Tsubota M 2012 *Phys. Rev.* A **85** 033642
[20] Fujimoto K and Tsubota M 2012 *Phys. Rev.* A **85** 053641
[21] Aoki Y and Tsubota M 2013 *J. Low Temp. Phys.* **171** 382
[22] Tsubota M, Aoki Y and Fujimoto K 2013 *Phys. Rev.* A **88** 061601(R)
[23] Fujimoto K and Tsubota M 2013 *Phys. Rev.* A **88** (in press)
[24] Aoki Y, Fujimoto K and Tsubota M 2013 *J. Low Temp. Phys.* DOI 10.1007/s10909-013-0996-7
[25] Frisch U 1995 *TURBULENCE* (Cambridge: Cambridge University Press)
[26] Davidson P A 2004 *turbulence* (Oxford: Oxford University Press)
[27] Ohmi T and Machida K 1998 *J. Phys. Soc. Jpn.* **67** 1822
[28] Ho T-L 1998 *Phys. Rev. Lett.* **81** 742
[29] Yukawa E and Ueda M 2012 *Phys. Rev.* A **86** 063614
[30] Lamacraft A 2008 *Phys. Rev.* A **77** 063622
[31] Kudo K and Kawaguchi Y 2010 *Phys. Rev.* A **82** 053614
[32] Kudo K and Kawaguchi Y 2011 *Phys. Rev.* A **84** 043607
[33] Saito H, Kawaguchi Y and Ueda M 2007 *Phys. Rev.* A **75** 013621
[34] Watanabe T, Fujisaka H and Iwayama T, *Phys. Rev.* E **55** 5575
[35] Binder K and Young A P 1986 *Rev. Mod. Phys.* **58** 801
[36] Sherrington D and Kirkpatrick S 1975 *Phys. Rev. Lett.* **35** 1792
[37] Vengalattore M, Leslie S R, Guzman J and Stamper-Kurn D M 2008 *Phys. Rev. Lett.* **100** 170403
[38] Nazarenko S 2011 *Wave turbulence* (Heiderberg: Springer)
[39] Zakharov V E, L'vov V S and Falkovich G 1992 *Kolmogorov Spectra of Turbulence I: Wave Turbulence* (Berlin: Springer)

Ultrabroadband Airy light bullets

Peeter Piksarv[1], Andreas Valdmann[1], Heli Valtna-Lukner[1,2] and Peeter Saari[1,3]

[1]Institute of Physics, University of Tartu, Riia 142, Tartu, 51014 Estonia
[2]Max Planck Institute for the Science of Light, Günther-Scharowsky Str. 1, 91058 Erlangen, Germany
[3]Estonian Academy of Sciences, Kohtu 6, Tallinn, 10130 Estonia

E-mail: peeter.piksarv@ut.ee

Abstract. We present the measurements of the spatiotemporal impulse responses of two optical systems for launching ultrashort Airy pulses, incl. ultrabroadband nonspreading Airy beams whose main lobe size remains invariantly small over propagation. First, a spatial light modulator and, second, a custom refractive element with continuous surface profile were used to impose the required cubic phase on the input field. White-light spectral interferometry setup based on the SEA TADPOLE technique was applied for full spatio-temporal characterization of the impulse response with ultrahigh temporal resolution approaching a single cycle of the light wave. The results were compared to the theoretical model.

Keywords: Airy pulses, ultrafast measurement, non-diffractive waves

PACS numbers: 42.25.Bs, 42.65.Re, 42.79.-e

1. Introduction

The family of non-diffracting or localized waves [1] has grown recently thanks to discovery of several wave-fields whose intensity maximum not only is transversely confined, but also moves along a curved trajectory during propagation in free space. The first and most well-known of such wavefields—the Airy beam—has attained enduring attention since its introduction in optics by Siviloglou and Christodoulides [2, 3]. Pulsed versions of the Airy beam have also been discussed and experimentally realized [4–7]. Although direct light-field modulation can be used to create Airy beams [8], a more common experimental setup involves imposing a cubic phase profile on the input Gaussian beam and performing an optical Fourier transform with a lens. The desired spatial phase modulation can be acquired, *e.g.*, with spatial light modulators (SLM) [2], lithographically etched micro-optical phase plates [9], combinations of cylindrical lenses [10] and even by a purposeful use of lens aberrations [6]. Airy beams have been put to use for generation of curved plasma channels [11] and optical micromachining [12].

For many possible applications, it is important, that the main lobe of the beam remains laterally confined. In the monochromatic case, even the truncated version of Airy beam resists diffraction over propagation distances much larger than the Rayleigh length and its main intensity lobe retains its lateral confinement [2]. However, in the case of ultrashort pulses with broadband spectrum spreading over tens or hundreds of nanometres or spanning even over more than an octave, the dispersion becomes unavoidable. Therefore, the wavelength dependence of the system for creating ultrashort Airy pulses must be taken into consideration to understand the spatial and temporal properties of the main intensity lobe.

In this paper we study the impulse responses of two optical systems for launching ultrashort Airy pulses. Ultrashort Airy pulses shaped by an SLM and an optical element that has a continuous surface profile following a cubic function are compared. A spatial-spectral interferometry technique based on the SEA TADPOLE method [13] was used to fully characterize the created Airy wave field with high spatial and temporal resolution.

2. Theoretical models

In general, a broadband 1-dimensional Airy beam can be expressed in the paraxial approximation as a superposition of its monochromatic constituents of wavenumber k over spectrum $S(k)$ [4, 5, 14]

$$\Psi(x,z,t) = \int_0^\infty \mathrm{d}k S(k)\,\Phi(x,z,k)\,\mathrm{e}^{\mathrm{i}k(z-ct)}, \tag{1}$$

where

$$\Phi(x,z,k) = \mathrm{Ai}\left[\frac{x}{x_0} - \left(\frac{z}{2z_0}\right)^2 + \mathrm{i}a\frac{z}{z_0}\right]$$

$$\times \exp\left[a\frac{x}{x_0} - \frac{a}{2}\frac{z^2}{z_0^2} + \mathrm{i}\varphi\left(x, z, k\right)\right], \tag{2}$$

$$\varphi\left(x, z, k\right) = \frac{x}{2x_0}\frac{z}{z_0} - \frac{1}{12}\left(\frac{z}{z_0}\right)^3 + \frac{a^2}{2}\frac{z}{z_0}. \tag{3}$$

Here, Ai(...) is the Airy function and a is small parameter describing the lateral exponential decay of the infinite Airy trail. Scale of the beam is determined by lateral and longitudinal characteristic lengths x_0 and z_0. The parabolic wave-equation requires that $z_0 = kx_0^2$. Additionally, the parameter $b_0 = x_0/4z_0^2$ characterizes the parabolic trajectory of the main lobe of the Airy beam. The wave function of the 2D Airy beams is obtained as the product of two 1D wave functions $\Psi\left(x, y, z, t\right) = \Psi_x\left(x, z, t\right)\Psi_y\left(y, z, t\right)$.

The Airy beam is typically created with a lens performing an optical Fourier transform of the initial field [3]

$$\Phi_0\left(\xi, k\right) = \mathrm{e}^{-ac_0^2\xi^2}\mathrm{e}^{\mathrm{i}\frac{1}{3}\left(c_0^3\xi^3 - 3a^2c_0\xi - \mathrm{i}a^3\right)}, \tag{4}$$

were ξ is the coordinate in the input plane of the lens. The lateral scale of the cubic phase function in the input plane is described by parameter c_0. The focal length f of the lens imposes a relation $c_0 = {}^{x_0k}/_f$ on the input and output fields. Four types of Airy pulses have been defined according to the dependence of c_0 on the wave-number [4, 5, 7]. Wavelength constituents of an Airy pulse can spatially overlap only in certain regions, which is determined by the four parameters of the Airy beam: x_0, z_0, c_0 and b_0. Type I Airy pulses have $x_0\left(k\right) = const$, for type II $c_0\left(k\right) = const$, type III pulses have $z_0\left(k\right) = const$ and for type IV $b_0\left(k\right) = const$. The type of an Airy pulse is determined by properties of the optical system for creating it. It was previously shown that an SLM with a cubic phase mask containing discontinuities will produce type II Airy pulses that exhibit strong lateral dispersion [14].

For applications that require a spatially and temporally localized ultrashort pulse, type IV Airy beams would be most suitable as all its spectral components propagate along the same trajectory and their main intensity lobe is not subjected to lateral dispersion. The cubic phase term in (4) is proportional to c_0^3 and therefore for the nondispersing type IV Airy pulses with $c_0 \propto k^{1/3}$, the phase modulation amplitude for different wavelengths in the Fourier plane is proportional to k. This could be achieved by using a suitably curved mirror as the cubic phase element. A refractive phase plate with a continuous cubic surface profile can also be used, but dispersion in the substrate material must be taken into account.

3. Methods and experimental setup

The underlying principle of our technique for measuring spatiotemporally resolved impulse responses is the well known spectrally resolved white-light interferometry. However, since the output field of an optical element generally depends not only on time and/or the longitudinal coordinate, but also on the transverse coordinates, the interferometry has to be generalized into three dimensions as follows.

Let $E_0(t)$ be the electric field of the homogeneous, transversely coherent light beam in the input plane $z = 0$ of the optical system. The field $E(\mathbf{r}, t)$ in the output half-space $z > 0$, which generally has acquired a 3D dependence on the field point coordinates $\mathbf{r} = (x, y, z)$ and a specific temporal dependence in every point, is given as temporal convolution

$$E(\mathbf{r}, t) = \int_0^\infty h(\mathbf{r}, t') E_0(t - t') \, \mathrm{d}t' , \tag{5}$$

where $h(\mathbf{r}, \tau)$ is the impulse response of the optical system for a location $\mathbf{r}$ behind it.

Let us suppose that somehow we have measured the cross-correlation function $K(\mathbf{r}, \tau)$ between the input and output fields $K(\mathbf{r}, \tau) \equiv \langle E(\mathbf{r}, t) E_0(t - \tau) \rangle$. Thereby we have also found out the impulse response function sought, since

$$\begin{aligned} K(\mathbf{r}, \tau) &= \left\langle \int_0^\infty h(\mathbf{r}, t') E_0(t - t') \, \mathrm{d}t' E_0(t - \tau) \right\rangle = \\ &= \int_0^\infty h(\mathbf{r}, t') \, \langle E_0(t - t') E_0(t - \tau) \rangle \, \mathrm{d}t' = \\ &= const \times h(\mathbf{r}, \tau) \ , \end{aligned} \tag{6}$$

as far as the temporal dependence of the input field can be considered as a delta-correlated stationary noise in which case $\langle E_0(t - t') E_0(t - \tau) \rangle \propto \delta(t' - \tau)$ and the integration over t' drops out. In a general case, $K(\mathbf{r}, \tau)$ is the convolution of $h(\mathbf{r}, \tau)$ with the autocorrelation function of the input laser field and the impulse response can be recovered by a deconvolution method. As a matter of fact, the proportionality relation $h(\mathbf{r}, t) \propto K(\mathbf{r}, t)$ practically holds always whenever the spectrum of the input field is sufficiently "white", *i.e.*, wider than the spectrum of $h(\mathbf{r}, t)$—in other words, wider than the support of the transfer function. Our setup follows the scheme of SEA TADPOLE, where the cross-correlation $K(\mathbf{r}, \tau)$ has been obtained through three procedures: (i) quadratic detection of the sum $E(\mathbf{r}, t) + E_0(t - \tau)$, where the τ-dependence appears in the overlap region of the two beams tilted with respect to each other; (ii) temporal averaging due to the "slowness" of the detector (a CCD camera); (iii) spectral-domain separation of the cross-correlation term from the two non-oscillating autocorrelation terms. The spatial dependence of $h(\mathbf{r}, t) \equiv h(\mathbf{r}, \tau)$ is obtained by simply scanning the light-collecting tip of the signal-carrying fibre in the output half-space.

It should be pointed out that (6) in the case of time averaging is actually insensitive to the nature of the temporal dependence of the input light—be it deterministic ultrashort pulses or stationary noise or periodical bursts of temporally incoherent noise (which is the case of our set-up with a supercontinuum fibre laser). This favourable independence from the spectral phase of the light source follows from the basic principle of the spectrally resolved white-light interferometry, or—more generally—from the properties of time-averaged responses of linear systems.

The experimental setup for generating and measuring ultrabroadband Airy pulses is shown in figure 1. The measurement setup follows the SEA TADPOLE arrangement for spatiotemporal ultrashort pulse characterization using spatial spectral

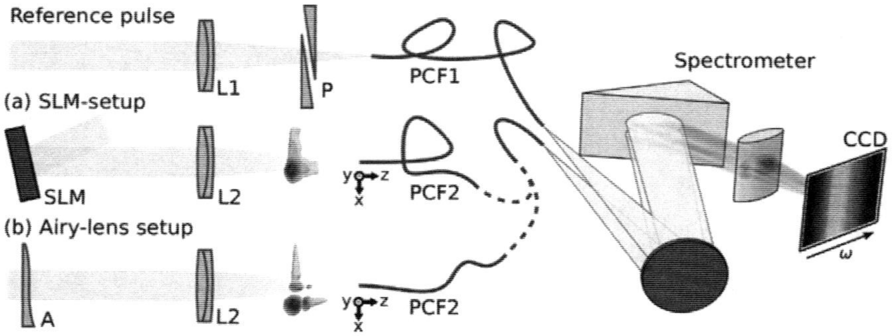

Figure 1. Experimental setup. Airy pulses were generated and measured in two setups: (a) using an SLM with wrapped cubic phase mask, (b) using a custom-made continuous phase element. An endlessly single mode photonic crystal fibre PCF2 samples a small region of the generated Airy fields. The reference is coupled into an identical fibre PCF1. The coupling lens L1 and the Fourier transforming lens L2 are of the same type. A wedge prism pair P was used to fine tune the dispersion between two arms of the inferferometer. The fibre outputs are placed on the entrance slit of the spectrometer at a small displacement resulting them to cross. A two-dimensional spectral interferogram is recorded, and $E(\omega)$ is retrieved [13]. The fibre inputs on the measurement arm are scanned in three dimensions in order to measure the spatiotemporal field.

interferometry [13]. A more detailed description of the SEA TADPOLE method using supercontinuum laser source (Fianium SC400-2-PP) can be found in [15] and references therein. In short, the device involves sampling a small spatial region of the field under study with a single-mode photonic crystal fibre and recording its interference pattern with a reference pulse in a spectrometer from which the complex spectral response and hence also the impulse response is reconstructed at every point in space. The fibre is scanned both transversely along x and y axis and longitudinally along z. For longitudinal scanning, the delay on the reference arm was adjusted correspondingly to keep the relative path lengths of the interferometer arms constant. Our measurable spectral range was from $428\ldots1088\,\mathrm{nm}$ yielding a near-cycle temporal resolution of $2.5\,\mathrm{fs}$. The spatial resolution is determined by the mode size of the fibres and is approximately $3\,\mu\mathrm{m}$.

Type II Airy beam profile was created by a reflective SLM (Holoeye LC-R 2500) with an imposed cubic phase mask in both transverse orthogonal directions [14]. The SLM was placed in the focal plane of an achromatic lens with focal length of $f = 500\,\mathrm{mm}$. A small uncompensated spectral phase term introduced by the thickness of the SLM and by the small inequality of the fibre lengths was determined with a zero phase mask on the SLM and subtracted from the measured spectral phase.

For generating type IV Airy pulses we chose to use a custom-made refractive phase element which can be directly placed into the beam path. The phase plate was fabricated using the PowerPhotonic LightForge service. The bulk dispersion of the fused silica substrate was compensated in the setup. Some material dispersion still remained as the phase element thickness varies, but this small perturbation does not cause any significant deviation of the created pulse from a pure type IV Airy pulse as it will be confirmed below.

In order to improve the transverse beam quality the laser beam was spatially filtered trough a single mode photonic crystal fibre and expanded using off-axis parabolic mirrors for type IV Airy pulse measurements. In addition, a pair of fused silica wedge prisms were added into the reference arm to precisely control the dispersion in the reference arm and compensate for the residual dispersion arising from the possible inequalities in the length of fibre and from the substrate of the phase element. Therefore any additional calibration of the device, *i.e.*, subtracting a spectral phase of a blank substrate from the measurement results, was not necessary.

Two sets of achromatic lenses ($f = 500\,\mathrm{mm}$ and $f = 300\,\mathrm{mm}$) were used in the setup, effectively allowing to obtain desired parameters of the generated Airy pulses even with the given single phase element. As the linear dimensions of the generated Airy beam scale inversely proportionally to the focal length of the Fourier transforming lens, it is a convenient way to generate beam with a desired size of the focal spot.

The simulations were carried out by evaluating (1) using the measured spectrum. The parameters x_0 and y_0 were calculated from the phase mask displayed on the SLM and from the phase profile of the cubic phase element by taking into account the dispersion relation of the fused silica substrate of the element. It was assumed that the Fourier transform was performed by an ideal achromatic thin lens. The parameter a for each wavelength was determined by measuring the beam profile with a separate CCD camera and a set of interference filters. These data were thereafter inter- and extrapolated to cover the full spectral range of the SEA TADPOLE.

4. Results and discussion

The measured spatiotemporal response of the systems with a lens of $f = 500\,\mathrm{mm}$ for generating type II Airy pulses and type IV Airy pulses are shown and compared with simulations in figures 2 and 3, respectively. By comparing those figures it is clearly seen that the type II Airy pulse generated using an SLM exhibits strong lateral dispersion of the Airy maximum as the shorter wavelength constituents of the pulse are deflected more from the linear path. In contrast, the comparable size Airy maximum of the type IV pulse retains its spot size over the same propagation distance. With a lens of $f = 500\,\mathrm{mm}$, the pulse cross section retained a distinct pattern typical to Airy beams throughout the longitudinal measurement range of 200 mm for both types of Airy pulses. However, near the Fourier plane of the lens at the distance $z = 0\,\mathrm{mm}$ the Airy profile of the type II pulse was obstructed by the so-called zero-order of the SLM.

The propagation characteristics depend also on the focal length of the Fourier lens. The results of type IV Airy pulses generated with the continuous cubic phase mask and $f = 300\,\mathrm{mm}$ lens are shown in figure 4. With $f = 300\,\mathrm{mm}$, the pulse exhibited stronger lateral acceleration, but the pattern on the cross section began to deteriorate

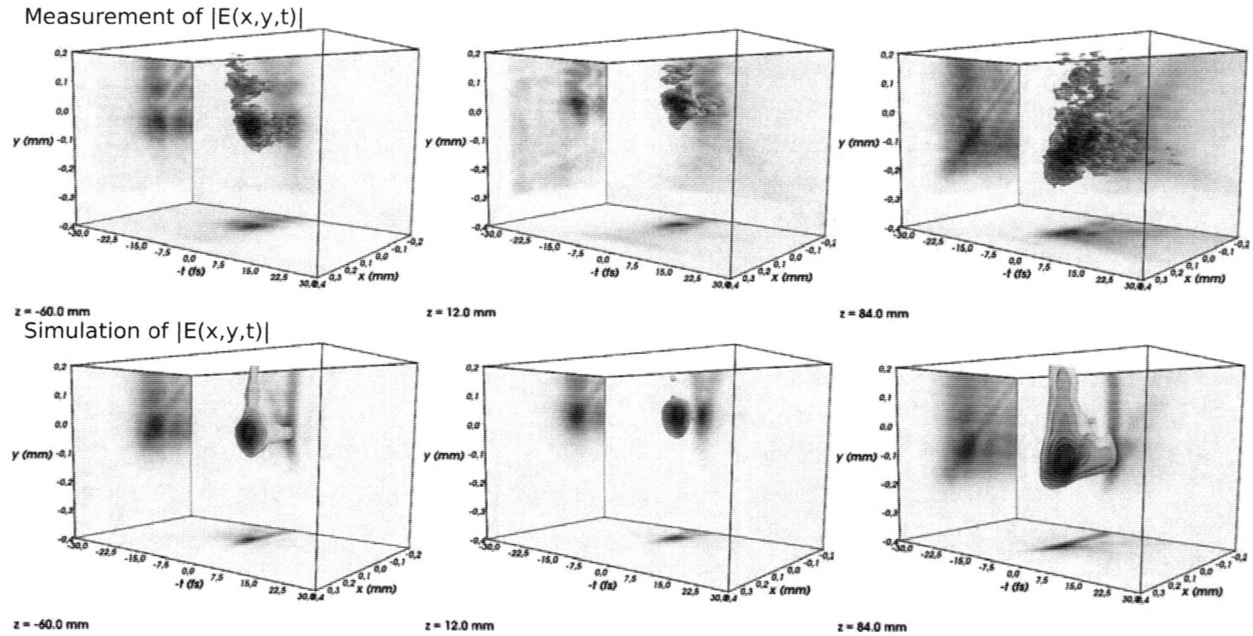

Figure 2. Measured (top) and simulated (bottom) impulse response of type II Airy pulse generator consisting of an SLM with wrapped cubic phase and a lens $f = 500\,\mathrm{mm}$ at three distances. Colour represents the relative amplitude of the electric field isosurfaces. The images show projections of the pulse to the corresponding planes.

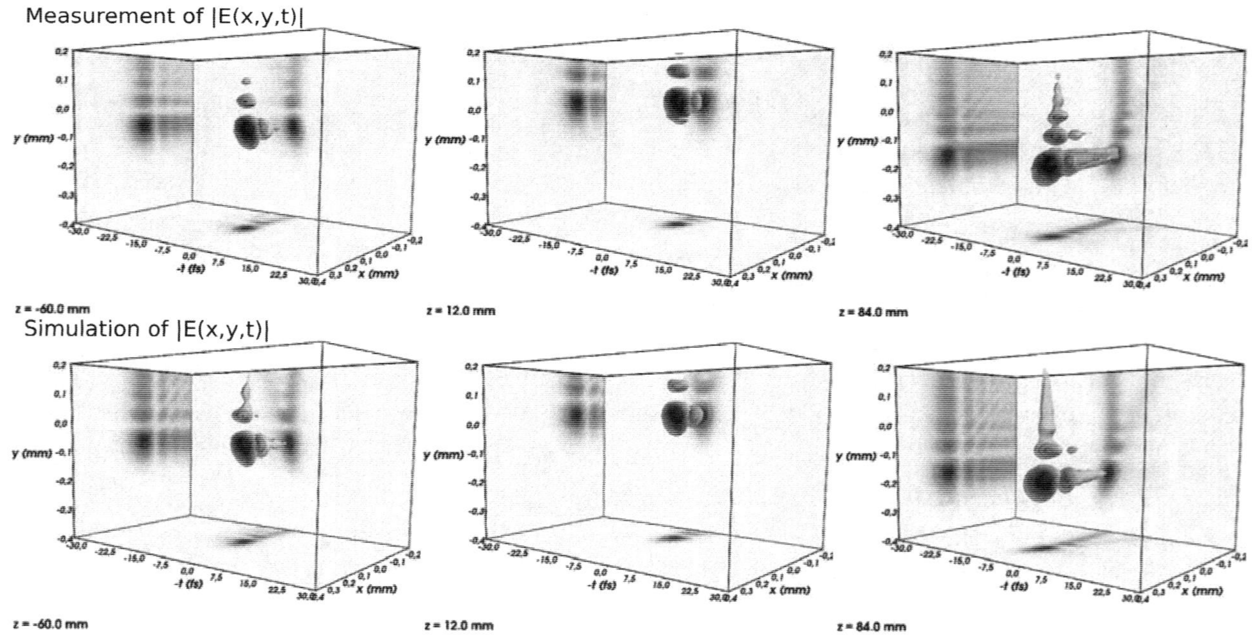

Figure 3. Measured (top) and simulated (bottom) impulse response of type IV Airy pulse generator consisting of an continuous cubic phase element and a lens $f = 500\,\mathrm{mm}$ at three distances.

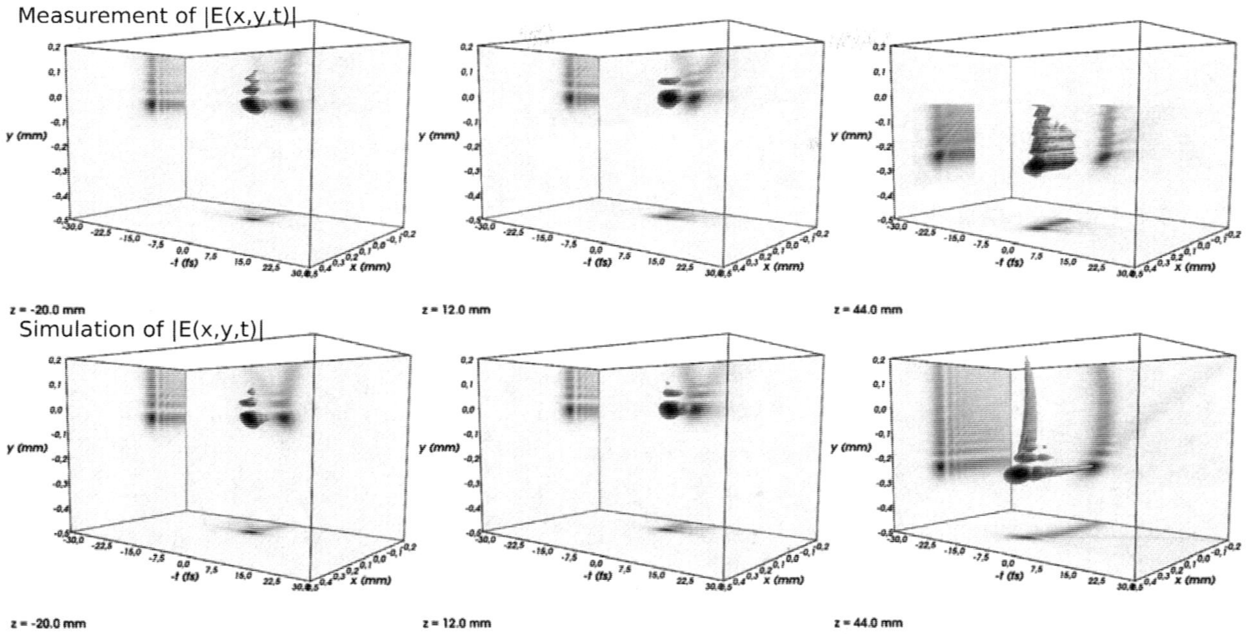

Figure 4. Measured (top) and simulated (bottom) impulse response of type IV Airy pulse generator consisting of the continuous cubic phase element and a lens $f = 300\,\text{mm}$ at three distances. The measurements were carried out only in the range $-225\ldots125\,\mu\text{m}$ respect to the intensity maximum with a $5\,\mu\text{m}$ step in both lateral directions which is two times smaller than in the previous measurements presented in this paper.

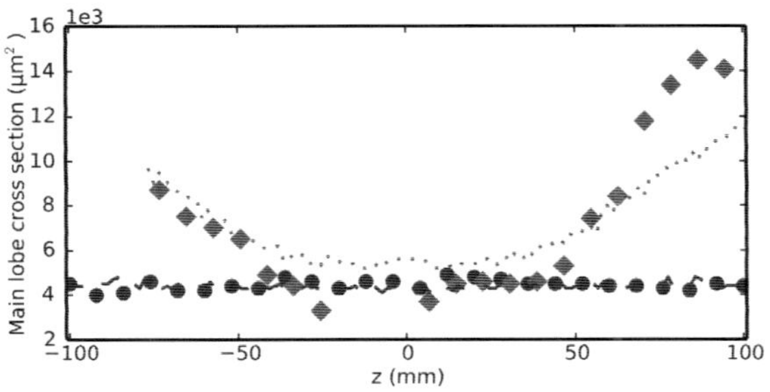

Figure 5. Measured (markers) and simulated (lines) main lobe cross section area of the ultrashort Airy pulse. Rhombuses and dotted line: Airy pulse generated by an SLM and Fourier lens with $f = 500\,\text{mm}$; circles and dashed line: nondispersing Airy pulse generated by special phase element and Fourier lens with $f = 500\,\text{mm}$.

when moving farther away from the apex of the parabolic trajectory along the z-axis. The main lobe was distinguishable throughout a propagation range of 100 mm.

Spatial and temporal localization of the main intensity maximum during propagation was examined to demonstrate the nondispersive characteristics of the created pulse. The pulse intensity profile was averaged over time for each longitudinal measurement point and the transverse size of the main lobe was calculated as an area surrounded by the half maximum intensity contour. The measured and simulated cross section areas are shown in figure 5. For comparison, the main lobe cross section area of an Airy wave field created with an SLM was calculated. The main lobe size of the newly created pulse remains constant throughout the whole propagation range. This was not the case for type II pulses shaped with an SLM, where the main lobe quickly spreads out while moving away from the focus of the Fourier lens. This was caused by strong lateral dispersion as the wavelength constituents of the pulse showed a varying parabolic deflection rate during propagation.

The temporal duration of the pulse intensity maximum over the measured propagation range remained almost constant at 3.6 fs and 3.3 fs (FWHM) for type II and type IV Airy pulses, respectively. This, along with the aforementioned lateral confinement of the main lobe, suggests that creation of curvilinearly propagating localized ultrashort pulses, *i.e.*, "light bullets", is feasible with the refractive phase element having a continuous cubic surface profile. When using a femtosecond laser to create the pulses, bulk dispersion of the cubic phase element can be compensated by adjusting the pulse compressor, which is typically present in the setup. It should be pointed out that the damage threshold of the used fused silica phase plate is much higher than of an SLM, which is promising for applications where high intensity pulses are needed.

5. Conclusion

In conclusion, we have measured the spatiotemporal impulse responses of optical systems for generating ultrabroadband Airy pulses. It was shown that an SLM with an wrapped cubic phase mask creates type II Airy pulses, which have identical spectral phase distribution over the beam cross section after the SLM. Furthermore, it was shown that a continuous cubic phase plate creates type IV Airy pulses, which have laterally nondispersing main lobe that does not spread out over its propagation range. The temporal duration and cross section area of the main lobe of the pulse remained constant during propagation. Properties of the created pulse can be adjusted by selecting a suitable Fourier lens. Choosing a lens with a shorter focal length will result in Airy pulses with stronger later acceleration and smaller main lobe, but aberrations are more noticeable at shorter propagation distances. The presented results are important for real generation of near-single-cycle pulses that follow a curved trajectory, which can be used, *e.g.*, for exploring nonlinear optical processes, high-precision micromachining, fluorescence microscopy, and particle manipulation. Finally, this work emphasises the important differences between the responses of diffractive and refractive optical elements under broadband illumination.

Acknowledgments

This work has been supported by the Estonian Science Foundation grant 7870 and by the Estonian Centre of Excellence in Research "Centre for Nonlinear Studies" receiving funding from the European Regional Development Fund under project 3.2.0101.11-0037. A. V. has been partially supported by the graduate school "Functional Materials and Technologies" receiving funding from the European Social Fund under project 1.2.0401.09-0079.

References

[1] H. E. Hernández-Figueroa, E. Recami, and M. Zamboni-Rached, editors. *Non-Diffracting Waves.* Wiley-VCH, 2013.

[2] G. A. Siviloglou and D. N. Christodoulides. Accelerating finite energy Airy beams. *Opt. Lett.*, 32(8):979–981, 2007.

[3] G. A. Siviloglou, J. Broky, A. Dogariu, and D. N. Christodoulides. Observation of accelerating Airy beams. *Phys. Rev. Lett.*, 99(21):213901, 2007.

[4] P. Saari. Laterally accelerating Airy pulses. *Opt. Express*, 16(14):10303–10308, 2008.

[5] P. Saari. Airy pulse—a new member of family of localized waves. *Laser Physics*, 19(4):725–729, 2009.

[6] D. Abdollahpour, S. Suntsov, D. G. Papazoglou, and S. Tzortzakis. Spatiotemporal Airy light bullets in the linear and nonlinear regimes. *Phys. Rev. Lett.*, 105(25):253901, 2010.

[7] Y. Kaganovsky and E. Heyman. Airy pulsed beams. *J. Opt. Soc. Am. A*, 28(6):1243–1255, 2011.

[8] D. M. Cottrell, J. A. Davis, and T. M. Hazard. Direct generation of accelerating airy beams using a 3/2 phase-only pattern. *Opt. Lett.*, 34(17):2634–2636, 2009.

[9] J. Wang, J. Bu, M. Wang, Y. Yang, and X. Yuan. Generation of high quality Airy beams with blazed micro-optical cubic phase plates. *Appl. Opt.*, 50(36):6627–6631, 2011.

[10] B. Yalizay, B. Soylu, and S. Akturk. Optical element for generation of accelerating Airy beams. *J. Opt. Soc. Am. A*, 27(10):2344–2346, 2010.

[11] P. Polynkin, M. Kolesik, J. V. Moloney, G. A. Siviloglou, and D. N. Christodoulides. Curved plasma channel generation using ultraintense Airy beams. *Science*, 324(5924):229–232, 2009.

[12] A. Mathis, F. Courvoisier, L. Froehly, L. Furfaro, M. Jacquot, P. A. Lacourt, and J. M. Dudley. Micromachining along a curve: Femtosecond laser micromachining of curved profiles in diamond and silicon using accelerating beams. *Appl. Phys. Lett.*, 101(7):071110, 2012.

[13] P. Bowlan, P. Gabolde, A. Shreenath, K. McGresham, R. Trebino, and S. Akturk. Crossed-beam spectral interferometry: a simple, high-spectral-resolution method for completely characterizing complex ultrashort pulses in real time. *Opt. Express*, 14(24):11892–11900, 2006.

[14] P. Piksarv, A. Valdmann, H. Valtna-Lukner, R. Matt, and P. Saari. Spatiotemporal characterization of ultrabroadband Airy pulses. *Opt. Lett.*, 38(7):1143–1145, 2013.

[15] P. Piksarv, H. Valtna-Lukner, A. Valdmann, M. Lõhmus, R. Matt, and P. Saari. Temporal focusing of ultrashort pulsed Bessel beams into Airy–Bessel light bullets. *Opt. Express*, 20(15):17220–17229, 2012.

Journal of Physics: Conference Series **497** (2014) 012004 doi:10.1088/1742-6596/497/1/012004

Measuring vibrational coherence lifetimes in liquid methanol using transient coherent Raman scattering

S Meiselman[1], O Cohen[2], M F DeCamp[1] and V O Lorenz[1]

[1] University of Delaware, Department of Physics and Astronomy, Newark, Delaware 19716, USA

[2] Joint Quantum Institute, National Institute of Standards and Technology & University of Maryland, Gaithersburg, MD 20849, USA

E-mail: vlorenz@udel.edu

Abstract. We demonstrate the measurement of vibrational coherence in liquids using transient coherent Raman scattering. This technique measures the coherence lifetime of vibrational states through the interference of time-delayed coherent Raman-scattered photons using low-power, non-resonant optical pulses. We measure the vibrational lifetime of the 1033 cm^{-1} mode in liquid methanol. The resulting lifetime agrees with frequency-domain lineshape measurements. This technique is a complementary and in some cases simpler alternative to standard nonlinear spectroscopy techniques.

1. Introduction

Vibrational spectroscopy is a useful tool to study molecular structure in the condensed phase, as the vibrational resonances of molecules are quantized and bond-specific. The study of vibrational modes in liquids is often undertaken using inelastic-scattering of optical photons, such as Raman scattering [1]. In the process of Raman scattering (see Figure 1(a)) an incident photon causes a molecule to make a transition from its ground state to a vibrational excited state by interacting non-resonantly with an electronic excited state. The scattered photon has an energy equal to the incident photon energy minus the resulting vibrational state energy. By spectrally resolving these so-called Stokes photons, it is possible to measure directly the vibrational modes of a complex system in the condensed phase.

A typical Raman Stokes spectrum of methanol, taken using a continuous-wave (CW) laser, is shown in Figure 2. The vibrational resonances appear as peaks in the spectrum. Their frequencies and lineshapes contain structural and dynamical information about the molecular bonds. In particular, the resonances at ~3000 cm^{-1} correspond to O-H and C-H vibrations, while the 1000-1500 cm^{-1} peaks correspond to C-O and C-H vibrations [2]. The results shown in Fig. 2 are in the spontaneous regime, in which the intensity of the scattered photons is linear with the incident light intensity and the photons are scattered isotropically. Nonlinear optical techniques [3] can provide greater signal-to-noise ratio and selectivity of which bonds are excited by using a second incident photon at the Stokes photon energy to stimulate the process (see Figure 1(b)). As a result a so-called anti-Stokes photon is emitted as a fourth field in a phase-matched direction. The directionality and enhanced emission rate of the stimulated process results in better signal-to-noise ratios compared to the spontaneous regime. Using ultrashort

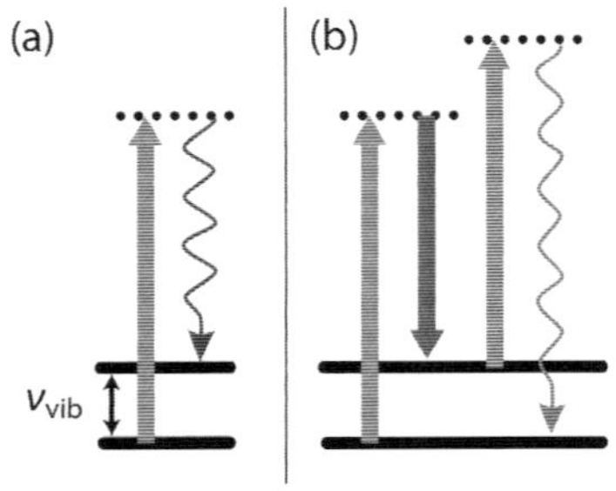

Figure 1. Energy level diagrams for (a) spontaneous Raman Stokes scattering and (b) stimulated Raman anti-Stokes scattering. ν_{vib} is the frequency separation between the ground and excited vibrational state.

pulses, one can obtain directly the temporal dynamics of the excited vibrational state. Thus nonlinear techniques such as coherent anti-Stokes Raman spectroscopy (CARS) are often used to reveal the underlying vibrational dynamics associated with complex spectra [4, 5, 6, 7, 8, 9]. Nevertheless, in the case of molecular liquids where inhomogeneous broadening is substantial, the full interpretation of nonlinear spectra can be difficult. When the incident pulses overlap in time, coherent artifacts can arise due to overlapping transition pathways. In addition, the choice of Stokes pulse frequency limits the frequency range of excitations that can be probed simultaneously and requires a separate light source.

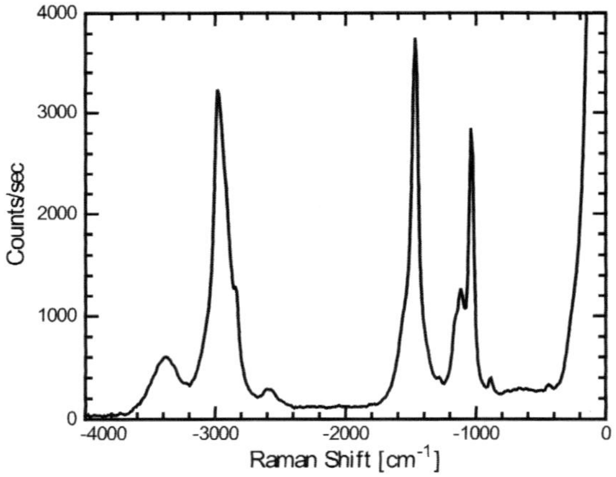

Figure 2. The CW Raman Stokes spectrum of methanol.

Recently it was demonstrated that the interference of spontaneously-scattered Stokes photons can be used to measure vibrational coherence dynamics in solid-state systems [10, 11, 12]. This technique uses two ultrashort pulses with a time delay between them to set up a kind of two-slit experiment. The pulses are in the same spatial mode and are low-intensity such that the probability of both pulses generating Stokes photons at the same time is negligible. Thus any emitted Stokes photon will exist in a superposition of having been created by one pulse and by the other. Measuring the emitted Stokes photons on an optical spectrometer reveals an interference pattern whose visibility depends on the coherence remaining in the system when the second pulse arrives. Thus the interference visibility as a function of time delay provides a direct measure of the coherence lifetime. In this work we extend this technique to the study of vibrational coherence in liquids, providing another tool to gain insight into the intermolecular dynamics that result in loss of coherence. Unlike frequency-domain spontaneous Raman scattering, the method works in the transient regime, in which the pulse duration is less than the coherence lifetime but greater than the period of the vibrational modes excited. Thus it provides a direct time-domain measurement of the dynamics. In contrast to nonlinear techniques, this method can probe the coherence dynamics of multiple vibrational states with widely separated frequency shifts simultaneously without the need for additional light sources. Coherent artifacts that are present in CARS are eliminated in the technique due to the low single-photon-level Stokes emission. The low single-photon-level excitation also eliminates excitation-induced line broadening that can be problematic in stimulated experiments [11]. Thus the technique, which we refer to in this paper as transient coherent Raman scattering, complements existing methods by providing an alternative and in some cases simpler setup that may be more suitable for applications in which available light sources are limited. In this work, we apply the technique to measure the coherence lifetime of an isolated vibrational resonance in methanol.

2. Experimental Setup

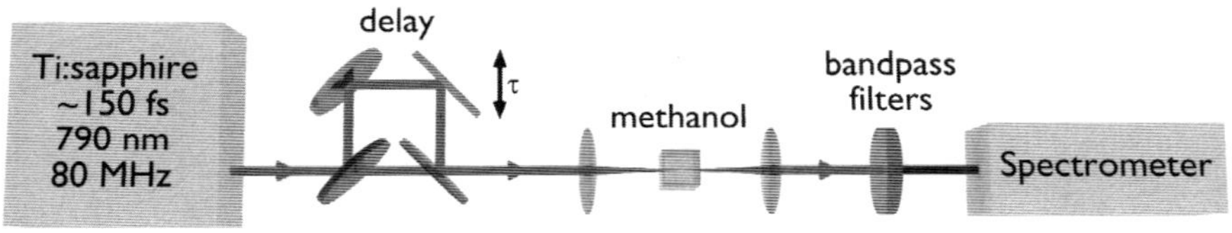

Figure 3. Schematic of the transient coherent Raman scattering experiment.

The experimental setup is shown in Figure 3. Optical pulses centered at 790 nm with a duration of ~150 fs are generated by a modelocked Ti:sapphire oscillator. The pulses are sent through an optical interferometer to create two identical time-delayed pulses, with the optical delay set by a computer-controlled mechanical stage. The average pump power in the two arms is approximately 280 mW. The time-delayed pulses are focused into a standard 1 cm thick quartz spectrometer cell filled with methanol, and the resulting scattered light is detected by an optical spectrometer with ~0.05 nm resolution. To isolate the Raman-scattered signal from the strong pump pulses, a series of bandpass filters are placed before the spectrometer. Ten independent time-delayed scans are collected for both the fundamental and Raman-scattered pulses, with an acquisition time of 2 seconds per spectrum.

In this experiment, the two time-delayed pump pulses can generate Stokes photons via off-resonant spontaneous Raman scattering. The resulting measured spectrum will display interference at both the laser and Stokes wavelengths. The visibility of the Stokes interference decays as a function of delay due to spontaneous fluctuations in the phase of the excited vibrational state due to intermolecular collisions. The spectrum can be written as a superposition of two spontaneous Raman scattering processes, $I_{\rm sig}(t) = |R_{\rm S}E_{\rm P}(t) + R_{\rm S}E_{\rm P}(t-\tau)e^{i\phi}|^2$, where $E_{\rm P}$ is the incident pump field, $R_{\rm S}$ is the associated Stokes Raman tensor, and ϕ represents a continuously fluctuating phase, which when integrated over longer and longer delays results in greater loss of visibility [11]. Thus by measuring the spectral interference visibility of the emitted Stokes photons as a function of time delay, one can obtain information about the loss of coherence.

Journal of Physics: Conference Series **497** (2014) 012004 doi:10.1088/1742-6596/497/1/012004

3. Results

As a proof-of-concept experiment we measure the coherence dynamics in liquid methanol, on which experimental investigations and theoretical structural calculations have been made [13, 14, 15, 16, 17, 18, 19, 20, 21]. We capture both the Stokes spectrum at the 1033 cm^{-1} vibrational mode and the pump spectrum for a series of delays between the pumps. For each Stokes and pump spectrum we acquire a background signal by measuring the spectrum at a time delay at which the interference fringes are averaged out due to the finite resolution of the spectrometer. We take data across both positive and negative time delays, corresponding to one of the pulses arriving before the other and vice versa. Exemplary experimental results are shown in Fig. 4.

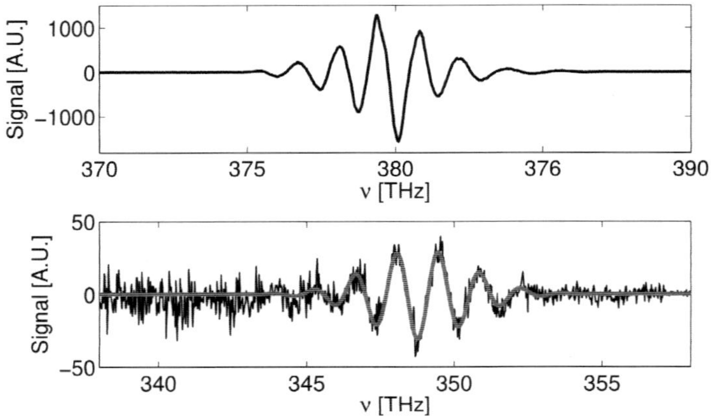

Figure 4. Exemplary background-subtracted spectra showing laser (top) and Stokes (bottom) interference at time delay $\tau = 0.7$ ps. Overlaid on the Stokes spectra is a numerical fit (in red). The separation between interference fringes corresponds to $1/\tau$.

We analyze the data by assuming that the pump pulse and subsequent Raman line have a Gaussian spectral profile. We subtract the long-time-delay background from the signal and fit the background-subtracted data via a nonlinear least squares method with a function of the form

$$A_1 \exp\left[-\frac{(x - A_2)^2}{2\sigma^2}\right] \sin(A_3 x + A_4) \tag{1}$$

in which σ is fixed as the bandwidth of the laser pulse, and the A_i are the fitting parameters. Figure 4 shows a fit result on exemplary background-subtracted Stokes data. After fitting the background-subtracted pump and Stokes spectra, we take a numerical weighted mean of the amplitude fitting parameter A_1 over the collected data sets at each delay for both the pump and Stokes spectra. We normalize the Stokes amplitudes by the pump amplitudes to address loss of visibility due to the finite resolution of the spectrometer [11]. The normalized amplitude of the Stokes fits as a function of time delay is then representative of the loss of visibility. We fit the amplitudes for both positive and negative delay with a simple exponential function of the form $B_1 \exp \; -\frac{|x-B_2|}{B_3}$, where the B_i are the fitting parameters. The fit, shown in Fig. 5, yields a coherence lifetime of 449 ± 16 fs. This result is in agreement with the CW Raman frequency-domain linewidth shown in Fig. 2, providing proof-of-principle that transient coherent Raman scattering enables the direct time-domain measurement of coherence lifetimes in liquids.

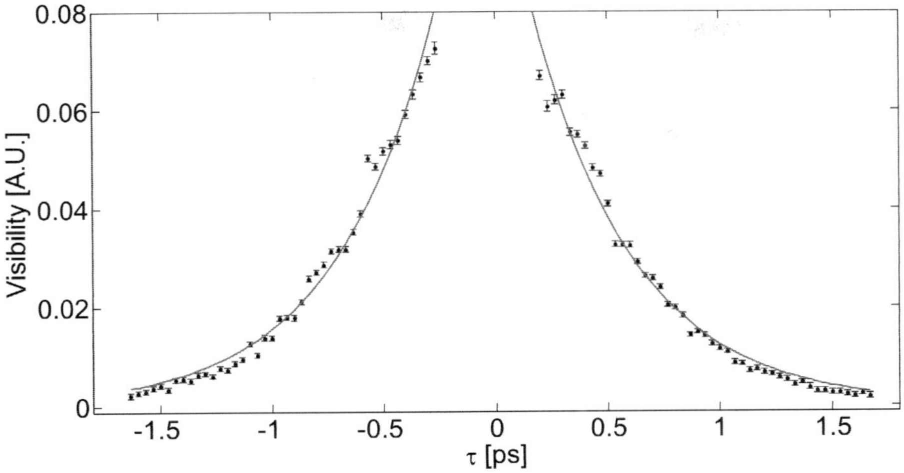

Figure 5. Coherence lifetime measurements of the 1033 cm^{-1} line of methanol. Plotted is the visibility of the Stokes interference fringes relative to the laser visibility, as a function of time delay τ. The points are experimental data and the solid line is an exponential fit.

4. Conclusion

We have demonstrated that transient coherent Raman scattering is a useful spectroscopic tool for measuring vibrational coherence in liquids. This technique can be used at the single-photon level with low pump powers, does not require more than one laser, and is free of nonlinear artifacts. In addition, using this method it is possible to probe the coherence lifetimes of multiple vibrational excitations simultaneously by measuring the interference visibilities of each Stokes line on an optical spectrometer. This technique provides a complementary perspective and in some cases simpler setup for measuring coherence dynamics in molecular liquids.

5. Acknowledgements

This work was supported in part by the University of Delaware Research Foundation and the NSF Physics Division, Grant No. 1205812.

References

[1] Boyd R 2008 *Nonlinear Optics* 3rd ed (Elsevier Inc.)
[2] Shimanouchi T 1972 *Tables of Molecular Vibrational Frequencies: Consolidated Volume I* (Washington, D.C.: U.S. Department of Commerce, National Bureau of Standards)
[3] Mukamel S 1995 *Nonlinear Optical Spectroscopy* (Oxford University Press, Inc.)
[4] Tolles W, Nibler J, McDonald J and Harvey A 1977 *Applied Spectroscopy* **31** 253–271
[5] Silberberg Y 2009 *Annual Review of Physical Chemistry* **60** 277–292 pMID: 18999997
[6] Laubereau A and Kaiser W 1978 *Rev. Mod. Phys.* **50**(3) 607–665
[7] Mukamel S, Piryatinski A and Chernyak V 1999 *Accounts of Chemical Research* **32** 145–154
[8] Mukamel S 2000 *Annual Review of Physical Chemistry* **51** 691–729
[9] Fayer M 2001 *Ultrafast Infrared and Raman Spectroscopy* (*Practical Spectroscopy* vol 26) (New York: Marcel Dekker)
[10] Lee K C, Sussman B J, Sprague M R, Michelberger P, Reim K F, Nunn J, Langford N K, Bustard P J, Jaksch D and Walmsley I A 2012 *Nat Photon* **6** 41–44 ISSN 1749-4885
[11] Waldermann F C, Sussman B J, Nunn J, Lorenz V O, Lee K C, Surmacz K, Lee K H, Jaksch D, Walmsley I A, Spizziri P, Olivero P and Prawer S 2008 *Phys. Rev. B* **78** 155201
[12] Lee K, Sussman B J, Nunn J, Lorenz V, Reim K, Jaksch D, Walmsley I, Spizzirri P and Prawer S 2010 *Diamond and Related Materials* **19** 1289 – 1295

[13] Iwaki L K, Deàk J C, Reah S T and Dlott D D 2000 *Ultrafast Infrared and Raman Spectroscopy* (New York: Marcel Dekker)
[14] Buck U and Huisken F 2000 *Chem. Rev.* **100** 3863–3890
[15] Chelli R, Ciabatti S, Cardini G, Righini R and Procacci P 1999 *J. Chem. Phys.* **111** 4218
[16] Garberoglio G and Vallauri R 2001 *J. Chem. Phys.* **115** 395
[17] Jedlovszky P 2000 *J. Chem. Phys.* **113** 9113–9121
[18] Liddel U and Becker E D 1957 *Spectrochim. Acta, Part A* **10** 70–84
[19] Matsumoto M and Gubbins K E 1990 *J. Chem. Phys.* **93** 1981
[20] Torii H and Tasumi M 1993 *J. Chem. Phys.* **99** 8459
[21] Wolfrum K, Graener H and Laubereau A 1993 *Chem. Phys. Lett.* **213** 41

Two-quantum photon-phonon laser

A A Zadernovsky

Moscow State Technical University of Radioengineering, Electronics and Automation, Department of Physics, 78 Vernadsky Ave., 119454 Moscow, RUSSIA

E-mail: zadernovsky@mirea.ru

Abstract. We discuss a new approach to the implementation of laser action in silicon and other indirect band gap semiconductors – two-quantum photon-phonon laser. Dynamics of the photon-phonon laser generation in indirect band gap semiconductors with electron population inversion is investigated in detail with aid of the rate equations. Numerical estimates are made for Si and Ge.

1. Introduction

The possibility to exploit indirect band gap semiconductors as an active medium for lasers is discussed by researchers for many years. Already in the first publication on this subject [1], which appeared shortly after demonstration of the first semiconductor laser, it has been concluded that indirect band gap semiconductors are not suitable for light emitting devices and realizing a laser. In such semiconductors the bottom of the conduction band is not located directly above the top of the valence band in the momentum space of a crystal. This means that an electron in the conduction band can not to have the same momentum as a hole in the valence band. As a result, the near-to-gap radiative electron-hole recombination can only occur through two-quantum transitions with emission of a photon and simultaneous emission or absorption of a phonon to satisfy the laws of energy and momentum conservation. The rate of such two-quantum transitions is very low and therefore, indirect band gap semiconductors are highly inefficient as a host material for light sources.

Despite this disappointing conclusion researchers do not leave the efforts to find a solution to this problem [2–5]. The reason for such persistence is silicon – one of the most famous and attractive indirect band gap semiconductors. Silicon is the fundamental material used in the electronics industry. It is exploited as a substrate for most integrated circuits. Therefore, creation of silicon-based light sources (either light emitting diodes or lasers) would pave the way for closer integration of photonics and electronics.

The advent of nanotechnology opened up entirely new possibilities for researchers. Optical properties of a tiny crystal with the size that are comparable to the electron de Broglie wavelength (for instance, less than 5 nm at room temperature) is fundamentally differ from its bulk counterpart. When the dimensions are so small, quantum effects start to play an important role and can fully alter the properties of the original material.

The quantum confinement effects may offer a solution to the problem of silicon and other indirect band gap semiconductors. According to the Heisenberg uncertainty principle, when a particle is spatially localized, its momentum becomes uncertain. This means that in low-dimensional indirect band gap semiconductors (nanocrystals, thin layered structures, porous materials), the tails of the wavefunctions of an electron and a hole can now partially overlap, allowing the quasi-direct transitions to occur and thus increasing the probability of radiative recombination.

A number of groups have presented the experimental data on the observation of an enhanced luminescence and even an optical gain in nanostructured silicon, namely, in nanocrystals [6, 7], in silicon on insulator superlattices [8] and in nano-porous silicon [9–11]. Soon afterwards, the efficient silicon-based light emitting diodes [12, 13] have been demonstrated.

Silicon laser is the ultimate goal for researchers and a number of important breakthroughs have been made in the past decade [14]. Among the impressive achievements are the first demonstration of a pulsed silicon Raman laser [15] and the first successful demonstration of a CW silicon Raman laser shortly thereafter [16].

Raman laser is a specific type of laser where the light-amplification mechanism is stimulated Raman scattering. The optical pumping in Raman lasers does not produce a population inversion. The pumping radiation is rather converted into stimulated radiation in nonlinear laser medium. In contrast, most of the conventional lasers rely on stimulated electronic transitions between inversely populated quantum states. Today all the conventional semiconductor lasers exploit the materials with direct band gap structure while the lasers on indirect band gap semiconductors remain still elusive despite the efforts of many researchers [17–22].

In this paper we discuss a new approach to the implementation of laser action in indirect band gap semiconductors – two-quantum photon-phonon laser. Dynamics of the photon-phonon laser generation in a bulk indirect band gap semiconductor with electron population inversion is investigated in detail with aid of the rate equations. Numerical estimates are made for Si and Ge.

2. Principles of photon-phonon laser action

Inverted electronic population, maintained in an indirect band gap semiconductor by a pumping, enables the band-to-band electron transitions with emission of a photon. If such transitions occur between the states close to the energy band edges, the emission of a photon must be assisted by simultaneous emission or absorption of a phonon to conserve the momentum (see figure 1).

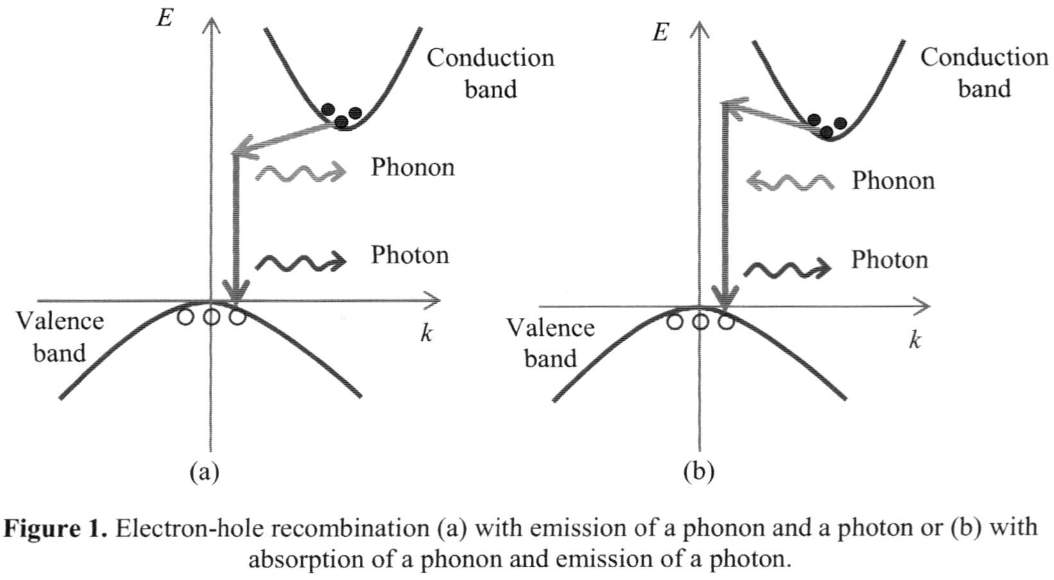

Figure 1. Electron-hole recombination (a) with emission of a phonon and a photon or (b) with absorption of a phonon and emission of a photon.

The rate of such two-quantum transitions and the corresponding photon gain in indirect band gap semiconductors is much less than in semiconductors with direct band gap structure, where the similar radiative transitions are single-quantum. Moreover, as it was pointed out [1] at the very beginning of the study of this problem, the free carrier photon absorption in indirect band gap semiconductors is increased with pumping more rapidly than the photon gain and, thus, in such semiconductors the photon laser action seems to be impossible.

The photon emission rate, however, could be increased by stimulating the phonon part of the two-quantum transitions with the help of a phonon flux injected through a crystal surface from an external acoustic source. At a high enough intensity of this flux and at a high enough level of inverted electronic population, the photon gain could exceed the photon losses and we might expect the photon laser action.

In typical indirect band gap semiconductors such as the crystals of Si and Ge, the phonons that involved in the two-quantum transitions are in THz-frequency range, in which the effective acoustic sources are not available. As a way of getting around this problem, we propose to generate the required phonons directly inside the crystals.

The phonons generated in the two-quantum transitions with emission of a phonon and a photon (figure 2a) can be absorbed in the two-quantum transitions with absorption of a phonon and emission of a photon (figure 2b). Such transitions are known as the Stokes and anti-Stokes vibronic sidebands. Since the phonon energy is much less than that of the photon, the probabilities of these processes are similar. The phonon emission rate, however, can be increased if the photon part of the two-quantum transition with simultaneous emission of a photon and a phonon is stimulated by the light from an appropriate external laser source. Since the energies of the photons radiated in Stokes and anti-Stokes sidebands differ by twice the phonon energy, the separate stimulation of the two-quantum process with phonon emission only is possible within a certain range of laser light frequencies.

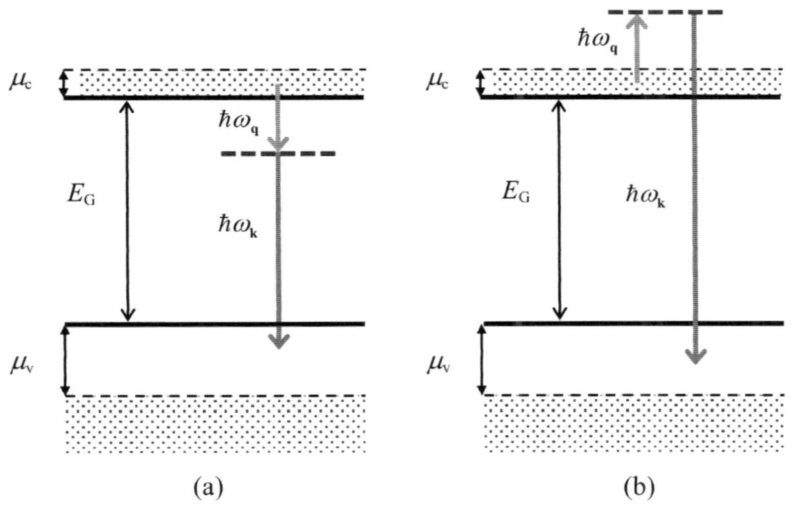

Figure 2. Two-quantum transitions (a) with emission of a phonon and a photon or (b) with absorption of a phonon and emission of a photon.

Moreover, the laser light frequency can be fitted in such a way to promote emission of phonons of only a specific type. Indeed, the luminescence spectrum of indirect electron-hole recombination has the form of a series of peaks corresponding to the different types of phonons involved in the process [23]. If the laser light frequency is close to one of these peaks, the emission rate of the phonons of the specific type is increased. We are interested in transitions involving emission of the transverse acoustic (TA) phonons, since at low temperatures of the crystals (temperature of liquid helium and lower) the phonons of this type are long-lived with respect to anharmonic phonon-phonon interactions (the selection rules forbid the TA phonon decay into two other phonons with lower energies) [24].

At a high enough intensity of the external stimulating radiation and at a high enough level of pumping, the phonon gain can exceed the phonon losses and the phonon laser action appears. The produced phonons can either couple to the anti-Stokes sideband, thus allowing the spectroscopic detection of the beginning of the phonon laser action, or can further stimulate the phonon part of the same transition, leading to a growth of the photon emission rate in the Stokes sideband. This permits us to sustain the phonon laser action by gradual increasing the pumping rate with simultaneous decreasing the intensity of the external stimulating radiation. Finally, the source of stimulating radiation can be switch off and the system continues to operate with simultaneous lasing of both photons and phonons.

3. Photon and phonon emission rates

In this section we consider the emission rates of photons and phonons via the two-quantum photon-phonon transitions in which either the optical or the acoustic components, or both may be stimulated. We follow our article [25], where such transitions have been first examined.

3.1. The rate of stimulated-stimulated photon-phonon transitions

The number of transitions per unit volume per unit time W_{kq} with simultaneous stimulated emission of photons and TA phonons is determined by the expression

$$W_{kq} = R_t N_k N_q \sum_{\mathbf{k}_1\mathbf{k}_2} \delta_{0,\mathbf{k}_1+\mathbf{q}-\mathbf{k}_2} \delta(E_{c\mathbf{k}_2} - E_{v\mathbf{k}_1} - \hbar\omega_{\mathbf{k}} - \hbar\omega_{\mathbf{q}t})[f_c(E_{\mathbf{k}_2}) - f_v(E_{\mathbf{k}_1})] \quad (1)$$

where N_k is the photon concentration in an optical mode including the photons with energy $\hbar\omega_{\mathbf{k}}$ and wave vectors $\mathbf{k}$ of opposite sign, N_q is the phonon concentration in an acoustic mode representing the total concentration of the TA phonons with energy $\hbar\omega_{\mathbf{q}t}$ and wave vectors $\mathbf{q}$ taking into account all the equivalent phonon propagating directions in a crystal. For example, in silicon there are 6 of such directions along the axes (100), while in germanium there are 8 equivalent directions along the axes (111).

The summation over the wave vectors $\mathbf{k}_1$ and $\mathbf{k}_2$ of electrons in the valence band v and the conduction band c, respectively, runs over the first Brillouin zone and includes the summation over the two electron spin states. The electron energies near the band extrema are

$$\begin{aligned} E_{c\mathbf{k}_2} &= E_G + \hbar^2(\mathbf{k}_2 - \mathbf{q}_0)^2/2m_c\,; \\ E_{v\mathbf{k}_1} &= -\hbar^2\mathbf{k}_1^2/2m_v \end{aligned} \quad (2)$$

where E_G is the energy gap width, m_c, m_v are the electron and the hole effective masses, $\mathbf{q}_0$ is the wave vector of the conduction band minimum pointing from the origin to one of equivalent valleys in the first Brillouin zone. The functions $f_c(E)$, $f_v(E)$ are the quasi-Fermi functions, which are defined at a temperature T by the familiar expression

$$f_{c,v}(E) = \frac{1}{\exp[(E - F_{c,v})/\kappa T] + 1} \quad (3)$$

where $F_c = \mu_c + E_G$, and $F_v = -\mu_v$ represent the quasi-Fermi energy levels in the bands.

The coefficient R_t contains the matrix element of the considered transition. Since the phonon energy is much less than that of the photon this matrix element can be replaced by another one corresponding to the two-quantum transition with photon absorption and TA phonon emission (the only difference between the matrix elements is the sign of the phonon energy). This permits us to determine R_t from data on measurement of the optical absorption coefficient in the region of the absorption edge spectrum of indirect band gap semiconductors. To a good degree of approximation R_t can be regarded as a constant in the vicinity of the band extrema. Taking into account that the transverse acoustic modes are degenerate and using the curves [26] for the optical absorption coefficient versus photon energy in Si and Ge at temperature T=4.2 K , we obtain [25] $R_t = 4\times10^{-31}$ eV cm^6 c^{-1} for Si and $R_t = 5\times10^{-30}$ eV cm^6 c^{-1} for Ge.

Due to the δ-function and the Kronecker symbol $\delta_{0,\mathbf{k}_1+\mathbf{q}-\mathbf{k}_2}$ the sum (1) is essentially reduced to a sum over all allowed vectors $\mathbf{k}_1$ and $\mathbf{k}_2$, consistent with the conservation of energy and momentum (the momentum of a photon can be neglected). For electron transitions between states close to the band edges, the wave vector $\mathbf{q}$ of the emitted phonons is near to $\mathbf{q}_0$. Therefore, summing in (1) over $\mathbf{k}_2$ at $\mathbf{q}=\mathbf{q}_0$ and then integrating over $\mathbf{k}_1$ (using the density of the electron states), we obtain in the range $\Delta_{\mathbf{q}_0} = \hbar\omega_{\mathbf{k}} + \hbar\omega_{\mathbf{q}_0 t} - E_G > 0$ the following expression

$$W_{kq} = AN_k N_q\left[\left(\exp\frac{\mu_0-\mu_c}{\kappa T}+1\right)^{-1} - \left(\exp\frac{\mu_v-\mu_0\, m_c/m_v}{\kappa T}+1\right)^{-1}\right] \qquad (4)$$

with

$$A = \frac{R_t}{2\pi^2}\left(\frac{2m_r}{\hbar^2}\right)^{3/2}\Delta_{\mathbf{q}_0}^{1/2} \qquad (5)$$

and $m_r = m_c m_v/(m_c+m_v)$, $\mu_0 = \Delta_{\mathbf{q}_0} m_r/m_c$.

One can see from (4) that in the range $0 < \Delta_{\mathbf{q}_0} < \mu_c + \mu_v$ there is a gain of phonons and photons. At low temperatures, $kT << \mu_0$, the gain maximum is reached at $\mu_c \approx \mu_0$ (taking into account the inequality $\mu_v > \mu_c m_c/m_v$ that follows from the condition of electric neutrality of a semiconductor). In this case we can neglect the second term in (4).

Once a level of μ_c , which is maintained by a pumping, is fixed, the optimum photon energy of externally applied stimulating radiation is determined from the condition $\mu_0 = \mu_c$. The pumping, in turn, is limited by the requirement of emission of TA phonons only. If the stimulating radiation promotes emission of phonons of several types, the quantum efficiency of the TA phonon generating process, which we are interested in, is decreased. Combined application of these requirements determines the optimum pumping condition $\mu_c^{opt} = 1.5\times10^{-2}$eV for Si and $\mu_c^{opt} = 0.8\times10^{-2}$eV for Ge and the optimum photon energy of stimulating radiation $\hbar\omega_{\mathbf{k}_0}$ =1.165 eV for Si and $\hbar\omega_{\mathbf{k}_0}$ = 0.745 eV for Ge. Using the data for Si and Ge [26] collected in table 1, we obtain for the rate constant A (5) at the optimum photon energy the following value $A = 0.4\times10^{-10}$ cm^3 s^{-1} for Si and $A = 1.9\times10^{-10}$ cm^3 s^{-1} for Ge.

Table 1.

Semi-conductor	Type of phonon	Phonon energy, eV	Effective mass m_c/m_0	Effective mass m_v/m_0	Energy gap width (T=0), eV	q_0, 10^{-8} cm^{-1}
Si	*Acoustic:*		0.33	0.61	1.16	0.94
	transverse (TA)	0.018				
	longitudinal (LA)	0.027				
	Optical:					
	longitudinal (LO)	0.091				
	transverse (TO)	0.122				
Ge	*Acoustic:*		0.22	0.35	0.74	0.96
	transverse (TA)	0.008				
	longitudinal (LA)	0.027				

3.2. The rate of stimulated-spontaneous photon-phonon transitions

The rate of stimulated-spontaneous transitions W_{k0} with stimulated photon emission into an optical mode and spontaneous phonon emission is determined by the expression

$$W_{k0} = R_t N_k N_c \sum_{\mathbf{q}} \sum_{\mathbf{k}_1 \mathbf{k}_2} \delta_{0,\mathbf{k}_1+\mathbf{q}-\mathbf{k}_2} \delta(E_{c\mathbf{k}_2} - E_{v\mathbf{k}_1} - \hbar\omega_{\mathbf{k}} - \hbar\omega_{\mathbf{q}t}) f_c(E_{\mathbf{k}_2})[1 - f_v(E_{\mathbf{k}_1})] \quad (6)$$

where N_c is the number of the equivalent conduction band valleys ($N_c = 6$ for Si and $N_c = 4$ for Ge). The summation over $\mathbf{q}$ runs over the wave vectors of TA phonons only, since in the range of the optimum pumping it is the only type of the phonons participating in the process.

At low temperatures the quasi-Fermi functions in (6) can be well approximated by the step function $f_{c,v}(E) = \Theta(F_{c,v} - E)$. Summing over $\mathbf{q}$, taking into account the two TA phonon polarizations, and then integrating over $\mathbf{k}_1$ and $\mathbf{k}_2$, we obtain

$$W_{k0} = C N_k f(\mu_c/\mu_0) \quad (7)$$

with

$$C = R_t \frac{N_c}{4\pi^3} \left(\frac{m_c}{\hbar^2}\right)^{3/2} \left(\frac{m_v}{\hbar^2}\right)^{3/2} \Delta^2_{\mathbf{q}_0} . \quad (8)$$

The function $f(\mu_c/\mu_0)$ in (7) is equal to 0 at $\mu_c/\mu_0 \leq (1+m_c/m_v)/(1+N_c^{2/3} m_c/m_v)$, is equal to 1 at $\mu_c/\mu_0 \geq 1+m_c/m_v$ and of the order of 1 at other cases. In particular, at the gain maximum, when $\mu_0 = \mu_c$, we have $f(1) \approx 0.7$ both for Si and Ge . At the optimum photon energy of stimulating radiation we obtain for the rate constant C (8) the following estimates $C = 2.1\times10^9$ s^{-1} for Si and $C = 1.3\times10^9$ s^{-1} for Ge.

3.3. The rate of spontaneous-stimulated photon-phonon transitions

The rate of spontaneous-stimulated transitions W_{0q} with spontaneous photon emission and stimulated TA phonon emission into an acoustic mode is determined by the expression

$$W_{0q} = R_t N_q \sum_{\mathbf{k}} \sum_{\mathbf{k}_1 \mathbf{k}_2} \delta_{0,\mathbf{k}_1+\mathbf{q}-\mathbf{k}_2} \delta(E_{c\mathbf{k}_2} - E_{v\mathbf{k}_1} - \hbar\omega_{\mathbf{k}} - \hbar\omega_{\mathbf{q}t}) f_c(E_{\mathbf{k}_2})[1 - f_v(E_{\mathbf{k}_1})] \quad (9)$$

where the summation over the photon wave vectors $\mathbf{k}$ includes the summation over the two photon polarizations. Summing in (9) over $\mathbf{k}_2$ at $\mathbf{q}=\mathbf{q}_0$ and then performing the integration over $\mathbf{k}_1$ and $\mathbf{k}$, taking into account that $\hbar\omega_{\mathbf{q}_0 t} << E_G$, we obtain

$$W_{0q} = D N_q N_e \quad (10)$$

where

$$D = \frac{R_t}{N_c} \frac{E_G^2}{\pi^2 (\hbar c)^3} \quad (11)$$

c is the velocity of light in the material ($c = 0.88\times 10^{10}$cm s^{-1} for Si and $c = 0.75\times 10^{10}$cm s^{-1} for Ge) and N_e is the electron concentration in the conduction band which is connected with μ_c by the relation

$$N_e = \frac{N_c}{3\pi^2} \left(\frac{2m_c}{\hbar^2}\right)^{3/2} \mu_c^{3/2} \quad (12)$$

The estimates for the rate constant D (11) give D=0.5×10^{16} cm^3 s^{-1} for Si and D=5.7×10^{16} cm^3 s^{-1} for Ge. The electron concentration (12) is estimated at the condition of $\mu_c = \mu_0$ by the value of $N_e = 8.9\times10^{18}$ cm^{-3} for Si and $N_e = 1.3\times10^{18}$ cm^{-3} for Ge.

3.4. *The rate of spontaneous-spontaneous photon-phonon transitions*

The rate of spontaneous-spontaneous photon-phonon transitions W_{00} can be expressed as follows [27]

$$W_{00} = BN_e^2 \tag{13}$$

with the rate constant B=0.2×10^{-14} cm^3 s^{-1} for Si and B=3.4×10^{-14} cm^3 s^{-1} for Ge.

4. Photon and phonon losses

4.1. *Phonon losses*

The TA phonons emitted into a particular acoustic mode can leave the mode (a) due to the phonon absorption in the two-quantum transitions with simultaneous photon radiation in the anti-Stokes sideband (see figure 2b), (b) due to anharmonic interaction with heat phonons (lattice absorption), (c) due to the phonon scattering by free carriers and (d) by the lattice imperfections (impurities, vacancies, dislocations).

(a) Since $\hbar\omega_{\mathbf{q}_{r}} << \hbar\omega_{\mathbf{k}}$, the rate of the phonon coupling to the anti-Stokes sideband is similar to the rate of spontaneous-stimulated transitions W_{0q} (9). Therefore, the inverse lifetime τ_{A}^{-1} of the TA phonon with wave vector $\mathbf{q}_0$ can be estimated by the expression $\tau_{\mathrm{A}}^{-1} = DN_e$, where N_e is the electron concentration (12). At the condition of the gain maximum μ_c=μ_0, we obtain τ_{A}^{-1}=4.5×10^2 s^{-1} for Si and τ_{A}^{-1} = 7.4× 10^2 s^{-1} for Ge.

(b) The inverse lifetime τ_{t}^{-1} of the TA phonon with wave vector $\mathbf{q}_0$ in phonon-phonon interactions can be estimated [24] by the value of τ_{t}^{-1} = 10^3÷10^4 s^{-1} for Si and τ_{t}^{-1} = 10^2÷10^3 s^{-1} for Ge at a temperature of the crystals of T= 1 K.

(c) Phonon absorption by free carriers obeys certain selection rules. It vanishes for TA phonons with wave vector $\mathbf{q}$ along any symmetry axis of the spheroidal Fermi surface [28]. In our case, the absorption of TA phonons with wave vector $\mathbf{q}_0$ is essentially forbidden by the selection rules. Thus for $\mathbf{q}$ near $\mathbf{q}_0$, the TA phonon absorption by free carriers can be neglected.

(d) Among all the above listed channels of TA phonon losses, elastic Rayleigh scattering by isotope impurities is the largest one in an otherwise perfect crystal [29]. For Si and Ge with natural isotopic abundances the inverse lifetime τ_{i}^{-1} of the TA phonon with wave vector $\mathbf{q}_0$ is estimated by [30] τ_{i}^{-1} =7.2×10^9 s^{-1} for Si and τ_{i}^{-1}= 5.2×10^8 s^{-1} for Ge. We will assume that the last two values can be reduced with the help of isotope separation methods by 3–4 orders of magnitude.

Summarizing all the above estimates we may conclude that the crucial contribution to the phonon losses from a particular acoustic mode gives the phonon scattering by isotope impurities. Thus, in sufficiently pure (including isotopic purity) and perfect crystals of Si and Ge, the total inverse lifetime of the TA phonon in an acoustic mode can be estimated by the value of τ_0^{-1} =7×10^5 s^{-1} for Si and τ_0^{-1}= 5×10^4 s^{-1} for Ge .

4.2. *Photon losses*

Photon losses from a particular optical mode are due to (a) the photon absorption by free carriers, (b) the photon scattering by optical inhomogeneities, (c) due to diffraction losses and (d) the losses of radiation through the partially reflecting crystal surfaces. We assume that the crystals under investigation are perfect enough to neglect the photon scattering by the optical inhomogeneities. As far as diffraction losses are concerned, one can expect that similar to the semiconductor lasers such losses can be generally reduced by waveguiding of the gain region. Thus, the bulk photon losses are dominated by the free carrier absorption with the rate

Journal of Physics: Conference Series **497** (2014) 012005 doi:10.1088/1742-6596/497/1/012005

$$W_e = \sigma c N_k N_e, \tag{14}$$

where σ is the photon capture cross section ($\sigma = 6\times10^{-18}$ cm^2 for Si and $\sigma = 4\times10^{-18}$ cm^2 for Ge). The rate of photon losses due to radiation through the end crystal surfaces of reflectivity R ($R = 0.30$ for Si and $R = 0.36$ for Ge) is equal to N_k/τ, where

$$\tau^{-1} = \frac{c}{L}\ln(1/R) \tag{15}$$

and L is the crystal length. Assuming $L = 250\ \mu$m, we obtain $\tau^{-1} = 3.5\times10^{11}$ s^{-1} for Si and $\tau^{-1} = 3\times10^{11}$ s^{-1} for Ge.

5. Rate equations

The dynamics of the photon-phonon generation in indirect band gap semiconductors can be described by the following set of the rate equations for the electron concentration N_e in the conduction band, the photon concentration N_k in an optical mode, and for the phonon concentration N_q in an acoustic mode

$$\frac{dN_e}{dt} = W_p - W_{kq} - W_{k0} - W_{0q} - W_{00} \tag{16}$$

$$\frac{dN_k}{dt} = \frac{1}{\tau}\left(\frac{1+R}{1-R}nN_0 - N_k\right) + W_{kq} + W_{k0} - W_e \tag{17}$$

$$\frac{dN_q}{dt} = W_{kq} - \frac{N_q}{\tau_0} \tag{18}$$

where W_p is the pumping rate of electrons into the conduction band, N_0 is the photon concentration of stimulating radiation outside the crystal, the term $nN_0(1+R)/(1-R)$, where n is the refractive index, represents the stationary photon concentration which is sustained in an optical mode inside the crystal with aid of the external simulating radiation.

In the equation (17) we neglect the spontaneous photon contribution into the optical mode, as well as in (18) we neglect the contribution of spontaneous phonons into the acoustic mode. This can be done due to the strict localization of the photons and the phonons in the modes. The contribution into the acoustic mode due to spontaneous-stimulated transitions W_{0q} (10) is balanced by the absorption of the phonons through coupling to the anti-Stokes sideband (see figure 2b). As a result, these terms cancel each other in the equation (18). Furthermore, the term W_{0q} could be omitted in equation (16), since for the steady state solutions, which we are interested in, we have $W_{kq} >> W_{0q}$.

To simplify the rate equations a conversion of variables to dimensionless units is made: $n_e = N_e/N$, $n_k = N_k/N$, $n_q = N_q/(N\tau_0/\tau)$, $\theta = t/\tau$, where the normalizing particle concentration N is determined by the condition $AN\tau_0 = 1$. Using the normalized variables, the rate equations (16–18) can be rewritten as follows

$$\frac{dn_e}{d\theta} = \frac{W_p}{W_0} - n_k n_q S - C\tau n_k f(\mu_c/\mu_0) - B\tau N n_e^2 \tag{19}$$

$$\frac{dn_k}{d\theta} = \frac{1+R}{1-R}nn_0 - n_k + n_k n_q S + C\tau n_k f(\mu_c/\mu_0) - \sigma c\tau N n_k n_e \tag{20}$$

$$\frac{dn_q}{d\theta} = \frac{\tau}{\tau_0}(n_k n_q S - n_q) \tag{21}$$

where $n_0 = N_0/N$; $W_0 = N/\tau$ and $S = \{\exp[(\mu_0 - \mu_c)/kT]+1\}^{-1}$. Equations (19–21) with the relation (12) rewritten through the normalized variables as

$$n_e = \frac{N_c}{3\pi^2 N}\left(\frac{2m_c}{\hbar^2}\right)^{3/2} \mu_c^{3/2} \qquad (22)$$

represent the complete set of equations for describing the dynamics of the photon-phonon generation in indirect band gap semiconductors.

6. Dynamics of photon-phonon laser action

We restrict ourselves to the steady state solutions of the rate equations. As it is seen from (21) at $dn_q/d\theta = 0$, there are two types of these solutions: below the phonon lasing threshold with $n_q = 0$ and above the phonon lasing threshold with $n_q \neq 0$. These solutions are shown in the figure 3 (curves of type I and type II, respectively) in the form of dependence of the normalized photon concentration n_k in an optical mode versus the normalized pumping rate W_p/W_0 with the normalized photon concentration n_0 of external stimulating radiation taken as a parameter of a family of curves. Numerical values in the figure 3 correspond to the silicon crystal.

It can be shown [31] with the help of Routh-Hurwitz criterion that only increasing parts of the curves II in figure 3 correspond to the stable steady state solutions. The intersection in the point A between the stable branch of a curve II and a curve I with the same n_0 determines the phonon lasing pumping threshold $(W_p/W_0)_{th}$ and the threshold value for the normalized photon concentration $(n_k)_{th}$ inside the crystal. If the pumping is further increased the system goes on the stable branch of the curve II and the phonon laser action occurs. This leads to a growth of the phonon and the photon concentrations in the modes. If the pumping rate W_p/W_0 exceeds some threshold value W_s/W_0 , the external source of stimulating radiation can be switched off and the system goes from the point C_1 to the working point C_2 on the stable branch of the curve II with $n_0 = 0$. Thus, the system continues to operate with simultaneous lasing of both photons and phonons.

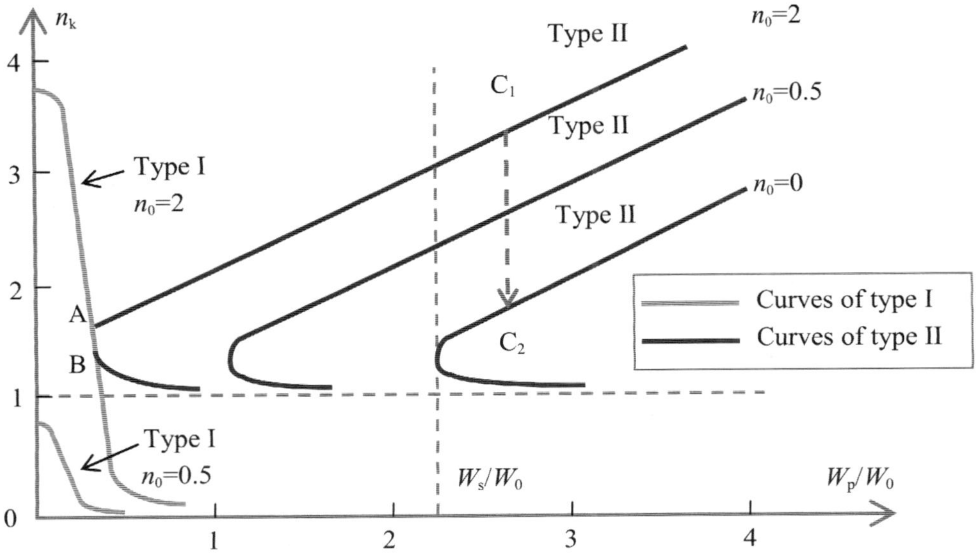

Figure 3. The steady state solutions of the rate equations for Si.

Numerical estimates [31, 32] are collected in the table 2.

Table 2. Numerical estimates

Semiconductor	N, 10^{14} cm^{-3}	W_0, 10^{25} cm^{-3}c^{-1}	$(W_p/W_0)_{th}$, 10^{-3}	$(n_k)_{th}$	W_s/W_0	n_{ks}	n_{qs}	n_0
Si	1.6×10^2	5.6×10^2	4.2	1.6	2.4	3.0	2.4	2.0
Ge	2.6	7.8	3.7	1.8	1.1	3.2	1.1	1.0

N the normalizing particle concentration

W_0 the normalizing pumping rate

$(W_p/W_0)_{th}$ the normalized threshold pumping rate for starting the phonon laser action

$(n_k)_{th}$ the normalized threshold photon concentration inside the crystal for starting the phonon laser action

W_s/W_0 the normalized threshold pumping rate for starting the simultaneous laser action of both photons and phonons

n_{ks} the normalized photon concentration inside the crystal for starting the simultaneous laser action of both photons and phonons

n_{qs} the normalized phonon concentration for starting the simultaneous laser action of both photons and phonons

n_0 the normalized photon concentration of external stimulating radiation

Returning from normalized to the physical variables, we obtain at the optimum photon energy of stimulating radiation ($\hbar\omega_{\mathbf{k}_0}$ =1.165 eV for Si and $\hbar\omega_{\mathbf{k}_0}$ = 0.745 eV for Ge) the following estimate for the threshold light intensity at the point A in figure 3 $I = c\hbar\omega_{\mathbf{k}_0} N(n_k)_{th}$ =4.2×10^7 W/cm^2 for Si and I = 4.0×10^5 W/cm^2 for Ge and the estimate for the intensity of external stimulating radiation $I = c\hbar\omega_{\mathbf{k}_0} Nn_0$= 5.2×10^7 W/cm^2 for Si and I = 2.2×10^5 W/cm^2 for Ge. For silicon these values are close to the radiative damage threshold of the semiconductor crystals which consist of $10^7\div10^8$ W/cm^2 [27].

7. Conclusions

We propose the use of photon-phonon interband electron transitions in indirect band gap semiconductors to obtain phonon lasing and simultaneous lasing of both photons and phonons.

To drive the process of photon-phonon laser generation, we start with stimulating the photon part of the two-quantum transitions with simultaneous emission of phonons and photons by the light from an appropriate laser source. This leads to a growth of the rate of phonon emission. At a high enough intensity of the external stimulating radiation and at a high enough level of pumping, the phonon gain turns out to be equal to the phonon losses in a particular acoustic mode and the phonon laser action appears. It is then possible to sustain the phonon laser action by gradual increasing the pumping rate with simultaneous decreasing the intensity of the external stimulating radiation. Finally, the source of stimulating radiation can be switch off and the system continues to operate with simultaneous lasing of both photons and phonons.

Dynamics of the photon-phonon laser generation has been investigated in detail with aid of the rate equations. Numerical estimates of the threshold intensity of stimulating radiation, the threshold rate of pumping for starting the phonon lasing and the threshold rate of pumping for starting the simultaneous lasing of both photons and phonons have been made for pure (including isotopic purity) and perfect crystals of Si and Ge.

The threshold intensity of stimulating radiation for Si is close to the radiative damage threshold of the semiconductor crystal. It should be noted, however, that the result of our calculations is strongly dependent on the concentration of isotope impurities in Si. We have assumed above that this concentration can be reduced with the help of isotope separation methods by four orders of magnitude as compared with the natural isotopic concentration. Otherwise, the threshold intensity of stimulating radiation must be increased in an appropriate number of times. On the contrary, in an isotopically pure silicon crystal, where the phonon losses from an acoustic mode is mainly due to lattice absorption and due to phonon absorption in the anti-Stokes vibronic sideband (see figure 2b), the intensity of stimulating radiation may be reduced by an order of magnitude.

Finally, a few remarks on a parasitic nonradiative recombination, which remains even in the pure and perfect semiconductor crystals. We are talking about well-known Auger recombination which has not been taken into account in the rate equation (16). This is a three-particle process where an excited electron recombines with a hole and the excess energy is transferred to another free carrier rather than emitted in the form of a photon. The rate of Auger recombination is proportional to the third degree of concentration of excited carriers, that is $W_{\mathrm{Auger}} = C_{\mathrm{n}} N_{\mathrm{e}}^3$ with $C_{\mathrm{n}} \sim 10^{-31}\mathrm{cm}^6\mathrm{c}^{-1}$ [33] for undoped Si at room temperature. Taken into account the electron concentration (12) at the gain maximum, $N_{\mathrm{e}} = 8.9\times10^{18}$ cm^{-3} for Si, we obtain for the rate of Auger recombination the value of $W_{\mathrm{Auger}} \sim 10^{26}$ cm^{-3}c^{-1} which is an order of magnitude greater than the rate of pumping required for the photon-phonon lasing (see table 2). These estimates show that Auger recombination may be the dominant mechanism at a high level of pumping in Si.

It should be noted, however, that the above estimate has been made at room temperature of the crystal. Temperature dependence of Auger recombination in silicon has been studied in a range from 243 to 473K [34]. To our knowledge, however, there is no data published on Auger recombination at temperatures of liquid helium and lower. Although the general trend is obvious – the Auger coefficient C_{n} should decrease with temperature.

Moreover, there is evidence [35] that some silicon solar cells are characterized by extremely long carrier recombination lifetimes of the order of some milliseconds. That is, for these solar cells, the nonradiative recombination lifetime is of the order of the radiative lifetime, hence their internal quantum efficiency of light emission is of the order of unity. These results are very encouraging for the use of silicon as an active medium for lasers. In any case, this issue requires further investigation.

Acknowledgements
This research was supported by the Ministry of Education and Science of the Russian Federation.

References

[1] Dumke W P 1962 Interband transitions and maser action *Phys. Rev.* **127** 1559–63
[2] Ossicini S, Pavesi L, Priolo F Light 2003 *Emitting Silicon for Microphotonics* (Berlin: Springer)
[3] Pavesi L and Lockwood D J 2004 *Silicon Photonics* (Berlin: Springer)
[4] Reed G T and Knights A P 2004 *Silicon Photonics: An Introduction* (Chichester UK: John Wiley)
[5] Reed G T and Knights A P 2008 *Silicon Photonics: The State of the Art* (Chichester UK: John Wiley)
[6] Pavesi L, Dal Negro L, Mazzoleni G, Franzo G and Priolo F 2000 Optical gain in silicon nanocrystals *Nature* **408** 440–4
[7] Pelant I 2011 Optical gain in silicon nanocrystals: current status and perspectives *Phys. Status Solidi* A **208** 625–30.
[8] Lu Z H, Lockwood D J and Baribeau J M 1995 Quantum confinement and light emission in SiO_2/Si superlattices *Nature* **378** 258–60
[9] Cullis A G and Canham L T 1991 Visible light emission due to quantum size effects in highly porous crystalline silicon *Nature* **353** 335–38
[10] Gösele U and Lehmann V 1995 Light-emitting porous silicon *Mater. Chem. Phys.* **40** 253–59
[11] Hirschman K D, Tsybeskov L, Duttagupta S P and Fauchet P M 1996 Silicon-based visible light–emitting devices integrated into microelectronic circuits *Nature* **384** 338–41
[12] Wai Lek Ng, Lourenço M A, Gwilliam R M, Ledain S, Shao G and Homewood K P 2001 An

efficient room-temperature silicon-based light-emitting diode *Nature* **410** 192–4
[13] Homewood K P and Lourenço M A 2005 Light from Si via dislocation loops *Matrials Today* **8** No 1 34–9.
[14] Di Liang and Bowers J E 2010 Recent progress in lasers on silicon *Nature Photonics* **4** 511–7
[15] Boyraz O and Jalali B 2004 Demonstration of a silicon Raman laser *Optics Express* **12** 5269–73
[16] Rong H, Jones R, Liu A, Cohen O, Hak D, Fang A and Paniccia M 2005 A continuous-wave Raman silicon laser *Nature* **433** 725–7
[17] Iyer S S and Xie Y H 1993 Light emission from silicon *Science* **260** 40–6
[18] Kimerling L C, Kolenbrander K D, Michel J, and Palm J 1997 Light emission from silicon *Solid State Physics* **50** 333.
[19] Trupke T, Green M A and Würfel P 2003 Optical gain in materials with indirect transitions *J. Appl. Phys.* **93** 9058–61
[20] Chen M J, Tsai C S and Wu M K 2006 Optical gain and co-stimulated emissions of photons and phonons in indirect bandgap semiconductors *Jpn. J. Appl. Phys.* **45** No 8B 6576–88
[21] Pavesi L 2008 Silicon-based light sources for silicon integrated circuits *Advances in Optical Technologies* **2008** 416926.
[22] Pavesi L 2005 Routes towards a silicon-based laser *Materials Today* **8** 18–25
[23] Haynes J R, Lax M and Flood W F 1959 *J. Phys. Chem. Solids* **8** 392
[24] Orbach R 1966 *Phys. Rev. Lett.* **16** 15
[25] Zadernovsky A A and Rivlin L A 1991 Stimulated two-quantum photon-phonon transitions in indirect-gap semiconductors *Sov. J. Quantum Electron.* **21** 255–260
[26] McLean T P 1960 The absorption edge spectrum of semiconductors *Progress in semiconductors* Vol. 5 (London: Heywood) p 53
[27] Pankove J I 1971 *Optical processes in semiconductors* (New Jersey: Englewood Cliffs) chapters 3–6
[28] Hensel J C and Dynes R C 1980 *Phonon scattering in condensed matter* ed Maris H J (New York: Plenum Press) p 395
[29] Bron W E 1985 *Nonequilibrium phonon dynamics (NATO ASI series:* ser B, vol 124*)* ed Bron W E (New York: Plenum Press) p 5
[30] Holland M C 1963 *Phys. Rev.* **132** 2461
[31] Zadernovsky A A and Rivlin L A 1993 Photon-phonon laser action in indirect-gap semiconductors *Quantum Electron.* **23** 300–308
[32] Zadernovsky A A and Rivlin L A 1993 Photon-phonon lasing in indirect gap semiconductors *Optics Communications* **100** 322–330
[33] Svantesson K G and Nilsson N G 1979 The temperature dependence of the Auger recombination coefficient of undoped silicon *J. Phys.* C: *Solid State Phys.* **12** 5111
[34] Wang S and Macdonald D 2012 Temperature dependence of Auger recombination in highly injected crystalline silicon *J. Appl. Phys.* **112** 113708
[35] Green M A, Zhao J, Wang A, Reece P J and Gal M 2001 Efficient silicon light-emitting diodes *Nature* **412** 805–808

Journal of Physics: Conference Series **497** (2014) 012006 doi:10.1088/1742-6596/497/1/012006

Magneto-optical effects and rf magnetic field detection in cold rubidium atoms

Krystian Sycz, Adam M. Wojciechowski, and Wojciech Gawlik
Jagiellonian University, Reymonta 4, 30-059 Kraków, Poland

E-mail: a.wojciechowski@uj.edu.pl

Abstract. We present the results of our latest experiments on atomic coherences in cold atoms. Interaction of atoms with a near-resonant, linearly polarized light leads to an effective creation of long-lived ground-state Zeeman coherences which is observed through the nonlinear Faraday effect or free induction decay signals of the Larmor precession. Both optically and radio-frequency induced Zeeman coherences are observed, with relaxation rates around a 100 Hz.

1. Introduction

Ever since the first experiments with laser cooling of neutral atoms, there is a growing interest in the field of cold and ultra-cold diluted gases, see, e.g., references [1, 2, 3]. Due to the slowed down motion and confinement, atom observation time can be as long as minutes, which results in the extraordinary stability of atomic clocks [4, 5], as well as the great sensitivity of atomic interferometers [6]. Moreover, cold atomic samples have been widely used for detection of electric and magnetic fields [7, 8, 9, 10, 11, 12, 13]. State-of-the-art experiments exhibit now quantum limited sensitivities, see, e.g., reference [14].

In this work we report on our latest experiments on nonlinear magneto-optical effects in laser-cooled rubidium samples released from a magneto-optical trap (MOT) [15]. Such experiments in a MOT present both unique opportunities and challenges compared to those performed in vapor cells at room temperatures. The most obvious difference is the greatly reduced velocity of atoms that brings the Doppler broadening below the homogeneous width of the hyperfine transitions. On the other hand, however, the trapping and cooling electromagnetic fields need to be turned off during the measurements because they significantly disturb atoms. Once released from the trap atom cloud spreads out and undergoes a freefall which limit the available observation time to about 20 milliseconds. The duty-cycle (measurement bandwidth) is also limited because the atoms need to be recaptured before the next measurement can be performed.

Interaction of atoms with a near-resonant, linearly polarized light leads to an effective creation of long-lived ground-state Zeeman coherences [16] which can be observed through the nonlinear, or paramagnetic Faraday effect [17]. Optical nonlinearity that results from existence of ground-state coherences allows for precision magnetometry of near-zero magnetic fields [18]. Additionally, by employing the amplitude-modulated optical rotation [19] it is possible to measure also higher magnetic fields up to the geophysical range [15]. Moreover, Zeeman coherences form a versatile tool for studying superposition states [20] which are vital to fundamental atomic physics and quantum information. In this work, we demonstrate the dynamics of coherent superposition states under the influence of laser and magnetic fields. Our

Journal of Physics: Conference Series **497** (2014) 012006 doi:10.1088/1742-6596/497/1/012006

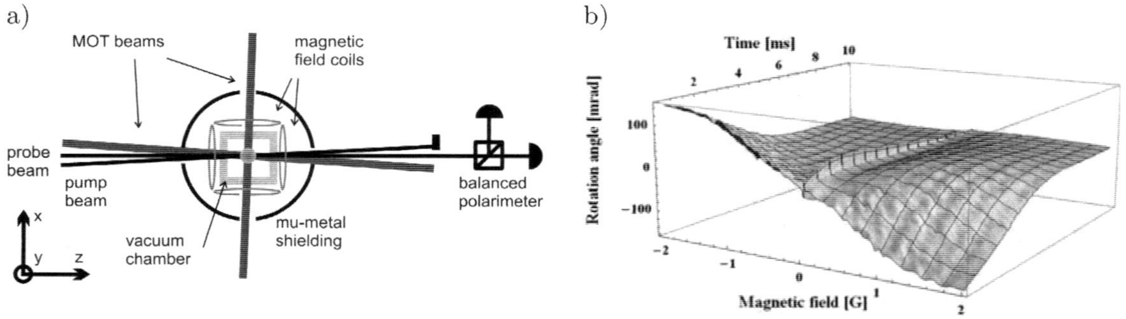

Figure 1. a) Schematic plot of the experimental setup. b) Faraday rotation in the atomic sample released from the MOT as a function of time and magnetic field B. Wide resonance is the linear, whereas the central, narrow structure around $B = 0$ is the nonlinear effect. Decay of the total signal corresponds to the loss of atoms due to the combined effect of their freefall, cloud expansion, optical pumping and magnetic field inhomogeneities.

experiments show that high rotation angles of above ten degrees and coherence lifetimes of a few milliseconds can be achieved with cold atoms released from the MOT in a simple magnetically shielded setup.

2. The experiment

In our experiments we prepare cold atomic samples of about 10^8 ^{85}Rb atoms at the temperature of the order of 50-100 μK in a standard vapor-pressure-loaded MOT. Atoms are then released from the trap by switching off the trapping laser beams and the quadrupole magnetic field. During the freefall of the cloud, a uniform magnetic field and a week probing beam are applied along the z axis [15]. Additionally, a resonant pumping beam along the z direction is used for preparation of the atomic state by optical pumping means. Polarization of the pumping beam is set either to the linear or the circular one, depending on the experiment. The probe beam is linearly polarized along y direction and the rotation of its polarization plane is recorded in time using a balanced polarimeter. Subsequently the trap is reloaded and the next measurement is performed for a different value of the Faraday magnetic field. In this way the magneto-optical rotation signals are recorded versus time and magnetic-field (Fig. 1).

To alleviate the effect of stray magnetic fields we shield the experimental vacuum chamber with a simple, single-layer μ-metal shielding. The magnetic shield consists of a cylinder with two end caps and a set of holes in three orthogonal directions x, y, z for the MOT, pumping and probing beams. Inside the shield the MOT-quadrupole field coils are mounted, as well as the set of coils that allows for application of the extra magnetic fields and their gradients in the three directions. The shield, despite its simplicity, provides a shielding factor of the order of 20 and, more importantly, reduces the time by which the quadrupole field can be switched off. This effect is attributed to the fact, that massive optical mounts in the vicinity of the MOT trap are shielded and the contribution of the generated eddy currents is diminished.

3. Experimental results

3.1. Nonlinear Faraday effect

When a linearly polarized, resonant laser beam is directed at the atomic cloud in a longitudinal magnetic field, its polarization plane undergoes rotation due to the para- and diamagnetic Faraday effects. After switching off the MOT, initial atomic population distribution in our experiments can be regarded as uniform over Zeeman substates with no Zeeman coherences in

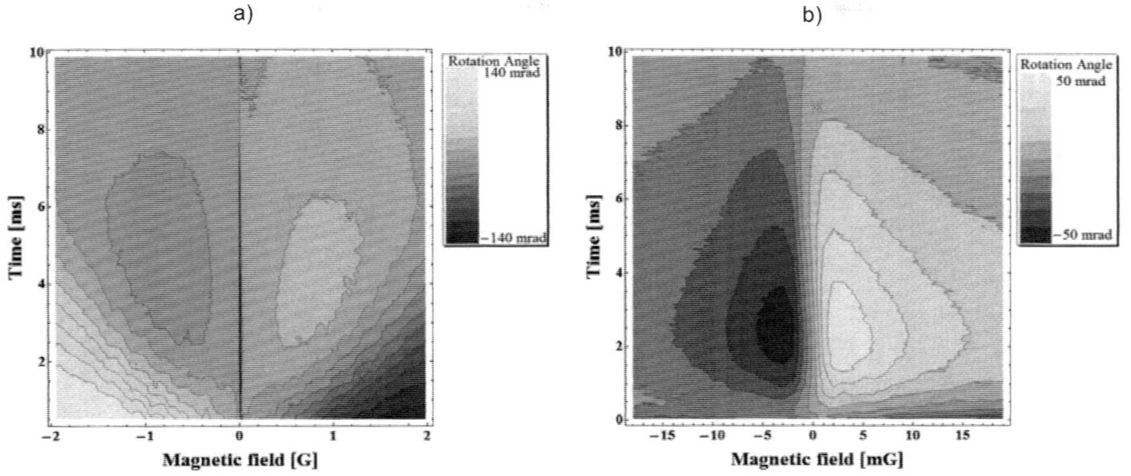

Figure 2. Contour plots of the nonlinear Faraday rotation signal with a single, probe beam of $2\,\mu\mathrm{W/mm^2}$ intensity in a wide (a) and narrow (b) magnetic field range. The width of the NFE resonance is 6.5 mG

the $F = 3$ manifold of the $5^2S_{1/2}$ ground state. However, application of the resonant laser beam may drastically change this distribution. The linearly polarized light creates an alignment in the ground state with no net magnetization, yet with Zeeman coherences evolving in the magnetic field. For a CW laser light, this results in a quasi-stationary (compared to the Larmor precession) rotation of the polarization angle, the nonlinear Faraday effect (NFE).

Typical magneto-optical rotation signal $\phi(B)$ obtained with a relatively weak probing beam ($\sim 2\,\mu\mathrm{W/mm^2}$), nearly resonant with the $F = 3 \rightarrow F' = 4$ transition of the ^{85}Rb D2 line is shown in Fig. 1. Signal decay with time is easily seen, and can be attributed to the combined effect of atom loss due to the freefall, cloud expansion (dilution) optical pumping, and inhomogeneous magnetic field. The wide dispersive structure around $B = 0$ is the linear Faraday effect (LFE) and has a width of $\sim$ 20 G. Additionally, the signal exhibits a much narrower resonance, also centered around $B = 0$, which corresponds to the NFE. This effect can be understood in terms of optical pumping, and thus requires time to build-up, which is easily seen in Fig. 2. The width of the NFE resonance, typically $\sim$ 1 mG, is much narrower than that of the LFE and corresponds to the lifetime of the $\Delta m = 2$ Zeeman coherences in the $F = 3$ ground state [17] (m being the magnetic quantum number).

The large amplitude of a few degrees and and narrow width of the NFE resonance allows for precision magnetic field measurements. The magnetometric sensitivity improves when the NFE resonance width is reduced [18]. It is thus advantageous to keep the resonance width as narrow as possible, i.e., prolong the coherence lifetime. Special care has to be taken to achieve long coherence life-times, since any stray magnetic fields can shorten it. In the following subsection we discuss the pump-probe method we used for assessment of magnetic-fields compensation.

3.2. Pulsed pump experiments

The nonlinear Faraday Effect allows one to measure an averaged response of an ensemble of a cold atom ensemble to the applied magnetic fields. It is, however, possible to observe the dynamics of the system by using a relatively short ($\tau << 1/\omega_L$, where ω_L is the Larmor precession frequency) strong light pulse for creation of the desired atomic population distribution. After such a pulse, the evolution of the system is interrogated by measuring the polarization rotation of a weak

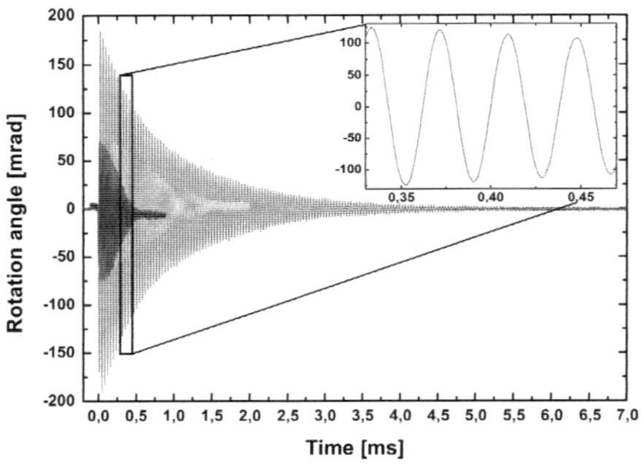

Figure 3. Comparison of the FID signals with and without magnetic shielding. The blue (shortest) and yellow (middle long) curves show data obtained in an unshielded chamber, with passive compensation of the ambient magnetic fields and additional compensation of the field gradients, respectively. The red (longest) curve was recorded in a shielded environment, where the measured decay time equals $\tau = 980\,\mu$s. Notice the change of the FID signal envelope shapes.

probing beam.

By means of a σ^+-polarized pump pulse propagating along the z direction we pump the atoms into the stretched state, $F = 3$, $m_F = 3$, assuming the quantization axis along the z-direction. We then apply a dc magnetic field B_x along the x axis which causes the atomic polarization to rotate around that field at a frequency proportional to its magnitude, i.e., Larmor frequency ω_L. The created distribution of atomic populations among the ground state Zeeman sublevels causes the polarization of a weak probe light to oscillate due to the paramagnetic Faraday effect, and this oscillation is recorded in the form of a free induction decay (FID) signal [7, 21, 22]. Such oscillations decay with a relaxation rate of Zeeman coherences, which is governed by several factors, including transverse magnetic fields, magnetic field instability and interaction with the probing light via absorption and AC Stark shifts.

The effect of stray magnetic field is shown in Fig. 3 where signals recorded with and without magnetic shielding are compared. Not only the coherence lifetime is prolonged in the shielded case, but also the envelope of the FID signal shows the exponential decay - a proof for the right magnetic field control. We were able to determine that after minimizing all other sources of decoherence we are left with the dephasing due to the spatial inhomogeneity of the applied magnetic field. Slight differences in the values of the B_x field across the atomic cloud are attributed to atoms from different parts of the sample experiencing precession at slightly deviated frequencies. The rate of such dephasing was measured to be $\gamma_x = 2\pi \cdot 160$ Hz.

3.3. FID signals of Zeeman coherences

In a standard observation of FID signals, the sample is magnetized along one direction and then subjected to the precession around a perpendicularly oriented magnetic field, which provides the highest amplitude of the rotation signal. It is, however, possible to observe the FID spin polarization dynamics in an arrangement where the pump beam is propagating along the B-field direction, i.e., the natural geometry for the Faraday rotation experiments. Since Zeeman substates are the eigenstates of the magnetic field, it is only Zeeman coherences which evolve in time in such a realization.

To record FID signals associated with Zeeman coherences, we use a short pump pulse of the linearly polarized beam, which prepares the $F = 3$ ground-state $\Delta m = 2$ Zeeman coherences, and monitor time dependence of the rotation angle of the probe beam polarization. Examples of such dependence of FID signals on the magnetic field are shown in Fig. 4. Clearly visible are the hyperbolas of the constant phase of Larmor precession. Frequency of the Larmor precession tends to zero around the zero field (long oscillation period) and increases with the magnetic field

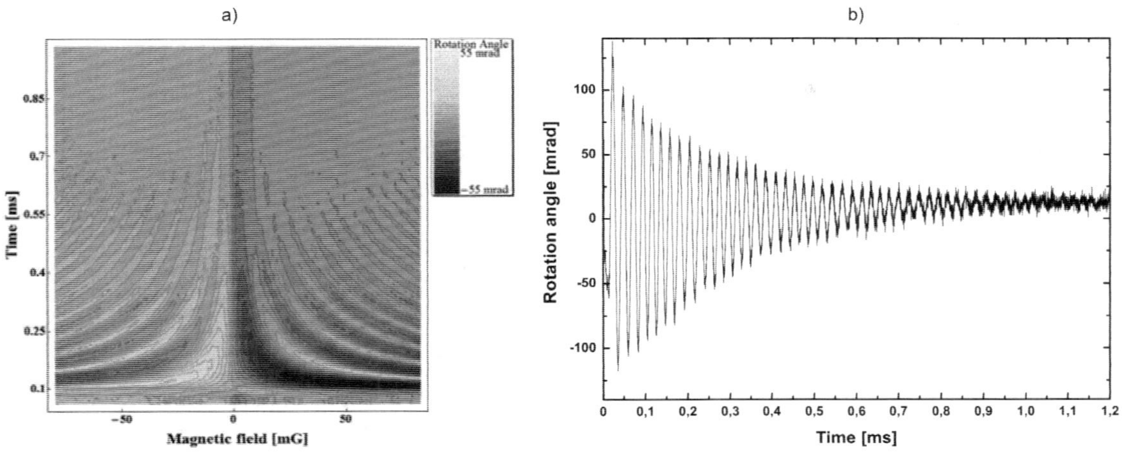

Figure 4. FID signals of the $\Delta m = 2$ atomic coherences prepared by a short pumping light pulse at $t = 0$. a) Contour plot of the rotation dependence on time and the magnetic field, b) coherences FID signal recorded for $B_z = 95$ mG.

magnitude.

3.4. Detection and measurements of the radio-frequency magnetic fields

By tuning the Zeeman splitting to resonance with an external radio-frequency field B_{rf} oscillating at the frequency ω_{rf} in the x direction we are able to observe resonant changes of the rotation angle. This happens provided that there is some initial inequality in atomic populations of given sublevels is established by optical pumping. These resonances can be applied for magnetometry in two ways: (i) the rf resonance frequency is taken as the measure of the dc magnetic field intensity B, (ii) the amplitude of resonance in the rotation amplitude can be used for measurement of the intensity of the rf field B_{rf} [23, 24].

For sufficiently strong amplitude B_{rf}, Rabi oscillations between neighboring Zeeman sublevels of the ground state manifold can be observed due to the paramagnetic Faraday rotation. By initially preparing atoms in the stretched state with a short σ^+ light pulse as described earlier, we are able to maximize the observed rotation angles.

Our present experimental setup is severely limited by a short lifetime of the MOT and cannot compete with other optical magnetometers in terms of sensitivity. Instead, we can use it for studies of a coherent interaction between cold atoms and various fields, like coherent evolution of atomic observables in the rf field. Figure 5 (a) depicts an example of Rabi oscillations recorded with rf field of 11 kHz frequency and 5.5 mG amplitude. The width of the rf rotation resonances is strongly power broadened. This allows observation of the dependence of the generalized Rabi frequency on the detuning of ω_{rf} from the exact Larmor frequency in the form of characteristic bending of the resonance features on a contour map [Fig. 5 (a)]. Rabi oscillations for lower rf field amplitude and with relaxation rate $\gamma_{rf}/2\pi \sim 50$ Hz are shown in Fig. 5 (b). It is worth noting that polarization rotation angle can easily reach hundreds of milliradians in an atomic cloud from a standard MOT.

4. Conclusions

We have demonstrated an ability to observe a coherent evolution of polarized atoms prepared in a magneto optical trap for times of the order of milliseconds and detection of dc and rf magnetic fields with high sensitivity. The present experimental setup is limited by a short

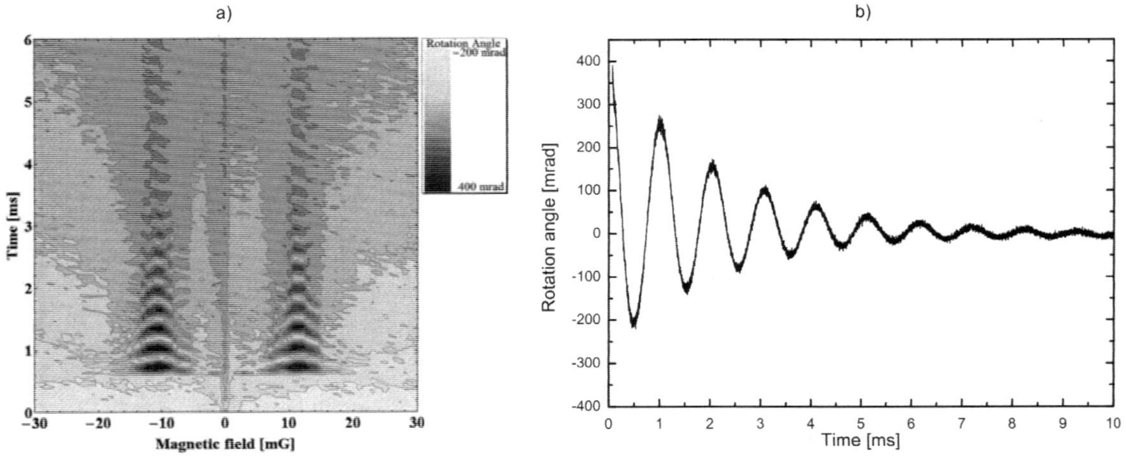

Figure 5. a) Contour plot of the polarization rotation dependence on time and a dc magnetic field showing Rabi oscillations induced by the rf magnetic field of frequency $\omega_{\mathrm{rf}} = 2\pi \cdot 11$ kHz and amplitude $B_{\mathrm{rf}} = 5.5$ mG switched on at $t \approx 0.6$ ms. b) Rotation signal for the resonance condition $\omega_L = \omega_{\mathrm{rf}} = 2\pi \cdot 8.2$ kHz, $B_{\mathrm{rf}} \approx 1.7$ mG.

lifetime of the MOT. The reported observations indicate, nevertheless, that the upgrade to a far-detuned optical dipole trap would significantly improve the lifetimes of the trapped atoms. Consequently, this would allow applications in sensitive magnetometry and studies of long-lived atomic superpositions states, including quantum degenerated samples.

Acknowledgments

Authors acknowledge financial support by the Foundation for Polish Science (Team Programme) and the National Science Centre (grant no. 2012/07/B/ST2/00251). AW acknowledges support by the Jagiellonian University within the SET project co-financed by the European Union.

References

[1] Ketterle W, Durfee D and StamperKurn D 1999 *Bose-Einstein condensation in atomic gases* Proceedings of the International School of Physics Enrico Fermi. Course CXL ed Inguscio M, Stringari S and Wieman C (IOS Press, Amsterdam) (*Preprint* `arXiv:cond-mat/9904034`)
[2] Lewenstein M, Sanpera A, Ahufinger V, Damski B, Sen(De) A and Sen U 2007 *Advances in Physics* **56** 243
[3] Ketterle W and Zwierlein M W 2008 *Ultracold Fermi Gases* Proceedings of the International School of Physics Enrico Fermi. Course CLXIV ed Inguscio M, Ketterle W and Salomon C (IOS Press, Amsterdam) (*Preprint* `0801.2500`)
[4] Katori H 2011 *Nature Photonics* **5** 203–210
[5] Le Targat R et al 2013 *Nature Communications* **4** 2109
[6] Dickerson S M, Hogan J M, Sugarbaker A, Johnson D M S and Kasevich M A 2013 *Phys. Rev. Lett.* **111**(8) 083001
[7] Isayama T, Takahashi Y, Tanaka N, Toyoda K, Ishikawa K and Yabuzaki T 1999 *Phys. Rev. A* **59** 4836–4839
[8] Franke-Arnold S, Arndt M and Zeilinger A 2001 *J. Phys. B* **34** 2527–2536
[9] Labeyrie G, Miniatura C and Kaiser R 2001 *Phys. Rev. A* **64** 033402
[10] Nash J and Narducci F A 2003 *Journ. Mod. Optics* **50** 2667
[11] Wildermuth S, Hofferberth S, Lesanovsky I, Groth S, Krüger P, Schmiedmayer J and Bar-Joseph I 2006 *Appl. Phys. Lett.* **88** 264103
[12] Vengalattore M, Higbie J M, Leslie S R, Guzman J, Sadler L E and Stamper-Kurn D M 2007 *Physical Review Letters* **98** 200801
[13] Terraciano M L, Bashkansky M and Fatemi F K 2008 *Opt. Express* **16** 13062–13069

[14] Sewell R J, Koschorreck M, Napolitano M, Dubost B, Behbood N and Mitchell M W 2012 *Phys. Rev. Lett.* **109**(25) 253605
[15] Wojciechowski A, Corsini E, Zachorowski J and Gawlik W 2010 *Physical Review A* **81** 053420–0533424
[16] Cohen-Tannoudji C 1999 *Frontiers in Laser Spectroscopy, Les Houches Summer School 1975* ed RBalian, Haroche S and Liberman S (North Holland) (*Preprint* `arXiv:cond-mat/9904034`)
[17] Budker D, Gawlik W, Kimball D F, Rochester S M, Yashchuk V V and Weis A 2002 *Reviews of Modern Physics* **74** 1153–11201
[18] Budker D, Kimball D F, Rochester S M, Yashchuk V V and Zolotorev M 2000 *Phys. Rev. A* **62**(4) 043403
[19] Gawlik W, Krzemień L, Pustelny S, Sangla D, Zachorowski J, Graf M, Sushkov A O and Budker D 2006 *Applied Physics Letters* **88** 131108–131110
[20] Gawlik W and Wojciechowski A 2011 *Optics and Spectroscopy* **111** 626–632
[21] Takahashi Y, Shimizu T, Tanaka N, Honda K, Toyoda K and Yabuzaki T 1999 *Phys. Rev. A* **59**(5) 3761–3765
[22] Behbood N, Martin Ciurana F, Colangelo G, Napolitano M, Mitchell M W and Sewell R J 2013 *Applied Physics Letters* **102**
[23] Savukov I M, Seltzer S J, Romalis M V and Sauer K L 2005 *Phys. Rev. Lett.* **95**(6) 063004
[24] Ledbetter M P, Acosta V M, Rochester S M, Budker D, Pustelny S and Yashchuk V V 2007 *Phys. Rev. A* **75**(2) 023405

Journal of Physics: Conference Series **497** (2014) 012007 doi:10.1088/1742-6596/497/1/012007

Light controlling light in a coupled cavity-atom system

Bichen Zou[1], Zheng Tan[1-2], and Yifu Zhu[1]

[1]*Department Physics, Florida International University, Miami, FL, USA*
[2]*Wuhan Institute of Physics and Mathematics, Chinese Academy of Sciences, Wuhan, China*

Abstract. We present a theoretical model of light controlling light in a multi-atom cavity QED system consisting of three-level atoms confined in a cavity and interacting with a control laser from free space. A signal laser is coupled into the cavity and provides two output light channels: the transmitted signal light through the cavity and the reflected signal light from the cavity. We show that with the cavity electromagnetically induced transparency manifested by the free-space control light, the amplitude and the phase of the intra-cavity signal light can be manipulated by the free-space control laser. We analyze the cavity transmitted signal light and the cavity reflected signal light, and show that the two output channels of the signal light are complementary and the cavity-atom system is a versatile system for studies of all-optical switching and cross-phase modulation at low control intensities.

1. Introduction

Cavity quantum electrodynamics (cavity QED) studies the coherent interactions of two-level atoms and photons in a cavity and has found a variety of applications in quantum physics and quantum electronics [1-4]. In recent years, studies of cavity QED have been extended to the interactions of the cavity mode and multi-level atoms. There were earlier studies of optical bistability in three-level atoms confined in an optical cavity [5]. Recent studies of atom-cavity interactions have been directed to a composite system of an optical cavity and coherently prepared multi-level atoms, in which the atomic coherence and interference such as electromagnetically induced transparency (EIT) [6] can be used to manipulate quantum states of the coupled cavity and atom system and selectively enhance the optical nonlinearities [7-11]. It has been shown that in a coherently coupled cavity and multi-atom system, the interplay of the collective coupling of the atoms and the cavity mode, and the atomic coherence and interference manifested by EIT may lead to interesting linear and nonlinear optical phenomena [12-21].

Here we present a theoretical study of an atom-cavity system consisting of N three-level atoms confined in an optical cavity and coherently coupled from free space by a control laser. The system forms a Λ-type standard EIT configuration with the cavity mode. We show that the free-space control laser induces EIT for the intra-cavity signal field, which can be used to control both the amplitude and phase of the cavity-reflected signal light and the cavity-transmitted signal light. Under appropriate conditions, all-optical switching and large cross-phase modulation (XPM) for the reflected signal field and transmitted signal field can be obtained with a weak control field. The cavity-atom system provides an example of resonant nonlinear optics at low light intensities and can be used to explore fundamental studies of light controlling light phenomena.

2. Theoretical analysis

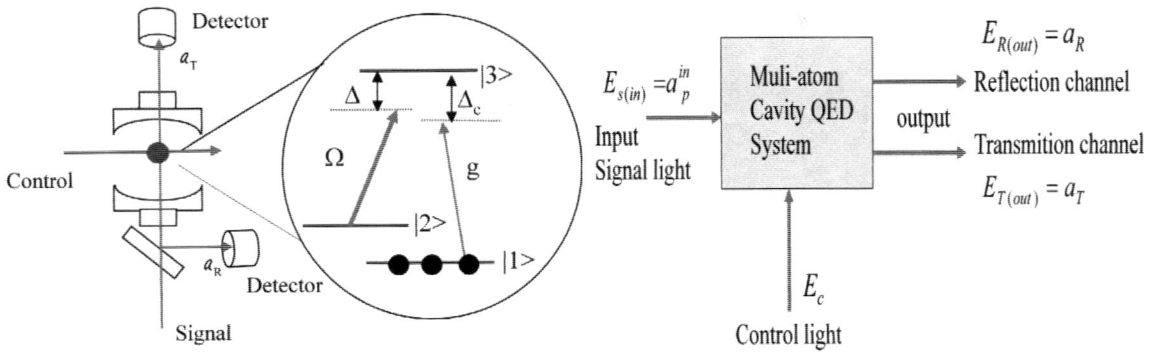

(a) (b)

Fig.1 (a) Schematic coupling scheme of coherently coupled three-level atoms in a cavity. A control laser drives |2> - |3> transition with Rabi frequency 2Ω. Δ is the control detuning. A cavity is coupled to the atomic transition |1> - |3> with the collective coupling coefficient $\sqrt{N}g$ (Δ_c is the cavity-atom detuning). A signal laser is coupled into the cavity and Δ_p is its frequency detuning from the atomic transition. The cavity transmitted signal light and the cavity reflected signal light are collected by two detectors. (b) Schematic diagram of the light input and output channels. The cavity-atom system provides two output signal channels: the reflection channel and the transmission channel.

Fig. 1 shows the composite atom-cavity system that consists of a single mode cavity containing N Λ-type three-level atoms interacting with a control laser from free space. The cavity mode couples the atomic transition |1>-|3> and the classical control laser drives the atomic transition |2>-|3> with Rabi frequency 2Ω. $\Delta = \nu - \nu_{23}$ is the control frequency detuning and $\Delta_c = \nu_c - \nu_{13}$ is the cavity-atom detuning. A signal laser is coupled into the cavity mode and its frequency is detuned from the atomic transition |1>-|3> by $\Delta_p = \nu_p - \nu_{13}$. The interaction Hamiltonian for the coupled cavity-atom system is

$$H = -\hbar\left(\sum_{i=1}^{N} \Omega\hat{\sigma}_{32}^{(i)} + \sum_{i=1}^{N} g\hat{a}\hat{\sigma}_{31}^{(i)}\right) + H.C. , \qquad (1)$$

where $\hat{\sigma}_{lm}^{(i)}$ (l, m=1-4) is the atomic operator for the ith atom, $g = \mu_{13}\sqrt{\omega_c / 2\hbar\varepsilon_0 V}$ is the cavity-atom coupling coefficient, and $\hat{a}$ is the annihilation operator of the cavity photons. The resulting equation of motion for the expectation value the intra-cavity light field (two-sided cavity, one input) is given by [22] is

$$\dot{a} = -((\kappa_1 + \kappa_2)/2 - i\Delta_c)a + \sum_{i=1}^{N} ig\sigma_{31}^{(i)} + \sqrt{\kappa_1 / \tau}a_p^{in} , \qquad (2)$$

where a_p^{in} is the input signal field, $\kappa_i = \frac{T_i}{\tau}$ (i=1-2) is the loss rate of the cavity field of the mirror i (Ti is the mirror transmission and τ is the photon round trip time inside the cavity). For simplicity, we consider a symmetric cavity such that $\kappa_1 = \kappa_2 = \kappa/2$. With g<<Ω, the atomic population is concentrated in |1> and the steady-state solution of the intra-cavity signal field is given by

$$a = \frac{\sqrt{\kappa / \tau} a_p^{in}}{\kappa - i\Delta_c - i\chi}, \qquad (3)$$

where χ is the atomic susceptibility given by $\chi = \frac{ig^2 N}{\Gamma_3 - i\Delta_p + \frac{\Omega^2}{\gamma_{12} - i(\Delta_p - \Delta)}}$. Here Γ_3 is the spontaneous decay rate of the excited state |3> and γ_{12} is the decoherence rate between the ground states |1> and |2>. The transmitted signal field is then given by $a_T = \sqrt{\kappa\tau} a$ and the reflected signal field from the cavity is $a_R = \sqrt{\kappa\tau} a - a_p^{in}$. We are interested in the regime of parameters near cavity EIT in which the laser fields are near or at resonance with the atomic transitions, and under the conditions of low light intensities in which the intra-cavity field is very weak and the control field is well below the saturation level. We show that a weak control light can induce large nonlinearities in the cavity-atom system, which then may be used to control the amplitude and phase of the cavity transmitted signal field and the cavity reflected signal field [23]. The primary control parameters of the light-control-light system are the frequency and intensity of the control laser that are characterized by the control detuning Δ and control Rabi frequency Ω respectively.

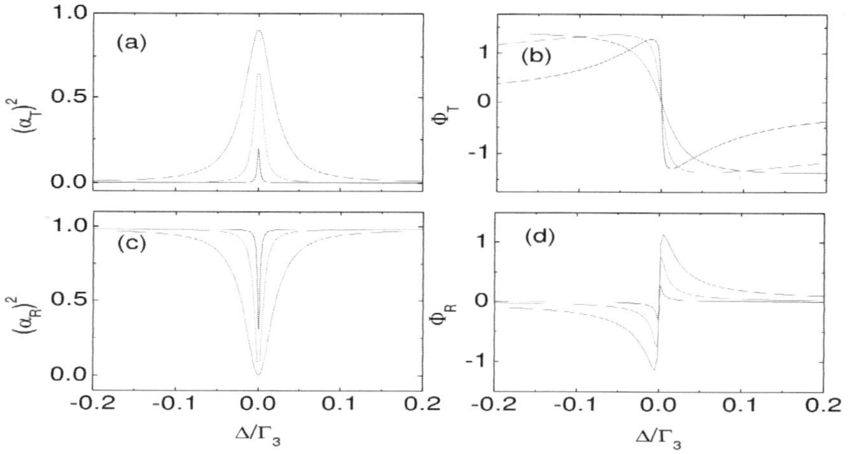

Fig. 2 (a) The intensity ratio α_T^2 and the phase shift Φ_T of the cavity transmitted field versus the control frequency detuning Δ/Γ_3. (c) The intensity ratio α_R^2 and (d) the phase shift Φ_R of the cavity reflected field versus the control frequency detuning Δ/Γ_3. The parameters used in the calculations are $g\sqrt{N} = 10\Gamma$, $\kappa = 2\Gamma_3$, γ_{12}=0.001Γ, and $\Delta_c = \Delta_p = 0$. The control Rabi frequency Ω=0.2Γ,0.5Γ, and Γ for the black lines, red lines, and blues lines respectively.

Fig. 2(a) and 2(b) plot the intensity ratio of the cavity transmitted field, $\frac{I_T}{I_{in}} = \alpha_T^2$ and the phase shift Φ_R versus the control frequency detuning Δ/Γ_3 respectively. Fig. 2(c) and 2(d) plot the intensity ratio of the cavity reflected field, $\frac{I_R}{I_{in}} = \alpha_R^2$ and the phase shift Φ_T versus Δ/Γ_3 respectively. The relevant parameters are $g\sqrt{N} = 10\Gamma_3$, $\kappa = 2\Gamma_3$, $\gamma_{12}=0.001\Gamma_3$, and $\Delta_c = \Delta_p = 0$. The transmission spectrum in Fig. 2(a) and 2(b) exhibit the standard cavity EIT spectral peak at $\Delta=0$ with a peak linewidth that is substantially smaller than the decay width Γ_3 and increases with the increasing Ω of the control laser. The phase shift of the transmitted field varies rapidly across the resonance at $\Delta=0$ and has a line shape of the anomalous dispersion. Concomitantly, the reflected field from the cavity exhibits a dip with the linewidth matching the peak linewidth of the transmitted field and its phase shift has a lineshape of the normal dispersion with a steep slope across $\Delta=0$. The calculations of Fig. 2 are done with the signal laser tuned to the resonance $\Delta p=0$ and show that the transmitted field and the reflected field can be controlled by varying the frequency of the control laser. For example, Fig. 2(a) and 2(c) show that with a resonant control laser ($\Delta=0$), the transmitted light field can be turned on or off by turning on or off of the control light, and at the same time and in contrast, the reflected light field is turned off or on. Thus the coupled cavity-atom system effectively performs all-optical switching with reciprocal functions for the transmission and reflection with a weak control laser. When the control laser is tuned slightly away from the resonance ($\Delta\neq0$ but $\Delta<<\Gamma_3$), large phase shifts are induced on the transmitted field and the reflected field (Fig. 2(b) and 2(d)), which effectively performs the cross-phase modulation (XPM) on the transmitted/reflected field with a weak control light near the atomic resonance.

2-1 All-optical switching

Fig. 2 shows that a weak control laser can be used to control the amplitude and phase of both the transmitted and reflected light fields from the cavity. In particular, when $\Delta=0$ and $\Delta_c = \Delta_p = 0$, cavity EIT is established, and the coupled cavity-atom system can perform the all-optical switching function with two complementary output channels. Under the resonance condition ($\Delta=0$ and $\Delta_c = \Delta_p = 0$), the reflected signal intensity is

$$I_R = \frac{g^4 N^2 \gamma_{21}^2 I_{in}}{(\kappa(\Omega^2 + \Gamma_3\gamma_{21}) + g^2 N\gamma_{21})^2}, \tag{4}$$

and the transmitted signal intensity is

$$I_T = \frac{\kappa^2 (\Omega^2 + \Gamma_3\gamma_{21})^2 I_{in}}{(\kappa(\Omega^2 + \Gamma_3\gamma_{21}) + g^2 N\gamma_{21})^2}. \tag{5}$$

The performance of an all-optical switch can be characterized by the switching efficiency defined as $\eta = \frac{I(|1>) - I(|0>)}{I_{in}}$, where I(|1>) is the signal output intensity when the switch is closed and I(|0>) is the signal output intensity (leakage) when the switch is open (I(|1>) > I(|0>)), and I_{in} is the input signal intensity. Eq. (4) indicates that $I_R(\Omega=0) > I_R(\Omega\neq0)$ so for the all-optical switching operating on the reflection output channel, we designate $I_R(\Omega\neq0)$ =I(|0>) for the open state |0> of the switch and $I_R(\Omega=0)$=I(|1>) for the closed state |1> of the switch, then the switching efficiency for the reflection output channel is

$$\eta_R == \frac{g^4N^2}{(\kappa\Gamma_3 + g^2N)^2} - \frac{g^4N^2\gamma_{21}^2}{(\kappa(\Omega^2 + \Gamma_3\gamma_{21}) + g^2N\gamma_{21})^2}. \quad (6)$$

Eq. (5) indicates that $I_T(\Omega\neq0)>I_T(\Omega=0)$. Therefore, for the all-optical switching with the transmission output channel, we designate $I_T(\Omega=0)=I_T(|0>)$ for the open state $|0>$ of the switch and $I_R(\Omega\neq0)=I_T(|1>)$ for the closed state $|1>$ of the switch. Then, the switching efficiency for the transmission output channel is

$$\eta_T = \frac{\kappa^2(\Omega^2 + \Gamma_3\gamma_{21})^2}{(\kappa(\Omega^2 + \Gamma_3\gamma_{21}) + g^2N\gamma_{21})^2} - \frac{\kappa^2\Gamma_3^2}{(\kappa\Gamma_3 + g^2N)^2}. \quad (7)$$

Eq.(6) and Eq.(7) show that under appropriate conditions, the switching efficiency for the reflection channel as well as the transmission can be near unity. In order to provide a detailed picture for the study of all-optical switching in the coherently coupled cavity-atom system, we present numerical calculations in Figures 3-6 and show that practical parameters can be identified, in which the all-optical switching of the two output channels with the switching efficiency near unity can be obtained.

The switching efficiency η_R and η_T depend on the collective coupling coefficient $g\sqrt{N}$, the control laser Rabi frequency 2Ω, the cavity decay rate κ, and the deocherence rate γ_{21}. In order to clarity the performance characteristics of the coupled cavity-atom system, we plot in Fig. 3 to

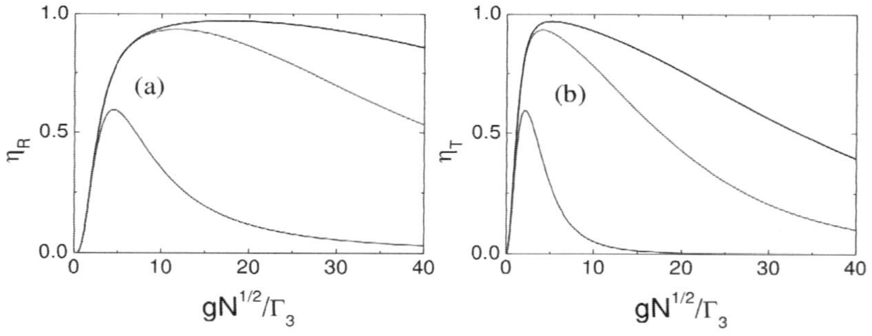

Fig. 3 (a) Switching efficiency of the reflected signal light η_R and (b) switching efficiency of the transmitted signal light η_T versus $g\sqrt{N}/\Gamma_3$ with $\gamma_{12}=0.0001\Gamma_3$ (black lines), $\gamma_{12}=0.001\Gamma_3$ (red lines) and $\gamma_{12}=0.01\Gamma_3$ (blues lines), respectively. Other parameters are $\Delta p=\Delta= g\sqrt{N}$. $\Omega=0.3\Gamma_3$, and $\kappa=3\Gamma_3$.

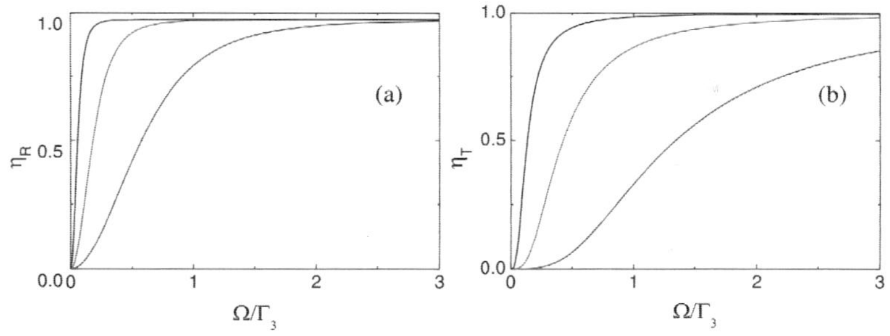

Fig. 4 (a) Switching efficiency of the reflected signal light η_R and (b) switching efficiency of the transmitted signal light η_T versus Ω/Γ_3 with $\gamma_{12}=0.0001\Gamma_3$ (black lines), $\gamma_{12}=0.001\Gamma_3$ (red lines) and $\gamma_{12}=0.01\Gamma_3$ (blues lines), respectively. Other parameters are $\Delta p=\Delta= g\sqrt{N}=20\Gamma_3$, $\Omega=0.3\Gamma_3$ and $\kappa=3\Gamma_3$,

Fig. 6 separately the switching efficiency η_R and η_T under the resonance conditions $\Delta=0$ and $\Delta_c=\Delta_p=0$, versus these parameters with practical values obtainable experimentally [9]. Fig. 3

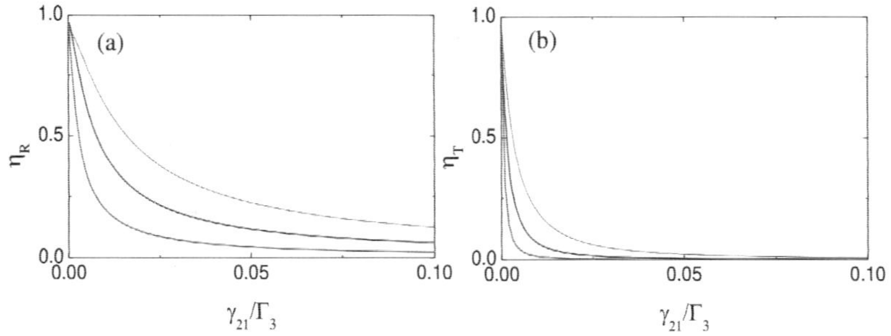

Fig. 5 (a) Switching efficiency of the reflected signal light η_R and (b) switching efficiency of the transmitted signal light η_T versus the decoherence rate γ_{21}/Γ_3 with $g\sqrt{N}=20\Gamma_3$ (red lines), $g\sqrt{N}=15\Gamma_3$ (black lines) and $g\sqrt{N}=10\Gamma_3$ (blues lines), respectively. Other parameters are $\Omega=0.5\Gamma_3$ and $\kappa=3\Gamma_3$.

shows that η_R and η_T increases initially with the collectively coupling coefficient $g\sqrt{N}$ and are maximized at moderate $g\sqrt{N}$ value. Fig. 4 shows that the all-optical switching of the cavity-atom system can be done with a weak control laser. The switching efficiency increases rapidly with the increasing control field, but saturates as Ω values near the decay rate Γ_3. Fig. 5 shows that η_R and η_T decreases monotonically with increasing γ_{21}, which indicates that the all- optical switching is a coherent process and the switching efficiency (particularly η_T) depends sensitively

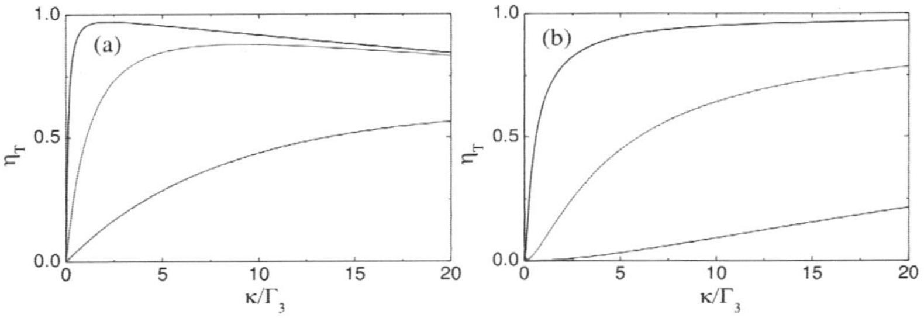

Fig. 6 (a) Switching efficiency of the reflected signal light η_R and (b) switching efficiency of the transmitted signal light η_T versus κ/Γ_3 with γ_{12}=0.0001Γ_3 (black lines), γ_{12}=0.001Γ_3 (red lines) and γ_{12}=0.01Γ_3 (blues lines), respectively. Other parameters are $g\sqrt{N}$ =15Γ_3, and Ω=0.2Γ_3.

on γ_{21}. For the high efficiency operation of the all-optical switching, it is desirable to have an atomic system with small decay rate of the ground state coherence. It has been shown [36] that in cold Rb atoms, the decoherence rate as small as $\gamma_{12}=10^{-4}\Gamma_3$ has been observed. Therefore, it is possible to achieve high switching efficiencies in experiments with cold alkaline atoms as the optical medium.

Fig. 6 plots η_R and η_T versus the cavity decay rate κ and shows that it is more efficient to operate the all-optical switching of the cavity-atom system at a relatively high κ value. Therefore, a moderate to low Q value of the cavity provides a better switching efficiency. Overall, the highly efficient all-optical switching at low control intensities requires a moderately large collective coupling coefficient $g\sqrt{N}$, a sufficiently large cavity decay rate κ, and a small decoherence rate γ_{21}. These requirements can be readily fulfilled experimentally. As a numerical example, consider cold Rb atoms (Γ_3=3 MHz) confined in a 5 cm cavity with a finesse of 150 (κ=10 MHz), with $g\sqrt{N}$=50 MHz (N≈10^4 atoms), γ_{21}=10 KHz(~0.003Γ_3), and Ω=1.5 MHz (corresponding to a control intensity $I_c = c\varepsilon_0 E^2 = c\varepsilon_0 (h\Omega/\mu_{23})^2$≈0.3 mW/cm^2 that is about 5 times smaller than the Rb saturation intensity of 1.6 mW/cm^2), the switching efficiency is derived to be η_R=0.83 and η_T=0.56. Fig. 3-6 also show that the switching efficiencies for the two output channels are different. Under the normal operating conditions discussed here, $\eta_R > \eta_T$: it is more efficient to operate the all-optical switch of the cavity-atom system in the reflection mode.

2-2 Cross-phase modulation

When the control laser is tuned away from the resonance ($\Delta \neq 0$, but keeping $\Delta_c = \Delta_p = 0$), the phase shift of the intra-cavity signal field is introduced by the control laser. We first derive the analytical results with γ_{12}=0 (neglecting the decoherence. The effect of decoherence with $\gamma_{12} \neq 0$ will be considered in the numerical calculations) for the two output channels of the signal light. Specifically, the light field transmitted through the cavity is $a_T = \alpha_T e^{i\Phi_T} a_p^{in}$. The amplitude ratio of the transmitted signal field to the input signal field is given by

$$\alpha_T = \frac{\kappa((\Omega^4 + (\Delta\Gamma_3)^2)^{1/2}}{((\kappa\Omega^2)^2 + ((g^2N + \kappa\Gamma_3)\Delta)^2)^{1/2}}, \quad (8)$$

and the phase shift Φ_T is given by

$$\Phi_T = \tan^{-1}\frac{(\Gamma_3\kappa - g^2N)\Omega^2\Delta}{g^2N\Delta^2\Gamma_3 + \kappa\Omega^4}. \quad (9)$$

Under the condition of the strong collective coupling ($g^2N >> \kappa\Gamma_3$) and for a weak control field ($\Omega^2 << \Delta\Gamma_3$. For the near resonant nonlinear optics studied here, $\Delta < \Gamma_3$, the phase shift of the signal field is $\Phi_T = -\frac{\Omega^2}{\Delta\Gamma_3}$, and the amplitude ratio of the transmitted field to the input signal field is $\alpha_T = \frac{\kappa\Gamma_3}{g^2N}$. The phase shift is proportional to the control laser intensity and corresponds to the standard Kerr nonlinearity. Under the strong collective coupling condition, a large phase shift of the transmitted field can be obtained but the transmitted field amplitude is very small.

The reflected signal field from the cavity is $a_R = \alpha_R e^{i\Phi_R} a_p^{in}$. The amplitude ratio of the reflected signal field to the input signal field is given by

$$\alpha_R = \frac{g^2N\Delta}{((\kappa\Omega^2)^2 + (g^2N + \kappa\Gamma_3)^2\Delta^2)^{1/2}}, \quad (10)$$

and the phase shift Φ_R of the reflected signal field is

$$\Phi_R = \tan^{-1}\frac{\kappa\Omega^2}{(g^2N + \kappa\Gamma_3)\Delta}. \quad (11)$$

As an example, with a weak, near resonant control (Ω=0.5Γ_3 and Δ=0.01Γ_3), $g\sqrt{N} = 10\Gamma_3$, and κ=2Γ_3, the amplitude ratio of the reflected signal field to the input signal field is ~ 79% and the

XPM phase shift of the reflected field from the control laser is ~0.5 rad. In the limit of $\kappa\Omega^2 << |(g^2N + \kappa\Gamma_3)\Delta|$, one derives $\alpha_R = \dfrac{g^2N}{g^2N + \kappa\Gamma_3}$ and $\Phi_R \approx \dfrac{\kappa\Omega^2}{(g^2N + \kappa\Gamma_3)\Delta}$, which is the phase shift from the Kerr nonlinearity induced by the weak control field on the reflected cavity field. The reflected field amplitude can be nearly equal to the input signal field when the condition of strong collective coupling is satisfied ($g^2N >> \kappa\Gamma_3$), but the phase shift Φ_R approaches zero. Therefore the optimal performance of the cavity system for XPM is achieved for moderate g^2N values. As it will be further clarified in the numerical calculations that by inducing cavity EIT in the coherently coupled cavity-atom system, the XPM can be obtained near the atomic resonance with the control detuning $\Delta << \Gamma$, which leads to a larger Φ_R value, but at the same time still maintains a sufficiently large amplitude of the transmitted/reflected field. This is in sharp contrast with the conventional XPM systems in which the laser fields have to be tuned far away from the atomic resonance to minimize the absorption loss.

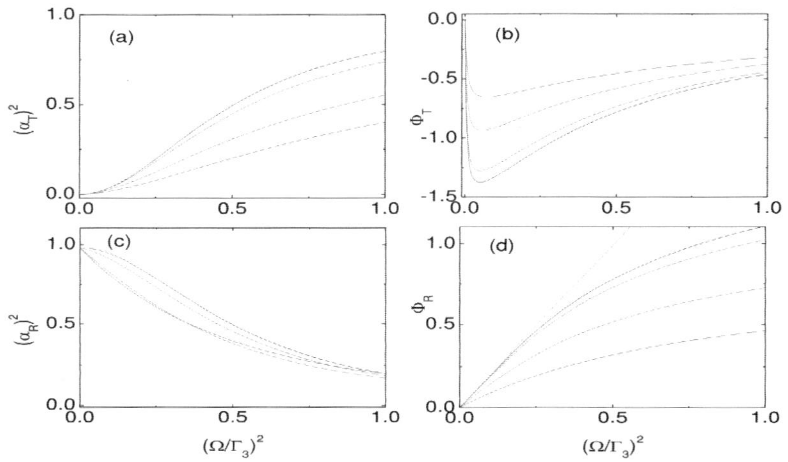

Fig. 7 (a) The intensity ratio α_T^2 and (b) phase shift Φ_T of the cavity transmitted signal field versus $(\Omega/\Gamma_3)^2$. (c) The intensity ratio α_R^2 and (d) phase shift Φ_R of the cavity reflected field versus $(\Omega/\Gamma_3)^2$. The parameters used in the calculations are $g\sqrt{N} = 10\Gamma_3$, $\kappa = 2\Gamma_3$, $\Delta=0.01\Gamma_3$, and $\Delta_c = \Delta_p = 0$. The decoherence rate γ_{12}=0, $0.001\Gamma_3$, $0.005\Gamma_3$, and $0.01\Gamma_3$ for the black lines, red lines, purple lines, and blues lines respectively.

It is necessary to quantify the XPM performance of the cavity-atom system by comparing the phase shit and the field amplitude. A useful system has to be able to produce large XPM phase shits and at the same time, provides sufficiently large field amplitudes. For studies of nonlinear optics at low light intensities, such requirements have to be obtained at the condition of low control field intensities. Fig. 7 plots $\dfrac{I_T}{I_{in}} = \alpha_T^2$, $\dfrac{I_R}{I_{in}} = \alpha_R^2$, Φ_T, and Φ_R versus the square of the control Rabi frequency Ω^2, which is proportional to the control light intensity $I = c\varepsilon_0(h\Omega/\mu_{23})^2$.

It can be seen that for the reflected field, the XPM phase shift increases with the control light intensity and exhibits saturation at high control intensities while the reflected light intensity decreases with the control intensity and approaches zero at high control intensities. For comparison, the green curve in Fig. 4(c) plots the phase shift given by $\Phi_R = \frac{\kappa\Omega^2}{(g^2N + \kappa\Gamma_3)\Delta}$, (Eq. (10) with $\kappa\Omega^2 << |(g^2N + \kappa\Gamma_3)\Delta|$). For the transmitted field, the XPM phase shift exhibits rapid change at low control intensities near zero and also saturates at high control intensities while the transmitted intensity increases with the control intensity.

The ground state decoherence degrades the system performance of the light control light. γ_{12} ultimately determines the minimum linewidth of the cavity EIT and also the lower limit of the control Rabi frequency Ω. As the decoherence rate γ_{12} increases, the amplitudes of the transmitted field and the reflected field decrease; the phase shifts of the transmitted field and the reflected field also decrease. It is preferable to select an atomic system with a small γ_{12}. We note that in cold Rb atoms, the decoherence rate γ_{12}=$10^{-4}\Gamma_3$ has been observed [24]. Therefore, a possible experimental system may consist of cold Rb atoms ($\Gamma_3 \approx 6$ MHz) and a cavity with a moderate finesse (with $\kappa = 2\Gamma_3$ and a cavity length of 5 cm, the required finesse is F=124). With a weak control laser, the linewidth of the cavity EIT is much smaller than Γ_3 (with Ω=0.5Γ_3 and γ_{12}=0.001Γ_3, the linewidth of the cavity EIT is ~ 0.01Γ_3), it is necessary to have the control laser and the signal laser with a linewidth much smaller than 0.01Γ_3. For the Rb atoms, the required laser linewdth is in the KHz range, which can be obtained with the frequency stabilized diode lasers.

Fig. 7(a) and 7(c) show that at low control intensities, the reflected field amplitude is greater than the transmitted field amplitude. To achieve a large field amplitude, it is desirable to operate the cavity-atom system with the reflected light field. At low control intensities, the XPM phase shift is proportional to the control intensity, which is in the Kerr nonlinearity regime. A large φ_R value can be obtained at a sufficiently large α_R^2 value with a weak control light. As a numerical example, In an cavity-atom system with γ_{12}=0.001Γ_3, the XPM phase shift Φ_R ~0.5 rad. and the reflected intensity ratio α_R^2 ~ 70% can be obtained with practical parameters Ω=0.5Γ_3 $g\sqrt{N} = 10\Gamma_3$, $\kappa = 2\Gamma_3$, Δ=0.01Γ_3, and $\Delta_c = \Delta_p = 0$.

The nonlinear optics at low light intensities requires a weak control field below the saturation intensity ($\Omega<\Gamma$). The presented calculations are valid under the condition ρ_{11}~1, i.e., the atomic population is concentrated in |1>, which requires g<<Ω. Therefore, the cavity-atom coupling coefficient g<<Γ, consequently, this is the weak coupling regime of the cavity QED with single atoms (the bad cavity regime).

It is interesting to compare the light-control-light scheme based on the cavity EIT here with the earlier published scheme based on the cavity polariton interference [19-20]. The cavity polariton scheme works also with three-level atoms, but the free-space control laser and the signal laser are tuned to the polariton resonances (the normal mode) that are separated from the respective atomic resonances by the vacuum Rabi splitting. The control laser induces the destructive interference for the polariton excitation and renders the cavity-atom system opaque to the signal light [25]. For the cavity EIT scheme presented here, the control laser and the signal laser are tuned to the atomic resonant transitions of the three-level system and create the EIT condition. EIT renders the medium transparent to the signal laser. Therefore, the scheme in Ref. 18 and 19 is based on

Journal of Physics: Conference Series **497** (2014) 012007 doi:10.1088/1742-6596/497/1/012007

electromagnetically induced opaque in the normal mode excitation while the scheme here is based on electromagnetically induced transparency. The two schemes result in comparable phase shifts and amplitudes for the transmitted field under the conditions of a weak control laser and moderate optical density.

Our earlier study of the cavity-atom polariton system only analyzed the cavity-transmitted field. Here the analysis is carried out for both the cavity-transmitted field and the cavity-reflected field. The phase shift and amplitude of the cavity-reflected field as well as the cavity-transmitted field are analyzed and compared. The results show that the cavity reflected field and cavity transmitted field are complementary in their performance merits. The cavity system provides the versatility of choosing either the reflected field or transmitted field for the studies of all-optical switching and the cross-phase modulation according to their respective performance merits in a given parameter regime.

There are several studies of the EIT enhanced Kerr nonlinearities and XPM at low light intensities in multi-level atomic systems in recent years [26-33]. Most of these studies were done in four-level N-type systems coupled by three laser fields in free space. They can be divided into 4 types: a four-level N type coupled by three laser fields (Ref. 26-28), a five-level M type coupled by four laser fields (Ref. 26), a four-level tripod type coupled by three laser fields (Ref. 30-31), and a modified five-level tripod type coupled by 4 laser fields (Ref. 32). Ref. 28 proposed to inject a signal light (that induces the Kerr nonlinearity on a free-space signal light) into a Michelson-type double cavity containing the four-level atoms and therefore increase its coupling time with the free-space signal light. The double cavity is a passive device used solely for the signal light. Ref. 33 explores basically the N-type in the time domain through the EIT light storage process. Ref. 34 deals with the self Kerr nonlinearity (self phase modulation) in the three-level Λ type system confined in a travelling-wave cavity. In contrast with the EIT schemes in free space with multi-level atoms coupled by at least three laser fields, our scheme requires only two laser fields coupled with three-level atoms, but needs to operate in a cavity (it differs from Ref. [28] in that the signal light is coupled into the cavity and the cavity QED effect measured by the vacuum Rabi splitting plays an essential role). Due to the cavity feedback enhancement, our scheme produces a large XPM phase shift (~ 0.5 rad. and with a transmitted light intensity about 50% of the input light intensity under practical conditions). In the fee-space EIT schemes, in order to produce the comparable XPM phase shift, a much greater optical depth of the atomic medium is required (a few orders of magnitude greater than our scheme, see Ref. 26, 28 and 32).

The XPM phase shift in our scheme is very small when the medium optical depth becomes very large ($\sqrt{N}g >> \kappa\Gamma_3$). There is no such limiting factor for the EIT schemes. If very large optical depths of the atomic medium are available, the EIT scheme may produce a greater XPM phase shift (maybe even reaching a value of π as suggested in Ref. 31 and 32). Therefore, our scheme performs better with an atomic medium of low to moderate optical depths while the free-space EIT schemes perform better with an atomic medium of very high optical depths.

3. Conclusion

In conclusion, we have shown that coherently coupled cavity-atom system can be used to study resonant nonlinear optics at low light intensities. The system is well suited for studies of all-optical switching and XPM at low control light intensities. In the weak coupling regime of the cavity QED, the interplay of the EIT and the collective coupling of atoms with the cavity mode enable the nonlinear control of the intra-cavity light field through a free-space control laser. The analytical results and numerical calculations show that large optical nonlinearities can be induced by the control laser near the narrow resonance of cavity EIT, which can be used to control the amplitude and phase of the cavity transmitted and reflected fields.

Acknowledgement

This work is supported by the National Science Foundation under Grant No. 1205565.

References

[1] *Cavity Quantum Electrodynamics*, 1994 edited by P. R. Berman (Academic, San Diego).
[2] J. M. Raimond, M. Brune, and S. Haroche 2001 Rev. Mod. Phys. **73** 565.
[3] H. Tanji-Suzuki, W. Chen, R. Landig, J. Simon, V. Vuletić 2011 Science **333** 1266.
[4] G. Rempe, R. J. Thompson, R. J. Brecha, W. D. Lee, and H. J. Kimble 1991 Phys. Rev. Lett. **67** 1727.
[5] P. Grangier, J. F. Roch, J. Roger. L. A. Lugiato, E. M. Pessina, G. Scandroglio, P. Galatola 1992 Phys. Rev. A **46** 2735.
[6] S. E. Harris 1997 Phys. Today **50** 36.
[7] M. D. Lukin, M. Fleischhauer, M. O. Scully, and V. L. Velichansky 1998 Opt. Lett. **23** 295.
[8] H. Wang, D. J. Goorskey, W. H. Burkett, and M. Xiao 2000 Opt. Lett. **25** 1732.
[9] G. Hernandez, J. Zhang, and Y. Zhu 2007 Phys. Rev. A **76** 053814.
[10] H. Wu, J. Gea-Banacloche, and M. Xiao 2008 Phys. Rev. Lett. **100** 173602.
[11] M. Mücke, E. Figueroa, J. Bochmann, C. Hahn, K. Murr, S. Ritter, C. J. Villas-Boas, and G. Rempe 2010 Nature **465** 755.
[12] A. Joshi and M. Xiao 2003 Phys. Rev. Lett. **91** 143904.
[13] T. Kampschulte, W. Alt, S. Brakhane, M. Eckstein, R.Reimann, A. Widera, and D. Meschede 2010 Phys. Rev. Lett. **105** 153603.
[14] J. K. Thompson, J. Simon, H. Loh, and V. Vuletic 2006 Science **313** 74.
[15] J. Sheng, H. Wu, M. Mumba, J. Gea-Banacloche, and M. Xiao 2011 Phys. Rev. A **83** 023829.
[16] M. Albert, A Dantan, and M. Drewsen 2011 Nature Photonics **5** 633.
[17] A. E. B. Nielsen, J. Kerckhoff 2011 Phys. Rev. A **84** 043821.
[18] Y. Zhu 2010 Opt. Lett. **35** 303
[19] X. Wei, J. Zhang, Y. Zhu 2010 Phys. Rev. A **82** 033808.
[20] Y. Wang, J. Zhang, and Y. Zhu 2012 Phys. Rev. A **85** 013814.
[21] W. Chen, K. M. Beck, R. Bücker M. Gullans, M. D. Lukin, H. Tanji-Suzuki, V. Vuletić, Science 2013 **341** 768.
[22] D. F. Walls and G. J. Milburn 1994 *Quantum Optics*, (Springer-Verlag, Berlin, Heidelberg).
[23] B. Zou and Y. Zhu 2013 Phys. Rev. A **87** 053802.
[24] R. Zhao, Y. O. Oudin, S. D. Jenkins, C. J. Campbell, D. N. Matsukevich, T. A. B. Kenndedy, and A. Kuzmich 2009 Nature Phys. **5** 100.
[25] J. Zhang, G. Hernandez, and Y. Zhu 2008 Optics Express **16** 7860.
[26] H. Schmidt and A. Imamoglu 1996 Opt. Lett. **21** 1936.
[27] H. Kang and Y. Zhu 2003 Phys. Rev. Lett. **91** 93601.
[28] T. Opatrny and D. G. Welsch 2001 Phys. Rev. A **64** 023805.

[29] A. B. Matsko, I. Novikova, G. R. Welch, and M. S. Zubairy 2003 Opt. Lett. **28** 96.
[30] D. Petrosyan and G. Kurizki 2002 Phys. Rev. A **65** 033833.
[31] D. Petrosyan and Y. P. Malakyan 2004 Phys. Rev. A **70** 023822.
[32] Z. B. Wang, K. P. Marzlin, B. C. Sanders 2006 Phys. Rev. Lett. **97** 063901.
[33] Y. F. Chen, C. Y. Wang, S. H. Wang, and I. A. Yu 2006 Phys. Rev. Lett. **96** 043603.
[34] H. Wang, D. Goorskey, M. Xiao 2001 Phys. Rev. Lett. **87** 073601.

Nonlinear and quantum optics with liquid crystals

Svetlana G. Lukishova

E-mail: sluk@lle.rochester.edu

Abstract. Thermotropic liquid crystals' usual application is display technology. This paper describes experiments on light interaction with pure and doped liquid crystals under *for these materials unconventional* incident light powers: (1) under *high-power* laser irradiation, and (2) at *the single-photon level*. In (1), I will outline several nonlinear optical effects under *high-power, nanosecond laser irradiation* which should be taken into account in the design of lasers with liquid crystal components and in fabrication of optical power limiters based on liquid crystals: (1.1) athermal helical pitch dilation and unwinding of cholesteric mirrors (both in free space and inside laser resonators); (1.2) some pitfalls in measurements of refractive nonlinearity using z-scan technique under two-photon or linear absorption of liquids; (1.3) the first observation of thermal lens effects in liquid crystals under several-nanosecond, low-pulse-repetition rate (2–10 Hz) laser irradiation in the presence of two-photon absorption; (1.4) feedback-free kaleidoscope of patterns (hexagons, stripes, etc.) in dye-doped liquid crystals. In (2), at *the single-photon level*, it will be shown that with a proper selection of liquid crystals and a single-emitter dopant spectral range, liquid crystal structures can be used to control emitted *single* photons (both polarization and count rate). The application of the latter research is absolutely secure quantum communication with polarization coding of information. In particular, in (2.1), definite handedness, circular polarized cholesteric microcavity resonance in quantum dot fluorescence is reported. In (2.2), definite linear polarization of single (antibunched) photons from single-dye-molecules in planar-aligned nematic host is discussed. In (2.3), some results on photon antibunching from NV-color center in nanodiamond in liquid crystal host and circularly polarized fluorescence of definite handedness from nanocrystals doped with trivalent ions of rare-earths dispersed in liquid crystal host are presented.

Introduction

The development of liquid-crystal optics for high-power lasers began in 1979 at the Laboratory for Laser Energetics (LLE), University of Rochester, when it became apparent that some optical-component requirements of large-aperture solid-state lasers for thermonuclear fusion could not easily be met by solid-crystalline materials [1-3]. One such component, a high-quality, laser-beam apodized aperture has been the goal of solid-state laser fusion research programs both in the US and USSR since the early 1970's [1,4]. I worked at the Soviet/Russian Academy of Sciences (Moscow) on high-damage threshold apodized apertures and graded reflectivity mirrors [4-8] for reducing small-scale self-focusing effects [4,9] in high-power laser fusion systems and for improving the beam quality of industrial lasers [8]. To learn from the experience in *high-damage threshold* liquid crystal (LC) apodizing devices [1] and cholesteric mirrors [10-15], I visited LLE in 1993. Crucial in the preparation of LC materials with *high-damage threshold* under high-power laser irradiation is the purification of "as received" liquid crystals (LCs) through filtration to remove condensation centers for laser-induced bubble formation on the surface of the LC cell. It should be pointed out that high-power laser applications of LCs are not the principal sector among all LC applications. Commercial vendors of LCs are therefore unlikely to provide materials optimized for high-power-laser use. That is why two types of possible contributors to laser damage of LC materials must be avoided. They are undissolved impurities that can be removed using particle filters and dissolved gaseous components that can be removed by degassing as well as synthesis byproducts or catalyst traces that can be removed by chromatography. In Refs. 1-3 it was shown that after purification of the "as received" LC samples the laser-damage threshold for massive material modification (visible burst of stable bubbles and glass chips in the cell surface and/or a web of scatter lines) improved sharply. For purified (using 0.2-μm particle filter) and degassed, isotropic 5CB, thresholds for permanent photolysis at ~ 9.6 J/cm^2 and ~ 4.4 J/cm^2 were observed under single and multiple shots respectively (1 ns pulse duration, 1-mm beam diameter, wavelength λ = 1.053 μm). Under similar conditions, transient-bubble "damage" observable under 110 x dark-field magnification in "as received" LCs occurred at ~ 0.76 J/cm^2 for isotropic 5CB and ~ 0.89 J/cm^2 for isotropic E7 (single-shot threshold) [1,3,16].

A similar problem in purification of "as received" LC materials arises when the LC is used as the host for single-emitters. In this case, single-photon counting detectors can detect fluorescence from impurities in the LC material itself [17] instead of fluorescence from a particular single emitter. In addition, for some types of single emitters it is desirable to avoid oxygen, the singlet-state of which can contribute to single-emitter bleaching. Removing the oxygen from a monomeric LC by saturation with helium or argon gas permitted us to record cw-excitation fluorescence from single dye molecules without bleaching for more than 1 hour [18,19].

The structure of this paper is as follows. It consists of two parts. Part 1 is an overview of my research on LCs under high-power laser irradiation. Chapter 1.1 of this part is devoted to my Moscow research at the Institute of the Radio Engineering and Electronics of the Russian Academy of Sciences on athermal cholesteric pitch dilation and unwinding of cholesteric liquid crystal (CLC) mirrors by the field of a light wave. Several groups tried to achieve this goal, but my group was the first who succeeded in it. These experiments were carried out both in free space and with the CLC mirror as the output coupler of a laser resonator. Special conditions of experiments permitted to prove that this effect is not connected with heating. I used purified CLC material provided by A.W. Schmid (LLE, University of Rochester). CLC mirrors were prepared by S.V. Belyaev's group of the Moscow Organic Intermediates and Dyes Institute NIOPIK. Two of my students of the Moscow Institute of Physics and Technology (FizTech), K. Lebedev and E. Magularia participated in this project. The research was supported by Grants of the International Science (Soros) Foundation, the Russian Government, and the Russian Foundation for Basic Research.

Chapter 1.2 is devoted to some of my results obtained at the Liquid Crystal Institute, Kent State University, on nonlinear absorption and refraction of cyanobiphenyl liquid crystals to high-power, several-nanosecond laser irradiation, using z-scan measurements. Two effects were observed under conditions of two-photon absorption: dependence of nonlinear refraction on laser-beam diameter (thermal-density nonlinearity) and the first-time observation of thermal-lens in LCs under several nanosecond irradiation and low-pulse-repetition rate. This work was made possible by the hospitality of P. Palffy-Muhoray and with the help of T. Kosa and B. Taheri.

Chapter 1.3 is devoted to feedback-free pattern formation using highly absorbing dye-doped nematic liquid crystals. This work was carried out at the Institute of Optics, University of Rochester, owing to the hospitality of R.W. Boyd and with the help of K. Marshall and N. Lepeshkin.

Part 2 of this paper describes my investigation, at the single-photon level, of LCs doped with single emitters (dye molecules, nanocrystal quantum dots, NV-color centers in nanodiamonds), and rare-earth doped nanocrystals. To my knowledge, this is the first research on LCs at the single-photon level. Chapter 2.1 describes the results on circularly polarized microcavity resonance in quantum dot fluorescence in a CLC host with 4.9 times intensity enhancement in comparison with polarization component of opposite handedness. Chapter 2.2 is devoted to a single-photon source with definite linear polarization using dye-doped nematic LC. Chapter 2.3 discusses using other type of emitters in liquid crystal hosts – nanodiamonds and nanocrystals doped with trivalent ions of rare-earths.

These projects supported by U.S. ARO and three NSF grants as well as Air Force and NASA graduate student fellowships, were carried out at the Institute of Optics, University of Rochester. My Ph.D. students L.J. Bissell [20] and J.M. Winkler as well as undergraduates C. Supranovich, R. Knox and summer research associate P. Freivald contributed to this project. I also used help and advice of K. Marshall, A.W. Schmid, L. Novotny, and A. Lieb. Some oligomeric CLC material was synthesized by S.-H. Chen's group, some nanocrystal quantum dots were synthesized by T. Krauss' group, nanocrystals with rare-earth ions were prepared at the University at Buffalo (Prasad's Institute for Lasers, Photonics, and Biophotonics). C.R. Stroud and R.W. Boyd contributed in discussions.

1. Liquid Crystals under High-Power Laser Irradiation

1.1. Athermal Cholesteric Pitch Dilation and Unwinding by the Field of a Light Wave

In 1982, Winful determined that athermal, light-induced changes in the pitch of the cholesteric helix lead to a reflectivity drop of a CLC mirror in the region of selective reflection [21]. However, later attempts [22,23] at detecting this effect in laser fields remained unsuccessful, in spite of the fact that chiral unwinding under the influence of both static and time-varying *electric* and *magnetic* fields had been observed. The challenge in observing *athermal* helix pitch dilation by *optical* fields arises from the larger than E ~ 10^4 V/cm electric-field strength requirement on the optical wave that needs to act on the CLC for time periods commensurate with the characteristic unwinding time of the helix t_h ~ several milliseconds. Therefore, neither short-pulse intensities [22,23] exceeding by an order of magnitude Winful's estimation nor rather low intensities applied cw [10-13], succeeded in triggering nonlinear changes in the reflectivity of nonabsorbing CLC mirrors. In reports [10-13], only changes in curvature and improvement of a beam profile of CLC mirrors were demonstrated in a cw-laser beam.

The first athermal drop in CLC mirror reflectivity in response to strong, circular polarized laser radiation was observed by my group in 1995 [24-28]. A special laser operating mode was selected using a pulsed, 4.5 kHz pulse-repetition-rate laser with an accumulation effect from many pulses. The effect was observed both in free space [24-25, 27-28] and inside a laser resonator [24,26,27-28] with a CLC mirror as the output coupler. To distinguish a field-induced orientational effect from thermal changes of the cholesteric pitch, the laser was switched to cw-operation with the same average intensity as in the pulsed regime. The effect was only observed at pulsed irradiation and depended only on peak intensity, but not on average intensity. The effect was demonstrated in four CLC mirrors with different substrate-surface-treatment methods for planar alignment. Sections of this Chapter 1.1 are devoted to the details of the experiments and their discussions.

It should be noted, that the authors of Ref. [29] also observed nonlinear selective reflection effects using the property of CLCs to achieve faster orientational response time to high-intensity beams. At incident, single-pulse peak intensities of ~ GW/cm^2, the CLC mirror reflectivity dropped even for single, nanosecond pulses. In paper [29], the light-induced reflectivity drop of CLC mirror was used in a Q-switched laser for passive cavity-dumping and an improved transverse beam profile.

1.1.1. Cholesteric mirrors in free space

For these experiments, we used a Nd:YAG laser with 1.064 μm wavelength. In this spectral region, *no two-photon absorption* effects in the CLC material should be observed. The laser operated in two regimes: (1) in cw mode and (2) in an acousto-optically Q-switched mode, emitting high-repetition-rate (4.5 kHz) pulse trains, each pulse being 500 ns long. In either mode, the average power reaching the CLC samples was between 0.3 and 1 W. Beam cross-sections before focus were monitored by a CID camera and video processing setup. From the recorded beam profiles, 1/e^2 intensity beam diameters of between 50 and 200 μm were derived, corresponding to sample-plane peak-intensities (for 500-ns pulses) of 10^6-10^7 W/cm^2.

Upon focusing of free-space-propagating, 500-ns, 4.5-kHz-repetition-rate pulses into the CLC layer, we observed an increase in CLC-mirror transmittance of between 5% and 30-80%. This reflectivity drop appeared within 1 to 10 min from the onset of irradiation (Figure 1), depending on the

Journal of Physics: Conference Series **497** (2014) 012008 doi:10.1088/1742-6596/497/1/012008

specific mirror. Interestingly, this effect did not depend on the magnitude of the average-intensity. Under cw irradiation conditions, we did not observe any effect even at average intensities twice as large as the average intensities in the pulsed, high-repetition rate mode. Furthermore, the effect appeared only, even in the case of high-repetition-rate pulse irradiation, under reflection of the incident light from the *strong-anchoring* side of the CLC mirror. Under 180° reversal, i.e., with the weak-anchoring side facing the incident beam, we did not observe any effect.

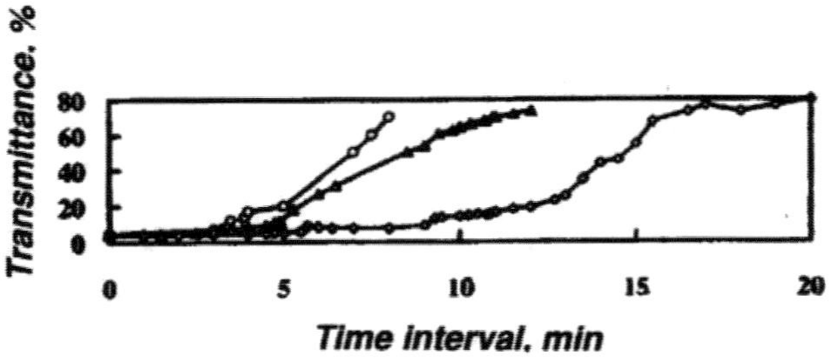

Figure 1. Dependence of a CLC mirror transmittance on irradiation time under high repetition rate, pulsed irradiation at different incident intensities.

The temperature drift in selective reflection for our given CLC mixture is estimated at +1 nm/C°, requiring a temperature excursion of more than 50 C° to account for the ~70% change in CLC-mirror transmittance at 1.064 μm. We did not observe such heating in the present CLC cells.

A recovery of the mirror reflectivity to its prior condition appeared after either terminating irradiation or switching it to cw-mode.

1.1.2. Cholesteric mirrors in laser resonators

For the laser-resonator studies, a plane-parallel cavity of the same laser used in section 1.1.1 was formed by a dielectric hard mirror and a CLC output coupler. An active Nd:YAG element was pumped by a continuously operating flashlamp filled with krypton (the voltage across this lamp was 100 V). Pumping of the active element generated a thermal lens whose optical power was +1 dioptre/kW. Switching to high pulse-repetition rate operation (500 ns pulse duration, 4.5 kHz pulse repetition rate) was performed by an acousto-optical switch. A quartz quarter waveplate was placed inside the resonator between the active element and the CLC output coupler. The diameter of the optical beam at the laser output was ~ 0.8 mm at an average output power of 1 W. No significant difference was observed between the slope efficiency of the laser with conventional dielectric or CLC output mirrors when the reflection coefficients were the same. The advantage of the CLC mirror was a lower sensitivity of the laser operation to cavity misalignment. At cw-operation and at low pumping rate in the pulsed high-repetition-rate regime, the slope efficiency and the lasing stability were the same for laser with CLC and dielectric mirrors. At high flashlamp currents, the average output power obtained in the high-repetition-rate mode decayed rapidly and disappeared after 0.5–5 min from switching, but only when the side of the CLC mirror with strong surface anchoring was facing the gain medium. The pump threshold (i.e., the current to the flashlamp) at which lasing was suppressed was different for each CLC mirror. It was 28.5–29.5 A for the output mirrors 1–3 with reflection coefficient R = 95%, and ~ 24.5 A for the CLC mirror 4 with R = 85%.

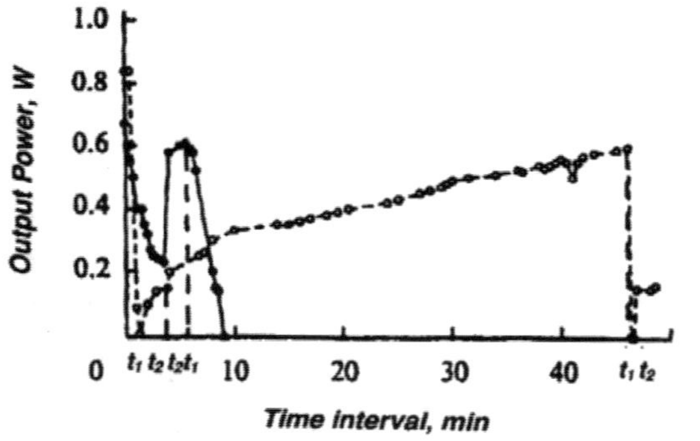

Figure 2. Time dependence of the average output power of a CLC mirror laser with changing the operating regime from cw to high-repetition rate (t_1) and back to (t_2). Solid and dashed curves present results for different mirrors.

Figure 2 depicts the time dependence of the laser output power obtained for two CLC mirrors when a change from the cw to the pulsed regime occurred at high flashlamp currents. Turning on the acousto-optical switch (at t_1) quenched the lasing action because of a drop in the reflection coefficient of the CLC mirror. Turning off the switch (at t_2) gradually restored the lasing action. A second change to the pulsed regime suppressed lasing again and it was restored when the laser was made to operate in the cw-mode again.

1.2. Thermal-Density Nonlinearity and Thermooptical Effects in Liquid Crystals under High-Power, Several Nanosecond Laser Irradiation

In this Chapter, I will briefly overview some of my results on nonlinear optical response of cyanobiphenyl LCs to high-power, nanosecond laser irradiation which I obtained at the Liquid Crystal Institute, Kent State University in 1997–1998. My full review on this subject including other group results is published in [28]. I would like to emphasize two important results which were overlooked in the past in z-scan measurements of LCs in the presence of nonlinear (or linear) absorption: (1) for isotropic LCs at several-nanosecond time scale and several tens-micrometers beam-waist-diameter the influence of coupled thermal and density effects on nonlinear refraction depends, through buildup time, on *the beam-waist diameter*; (2) for planar nematic layers, *cumulative effects* in heating (and in refractive nonlinearity) were observed even at low, 2–10 Hz pulse repetition rate, even for undoped (pure) liquid crystals. These results are important for optical power limiting and switching applications, and for intensity and beam-quality sensors of pulsed, high-power lasers.

Heating of undoped cyanobiphenyl LCs by short-pulse laser radiation in the visible range that drives photoacoustical (thermal-density) and thermooptical effects, is caused by two-photon absorption [30,31], concurrent or subsequent excited state absorption [30,32-34], and the efficient decay of the excited states through radiationless-recombination channels [30,32-34]. Strong nonlinear absorption in the visible was observed in optical power limiting studies of LCs [35-43] and in z-scan measurements [44-49]. For the nematic phase, the two-photon absorption coefficient has a several-times-higher value for incident polarization parallel to the LC molecular dipole direction than for perpendicular polarization [31].

As in any absorbing medium, in the presence of nonlinear absorption, the thermal mechanism causes refractive nonlinearity in nematic and isotropic LCs. The thermal nonlinearity coupled to density changes [35-36] (which is negative) competes with the orientational (positive) nonlinearity in changing the sign of the total refractive nonlinearity [28]. Another mechanism of density changes, electrostriction, with buildup time in the nanosecond range, is comparable to thermal/density contributions if medium absorption coefficients are low. At 0.532 μm a strong heating mechanism exists for the monomeric LCs as a result of two-photon and excited state absorption. That is why electrostriction is neglected in these experiments.

Refractive index changes by heating are the sum of two contributions:

$$\Delta n = (\partial n / \partial \rho)_T \Delta \rho + (\partial n / \partial T)_\rho \Delta T \quad , \qquad (1)$$

where ρ is the density, T is the temperature. It is very important in the current context that each term in (1) has its own turn-on time, and will contribute differently under transient and steady-state regimes [28,50-52].

1.2.1. Thermal-density nonlinearity $(\partial n / \partial \rho)_T \Delta\rho$

For several-nanosecond pulse duration and several tens of micrometers beam waist, the buildup time t_{ac} of the thermal-density nonlinearity ($\Delta n = (\partial n / \partial \rho)_T \Delta \rho$) can be close to the laser pulse duration t_o:

$$t_{ac} = r_o / V_s \ , \qquad (2)$$

where r_o is the beam radius and V_S is the velocity of sound (V_S (LCs) ~ 1500 m/s). For $t_{ac} > t_o$ the thermal-density-nonlinearity will not develop during the pulse. Experimental values of nonlinear refractive index n_2 for this transient regime were reported in Ref. 50 for absorbing organic liquid. It was found experimentally, that the transient absolute value of n_2 for t_o = 5 ns diminishes with r_o

increasing from 9 to 32 μm. (The value of n_2 is defined in our paper by $n = n_0 + 1/2\, n_2|E|^2$). In a series of papers [51,52] numerical modeling was carried out for this regime. Acoustic grating generation in LCs have been intensively studied both under subpico-, pico- and nanosecond excitation [30,32-36].

The typical z-scan curves of *transient* nonlinear refraction of room-temperature chiral additive CB15 samples with 2-mm thickness for 0.532 μm laser excitation [28,53] are presented in Figure 3 (a)-(c) for incident intensity I = 0.74 GW/cm^2. For comparison, a z-scan curve for CS_2 with 2-mm thickness is presented as well (Figure 3 (d)) for I = 0.73 GW/cm^2. The results from these curves yield the following values [28]: n_2(CB15) = 2.5 n_2 (CS_2). Beam-waist diameter was ~ 36 μm, and t_o = 6.8 ns, 10-Hz pulse repetition rate. The aperture factor for these z-scan measurements was S~0.1 [28].

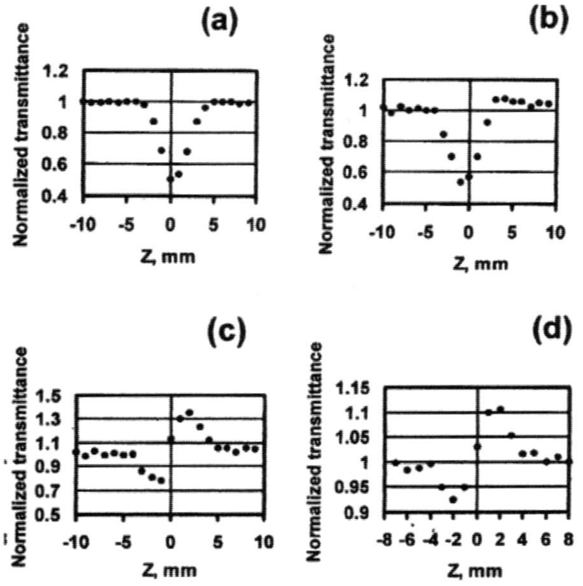

Figure 3. Z-scan curves for chiral additive CB15: (a) open aperture; (b) closed aperture; (c) closed/open aperture; and (d) closed aperture for CS_2.

For 5CB above phase transition from nematic to isotropic state with 2-mm-cell thickness under similar conditions in the temperature interval 35.6–50°C, the *transient* value of n_2(5CB) varied from 2.5 to 3.7n_2(CS_2). In both cases, isotropic 5CB and CB15, self-focusing (converging lens) was observed. The absolute value of n_2(CS_2) for these conditions is equal to 1.2 x 10^{-11} esu [54]. In papers [44,49] for 7-μm diameter measurements, negative values of n_2 (self-defocusing) were reported for isotropic 5CB using the same laser. This difference is explained by the dependence of the value of transient thermal-density nonlinearity with negative sign on the beam-waist diameter. For 7-μm diameter, t_{ac} is less than the pulse duration time and nonlinear refraction as a result of density changes prevails under the positive-sign, transient, orientational nonlinearity. (5CB (nematic) and CB15 have identical chemical composition, but CB15, in difference to 5CB, contains a chiral carbon [28]).

1.2.2. Thermooptical effects $(\partial n / \partial T)_\rho \Delta T$ *as the result of cumulative effects in heating of planar-aligned LC-cells*

The thermooptic coefficient dn/dT in nematic LCs is extraordinarily large, ranking among the largest of all known materials [55]. Not unexpected for these highly anisotropic molecules, it depends on the orientation of the molecules. E.g., dn_{par}/dT was reported [56] of ~ 2.5 x 10^{-3} grad^{-1} for 5CB in the temperature interval 26–31°C for incident linear polarization parallel to the molecular orientation direction. For "perpendicular" polarization, dn_{perp}/dT is of opposite sign and equals ~ -1/2 dn_{par}/dT [55]. In the proximity of the phase transition to the isotropic state, the slope of both dn_{par}/dT and dn_{perp}/dT steepens [56].

For nematic LCs, it is well-known that the thermooptical effect experiences a lengthening of its buildup time (tens-hundreds of nanoseconds), rendering it insignificant on a several-nanosecond time scale [35,36]. Similar buildup times are reported in Ref. 50 for an absorbing organic fluid. Thermooptical effects for r_o = 32 μm were found to be more significant for pulses longer than ~100 ns.

Because of the slow buildup time, thermooptical (thermal-lens) spatial self-phase-modulation rings in the beam cross-section were observed only for cw laser radiation [57-59]. LC heating was in this case due to absorbing coatings on cell glass-substrates, dissolving absorbing dyes in LCs, and/or irradiation of unpurified LCs by high-average power (~ 1W for 100-μm spot size [57]) beams.

For pulsed lasers the first report on thermal-lens spatial self-phase-modulation rings was published in my papers [28,60,61]. I irradiated a planar-aligned nematic cell with a 0.532-μm laser beam (7 ns pulse duration, 10 Hz pulse repetition rate), propagating perpendicular to the cell walls. Linear polarization of the beam was parallel to the alignment direction of the LC molecules. In this configuration, no orientational effects should be observed. Over time spans of ~ 0.5–ten seconds high-contrast, concentric, elliptical diffraction rings appeared in the far-field (Figure 4) whose major axes orient themselves perpendicular to the incident polarization direction. Once formed, each set of rings may remain stable for up to several minutes. Upon continued pulsed irradiation over several minutes, the number of these rings varies systematically with laser intensity (sometimes between 1–2 and (up to) 20 rings). This ring pattern (see Figure. 4) differs from that typically obtained from orientational spatial self-phase modulation. In our case, the ring pattern evolves under polarization conditions for which, in principle, no orientational effects may occur (polarization of laser light is parallel to the director orientation). Under cw-irradiation at twice the average intensity than the average intensity under pulsed irradiation, the effect fails to exist.

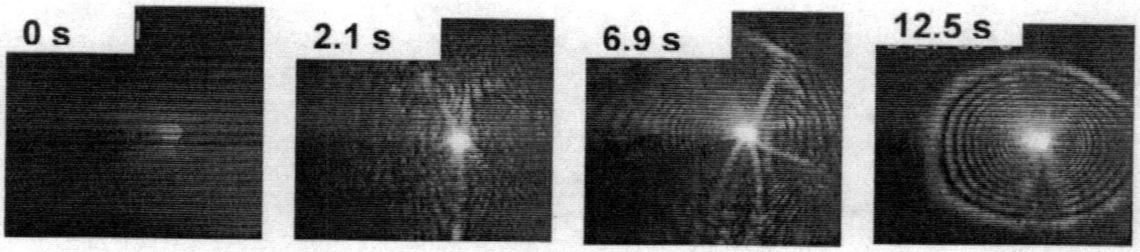

Figure 4. Time evolution of the elliptical ring pattern in the far-field (development of a thermal lens).

For explanation of the slow buildup time spatial self-phase modulation as well as its absence under cw irradiation of comparable average intensity, the heating of LC under *two-photon absorption* conditions was further scrutinized. The finite-element solutions were obtained by A.W. Schmid using the commercial ANSYS/Thermal code [28]. Numerical calculations of heating as cumulative action of many, 10-Hz repetition-rate pulses yield results on temperature increase and saturation at the end of the sequence. Evaluation of the phase shift caused in a laser beam by calculated temperature increase (ΔT ~ 5°C) gave number of self-modulation rings $N = 3$ [28]. A temperature increase after single pulse irradiation results in no diffraction rings in the beam. Only the cumulative action of many pulses approaching a steady-state value yields a stable spatial self-phase modulation elliptical ring pattern as displayed in Figure 4. It should be noted that some cumulative effects in LC nonlinearities at low pulse repetition rates were also reported in paper [63].

1.3. Feedback-Free Pattern Formation from Dye-Doped Liquid Crystals under Nanosecond Laser Irradiation

In this chapter, I will describe some experimental results on high-definition patterns observed under 10-Hz-pulse-repetition-rate, nanosecond laser irradiation of highly absorbing planar-nematic LC layers doped with dichroic dye. The patterns were observed at incident intensities I ~ 5–10 MW/cm^2 in a single beam and *without any feedback involved.* An incident polarization parallel to the nematic director was used. Under periodic pulsed laser irradiation, far-field beam patterns at the output of a dye-doped LC layer changed kaleidoscopically from rings and stripes to multiple hexagons. This pattern-formation regime had a buildup time of several seconds to minutes.

Journal of Physics: Conference Series **497** (2014) 012008 doi:10.1088/1742-6596/497/1/012008

1.3.1. Experimental setup and cell preparation

A 0.532-μm laser beam with a pulse duration of ~ 20–26 ns and pulse repetition rate 10 Hz was focused by a 24-cm-focal-length lens into the dye-doped, planar-aligned nematic LC cell. The incident polarization was parallel to the orientation of the LC molecular director. The beam diameter in the focus was ~ 150 μm at the 1/e level [64]. The far-field patterns were recorded from a screen by a video camera onto magnetic tape from which the spatial intensity distribution for each pulse was digitized afterwards.

The nematic LC mixture E7 doped with the dichroic azodye "Oil Red O" with a 1.5% weight-concentration was used for filling the LC cells. E7 was supplied by EM-Industries (E7 is the mixture of cyanobiphenyls with ~51% of 5CB [28]); Oil Red O was supplied by Sigma-Aldrich. Planar-aligned nematic LC-layers were prepared using buffing techniques on Nylon 6/6 alignment layers on Soda-Lime-glass plates of size 2.5 cm x 2.5 cm x 0.3 cm. Glass-bead-spacers mixed with an UV-epoxy provided cell thicknesses between 10 and 20 μm. Coating the substrates by Nylon 6/6, buffing, cell assembly, and filling the cells with LC were carried out in a clean room. UV-cured epoxy was used for sealing the cells. Cell transmittance at low incident intensities was ~ 1% for an incident polarization parallel to the nematic director, and ~ 10–15 % for the perpendicular polarization.

1.3.2. Pattern formation experiments

At incident intensities I ~ 1–5 MW/cm^2 several phenomena were observed during repetitive illumination of LC-cells by a focused laser beam with 10-Hz pulse-repetition rate, for time periods of *several seconds* to *several minutes*. These phenomena include the following [65]:

- A polarization component *perpendicular* to the nematic director appeared after the beam passed through the nematic layer. Simultaneously, a several-fold increase in cell transmission was observed.
- Following that, *stable* far-field patterns appeared in the beam cross-section, possibly as the result of heat-flow [66,67] and/or flow-reorientational [67-71] birefringence (Figure 5):
    (1) The *perpendicular* polarization component (Figure 5, b) took on the far-field form of an optical four-leaf clover (Maltese-like cross). The bright axes of the cross were oriented at 45° to the incident polarization;
    (2) The incident, *parallel* polarization component evolved into the far-field pattern in the form of a ring with a bright spot inside (Figure 5, c);
- *Stable* patterns existed for more than one hour of irradiation, but disappeared after switching the laser to a 5-Hz repetition rate.

At the higher-incident-intensities regime (I ~ 5–10 MW/cm^2), a new kind of high-definition patterns developed with a buildup time of several seconds to minutes and only for the polarization component *parallel* to the nematic director. Figure 6 shows the beam cross-sections of the two spatially separated polarization components both incident and induced (using a Glan-prism) overlapping in the center. The left side of the images depicts *the parallel-polarization* (incident) component; the right side – *the perpendicular-polarization* (induced) component. The optical four-leaf-clover of the *perpendicular-polarization* component (Figure 5, b) almost always became smeared into a random speckle pattern with a bright spot in the center (see the right side of the images). At the same time, *the parallel-polarization* component developed a high-definition pattern that kaleidoscopically changed from pulse to pulse from multiple hexagons (Figure 6, a), stripes (Figure 6, b), to rings (Figure 6, c) (see left sides of the Figure 6 images). The patterns disappeared after switching the laser to 5-Hz repetition rate. More details will be published elsewhere.

Near-field images of the laser-beam cross-section during the pulsed laser action showed kaleidoscopic changes from one to two and three spots patterns with dimensions ~ 5–15 μm and distance between spots ~ 35–70 μm.

The observation of high-definition hexagonal/stripes/rings patterns only for the "parallel" polarization component can be explained by the highly dichroic dye used for this experiment: cell transmittance was an order of magnitude higher for the "perpendicular" polarization component than for the "parallel" polarization component. Hexagonal/stripes/rings patterns' low contrast makes them invisible for the "perpendicular" polarization component. Only a four-leaf-clover pattern can be seen in this component because of its existence's connection with another mechanism — the aberrations of the laser beam in heat-flow and/or flow-reorientational birefringent media.

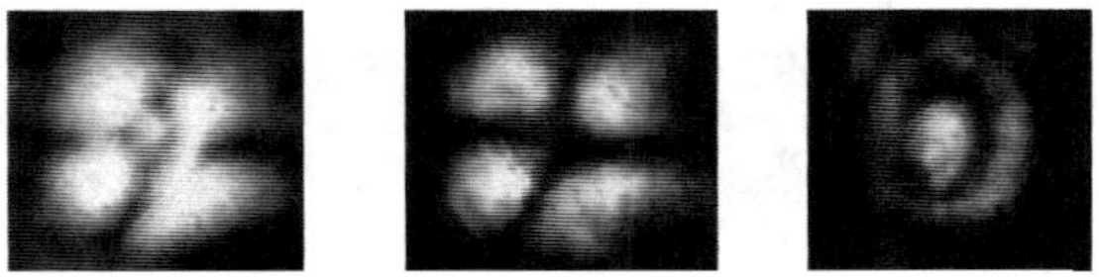

(a) (b) (c)

Figure 5. Far-field spatial patterns at low incident intensities: (a) no polarizer; (b) perpendicular-polarization (induced) component; (c) parallel (incident) polarization component.

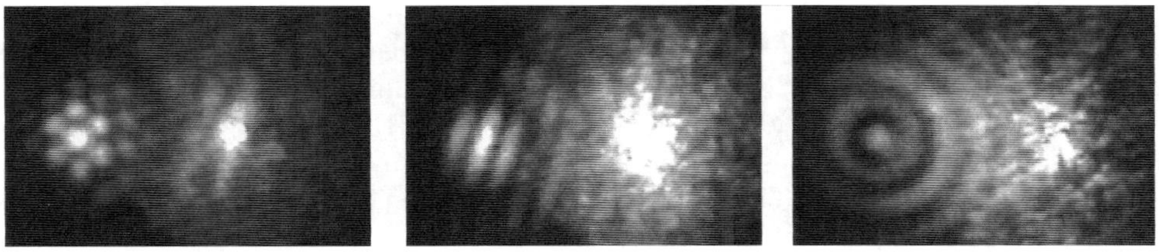

(a) (b) (c)

Figure 6. Representative selection of far-field spatial patterns at incident intensities ~5 – 10MW/cm^2 for the polarization component parallel to the director (left side of each image) and for the induced, perpendicular polarization component (right side of each image). A Glan prism was used to separate the two polarization components.

We explain the observed hexagonal/stripes/rings patterns in the far-field by the diffraction of the laser beam on light-induced one/two/three or more several-micron-size "drops" or "holes" with absorption and/or refraction properties different from the surrounding material. The patterns' ring structure can be attributed to the diffraction of laser light at the sharp boundary of the "drops". The variation in the "drop" numbers in the focal region, their size, the distance between them, and the gradient of transmittance inside the drop define the enormous variety of patterns we observed. A possible mechanism contributing to the creation of such "drops" is the instabilities in the presence of the Soret (thermal diffusion) effect, including phase separation of the dye from the liquid crystal. Phase separation of the dye from liquid crystal was also observed in Ref. 72 under cw-irradiation of a dye-doped liquid crystal.

2. Liquid Crystals at the Single-Photon Level

This chapter describes a new application for LCs — quantum information technology. A single-photon source (SPS) with definite polarization that efficiently produces photons exhibiting antibunching (separation of all photons in time) is a pivotal hardware element in absolutely secure quantum communication. For *single* photons, the second order correlation function $g^{(2)}(t) = \frac{<I(t)I(t+\tau)>}{<I(\tau)>^2}$ should have a minimum at $t = 0$ (in an ideal case $g^{(2)}(0) = 0$), indicating the absence of pairs, i.e., antibunching [73]. Here I is intensity. The critical issue in producing single photons exhibiting antibunching is the very low concentration of photon emitters dispersed in a host, such that, within an excitation-laser focal spot, only one emitter becomes excited which will emit only one photon at a time, because of its finite fluorescence lifetime.

In the BB84 quantum key distribution protocol, the sender (Alice) and receiver (Bob) employ the linear and circular polarization states of single photons. The linear and circular polarization bases can be used to provide two different quantum level representations of zero and one. So a desirable feature for a SPS is definite photon polarization, since, if the photon has unknown polarization, filtering it through a polarizer to produce the desired polarization for quantum key distribution will reduce by half the efficiency of a quantum cryptography system. In another implementation, a SPS becomes the key hardware element for quantum computers with linear optical elements and photodetectors [74].

In this chapter, experimental results of *room temperature*, robust SPSs with definite polarization using *single*-emitter fluorescence in either *cholesteric* or *nematic* LC hosts are discussed [17-20,75-79]. A desirable polarization state (either circular with definite handedness or linear with definite direction) of a fluorescence of the emitter in a LC host can be produced either by providing a chiral microcavity environment of CLC or by aligning emitters' dipole moments in a definite direction in a nematic LC. SPSs based on single emitters in LCs are the room-temperature alternatives to cryogenic SPSs based on semiconductor heterostructured quantum dots in microcavities prepared by molecular beam epitaxy (MBE), see review [80]. Definite linear polarization of single photons from heterostructured quantum dots both in elliptical pillar microcavities, and in a 2-D photonic crystal, was reported for resonance wavelength at *cryogenic* temperatures. In difference to expensive MBE, well developed LC alignment technology is relatively easy and fast. Different types of single emitters can be easily dissolved or dispersed both in monomeric (fluid-like) or oligomeric (solid) LCs. In addition to emitter alignment and the possibility of fabrication of photonic-bandgap structures from LCs, LC technology has another advantage. Special treatment of LCs (oxygen depletion) can protect the emitters from bleaching [18,19]. In Ref. [18,19], we reported on a significant diminishing of dye bleaching by saturation of LC with helium. Another remarkable advantage of LCs, e.g., changing its properties with temperature or by external-field variation, can provide SPS tunability. As single emitters, we used nanocrystal quantum dots (NQD), single NV-color centers in nanodiamonds, and single dye molecules.

The structure of this chapter is as follows. Section 2.1 describes antibunching and circularly polarized resonance with definite handedness in nanocrystal quantum dot fluorescence in a glassy CLC oligomer photonic bandgap microcavity. 4.9 times intensity enhancement was observed in comparison with a polarization component with opposite handedness. Section 2.2. is devoted to linearly polarized fluorescence with a definite polarization state from single dye molecules aligned in a glassy nematic LC oligomer. The last section 2.3. of this chapter describes photon antibunching from NV-color center in nanodiamond in liquid crystal host and circularly polarized fluorescence of definite handedness from nanocrystals doped with trivalent ions of rare-earths dispersed in liquid crystal host.

2.1. Circularly Polarized Microcavity Resonance in Quantum Dot Fluorescence in a Photonic Bandgap Cholesteric Liquid Crystal Host

We used left-handed cyclosiloxane oligomeric CLC powder from Wacker Chemie [81,82] and produced a planar-aligned glassy CLC structure doped with CdSeTe nanocrystal quantum dots (NQDs), Qdot 800 ITK organic, Invitrogen, fluorescence maximum at 790 nm [78]. Doping was accomplished by heating the CLC to ~135° C (the oligomer's melting temperature) and then mixing the melted CLC with quantum dots dispersed in toluene at a concentration of ~1 μM, with heating allowed to continue until the toluene evaporated. Subsequently, cells were prepared using two polyimide buffed glass coverslips. The CLC doped with quantum dots was placed on a buffed coverslip and heated beyond the oligomer clearing temperature of 180° C. After the sample was cooled to ~135° C, the second buffed coverslip was placed on the first and sheared along the direction of polyimide buffing. A slow cooling process back to a glassy (solid) state preserved the CLC order. This resulted in a photonic bandgap microcavity with a center wavelength of 910 nm.

Prepared samples were analyzed using our homemade confocal microscope setup. We excited the sample with cw, 633 nm laser light from a HeNe laser and observed the spectrum of the sample's fluorescence. By placing an achromatic quarter waveplate and linear polarizer in front of the spectrometer, we were able to filter for different handedness of circularly polarized (CP) fluorescence.

The resulting fluorescence spectra can be seen in Figure 7, left. Left-handed (LH) CP light experienced the photonic bandgap and therefore the black LHCP curve in Figure 7, left (curve 1) shows microcavity resonance, indicating that the LHCP light coupled to the cavity mode. The LHCP fluorescence had a center wavelength of 833 nm and a FWHM of 16 nm (Q ~ 50), as compared to a FWHM of 76 nm for RHCP fluorescence. The center wavelength of this resonance roughly matches the edge of the photonic stopband, centered at 910 nm, and shown in Figure 7, left by the blue curve 3. The observed right-handed (RH) CP fluorescence is shown in red in Figure 7, left (curve 2) and was less intense due to not experiencing the CLC microcavity, showing no sign of line narrowing. The maximum intensity ratio between LHCP and RHCP was a factor of 4.9. To characterize the degree of circular polarization, the circular polarization dissymmetry factor g_e is used:

$$g_e = 2\,(I_L - I_R)/(I_L + I_R)\,, \qquad (3)$$

where I_L and I_R are intensities of LHCP and RHCP light, respectively. At the wavelength of the resonance shown in Figure 7, left, $g_e = 1.3$ [78]. Note that for unpolarized light, $g_e = 0$, which we observed when NQDs were spin-coated on a bare glass slip.

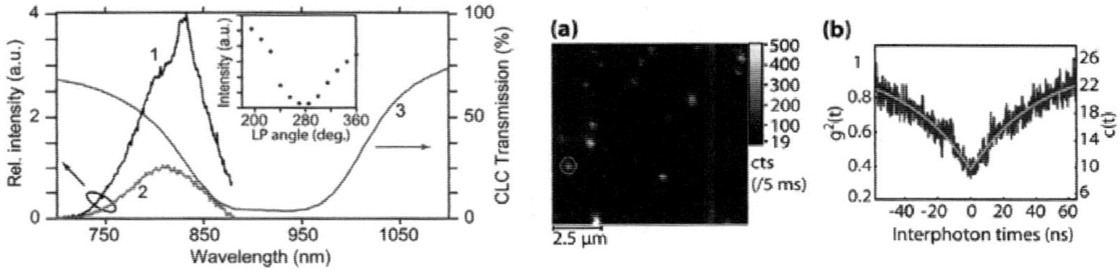

Figure 7, LEFT: Circularly-polarized fluorescence resonance from NQDs doped in a glassy CLC microcavity. Curve 1: LHCP fluorescence spectrum of the NQDs with a resonance at 833 nm. Curve 2: RHCP fluorescence spectrum for the same NQDs. Curve 3: selective transmission of LHCP light through CLC microcavity. Inset: Dependence of resonance peak intensity on rotation of a linear polarizer (LP) after a fixed quarter wave plate. RIGHT: (a) Confocal fluorescence microscopy image of single NQDs in a glassy CLC photonic bandgap microcavity. (b) Raw coincidence counts $c(t)$ (right-hand scale) and $g^{(2)}(t)$ (left-hand scale), showing antibunching (dip at $t = 0$).

Figure 7, right (a), shows a confocal fluorescence raster scan taken of a sample prepared using a relatively low concentration of quantum dots dispersed in toluene (~10 nM), with the higher intensity spots indicating the location of fluorescing quantum dots in the glassy CLC microcavity. Focusing on a spot (circled in white on Figure 7, right (a)), we checked for photon antibunching and obtained the coincidence histogram shown in Figure 7, right (b).

Figure 7, right (b) displays the histogram of coincidence counts $c(t)$ in blue, with $g^{(2)}(t)$ derived by normalizing $c(t)$. The measured $g^{(2)}(0)$ value from the fit shown in Figure 7, right (b) (green, solid curve) is $g^{(2)}(0) = 0.382 \pm 0.037$ [78]. As $g^{(2)}(0) < 0.5$, this indicated that we have managed to isolate the fluorescence of a single quantum dot, serving as a source of antibunched light in a glassy CLC microcavity.

It should be mentioned that the same cholesteric structures can be used for lasing [83]. Our results on dye-doped cholesteric lasers are reported in Refs [84-85].

2.2. Single-Photon Sources with Linear Polarization

Another intriguing application of liquid crystals as hosts for single emitters is using the planar-alignment of nematic liquid crystals to provide doped molecular dipoles definite alignment along a preferred direction for efficient excitation [17,75]. This allows for the creation of a single-photon source with fluorescence of definite linear polarization.

Samples were prepared using $DiIC_{18}(3)$ dye (DiI) molecules from Molecular Probes doped in a glassy nematic liquid crystal oligomer synthesized by S. H. Chen's group of University of Rochester [17, 75]. Planar-aligned layers of this doped liquid crystal host of ~100 nm thickness were prepared using photoalignment of the liquid crystal molecules.

Photoalignment was performed by first spin-coating a Staralign-2100 linearly photopolymerizable polymer (Rolic Technologies Ltd.) to a cleaned cover glass slip, which was then cured at ~135° C. This film was then irradiated by a polarized UV light for 10 to 15 minutes, with further irradiation used to bleach the polymer impurity fluorescence.

An oligomer solution doped with dye and diluted in chloroform was subsequently spin-coated onto these Staralign coated glass slips. After the chloroform evaporated, we heated the sample to ~80°C, slightly above where the oligomer transition to a nematic state occurs, after which the sample was slowly cooled to a glassy state, preserving the planar-aligned nematic order.

To characterize prepared samples, we used the Witec alpha-SNOM in a confocal mode, exciting the sample using a Nd:YAG laser for cw, 532 nm irradiation. Single-photon counting avalanche photodiodes were used as photodetectors for confocal fluorescence scans, with a polarizing beamsplitter used so that each photodetector collected light of orthogonal polarization. Molecules were identified by the peak pixel intensity values and a linear polarization measure ρ was determined for each molecule by comparison of the perpendicular and parallel polarization images. Our measure of linear polarization was

$$\rho = (I_{\text{par}} - I_{\text{perp}}) / (I_{\text{par}} + I_{\text{perp}}), \tag{4}$$

where I_{par} and I_{perp} are, respectively, the fluorescence intensities parallel and perpendicular to the direction of alignment. The only difference of ρ with the degree of polarization is that it allows inclusion of the direction of polarization as part of the measure, depending on whether ρ is positive or negative.

The different ρ values that we found are histogrammed in Figure 8. A clear asymmetry in Figure 8 demonstrates a preference towards fluorescence polarized perpendicular to sample alignment. This is in contrast to the expected fluorescence from an unoriented sample, which would yield a symmetric ρ histogram [86].

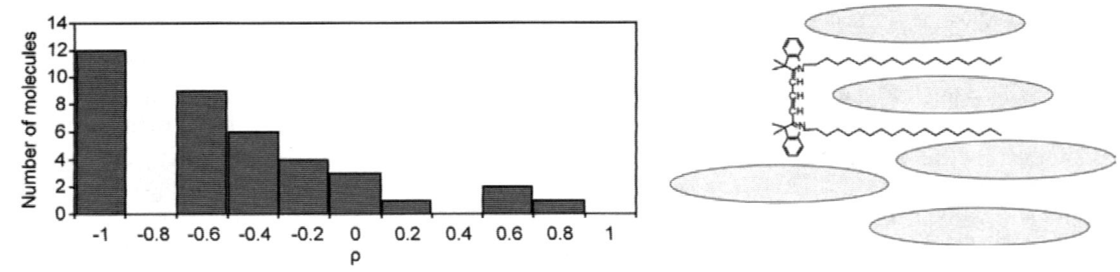

Figure 8. LEFT: Histogram of polarization measure ρ from 38 different DiI dye molecules in a planar-aligned nematic liquid crystal host [17,75]. RIGHT: Schematic view of a DiI dye molecule in a nematic liquid crystal host. The long axis (alkyl chains) of the DiI molecules tend to orient themselves along the rod-like nematic molecules, while the dipole, which is parallel to the bridge between the alkyl chains, orients perpendicular to the direction of liquid crystal alignment.

That DiI molecules doped in this planar-aligned liquid crystal host tend to fluoresce with polarization perpendicular to the alignment of the liquid crystal can be explained by the molecular structure of these molecules. As illustrated in Figure 8, right, it is likely that two alkyl chains orient themselves parallel to the rod-like molecules of the nematic liquid crystal host. The absorbing and emitting dipoles, however, are parallel to the bridge between these alkyl chains. Therefore, these dipoles end up oriented perpendicular to the direction of liquid crystal alignment, hence the fluorescence having a polarization orthogonal to the alignment of the liquid crystal.

To confirm these results, additional experiments were carried out both with monomeric nematic liquid crystal and nematic glassy liquid crystal oligomers. Figure 9, left, shows intensity changes of fluorescence of DiI molecules in a planar-aligned monomeric nematic liquid crystal (E7), as the linear polarization of the excitation 532-nm laser beam was rotated over 360° degrees. Figure 9, left, shows a clear dependence of the fluorescence intensity on the exciting angle of polarization, where the maximum fluorescence occurred when the DiI molecules were excited by light with linear polarization perpendicular to the alignment of the liquid crystal.

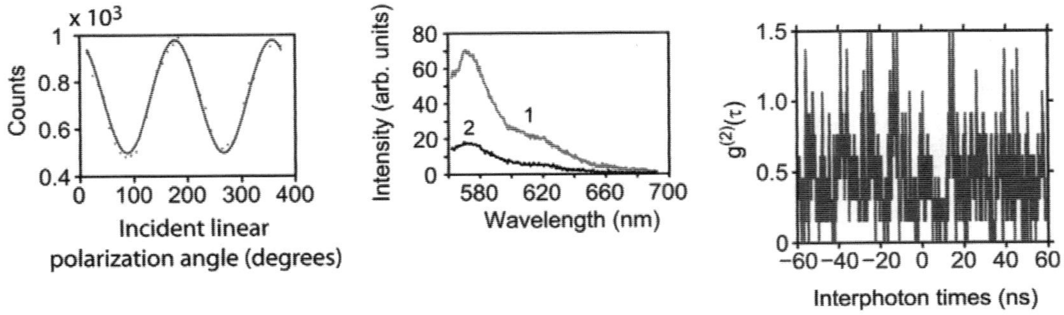

Figure 9. Polarized fluorescence of DiI dye in planar-aligned nematic liquid crystal hosts. Samples were excited using a 532 nm laser. LEFT: Intensity of DiI fluorescence in E7 as linear polarization of excitation laser light was rotated over 360°, with red dots showing experimental measurements and the blue solid curve providing a sinusoidal fit. The zero angle of polarization corresponds to the direction perpendicular to liquid crystal alignment. CENTER: Red (curve 1) and black (curve 2) show fluorescence spectrum with polarization perpendicular and parallel to the host alignment direction respectively. RIGHT: Antibunching histogram with a dip taken from the fluorescence of a single molecule of DiI dye in an E7 nematic host under pulsed, 532 nm excitation.

Figure 9, center, shows spectrofluorimeter measurements made for polarization perpendicular and parallel to the sample alignment, this time having used a sample of planar-aligned glassy nematic liquid crystal oligomer doped with DiI molecules of more than 1% concentration by weight. The value of ρ was measured from Figure 9, center, by comparing the peak intensities of the curves, yielding ρ = -0.5. The results of Figure 9, left and center, indicated a clear preference for fluorescence with definite linear polarization perpendicular to the sample alignment.

Figure 9, right, shows an antibunching histogram for a single molecule of DiI dye doped in an E7 nematic host, taken using pulsed, 532 nm excitation [20]. The value for $g^{(2)}(0)$ is $g^{(2)}(0) = 0.77 \pm 0.10$. This relatively high value of $g^{(2)}(0)$ may be due to the fact that the Raman spectrum of the E7 overlaps with the fluorescence spectrum of DiI dye.

2.3. Rare-Earth-Doped Nanocrystals and Color Centers in Nanodiamonds

Additional work with monomeric CLC based microcavities included doping with emitters such as nanodiamonds with nitrogen vacancy color centers and nanocrystals rare-earth ions. These alternative emitters can provide a source of single photons that is less likely to bleach.

Figure 10, left, shows photon antibunching from nitrogen vacancy color center in nanodiamond in a CLC (E7 and CB15) microcavity with a stopband centered at 725 nm, under cw, 514 nm excitation. The $g^{(2)}(0)$ value determined via fit (shown in Figure 10, left, as the dashed red line) was $g^{(2)}(0) = 0.74 \pm 0.08$. The fluorescence spectrum of nitrogen vacancy centers in nanodiamonds doped in this monomeric CLC when excited using cw, 514 nm excitation from an argon ion laser is shown in Figure 10, center.

A monomeric CLC microcavity made from a mixture of E7 and CB 15 was also doped with rare-earth Er^{3+} and Yb^{3+} ions in 20 nm to 30 nm sized $NaYF_4$ nanocrystals with 20% Yb and 2% Er. When these ions were excited using a cw, 976 nm diode laser at incident powers of ~500 μW, we were able to observe upconverted fluorescence of Er^{3+}, as shown in Figure 10, right. The emission lines observed were attributed to the transitions $^2H_{11/2}$, $^4S_{3/2} \rightarrow {}^4I_{15/2}$ (green) and $^4F_{9/2} \rightarrow {}^4I_{15/2}$ (red) of the Er^{3+} ions [87]. The populations of upper levels in Er^{3+}occur due to an efficient energy transfer from the Yb^{3+} to the Er^{3+}. This fluorescence was measured to have a circular polarization dissymmetry factor of g_e = -0.77 at 680 nm. These nanocrystals doped with rare-earth ions were prepared at the University at Buffalo (The Institute for Lasers, Photonics and Biophotonics).

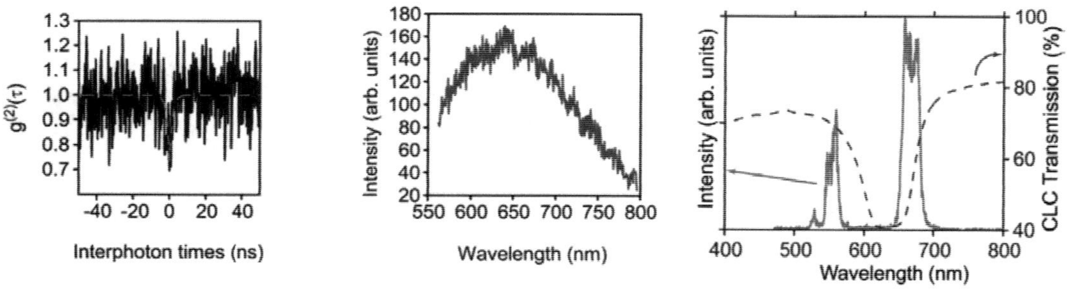

Figure 10. LEFT: Photon antibunching from nitrogen vacancy center nanodiamond in a CLC host. Black curve shows the raw data. Red dashed curve shows the fit. CENTER: Fluorescence spectrum of nitrogen vacancy centers in nanodiamonds in CLC host (cw, 514-nm laser excitation). RIGHT: Red solid lines: Fluorescence spectrum of Er^{3+} ions doped in $NaYF_4$ nanocrystals dispersed in a chiral CLC microcavity (E7 and CB15). The blue dashed curve shows the spectral transmission of CLC microcavity measured with unpolarized light.

Summary

An overview of some experiments and new applications of liquid crystals under two "extreme" levels of incident light powers are presented [79]. In many such applications for better device performance, "as received" liquid crystals materials must be purified and degassed. Several *nonlinear optical* effects under high-power, nanosecond laser irradiation are outlined: (1) athermal helical pitch dilation and unwinding of cholesteric mirrors by the field of a light wave; (2) dependence of refractive nonlinearity on the geometry of irradiation in the presence of two-photon (or linear) absorption; (3) the first observation of thermal lens effects in liquid crystals under several-nanosecond, low-pulse-repetition rate (2–10 Hz) laser irradiation in the presence of two-photon absorption; (4) feedback-free pattern kaleidoscope in dye-doped, highly absorbing liquid crystals.

At the single-photon level, definite linear and circular polarizations of single (antibunched) photons were obtained using single-emitter fluorescence in planar-aligned nematic and cholesteric hosts. Circular-polarized with definite handedness, cholesteric microcavity resonance was observed. These results are important for creation of room-temperature, single-photon sources, key devices for absolutely secure quantum communication. These single-photon sources can be used in fundamental *quantum optics* experiments.

Acknowledgement

The author acknowledge the support by the U.S. Army Research Office Award DAAD19-02-1-0285, National Science Foundation Awards ECS-0420888, DUE-0633621, DUE-0920500, and NASA NNX11AM82H Award. Research carried out in Russia was supported by the International Science Foundation Long-Term Award, Russian Government and Russian Foundation for Basic Research Awards.

The author thanks A.W. Schmid, S.V. Belyaev, V.F. Zolin, Ch.M. Briskina, V.M. Markushev, P. Palffy Muhoray, T. Kosa, B. Taheri, R. W. Boyd and his group members, L.J. Bissell, K. Marshal, L. Novotny and his group members, A. Lieb, J. Winkler, C.R. Stroud, S.-H. Chen and his group members, T. Krauss and his group members for help and/or support. These experiments were carried out in three institutions: the Institute of Radio Engineering and Electronics of the Russian Academy of Sciences, Moscow, Russia; the Liquid Crystal Institute (Kent State University), USA; the Institute of Optics, University of Rochester, USA. Some experiments were carried out at the liquid-crystal clean room and nanometrology facilities at the Laboratory for Laser Energetics, University of Rochester.

References

1. S. D. Jacobs, K. A. Cerqua, K. L. Marshal, A. W. Schmid, M. J. Guardalben and K. J. Skerrett, *JOSA. B* 5, 1962 (1988).
2. A. Schmid, S. Papernov, Z.-W Li, K. Marshal, T. Gunderman, J.-C. Lee, M. J. Guardalben and S. D. Jacobs, *Mol. Cryst. Liq. Cryst.* 207, 33 (1991).
3. S. D. Jacobs, K. L. Marshal and A. Schmid, in *Handbook of Laser Science and Technology, Supplement 2: Optical Materials*, ed. M. J. Weber (CRC Press, Boca Raton, FL, 1995), pp. 509-577.
4. S.G. Lukishova, Y.V. Senatsky, N.E. Bykovsky, A.S. Scheulin, Chapter 8, pp. 191-229, in *Self-focusing: Past and Present. Fundamentals and Prospects*, S.G. Lukishova, R.W. Boyd, Y.-R. Shen, Eds., Springer Series: Topics in Applied Physics, vol. 114, Springer NY, 2009, 605 p.
5. S.G. Lukishova, P.P. Pashinin, S.Kh. Batygov, V.A. Arkhangelskaya, A.E. Poletimov, A.S. Scheulin, B.M. Terentiev, *Laser and Particle Beams* 8, N 1-2, 349 (1990).
6. B.V. Gorshkov, V.K. Ivanchenko, V.K. Karpovich, I.K. Krasyuk, S.G. Lukishova, D.M. Margolin, P.P. Pashinin, EA. Simun, V.A. Sokolov, V.D. Terekhov, L.V. Chernysheva, *Kvantovaya Elektronika* 12, 1453 (1985). *Sov. J. Quant. Electron.* 15, 959 (1985).
7. S. G. Lukishova, N.R. Minhuey-Mendez, T.V. Tulajkova, *Kvantovaya Elektronika 21* 126 (1994). *Quant. Electron.* 24, 117 (1994).
8. S.G. Lukishova, S.A. Chetkin, N.V. Mettus, E.A. Magulariya, *Kvantovaya Elektronika* 23, 1040 (1996). *Quant. Electron.* 26, 1014 (1996).
9. S.G. Lukishova, R.W. Boyd, Y.-R. Shen, Preface, pp. V-XVIII, in *Self-focusing: Past and Present. Fundamentals and Prospects,* S.G. Lukishova, R.W. Boyd, Y.-R. Shen, Eds, Springer Series: Topics in Applied Physics, vol. 114, Springer NY, 2009, 605 p.
10. J.-C. Lee, S. D. Jacobs, T. Gunderman, A. Schmid, T. J. Kessler and M. D. Skeldon, *Opt. Lett.* 15, 959 (1990).
11. J.-C. Lee, S. D. Jacobs and R. J. Gingold, *Proc. SPIE* 824, 7 (1987).
12. J.-C. Lee, S. D. Jacobs and A. W. Schmid, *Mol. Cryst. Liq. Cryst.* 150b, 617 (1987).
13. J.-C. Lee, A. W. Schmid and S. D. Jacobs, *Mol. Cryst. Liq. Cryst.* 166, 253 (1989).
14. J.-C. Lee and S. D. Jacobs, J. Appl. Phys. 68, 6523 (1990).
15. J.-C. Lee, J. H. Kelly, D. L. Smith and S. D. Jacobs, IEEE J. Quant. Electron. 24, 2238 (1988).
16. A.W. Schmid, private communication.
17. S.G. Lukishova, A.W. Schmid, R.P. Knox, P. Freivald, A. McNamara, R.W. Boyd, C.R. Stroud, Jr., K.L. Marshall, *Molec. Cryst. Liq. Cryst.*, 454, 1-14 (2006).
18. S.G. Lukishova, A.W. Schmid, A. J. McNamara, R.W. Boyd, and C.R. Stroud, *IEEE J. Select. Top. Quantum Electron.* 9, No 6, 1512 (2003).
19. S.G. Lukishova, A.W. Schmid, Ch. M. Supranowitz, N. Lippa, A. J. McNamara, R.W. Boyd, C.R. Stroud, Jr., *J. Mod. Opt.* 51, No 9-10, 1535 (2004).
20. L.J. Bissell, *Experimental Realization of Efficient, Room-Temperature Single-Photon Sources with Definite Circular and Linear Polarizations*, Ph.D. Thesis, University of Rochester, Rochester, NY (2011).
21. H. G. Winful, *Phys. Rev. Lett.* 49, 1179 (1982).

22. R. S. Becker, S. Chakravorti and S. Das, *J. Chem. Phys.* 90, 2802 (1989).
23. H. Espinet, M. Lesieski, M. Ramsburg and D. Jenkins, *J. Appl. Phys.* 59, 1386 (1986).
24. S. G. Lukishova, E. A.Magulariya and K. S. Lebedev, *Izvestiya Akademii Nauk Seriya Fizicheskaya* 59, 130 (1995). *Bulletin Russian Acad. Sci., Phys.* 59, 2086 (1995).
25. S. G. Lukishova, K. S. Lebedev, E. A. Magulariya, S. V. Belyaev, N. V. Malimonenko and A.W. Schmid, *JETP Lett.* 63, 423 (1996).
26. S. G. Lukishova, S. V. Belyaev, K. S. Lebedev, E. A. Magulariya, A. W. Schmid and N. V. Malimonenko, *Kvantovaya Elektronika* 23, 817 (1996). *Quantum Electron.* 26, 796 (1996).
27. S. G. Lukishova, S. V. Belyaev, K. S. Lebedev, E. A. Magulariya, A. W. Schmid and N. V. Malimonenko, *Mol. Cryst. Liq. Cryst.* 303, 79 (1997).
28. S.G. Lukishova, *J. Nonl. Opt. Phys. & Mater.* 9, N 3, 365 (2000).
29. D. Grebe, R. Macdonald and H. J. Eichler, *Mol. Cryst. Liq. Cryst.* 282, 309 (1996).
30. F. W. Deeg and M. D. Fayer, *J. Chem. Phys.* 91, 2269 (1989).
31. S. D. Durbin and Y. R. Shen, *Phys. Rev.* A 30, 1419 (1984).
32. R. Macdonald and H. E. Eichler, *Appl. Phys.* B 60, 543 (1995).
33. R. Macdonald and H. E. Eichler, in *The Optics of Thermotropic Liquid Crystals*, eds. S. Elston and R. Sambles (Taylor & Francis, London, 1998), 137
34. H. J. Eichler, R. Macdonald and B. Trosken, *Mol. Cryst. Liq. Cryst.* 231,1 (1993)
35. I.-C. Khoo, *Liquid Crystals, Physical Properties and Nonlinear Optical Phenomena* (John Wiley & Sons, New York, 1995).
36. I.-C. Khoo and S.-T. Wu, *Optics and Nonlinear Optics of Liquid Crystals* (World Scientifc, Singapore, 1993).
37. I. C. Khoo, *J. Nonlin. Opt. Mat.* 8, 305 (1999).
38. M. J. Soileau, E. W. Van Stryland, S. Guha, E. J. Sharp, G. L. Wood and J. L. W. Pohlmann, *Mol. Cryst. Liq. Cryst.* 143, 139 (1987).
39. M. J. Soileau, S. Guha,W. E.Williams, E.W. Van Stryland, H. Vanherzeele, J. L. W. Pohlmann, E. J. Sharp and G. Wood, *Mol. Cryst. Liq. Cryst.* 127, 321 (1985).
40. A. Hochbaum, J. L. Fergason and J. D. Buck, *Proc. SPIE* 1692, 96 (1992).
41. A. Hochbaum, Y. Y. Hsu and J. L. Fergason, *Proc. SPIE* 2229, 48 (1994).
42. K. McEwan and R. C. Hollins, *Proc. SPIE* 2229, 122 (1994).
43. K. J. McEwan and R. C. Hollins, *J. Nonlin. Opt. Phys. Mat.* 4, 245 (1995).
44. H. J. Yuan, L. Li and P. Palffy-Muhoray, *Mol. Cryst. Liq. Cryst.* 199, 223 (1991).
45. P. Palffy-Muhoray, H. J. Yuan, L. Li, M. A. Lee, J. R. DeSalvo, T. H. Wei, M. Sheik-Bahae, D. J. Hagan and E. W. Van Stryland, *Mol. Cryst. Liq. Cryst.* 207, 291 (1991).
46. T. Kosa, A. Dogariu, P. Palffy-Muhoray and E. W. Van Stryland, *Opt. Soc. Am. Tech. Dig. Ser.* 21, 57 (1995).
47. W. Zhao and P. Palffy-Muhoray, *Appl. Phys. Lett.* 65, 673 (1994).
48. 70. P. Palffy-Muhoray, T. Wei and W. Zhao, *Mol. Cryst. Liq. Cryst.* 251, 19 (1994)
49. L. Li, H. J. Yuan, G. Hu and P. Palffy-Muhoray, *Liq. Cryst.* 16, 703 (1994).
50. P. Brochard, V. Grolier-Mazza and R. Cabanel, *J. Opt. Soc. Am.* B14, 405 (1997).
51. D. I. Kovsh, S. Yang, D. J. Hagan and E. W. Van Stryland, *Appl. Opt.* 38, 5168 (1999).
52. D. I. Kovsh, D. J. Hagan and E. W. Van Stryland, *Opt. Express* 4, 315 (1999).
53. G. Lukishova, *Mol. Cryst. Liq. Cryst.* 331, 2469 (1999).
54. M. Sheik-Bahae, A. A. Said, T.-H. Wei, D. J. Hagan and E. W. Van Stryland, *IEEE J. Quantum Electron.* 26, 760 (1990)
55. P. Palffy-Muhoray, in *Liquid Crystals: Applications and Uses*, Ed. B. Bahadur (World Scientific, Singapore, 1990),Vol. 1, 493.
56. R. G. Horn, *J. Physique* 39, 105 (1978).

57. V. Volterra and E. Wiener-Avnear, *Opt. Comm.* 12, 194 (1974).
58. P. Wang, H. Zhang and J. Dai, *Opt. Lett.* 13, 479 (1988).
59. F. Bloisi, L. Vicari, F. Simoni, G. Cipparone and C. Umeton, *J. Opt. Soc. Am.* B5, 2462 (1988).
60. S. G. Lukishova, *Proc. SPIE* 3798, 128 (1999).
61. S. G. Lukishova, *Opt. Soc. Am. Tech. Dig. QELS'99*, 126 (1999).
62. M. Marinelli, F. Mercuri, U. Zammit and F. Scudieri, *Phys. Rev.* E 58, 5860 (1998).
63. C. Umeton, G. Cipparrone and F. Simoni, *Opt. Quantum Electron.* 18, 312 (1986).
64. S.G. Lukishova, N. Lepeshkin, R.W. Boyd, K. Marshall, *Molec. Cryst. Liq. Cryst.* 453, 393, 2006.
65. S.G. Lukishova, R.W. Boyd, N. Lepeshkin, and K.L. Marshal, *J. Nonl. Opt. Phys. & Mater.* 11, 341 (2002).
66. D. R. Baals and S. Hess, *Z. Naturforsch.* 40a, 3 (1985).
67. R. Elscher, R. Macdonald, H. J. Eichler, S. Hess and A. M. Sonnet, *Phys. Rev.* E 60, 1792 (1999).
68. I. C. Khoo, R. G. Lindquist, R. R. Michael, R. J. Mansfield and P. LoPresti, *J. Appl. Phys.* 69, 3853 (1991).
69. H. J. Eichler and R. Macdonald, *Phys. Rev. Lett.* 67, 2666 (1991).
70. R. S. Akopyan, B. Ya. Zel'dovich and N. V. Tabiryan, *Tech. Phys.* 38, 493 (1994).
71. C. Glorieux, K. A. Nelson, G. Hinze and M. D. Fayer, *J. Chem. Phys.* 116, 3384 (2002).
72. D. Voloschenko, D. and O. Lavrentovich, *J. Appl. Phys.* 86, 4843 (1999).
73. D. F. Walls and G. F. Milburn, *Quantum Optics*, Springer Verlag: Berlin, NY (1995).
74. E. Knill, R. Laflamme, G. J. Milburn, *Nature*, 409, 46 (2001).
75. S.G. Lukishova, A.W. Schmid, R. Knox, P. Freivald, L. Bissell, R.W. Boyd, C.R. Stroud, Jr, K.L. Marshall, *J. Mod. Opt.* 54, iss. 2 & 3, 417 (2007).
76. S.G. Lukishova, L. J. Bissell, V.M. Menon, N. Valappil, M.A. Hahn, C.M. Evans, B, Zimmerman, T.D. Krauss, C. R. Stroud, Jr., R.W. Boyd, *J. Mod. Opt.* 56, is 2 & 3, 167 (2009).
77. S.G. Lukishova, L.J. Bissell, C. R. Stroud, Jr., R.W. Boyd, *Optics and Spectroscopy*, 108, 417 (2010).
78. S.G. Lukishova, L.J. Bissell, J. Winkler and C.R. Stroud, *Opt. Lett.* 37, 1259 (2012).
79. S.G. Lukishova, *Mol. Cryst. Liq. Cryst.* 559, 127–57 (2012).
80. Y. Yamamoto, Ch. Santori, J. Vuskovic, D. Fattal, E. Waks, E. Diamanti, *Progr. Informatics* 2005, 1, 5 (2005).
81. T.J. Bunning and F.-H. Kreuzer F-H., *Trends Polym. Sci.* 3, 318 (1995).
82. S.G. Lukishova and A.W. Schmid, *Molec. Cryst. Liq. Cryst.* 454, 15-21 (2006).
83. H. Coles and S. Morris, *Nature Photonics* 4, 678 (2010).
84. K. Dolgaleva, S. K. H. Wei, S.G. Lukishova, S.-H. Chen, K. Schwertz, R.W. Boyd *JOSA B* 25, 1496 (2008).
85. S. K .H. Wei, S. H. Chen, K. Dolgaleva, S. G. Lukishova, and R. W. Boyd, *Appl. Phys. Lett.* 94, 041111 (2009).
86. H. Huang, A. Dorn, V. Bulovic and M. Bawendi, *Appl. Phys. Lett.* 90, 023110 (2007).
87. Z. Wang, F. Tao, W. Cai, L. Yao and X. Li, *Solid State Commun.* 144, 255 (2007).

A comparison of the laser performance of Yb^{3+}:LuAG crystals with different doping levels

A Pirri [1], M Vannini [2*], V Babin [3], M Nikl [3] and G Toci [2]

[1] Istituto di Fisica Applicata "N. Carrara", Consiglio Nazionale delle Ricerche, IFAC-CNR, Via Madonna del Piano 10 C, 50019 Sesto Fiorentino, Florence, Italy

[2] Istituto Nazionale di Ottica, Consiglio Nazionale delle Ricerche INO-CNR, Via Madonna del Piano 10 C, 50019 Sesto Fiorentino, Florence, Italy

[3] Institute of Physics Academy of Sciences of the Czech Republic, Cukrovarnicka 10, 16200 Prague 6, Czech Republic

E-mail: matteo.vannini@ino.it

Abstract: We present a study and a comparison of the laser performance achieved by using 10*at.*%, 15*at.*% at. and 20*at.*% Yb^{3+}:LuAG crystals longitudinally pumped at 936 nm in quasi-continuous wave.

1. Introduction

Yb:LuAG [1-4] is one of the most attractive gain materials due to several reasons. First, it has similar optical properties of YAG as well as a well-defined absorbed peak around 940 nm, which allows it to be used in a laser system without specific adaptations of commercially available pump diodes. Second, it has a high thermal conductivity [5-6] which is independent on the concentration of dopant In fact, the similar atomic masses of Yb^{3+} and Lu^{3+}(173 and 175 g/mol [7], respectively) reduces the scattering at the mass defect of the phonons responsible of the heat transportation. Finally, this material has a broad and high emission cross section spectrum making it appealing for the generation of short pulses [8-9].

In this paper we reported a comparative investigation of the laser performances achieved by using three crystals with 10*at.*%, 15*at.*% and 20*at.*% concentration of Yb^{3+}. Longitudinally pumped in *quasi*-Continuous Wave at 936 nm, the crystals have shown different behavior in terms of laser output power and efficiency. Excellent performances are achieved by the 10*at.*%. Emitting at 1046 nm it delivers P_{out}=11.8 W with a slope efficiency of μ_s=82%. In the case of 15*at.*% doping, two different regimes are observed. At high ion excitation density, we measured a dramatic decrease of the output power due to the appearance of a non-linear and non-radiative loss mechanism [10-13]. Conversely, at low excitation density, the crystal releases the maximum output power at 1030 nm, *i.e.* P_{out}=9.7 W with a corresponding slope efficiency as high as μ_s=59%. Low laser performance are achieved with the 20*at.*%. Also for this sample two different regimes are observed, which, however, depend on the

[*] To whom any correspondence should be addressed.

Journal of Physics: Conference Series **497** (2014) 012009 doi:10.1088/1742-6596/497/1/012009

pumping level and not on the excitation density as the decrease of the delivered power occur with all OC mirrors. The maximum output power of P_{out}=5.3 W is reached by T=9%. The cause of the reduced performance is addressed to thermal lensing effects [10].

2 .Experimental set up

The experimental setup used to test the laser performance of the crystals is depicted in Fig. 1. The resonator is constituted by three mirrors placed in a V-shaped configuration. EM is a flat mirror with a dichroic coating with high transmission around the pump wavelength and high reflectivity in the laser emission band. The folding mirror, FM, has a curvature radius of 100 mm with high transmission at the pump wavelengths. The OC is the output coupler mirror. The uncoated 10*at.*%, 15*at.*% and 20*at.*% Yb:LuAG crystals have a thickness of 2 mm, 1 mm and 2 mm, respectively, while both face are flat and perpendicular of 5x5 mm^2. They are soldered with indium on a copper heat sink which in turn is water cooled at 18 °C; the crystals are placed as near as possible to the flat EM. A very carefully orientation of the facets perpendicular to the cavity axis allowed the re-injection of the Fresnel reflection, which are estimated around 8.5%.

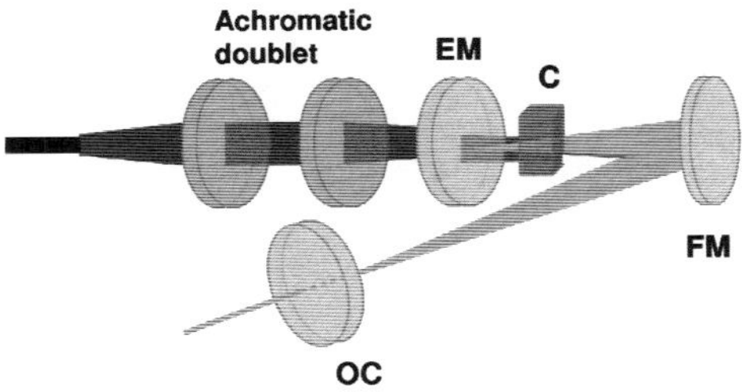

Figure 1. Experimental set up to test the laser performance of the samples. EM: End Mirror (flat); FM: Folding Mirror; OC: Output Coupler (flat).

For the pumping of these samples, a fibre-coupled laser diode is used. The emission from the fibre is refocused and injected into the sample by a pair of achromatic doublets with a magnification of 1:1. The pump emits at 936 nm (fiber core diameter Φ=200 μm, numerical aperture N.A.=0.22, maximum output power P_p=25 W) and it has an almost Gaussian intensity distribution in the focal plane of the doublets with 150 μm of radius at $1/e^2$ In the experiment Duty Factor, DF, of 20% and 40% are employed.

3. Experimental results and comments

Excellent performances, *i.e.* output power and slope efficiency, are achieved by the 10*at.*% Figure 2 (a)-(b) reports the experimental results obtained with several OCs (from T=1.9% to T=86%). The maximum powers, P_{out}=11.8 W, is delivered at 1046 nm with a corresponding slope efficiency of μ_s=82%. The lowest laser threshold is P_{th}=0.5 W. The shift of the emitted wavelength from 1049 nm with T~2% to 1030 nm measured with the other OC is ascribed to the behavior of the Yb^{3+}, which is a quasi three-level system. As a matter of fact, for increasing losses the increasing inversion population fraction needed to reach the lasing threshold determines a shift toward shorter wavelengths in the peak of the effective gain spectrum.

From the comparison of the performance obtained with 15*at.*% doping, see Fig. 3 (a)-(b) it is immediately evident that the sample has two different behavior depending on the ion excitation density involved in the laser action.

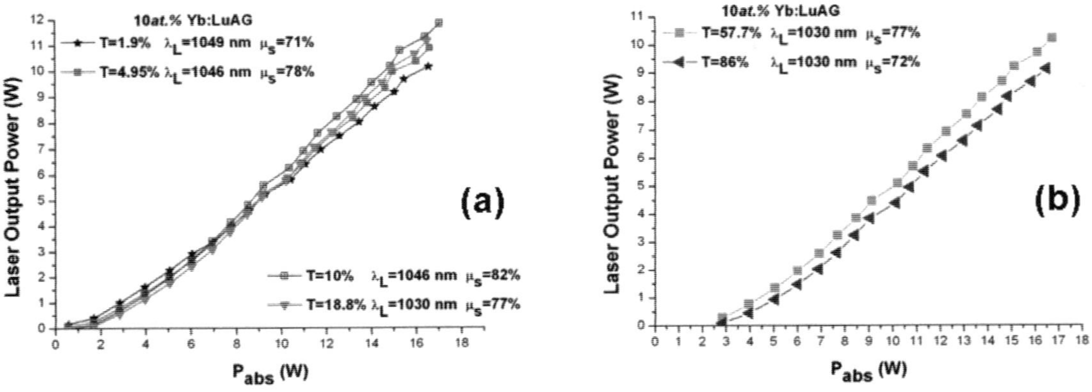

Figure 2. 10*at.*% LuAG laser output power as a function of the absorbed pump power. T: output coupler transmission; λ_L: laser wavelength; μ_S: slope efficiency. The crystals are pumped at 936 nm in quasi-CW (DF=20%, 10 Hz).

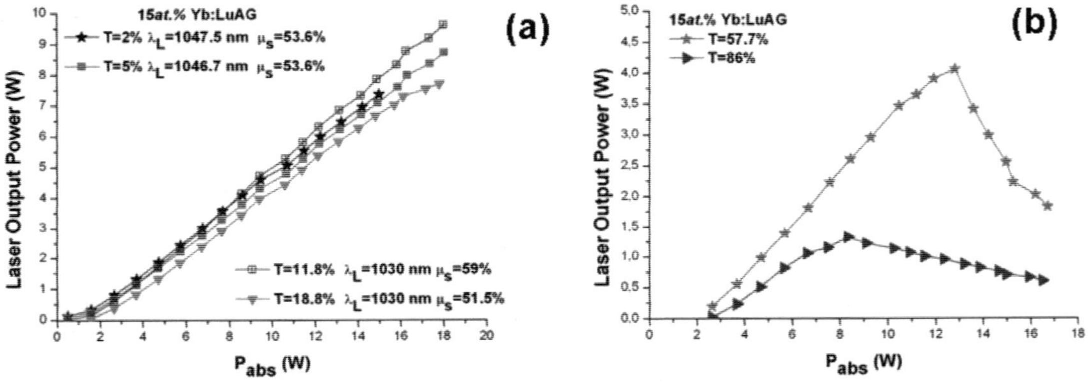

Figure 3. 15*at.*% LuAG laser output power versus the absorbed pump power. T: output coupler transmission; λ_L: laser wavelength; μ_S: slope efficiency. The crystals are pumped at 936 nm in quasi-CW (DF=20%, 10 Hz).

Good performances are achieved by the 15*at.*% sample with low ion excitation densities, *i.e.* with OC transmission ranging from 2% to 18.8%. At 1030 nm it releases a maximum output power of P_{out}=9.7 W with a corresponding slope efficiency of μ_s=59%. The threshold is P_{th}=0.57 W which is quite similar to the one obtained with the 10*at.*% in the same experimental conditions. Contrary, see Fig. 3 (b), when higher Yb-ion excitation densities in the laser action are involved (*i.e.* with T=86% and T=57.7%), the performance is dramatically reduced. The observed shift of the onset of the roll-over in the curves toward lower value of the absorbed pump power (from P_{abs}~13 W with T=57.7% to P_{abs}~8W by T=86%) demonstrates that it starts at a well-defined threshold of the excitation levels and its rate is enhanced by increasing the ions excitation density. According to our calculation based on the equation rates, we estimated a threshold of the density of $4.5x10^{20}$exc. ion/cm^3.

The role played by the temperature on this effect is studied by increasing the DF from 20% to 40% and closing the cavity with T=57.7%, see Fig. 4. Also in this case, two different regimes can be easily observed. At lower absorption levels, *i.e.* P_{abs}~13 W by T=57.7% and P_{abs}~7.5 W by T=86%, the increase of the temperature inside of the crystal does not affect the laser performance as the output power linearly increases with the pump power. The scenario totally changes as the P_{abs} exceeds the mentioned threshold with a dramatic decrease of the P_{out}. We point out that, at the maximum level of the absorption used in the experiment (P_{abs}~16 W), the laser cavity still emits 2 W when the crystal is pumped with DF=20%, whereas the laser action is practically hampered with DF=40%.

The experimental data are explained by the appearance of non-radiative channels which take place when higher ion excitation densities are involved in the laser action. The origin of these channels is still under investigation [10].

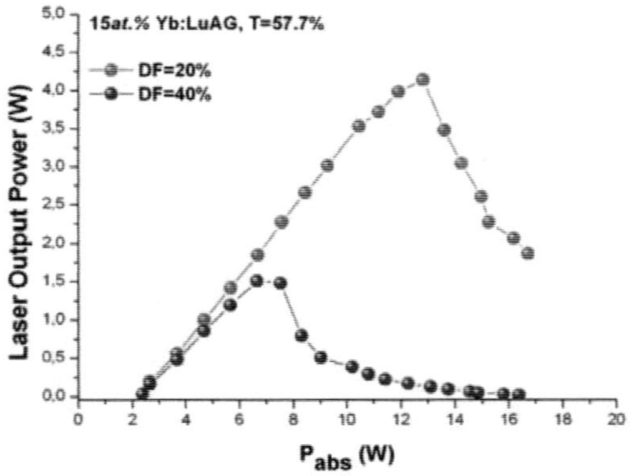

Figure 4.: Laser output power versus absorbed pump power at higher ion excitation density with two Duty Factor (20% and 40%).

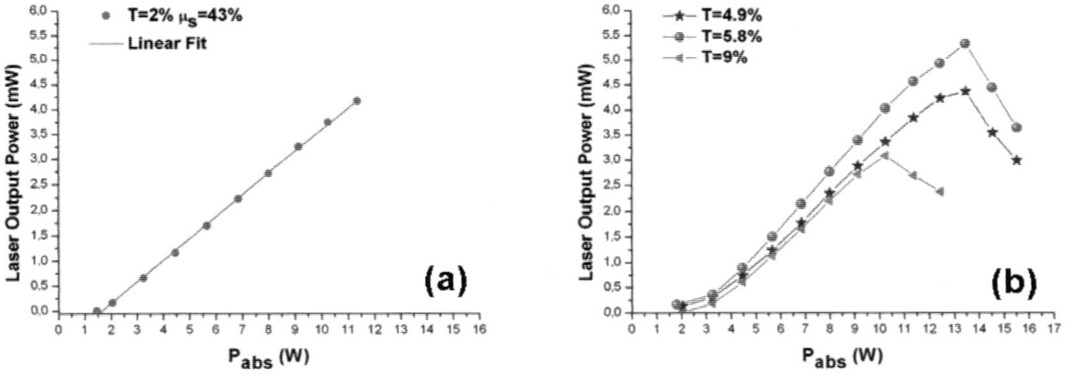

Figure 5.: 20*at.*% laser output power versus absorbed pump power. T: output coupler transmission; λ_L: laser wavelength; μ_S: slope efficiency. The crystals are pumped at 936 nm in quasi-CW (DF=20%, 10 Hz).

Low laser performance are achieved with the 20*at.*% doping as reported in Fig 5. Also for this sample two different behaviors are observed, which, however, depend on the pumping level and not on the excitation density as the decrease of the delivered powers occur with all OC mirrors. The maximum output power of P_{out}=5.3 W is reached by T=5.8%. The cause of the reduced performance is addressed to thermal lensing effects as reported in [10]. In other words, the thermal lens was so strong to drive the resonator out of the stability region, thus preventing the laser action.

4. Conclusions

In this paper we reported the laser performance achieved with a longitudinally diode-pumped laser based on 10*at.*%, 15*at.*% *at.* and 20*at.*% Yb^{3+}:LuAG crystals. In particular, we measured with the 10*at.*% doping an output power of 11.8 W with a slope efficiency as high as μ_s=82% at 1046 nm, which confirms the full potential of the Yb:LuAG to be employed in the development of diode-pumped solid state laser sources, at least for reasonable doping levels. A heavy doping of the matrix can trigger loss mechanisms which degrade the laser performance.

Acknowledgements

The research was supported by Regione Toscana, project "CTOTUS-Progetto integrato per lo sviluppo della Capacità Tecnologica e Operativa della Toscana per l'Utilizzo dello Spazio" (POR FESR 2007-2013 Attività 1.1 Linea d'intervento D); by the Consiglio Nazionale delle Ricerche, CNR-RSTL "Ricerca Spontanea a Tema libero", id. 959 and by the joint project by the joint project of ASCR and CNR and Czech GA AV project M100100910, 2012-2015.

References

[1] Brenier B, Guyot Y, Canibano H, Boulon G, Ròdenas A, Jaque D, Eganyan A and Petrosyan A G 2006 J. Opt. Soc. Am. B **23** 676
[2] Xu C W, Luo D W, Zhang J, Yang H, Qin H P, Tan W D and Tang D Y 2012 Laser Phys. Lett. **9** 30
[3] Luo D, Zhang J, Xu C, Yang H, Lin H, Zhu H, Tang D 2012 Opt. Mat. Express **2** 1425
[4] Dong J, Ueda K, Kaminskii A A 2010 Laser Phys. Lett. **7** 726
[5] Geballe T H and Hull G W 1958 Phys. Rev. **110** 773
[6] Nakao H, Shirakawa A, Ueda K, Yagi H and Yanagitani T 2012 Opt. Express **20** 15385
[7] Wieser M E and Coplen T B 2011 Pure Appl. Chem. 83 359
[8] He J, Liang X, Li J, Yu H, Xu X, Zhao Z, Xu J and Xu Z 2009 Opt. Express **17** 11537
[9] Beil K, Fredrich-Thornton S T, Tellkamp F, Peters R, Kränkel C, Petermann K and Huber G 2010 Opt. Express **18** 20712
[10] Pirri A, Toci G, Nikl M and Vannini M. 2013 Opt. Express. Submitted
[11] Pirri A, Toci G, Alderighi D and Vannini M 2010 Opt. Express **18** 17263
[12] Larionov M, Schuhmann K, Speiser J, Stolzenburg C and Giesen A 2005 Advanced Solid-State Photonics, Technical Digest (Optical Society of America, 2005), paper TuB49.
[13] Brandt C, Fredrich-Thornton S T, Petermann K and Huber G 2011 Appl. Phys. B **102**, 765

High power diode laser remelting of metals

H Chmelickova[1, 4], J Tomastik[1], R Ctvrtlik[1], J Supik[2], S Nemecek[3] and M Misek[3]
[1] Palacky University, Faculty of Sciences, Joint Laboratory of Optics of Palacky University and Institute of Physics of the Academy of Sciences of the Czech Republic, 17. listopadu 50a, Olomouc, 772 07, Czech Republic
[2] Palacky University, Faculty of Sciences, Department of Experimental Physics, 17. listopadu 12, Olomouc, 771 46, Czech Republic
[3] MATEX PM, Morseova 5, Plzeň, 301 00, Czech Republic

E-mail: hana.chmelickova@upol.cz

Abstract. This article is focused on the laser surface remelting of the steel samples with predefined overlapping of the laser spots. The goal of our experimental work was to evaluate microstructure and hardness both in overlapped zone and single pass ones for three kinds of ferrous metals with different content of carbon, cast iron, non-alloy structural steel and tool steel. High power fibre coupled diode laser Laserline LDF 3600-100 was used with robotic guided processing head equipped by the laser beam homogenizer that creates rectangular beam shape with uniform intensity distribution. Each sample was treated with identical process parameters – laser power, beam diameter, focus position, speed of motion and 40% spot overlap. Dimensions and structures of the remelted zone, zone of the partial melting, heat affected zone and base material were detected and measured by means of laser scanning and optical microscopes. Hardness progress in the vertical axis of the overlapped zone from remelted surface layer to base material was measured and compared with the hardness of the single spots. The most hardness growth was found for cast iron, the least for structural steel. Experiment results will be used to processing parameters optimization for each tested material separately.

1. Introduction

Laser remelting is one of the surface modification methods with aim to change a microstructure of highly stressed, ferrous machine parts and enhance their mechanical and chemical resistance (bearing races, gears, shafts, cylinders, camshafts) [1]. Nodular cast irons and low carbon steels might be suitable materials, because they are difficult hardened by solid state transformations due to extensive or insufficient carbon content. Heating and cooling rates in case of the laser melting are much higher then in case of the usually used conventional methods (oxy-acetylene flame, induction heating and TIG electric arc) that cause extensive heating of material and risk of deformations. Power density of the laser beam is controlled to ensure heating slightly above material melting point without vaporization. Laser beam with uniformly distributed intensity is created by various methods of beam shaping to ensure constant depth of the remelted zone [2], [3]. Main methods are defocusing by focal plane shift above surface, optical integration by segmenting the beam into a large number of portions and superposing the individual segments on the same focal plane using a beam integrator, scanning by two perpendicular galvo-mirrors respectively. Because dimensions of the homogenized laser beam is

[4] To whom any correspondence should be addressed.

about some millimetres, large work pieces have to be remelted by deposition of tens laser spots with optimal overlapping. Part of the once remelted material at the first spot boundary is heated again during the second spot deposition. The goal of our experimental work was to evaluate microstructure and hardness both in overlapped zone and single pass ones for three different kinds of ferrous metals, cast iron, non-alloy structural steel and tool steel.

2. Experiment

2.1. Material

Samples with approximate length 150 mm and width 80 mm were prepared from three kinds of the ferrous metal semi-product plates of the thickness 15 mm (figure 1), specifically cast iron EN-JS2070, non-alloy structural steel S355 JR and tool steel OH260 (table 1). Each of these materials has a defined position in the iron/carbon alloy phase diagram according a typical content of carbon. EN-JS2070 with carbon content from 3.4 % to 3.85 % is ductile cast iron with pearlitic structure containing nodular graphite inclusions that is used in production of the casts with thickness from 5 mm to 200 mm and mechanically stressed machine parts (gear wheels, drum brakes). S355 JR with carbon content less or equal to 0.24 % is hypo-eutectoid non-alloy structural steel with pearlitic and ferritic structure that is widely used for steel constructions in various industry branches (buildings, bridges, towers, ship desks etc.) [4], [5]. Both mentioned materials are difficult to be hardened due to extensively high and low content of carbon respectively, so laser melting can be suitable way for surface hardness improving.

Table 1. Specification of the experimental material.

material	carbon (%)	hardness (HV)	output structure	typical application
EN-JS2070	3.4 – 3.85	236 - 322	pearlite with nodular graphite	gear wheel, drum brake
S355 JR	$\leq$ 0.24	170 - 194	pearlite and ferrite	bridges, towers
OH260	0.77 – 0.85	243 - 253	oil quenched martensite	cutting and punched dies

OH260 (steel. no. 1.2369, DIN 81MoCrV42-16) with carbon content from 0.77 % to 0.85 % is cold work tool steel with hardness and wear resistance achieved by conventional hardening and oil quenching with martensitic structure to be used for cutting and punches dies.

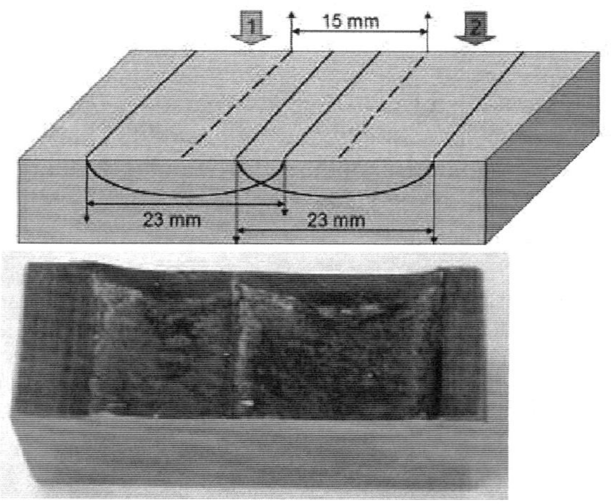

Figure 1. Scheme of the experiment set-up and remelted surface of the steel OH260.

Figure 2. Robotic guided laser processing head with beam shape homogenizer.

2.2. Laser parameters

The high power fibre coupled diode laser Laserline LDF 3600-100 with maximal power 3 600 W and a beam quality (beam radius times half angle divergence) 100 mm.mrad was used in experiment realized in MATEX PM Facility. Processing head with focus lens distance 200 mm is equipped by beam homogenizer that creates rectangular shape of the beam profile 5 mm x 23 mm with uniform intensity distribution. Connection between laser source and processing head was realized by the optical fibre with diameter 1 mm, relative motion between the beam focus plane and material surface was ensured by computer guided robotic arm (figure 2). Process parameters were kept on the same values, it means laser power P = 2800 W, speed of the processing head motion v = 1.5 m s^{-1} and the spot axis shift d = 15 mm, that results in 40% overlap.

2.3. Experiment set up

Two spots with preset distance d = 15 mm were applied consequently on each of the prepared samples with length 150 mm. Focus plane was positioned on the samples surface, which temperature was controlled by means of the pyrometer not to exceed temperature 1550 °C. Experiment was realized in the natural atmosphere without any shielding gas.

3. Results

The samples were cut in the plane perpendicular to the material surface and to the direction of the laser beam motion. These obtained cross sections are then prepared for microanalysis by the mechanical grinding, polishing with diamond paste and etching by 1% respective 3% solution of HNO_3. Dimensions and structures of the remelted zone, zone of the partial melting, heat affected zone and base material were detected and measured by means of the laser scanning microscope Olympus LEXT OLS3100 and optical microscope GX-51 [6]. Hardness was measured both along the second spot axis and the overlapped area by Vickers hardness test for each of the three samples to detect differences in the once and repeatedly remelted regions

3.1. Remelted cast iron sample microstructure

The remelted zone is created by ledeburite that is phase mixture of the cementite and austenite with presence of the martensite. Globular graphite was dissolved in the melt pool and during solidification it was excluded in the form of the dendritic structured cementite (figure 3). The heat affected zone is created by martensite and because the melting temperature was not achieved in this region, globular graphite is present in its original form observable as the black circular shapes as is seen on the figure 4. Transition between remelted and heat affected zone is gradual, presence of the both structures was observed in some places.

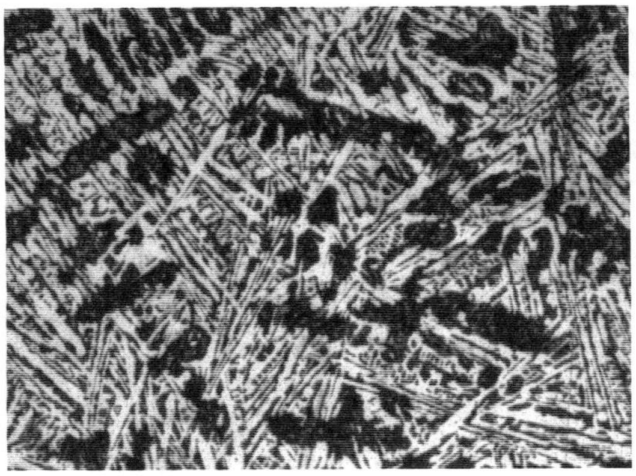

Figure 3. The remelted zone of the cast iron created by mixture of the cementite and austenite with martensite.

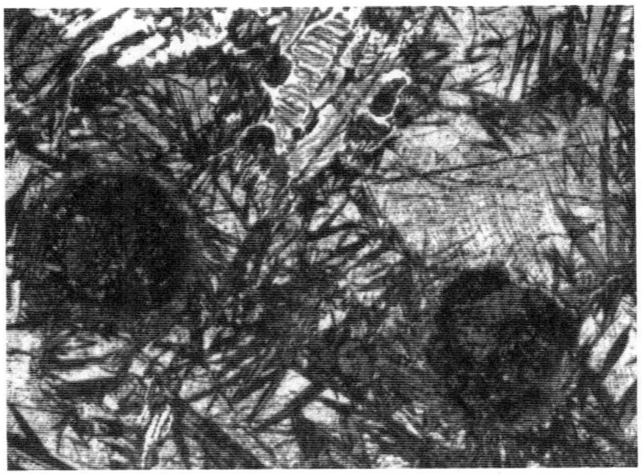

Figure 4. The heat affected zone of the cast iron created by martensite with non dissolved globular graphite.

3.2. Remelted structural steel sample microstructure

The remelted zone of this sample is created by martensite and ferrite (figure 5), where some inclusions of ferrous oxides are present probably due to material interaction with ambient atmosphere. These inclusions have form of the black dots and some group of them is highlighted in the figure by rectangle. The heat affected zone is created by martensite (figure 6), but there is present also ferrite with decreasing amount in direction to base material, so the heat affected zone has no discrete border and can be considered as a transition. The base material structure can be described as alternated layers of the pearlite and ferrite that was created during semi-product forming.

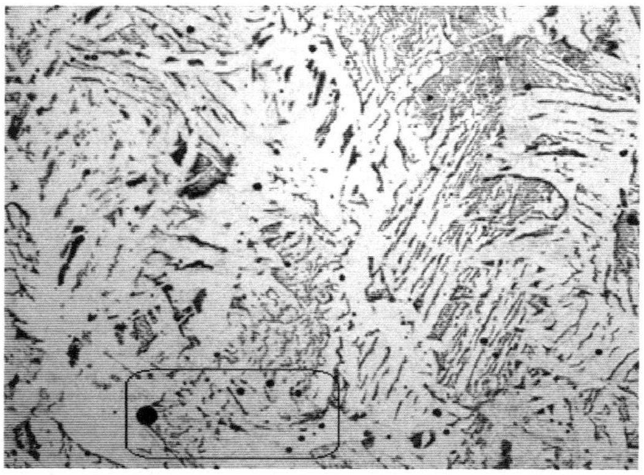

Figure 5. The remelted zone of the structural steel created by martensite and ferrite with ferrous oxides inclusions.

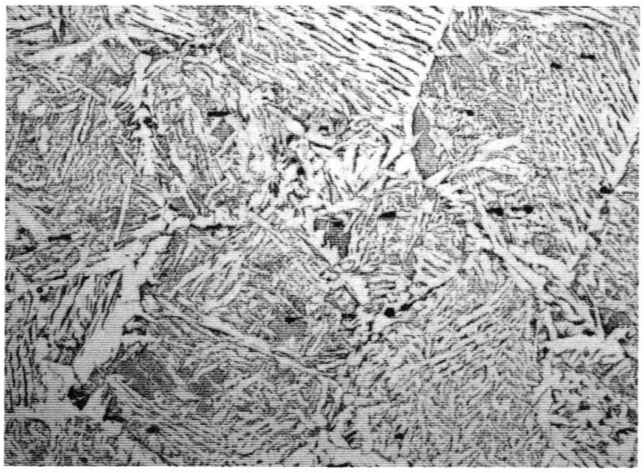

Figure 6. The heat affected zone of the structural steel created by martensite with occurrence of ferrite.

3.3. Remelted tool steel sample microstructure

Structure of this sample is oriented vertically. Martensite was created in the remelted zone and the original cellular structure of the austenite is observable too (figure 7). There are observed white and long segregation strips formed by cementite in the vertical direction going through martensite structure of the heat affected zone up to base material (figure 8). Intensity and broadness of these strips decreases in the direction from the surface to the inner volume of material. Base material has also martensitic structure due to previous heat treatment. Transition between the remelted and heat affected zone is strongly bounded in contrast with that of the structural steel and cast iron, what is observable on the macro photographs of the overlapped regions.

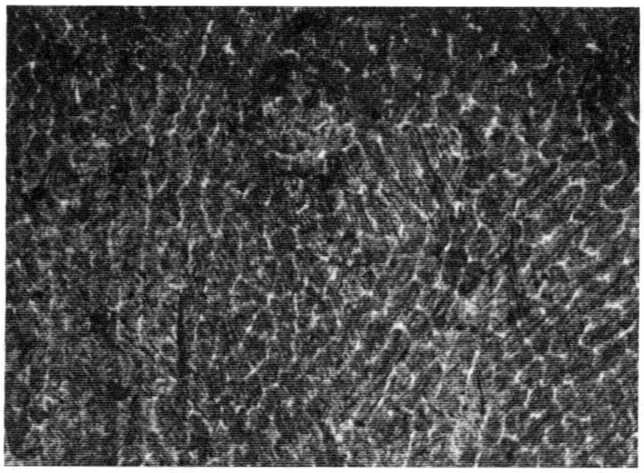

Figure 7. The remelted zone of the tool steel created by martensite and original cellular structure of the austenite.

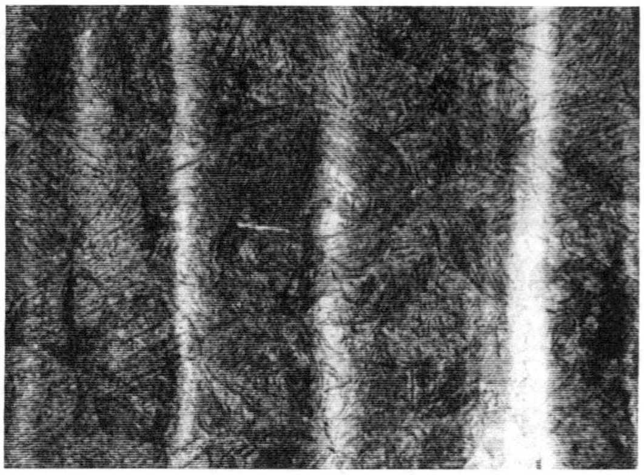

Figure 8. The heat affected zone of the tool steel created by martensite with vertical cementite white strips.

3.4. Overlapped zones microstructure

Microstructure of the overlapped zones is created by the same crystallographic phases as the particular spots in case of all three samples; differences were found only in the phase's boundaries and transitions. In case of the cast iron, the boundary of the remelted layer of the first spot was overlapped by zone of partial melting of the second one. Continuous transition between spots with mixing of structures, growing from two different directions was observed in the region of the spot overlap in case of the structural steel (figure 9). Apart of this remelted zone of the second spot overlies the first one with significant boundary where micro cracks were detected in case of the tool steel (figure 10).

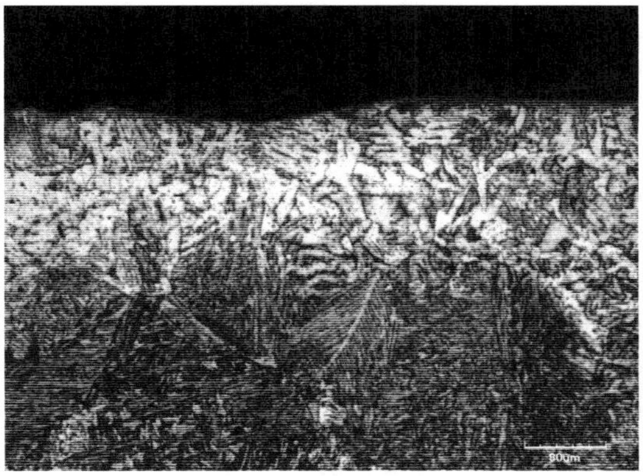

Figure 9. The overlapped region of the structural steel sample with continuous transition.

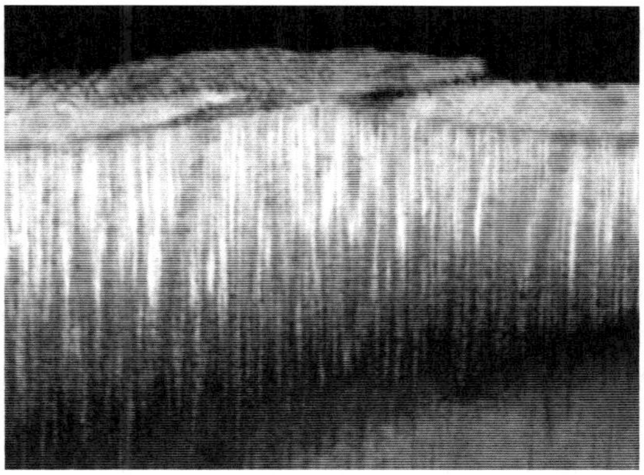

Figure 10. The overlapped region of the tool steel sample with significant boundary and micro cracks.

3.5. Hardness measurement

The Vickers hardness test was at the first time realized inside the second spot deepest region in the direction from the sample surface to the base material with load 9.81 N. Hardness increasing in comparison with the base material was achieved in the remelted and heat affected zones of the all three samples with some differences in the depth dependence progress (figure 11).

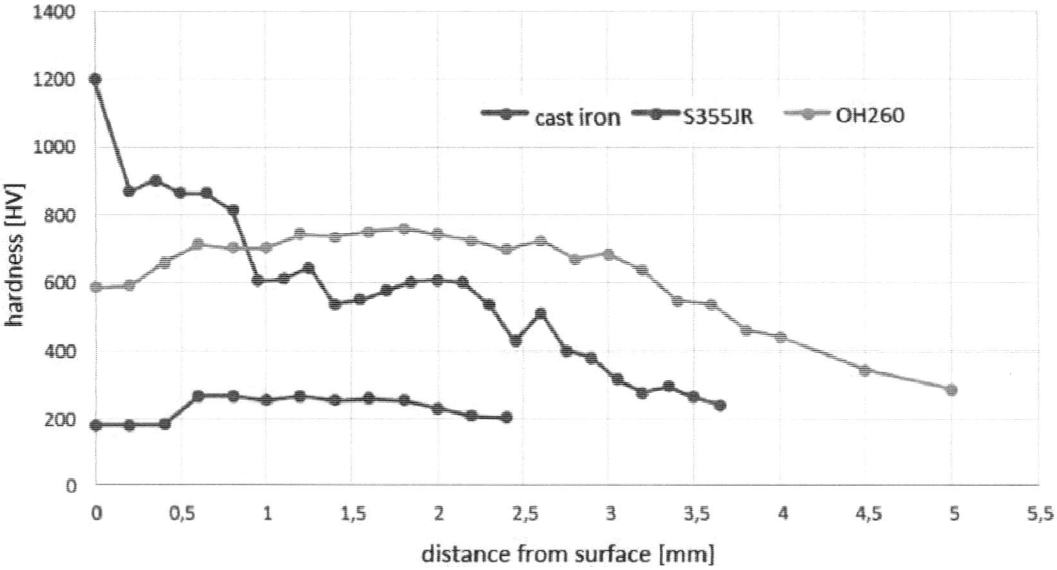

Figure 11. Hardness progress in the second spot vertical axis.

The highest hardness was measured in the remelted zone of the nodular cast iron (1200 HV) that is approximately four times greater than hardness in the base material (282 HV) and approximately twice greater than hardness in the heat affected zone (643 HV). In case of the construction steel hardness in remelted zone close to surface has less hardness (180 HV) then in heat affected zone (267 HV) due to ferrite phase presence and low content of carbon in the base material (204 HV). Hardness of the oil quenched tool steel with output martensitic structure initially grows from the sufficient surface value (591 HV) towards heat affected zone, where it achieves maximal value 760 HV and slowly flows down to base material (286 HV). Following measurement was realized in the vertical axis of the overlapped region to evaluate the influence of the first deposited spot repeated heating to melting temperature while the second one was deposited. Values of the cast iron overlapped region hardness in the various depths are outlined in the table 2.

Table 2. Cast iron hardness progress in the vertical axis of the overlapped region.

No.	1	2	3	4	5	6	7	8	9	10
depth (mm)	0.42	0.73	0.98	1.34	1.66	2.07	2.77	3.32	4.26	5.92
HV ($N.mm^{-2}$)	840.4	651.5	631.7	809.3	768.2	777.7	658.5	586.4	303.5	281.4
deviation (mm)	85.7	84.0	107.7	114.2	130.0	79.7	42.5	120.8	73.7	82.6

Strong boundary between the remelted and heat affected zone fully overlies the remelted zone of the first spot (measurement no. 2 and no. 3), where hardness falls down to the typical values for heat affected zone and its increasing is observable in the region below 1.34 mm where typical remelted structure is present again (figure 12). The analogues results were obtain in case of the other experimental materials, structural and tool steels. From obtained results implies, that cast iron is the most suitable material for the laser surface remelting due to hardness increasing above values achievable in the conventionally used hardening process. Structural steel due to low carbon content is material suitable rather for welding than for the surface treatment. Hardness about 775 HV (63HRC) is achieved by heat treatment of the tool steel both by use of the conventional and laser methods.

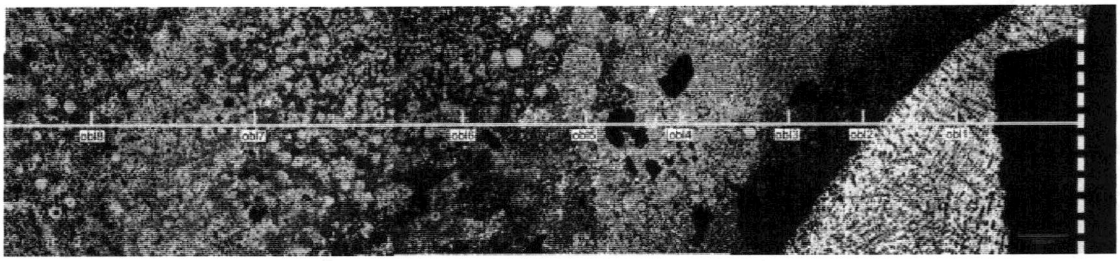

Figure 12. Cast iron hardness measurement in region of the spot overlap.

3.6. Surface roughness

The information about surface dimensional and volumetric changes after thermal treatment might be important data for the optimal function of the machine parts that undergone the laser surface remelting, when convection in the melt pool is caused by the surface tension gradient and surface deformations are observed after material re-crystallization [7]. Contact profile-meter Taylor Hobson was used to scanning across the complete remelted region of the two overlapped spots including surrounding non-treated surface in the direction perpendicular to the laser beam movement [8]. Record of five hundreds linear profiles separated of each other by 0.02 mm is used for various plane and three-dimensional displaying of the measured region according user requirements by means of the software tool TALYMAP. Black-and-white 3D axonometric display of the cast iron remelted spots in 10 mm long section is presented on the figure 13 with zoomed vertical axis.

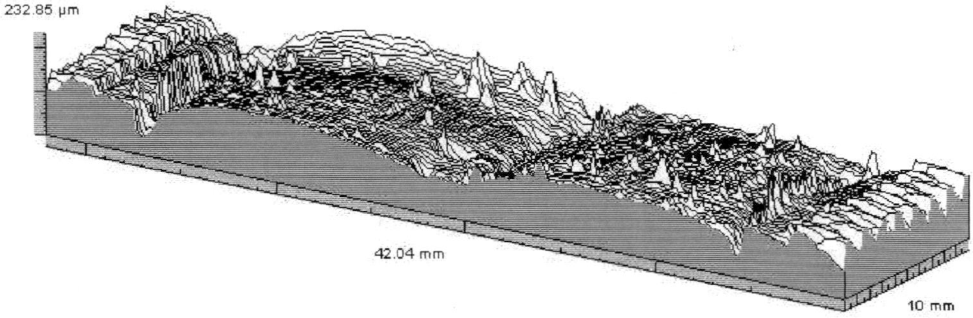

Figure 13. 3D axonometric display of the cast iron remelted region in 10 mm section.

Example of the average profile computing of this scanned area is shown on the figure 14, where result for the cast iron sample is presented. Black curve displays average profile while colored areas correspond to the surface shape changes with respect to non-treated sample plane and they are designed as Area of the hole (red areas) and Area outside (green areas). Maximum depth and height are extracted from measured data respectively and their values can be displayed below graph. The maximal positive difference in these areas dimensions 1.4534 mm^2 was found out in case of the tool steel, from this value volumetric growth was dedicated, maximal height 415.77 µm is approximately twofold than maximal depth 202.68 µm. The minimum positive difference 0.0004 mm^2 between areas was achieved in construction steel although its surface was noticeably modified by melt pool convection and maximum depth 177.94 µm is of 11.4 % greater than maximum high 159.71 µm. Negative areas difference 0.42358 mm^2 was computed in case of the cast iron, where the lowest value of maximum height 17.794 µm and depth 92.876 µm respectively was measured (figure 14).

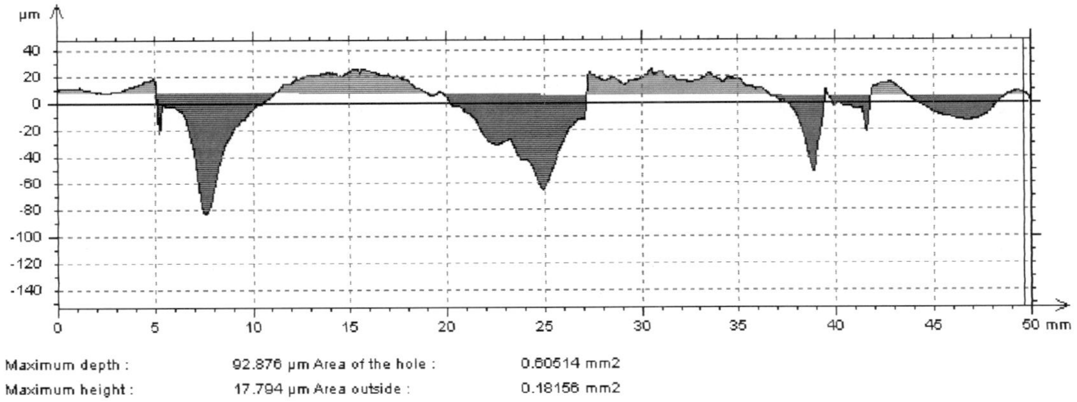

Figure 14. Average profile of the cast iron remelted region from data measured in 10 mm section.

4. Conclusions

This work has presented high power diode laser remelting of three various ferrous metal samples with different amount of carbon namely specific kind with cast iron, structural and tool steel. To evaluate material properties dependence on the expected hardness progress, processing parameters were kept constant. Two overlapped spots were applied on each sample surface to found changes in microstructure of the first spot after the second one treatment. Due to high heating and cooling rates martensitic microstructure was find in the heat affected zones, surface remelted layers are mixtures of martensite and other metallurgical phases from original structure such as austenite and ferrite. Hardness of the heat affected zone has higher values than in the base material of all three samples; the maximal increase was detected in case of the cast iron, the minimal one for structural steel respectively. Small hardness decrease at surface remelted zone in comparison with heat affected zone was detected for structural and tool steel. Only cast iron has maximal values of the hardness in the surface remelted layer. Differences between the first and second spot dimensions and structure roughness were also found, probably caused by sample pre-heating during the first laser passing (figure 15). Hardness progress was also measured in the vertical axis of the overlapped zone and it was compared with previous results. Information about surface and dimensional changes was obtained by means of measurement on the contact profile-meter.

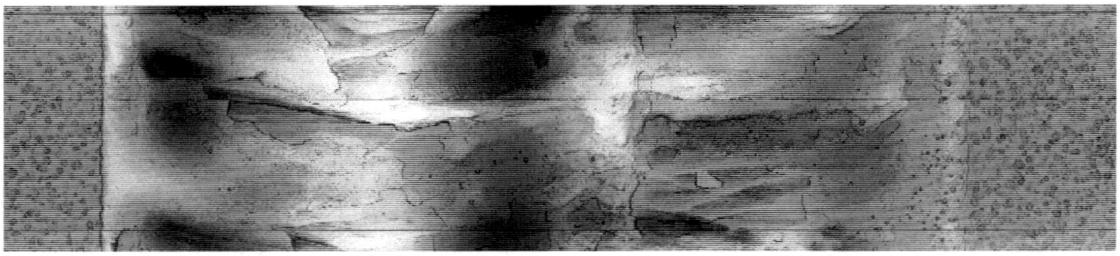

Figure 15. Pseudo-photo of the construction steel remelted surface at region 48 mm x 10 mm.

5. References

[1] Ion J C 2005 *Laser Processing of engineering materials* (Amsterdam: Elsevier) p 133
[2] Steen W M and Mazumder J 2010 *Laser Material Processing* (London: Springer) pp 302–4
[3] Ready J F and Farson F D 2001 *LIA handbook of laser material processing* (Orlando, FL Laser Institute of America) pp 227–32
[4] Fürbacher I, Macek K and Steidl J 1998 *Lexikon technických materiálu se zahranicními ekvivalenty* (Praha:Verlag Dashöfer)
[5] American Society for Metals 1989 *Metals handbook 9th Edition. Volume 1: Properties and Selection - Irons and Steels* (Materials Park, OH: ASM International)
[6] Bramfitt B and Benscoter A O 2001 *Metallographer's guide: practice and procedures for irons and steels* (Materials Park, OH: ASM International)
[7] Asibu E K 2009 *Principles of laser materials processing* (Hoboken, NJ: Wiley & Sons) pp 262– 4
[8] Hiklova H, Havelkova M, Chmelickova H and Lapsanska H 2010 Contact system of solid state surface mapping *Proc. 48th Int. Conf. Experimental Stress Anylysis (Velke Losiny, CZ, 31 May–3 Jun 2010)* ed(s) Smid P, Horvath P and Hrabovsky M (Olomouc: Palacky University) pp 89–93

Acknowledgments
This contribution was supported by the project "Otevřená síť partnerství na bázi aplikované fyziky" (Open partnership net based on applied physics) number CZ 1.07/2.4.00/17.0014.

Multiline generation capabilities of diode-pumped Nd:YAP and Nd:YAG lasers

Martin Fibrich[1,2,*], Tomáš Hambálek[1], Michal Němec[1], Jan Šulc[1], Helena Jelínková[1]
[1]Czech Technical University in Prague, Faculty of Nuclear Sciences and Physical Engineering, 11519 Prague 1, Czech Republic
[2]Institute of Physics ASCR, Department of Laser Systems
Na Slovance 1999/2, 182 21 Prague 8, Czech Republic
E-mail: *martin.fibrich@fjfi.cvut.cz

Abstract. Multiline generation capabilities of diode pumped Nd:YAP and Nd:YAG lasers within 1.3 – 1.5 μm wavelength range at room temperature are reported. Two optical resonators designed for 1.3 μm and 1.4 μm laser operation have been realized. Using a single quartz plate as a tuning element, six single emission lines within the 1.3 – 1.5 μm spectral range for both Nd:YAG and Nd:YAP lasers have been reached. Moreover, as also demonstrated, it was possible to obtain Nd:YAG/YAP dual frequency regime operation for some line combinations.

1. Introduction

So far, the neodymium ion Nd^{3+} has been one of the most frequently used laser ions. The success of this ion is due to its favorable energetic level structure and spectroscopic properties suitable for laser generation. Since 1961, when this ion was lasing first time using a $CaWO_4$ crystal [1], the Nd^{3+} ion has been successfully tested with more than 180 crystalline laser hosts [2]. Nevertheless, by far the most important solid-state laser for scientific, medical, industrial and military applications based on Nd^{3+} ion is Nd:YAG, because it has a high gain and good thermal and mechanical properties. Thanks to this, relatively cheap high-quality Nd:YAG rods and appropriate reliable highly-efficient high-power pumping laser diodes are affordable. Another attractive laser host for neodymium is the yttrium aluminium oxide $YAlO_3$ (YAP,YAlO), due to its natural birefringence combined with good thermal and mechanical properties similar to those of YAG [3]. Owing to the birefringent character of the YAP matrix, the thermally-induced birefringence does not degrade the laser performance [4].

Even though the available laser emissions of Nd:YAG and Nd:YAP cover tens of lines between 0.94 μm and 1.84 μm, the most frequently used lines are 1064 nm and 1070 nm for Nd:YAG and Nd:YAP, respectively, which corresponds to transition $^4F_{3/2} \rightarrow {}^4I_{11/2}$ of the Nd^{3+} ion [4, 5]. In the work presented, we focused on laser lines corresponding to transition $^4F_{3/2} \rightarrow {}^4I_{13/2}$ which covers laser emission from 1.32 to 1.44 μm. Although the stimulated emission cross-sections at this transition are more than ten times lower [6] in comparison with emission at 1064 nm, and the quantum defect is higher than 35 %, this wavelength can be obtained relatively easily with reasonable efficiency by using a low-loss resonator (multi-Watt output was demonstrated previously using diode pumped Nd:YAG laser operating even at 1444 nm [7]). Unlike 1064 nm

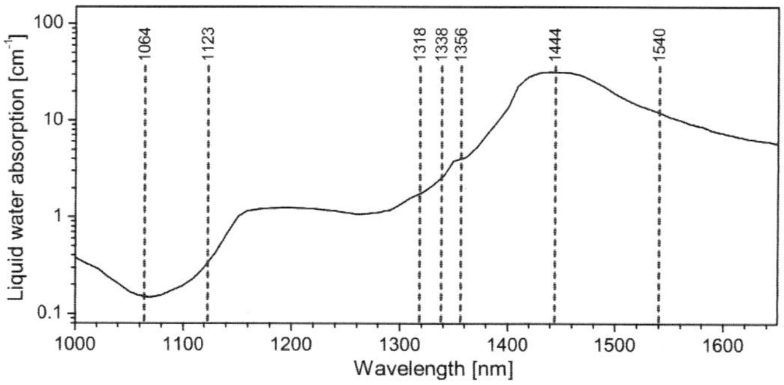

Figure 1: Liquid water absorption [9] versus wavelengths within 1000 – 1650 nm. Main Nd:YAG laser emission lines in this range as well as emission of Er:glass laser at 1.54 μm are marked.

wavelength, radiation in the 1.3 - 1.4 μm spectral range is much more absorbed by water – see Fig. 1. The highest absorption coefficient ($\sim 31\,\mathrm{cm}^{-1}$) of liquid water in this region corresponds to 1.44 μm, which is even higher in comparison with the "eye-save" 1.54 μm radiation. For 1.44 μm radiation, the soft tissue penetration depth is 320 μm. Going to shorter wavelengths down to 1.32 μm, the penetration depth into the soft tissue can change about one order. From our investigation [8] of radiation transmission through the eye structures, a difference has been observed in radiation attenuation for a particular eye tissues and for the measured wavelengths. The knowledge of energy loss distribution, e.g., inside the eye, is very important from the point of internal tissue injury.

The aim of this study was to investigate the diode-pumped Nd:YAG and Nd:YAP laser tuning capabilities in the spectral range of general interest 1.3 – 1.5 μm. The laser emission within this band is very much required for many applications in medicine, atmospheric physics, and spectroscopy due to high absorption of this radiation in liquid water and water vapor. In our experiment, six single emission lines within the desired spectral region have been observed for both Nd:YAG and Nd:YAP active media. Moreover, it was possible to operate both lasers in the dual frequency regime for some line combinations, which makes these lasers interesting for sum and different frequency generation.

2. Materials and methods

2.1. Active medium and pumping source

In our experiment, 1.3 at. % and 1.0 at. % doped Nd:YAG and Nd:YAP active media, respectively, grown by the Czochralski method, were used. The Nd:YAG and Nd:YAP samples $\phi 5\times 5$ mm and $\phi 5\times 8$ mm in dimensions, respectively, were of high quality and free of crack or twins.

As a pumping source, a fibre-coupled (core diameter 400 μm, numerical aperture 0.22) 808 nm laser-diode (HLU20F400, LIMO Laser Systems) was employed. The laser diode was operated in pulsed regime with low duty cycle (pulse length 10 ms, repetition rate 10 Hz, maximum pulse power amplitude 15 W), so it was not necessary to actively cool the laser crystals mounted in a copper heat sink. The diode radiation was focused into the active medium by two achromatic doublets with a focal length of 75 mm, resulting in the spot size of about 360 μm.

2.2. Spectroscopic characteristics

Emission spectra of Nd:YAG and Nd:YAP samples were measured by fiber coupled-spectrometer Ocean-Optics NIR 512 (0.9 – 1.7 nm wavelength range) along the crystal longitudinal axis together with a cut-off filter used for laser-diode excitation radiation at 808 nm. In the case of anisotropic Nd:YAP crystal, the polarization-resolved emission spectra were evaluated. To do this, an oriented film polarizer was placed in front of the spectrometer entrance. The Nd:YAG and Nd:YAP emission spectra within the range of our interest (1300 – 1450 nm) are shown in Fig. 2a) and Fig. 2b), respectively.

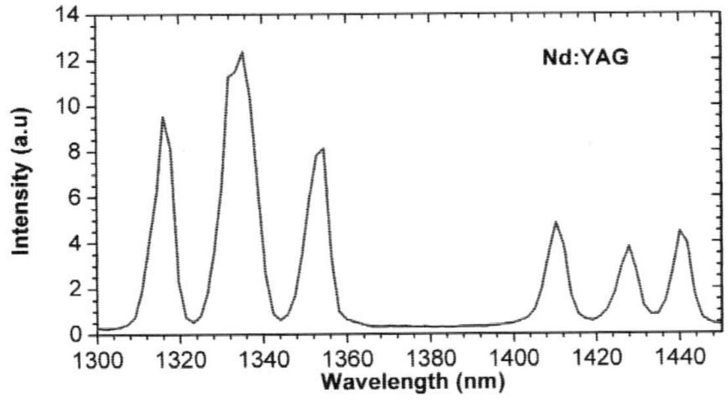

(a)

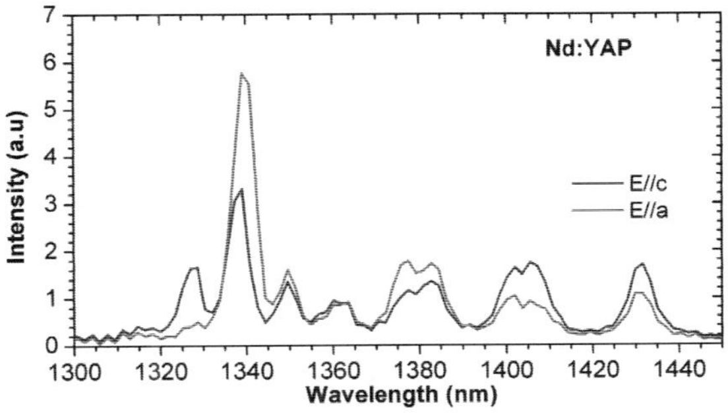

(b)

Figure 2: Nd:YAG emission spectra (a) and polarization-resolved Nd:YAP emission spectra (b) within the 1.30 – 1.45 μm spectral range

2.3. Measuring instruments

Laser output characteristics in terms of mean power, emission spectra, and spatial beam profile were measured by Molectron energy/power meter EMP2000 in connection with PowerMax probe PM3, fiber-coupled Spectrometer Ocean-Optics NIR 512 sensitive in the 0.9 – 1.7 μm spectral range, and Spiricon camera Pyrocam III, respectively.

3. Experimental arrangement and results

The laser experiment set-up is schematically illustrated in Fig. 3. Two semi-hemispherical laser resonators 110 mm long formed by a flat pump mirror and a curved output coupler (curvature radius 150 mm) were realized — the first one for 1.3 μm spectral region ($R_{oc} = 91$ % @ 1.3 μm) and the second one for 1.4 μm ($R_{oc} = 98$ % @ 1.4 μm). The Nd:YAG and Nd:YAP active media mounted on a copper heat sink were situated closest to the pump mirror, i.e., the location of the cavity beam waist. For laser line selection, a single quartz plate 1.5 mm thick was placed inside the optical resonator at Brewster's angle between the output coupler and laser active medium.

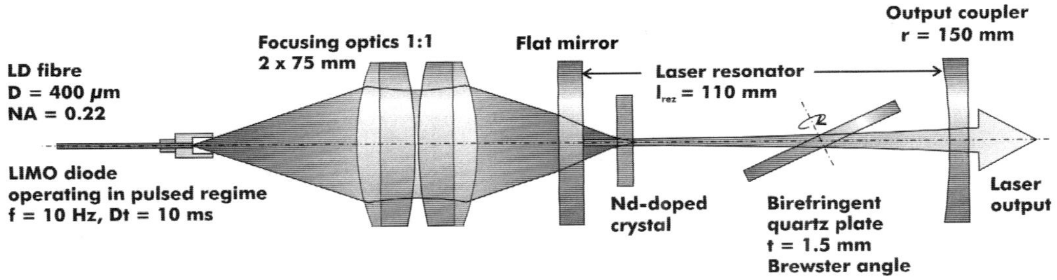

Figure 3: Layout of experimental set-up for switchable Nd-doped-crystal laser-operation within the 1.3 or 1.4 μm spectral bands

In the case of laser systems without a tuning element inside the resonator, the spectral line shapes for Nd:YAG and Nd:YAP active media together with beam profiles are displayed in Fig. 4a) and Fig. 4b), respectively. It is worth mentioning that for the Nd:YAG crystal designed for emission at 1.3 μm wavelength, the laser inherently operated in the dual frequency regime (see Fig. 4a). The corresponding maximal mean powers P_{mean}, oscillation thresholds P_{thr}, and slope efficiencies η_{diff} for the Nd:YAG and Nd:YAP lasers are summarized in Table 1.

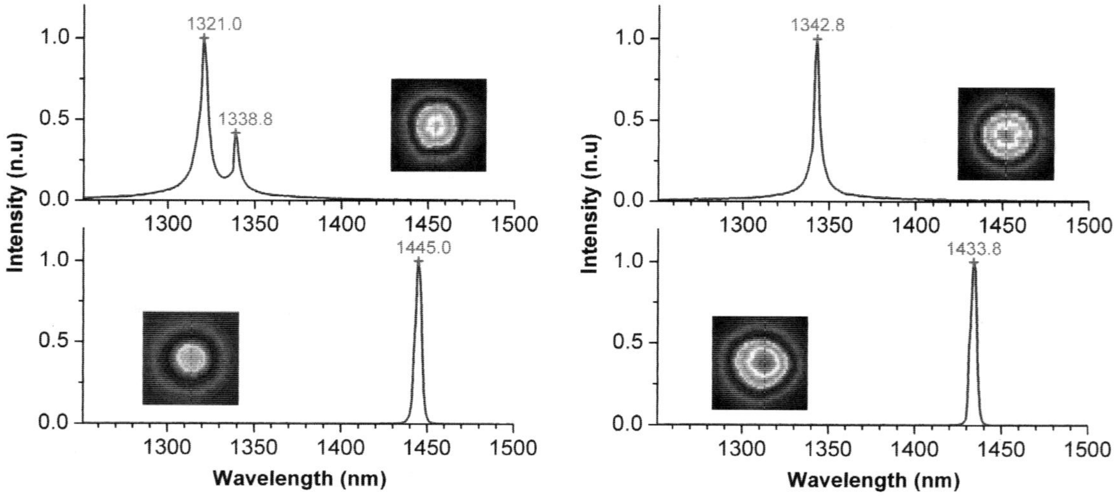

(a) Nd:YAG (b) Nd:YAP

Figure 4: Spectral lines of Nd:YAG (a) and Nd:YAP (b) lasers designed for operation at 1.3 μm and 1.4 μm spectral regions without tuning element inside the resonator. Inset: spatial beam profiles at max. pumping

Table 1: Summary of laser results reached without tuning element inside the resonator

Laser	λ [μm]	P_{mean} [mW]	P_{thr} [mW]	η_{diff}
Nd:YAG	1.3	374	160	0.33
	1.4	167	230	0.22
Nd:YAP	1.3	420	167	0.33
	1.4	121	560	0.14

If the birefringent quartz plate serving as a tuning element was inserted inside the laser resonator, six single emission lines within the spectral range 1.3 – 1.5 μm were reached for both Nd:YAG and Nd:YAP lasers. In addition, for some spectral line combinations, it was possible to operate the lasers in the dual frequency regime. The individual spectral lines obtained for Nd:YAG and Nd:YAP materials are shown in Fig. 5a) and Fig. 5b), respectively.

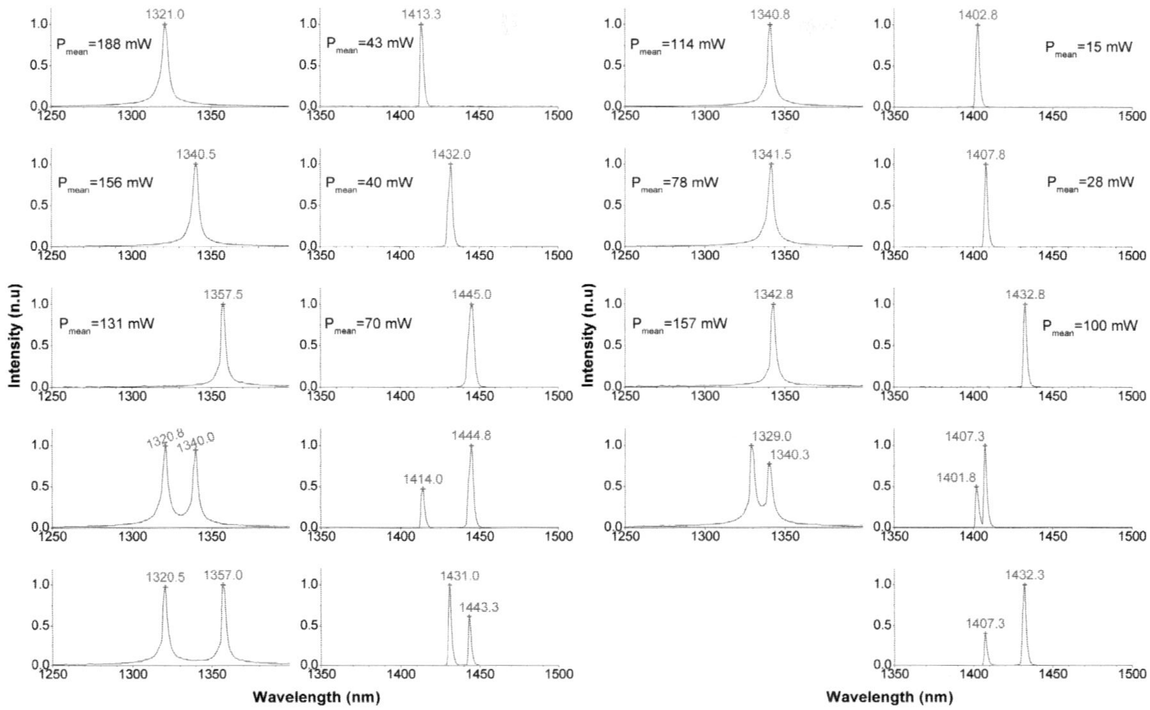

(a) Nd:YAG (b) Nd:YAP

Figure 5: Spectral lines of Nd:YAG (a) and Nd:YAP (b) lasers designed for operation at 1.3 μm and 1.4 μm spectral regions reached by birefringent quartz plate tuning

4. Conclusion

To conclude, multiline generation capabilities of diode-pumped Nd:YAP and Nd:YAG lasers within the range of general interest 1.3 – 1.5 μm have been investigated. Using pulsed 808 nm laser-diode pumping (10Hz, 10 ms, P_{mean} = 1.5 W) together with a single 1.5 mm thick quartz plate placed in the resonator at Brewster's angle, six single emission lines (max. mean power in parentheses) within the desired spectral range were obtained for each laser; specifically – 1321.0 nm (188 mW), 1340.5 nm (156 mW), 1357.5 nm (131 mW), 1413.3 nm (43 mW), 1432.0 nm (40 mW), 1445.0 nm (70 mW) for the Nd:YAG laser, and 1340.8 nm (114 mW), 1341.5 nm (78 mW), 1342.8 nm (157 mW), 1402.8 nm (15 mW), 1407.8 nm (28 mW), 1432.8 nm (100 mW) for the Nd:YAP laser in the same resonator configuration. Moreover, for some spectral line combinations, it was possible to operate both lasers in the dual frequency regime, which makes these lasers interesting for sum and different frequency generation. Further improvement in terms of generation of other laser lines is expected by using a tuning element with a smaller transmission bandwidth. For example, a sequence of birefringent quartz plates and polarizers forming so called Lyot filter, exhibiting a sharper filter function, could be a proper solution.

5. Acknowledgement

We gratefully acknowledge the grant of Technology Agency of the Czech Republic registered under No. TA03011141.

References

[1] Johnson L F and Nassau K 1961 *Proc. of Inst. of Radio Eng.* **49** 1704–6.
[2] Kaminskii A A 1996 *Crystalline lasers: Physical processes and operating schemes* (Florida: CRP Pres, Inc.)
[3] Kaminskii A A 1981 *Laser crystals. Their physics and properties* (Berlin: Springer-Verlag)
[4] Koechner W 1999 *Solid state laser engineering* (Berlin: Springer-Verlag, 5th ed.)
[5] Marling J B 1978 *IEEE J. of Quant. Electron.* **14** 56–62
[6] Moncorgé R, Chambon B, Rivoire J, Gamier N, Descroix E, Laporte P, Guillet H, Roy S, Mareschal J, Pelenc D, Doury J and Farge P 1997 *Opt. Materials* **8** 109–19
[7] Kretschmann H M, Heine F, Ostroumov V G and Huber G 1997 *Opt. Lett.* **22** 466–8
[8] Jelínková H, Pašta J, Šulc J, Němec M and Koranda P 2005 *Laser Phys. Lett.* **2** 603–7
[9] Segelstein D J 1981 *The complex refractive index of water* (Kansas City: Master's thesis, University of Missouri)
[10] Elder I F and Payne M J P 1998 *Opt. Communications* **148** 265–9

Increasing the laser damage threshold of the Nd:YAG crystal by the color center annihilation

S Panahibakhsh[1,2], S Jelvani[1], M Jaberi[3], M Mollabashi[4] and M H Maleki[1]
[1]Laser and Optics Research school, Tehran, Iran
[2]Department of Energy Engineering and Physics, Amirkabir University of Technology, Tehran, Iran
[3]Department of Physics, the University of Guilan, Rasht, Iran
[4]Department of Physics, Iran University of Science and Technology, Tehran, Iran

E-mail: panahi.spb@gmail.com

Abstract. This paper presents increase in laser damage threshold of the Nd:YAG crystal subsequent to annihilation of oxygen vacancies in the crystal structure at low dose of XeCl laser irradiation. A Nd:YAG crystal was irradiated with 500 pulses of XeCl laser (λ=308nm, τ=20 ns) at the fluence of ~100 mJ/cm^2 and 25 Hz repetition rate. The optical absorptions of the Nd:YAG crystal reduced in ultraviolet and visible regions due to elimination of some color centres indicated by UV-Vis-NIR spectra. We measured the laser damage threshold of the irradiated and un-irradiated crystals using 30 ns, 1064 nm single longitudinal mode, TEM_{00} pulses of a passively Q-switched Nd:YAG laser. The laser damage threshold of the irradiated crystal (after dividing by $\sqrt{3}$ to scale for pulse duration of 10 ns) was obtained to be 501±50 mJ/cm^2. It is nearly 5 times as high as that of un-irradiated Nd:YAG crystal, which was found to be 102±15 J/cm^2. The present results are very important for using Nd:YAG crystals as the laser amplifier media.

1. Introduction

Damage is a problem that optical materials are encountered when exposed to laser irradiation. As a result, solid state laser materials and optical components of laser systems are damaged due to the laser irradiation effects. By increasing the application of high peak power lasers, it is necessary to high quality laser components and active media be used in the laser systems.

Experimentally measured laser induced damage threshold (LIDT) of the most of the optical and laser active materials is much less than the intrinsic LIDT of them [1]. The parameters involved in damage phenomena have been studied extensively and some responsible factors have been mentioned. They include inclusions [2], impurities [3] and intrinsic point defects [1].

For high purity good polished laser crystals, point defect factors can be important due to the uncontrolled and unintentional incorporation of some definite point defects during the growth process. The most important point defects of the Nd:YAG crystal structure are oxygen vacancy and its related defects. With growing Nd:YAG crystal in vacuum or argon and nitrogen gas atmosphere, many oxygen vacancy centers are formed in the ordered crystal structure due to the absence of sufficient oxygen ions. Each oxygen vacancy by trapping an electron can be changed to F^+ center. Similarly, each F^+ center defect can capture another electron to form a neutral center named as F center. These defects have several absorption bands in the UV-Vis-NIR spectral region. In general, these defects are named as colour centers due to absorbtion of light in the visible spectrum such that a material that is usually transparent becomes coloured. With having energy states in the band gap, they couple to the laser irradiation, strongly absorb laser energy, enhance laser materials interaction and finally lead to damage of the materials.

In this paper, a Nd:YAG crystal sample was irradiated by low dose XeCl laser irradiation in the ambient air for removing some oxygen vacancies and related defects of the crystal structure. Modification of the defect structure is studied by additional absorption spectrum of the irradiated sample. Then, LIDT of un-irradiated and the irradiated samples are measured by a1064 nm polarized Gaussian single longitudinal mode of a passively Q-switched Nd:YAG laser system and compared with each other. Nearly 5 times higher LIDT of low dose XeCl laser pre-irradiated sample was obtained in the present work that is an important result for using Nd:YAG crystals as an amplifier media.

2. Experiments

A Nd:YAG single crystal (1 at.% Nd^{3+}) was grown in an iridium crucible along the [111] crystallographic direction by the Czochralski method under Ar atmosphere. The crystal was cut into thin discs of 12.5 mm in diameter and 2.4 mm in thickness. The disc samples were polished on both sides ($\leq \lambda/4$@632.8 nm flatness) and cleaned by sulfochromic acid and deionized water at ultrasonic chamber.

A Nd:YAG crystal sample was irradiated by 500 pulses of XeCl excimer laser system (Lambda Physik X200, λ=308nm, τ=20 ns) at the fluence of ~100 mJ/cm^2 and 25 Hz repetition rate. Irradiations were performed perpendicular to the crystal surface in the environmental condition with the 300 mJ laser output energy and 30×10 mm^2 spot size

The optical absorption spectra of the samples were recorded by a (Cary 500) UV-VIS-NIR Spectrophotometer (200-900 nm) before and after the irradiation. Defects in oxide crystals can be studied by optical absorption spectrum. Change of the defect energy state leads to new optical absorption[4].

To determine the type and amount of changes in defect structure, the value of laser induced additional absorption (AA) is calculated by [5]:

$$\Delta k = \ln(T_1 / T_2) / d \qquad (1)$$

Where k is the absorption coefficient, d is the sample thickness; T_1 and T_2 are the crystal transmission before and after the laser irradiation, respectively.

After XeCl laser irradiation, effect of the irradiation on the LIDT of the Nd:YAG crystal sample was examined experimentally. The laser induced damage threshold experimental setup is illustrated schematically in Figure 1. Laser system delivered a passively Q-switched single longitudinal mode, TEM_{00} Gaussian-shaped laser beam at a wavelength of 1064 nm with a pulse width of 30 ns FWHM. The beam profile was nearly perfect TEM_{00}, and the polarization was linear. The experiments were performed according to the ''1-on-1'' standard testing method for single shot damage measurement in laboratory environmental conditions. Then, each site received only one pulse, for each of which the pulse width and pulse energy were monitored.

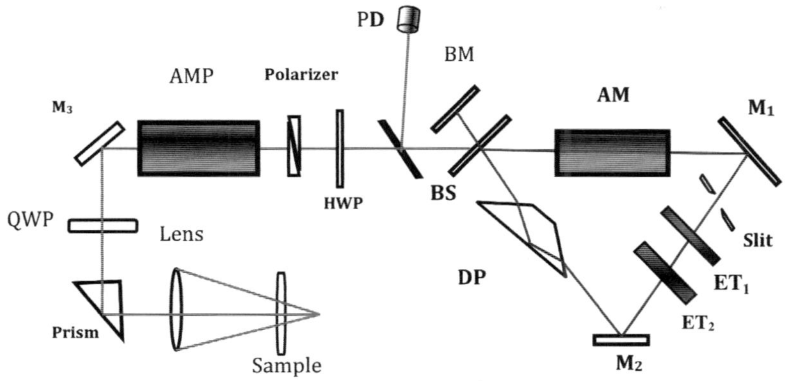

Figure 1. The experimental setup of 1064 nm Nd:YAG laser induced damage threshold measurement, where M_i are mirrors, ET_i are etalons, PD is a photodiode, QWP is a quarter wave plate, HWP is a half wave plate, BS is a beam splitter, DP is a Dove prism, BM is a back mirror, AM is the active medium and AMP is the amplifier.

Damage was identified by visually observing a laser spark and scattering of a coaxial He–Ne laser probe beam focused on the irradiated site that is collinear with the path of the 1064 nm beam. Onset of the damage was clearly visible due to the scattering of the probe beam. Occurrence of damage was confirmed afterward by observing the irradiated sites under an optical microscope.
Increasing the lens focal length reduces statistical variability and allows more accurate determination of the LIDT of materials. Thus, we used a 400 mm focal length lens with antireflection coatings on the both surfaces. The spot size of the beam incident on the sample was 90±10 μm diameters at $1/e^2$ of the maximum intensity. The sample front face was placed on 12 mm behind the focal point of the lens. Damage locations were laterally separated by about 1 mm.

3. Results and discussion

Figure 2 shows the UV-Vis optical transmission spectra of the Nd:YAG sample before and after the irradiation. As seen, the optical transmission has increased after XeCl laser irradiation. It has been shown that increase and decrease of optical absorption in the entire spectral range are the characteristic features of generation and annihilation of point defects in the materials, respectively [6]. Thus, it seems some point defects can be annihilated during the laser irradiation.

To determine the type and value of XeCl laser induced annihilated defects, additional absorption (AA) spectrum of the Nd:YAG sample was obtained. Figure 3 shows the AA spectrum in which a broad negative AA band at 236 nm is observed. There are also some tiny negative bands that are corresponded to Nd^{3+} absorption bands. It is known that oxygen vacancy is the main defect of the Nd:YAG crystal structure[7], so it is proposed that the 236 nm big negative band can be related to the annihilation of many oxygen vacancy centers.

On the other hand, it is known that annealing at high temperatures improves optical features of oxide crystals by removing some oxygen vacancy centers[8]. Thus, in order to find the oxygen vacancy related absorption band we annealed another Nd:YAG sample in the air for 3 hours at 400°C for thermal relaxation and then 3 hours at 1100°C for changing the defect structure. The optical spectrum was recorded before and after annealing and then the AA spectrum of the annealed sample was obtained.

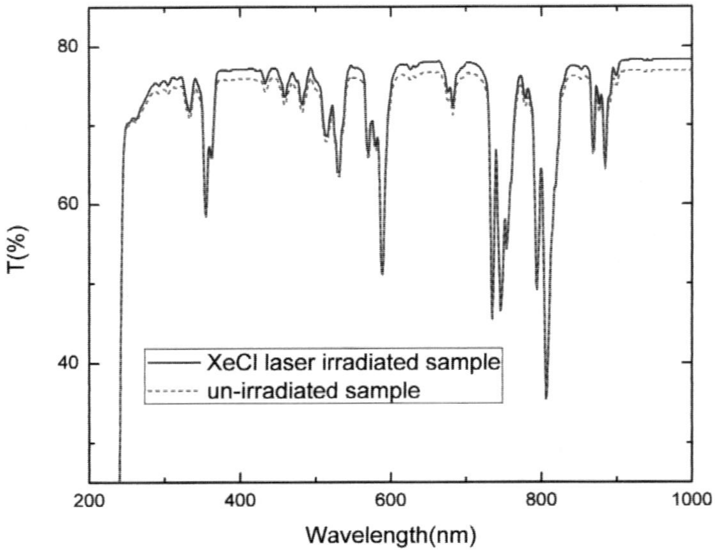

Figure 2. Transmission spectra of the irradiated (100 mJ/cm^2, 500 pulses, 25Hz) and un-irradiated Nd:YAG crystal sample.

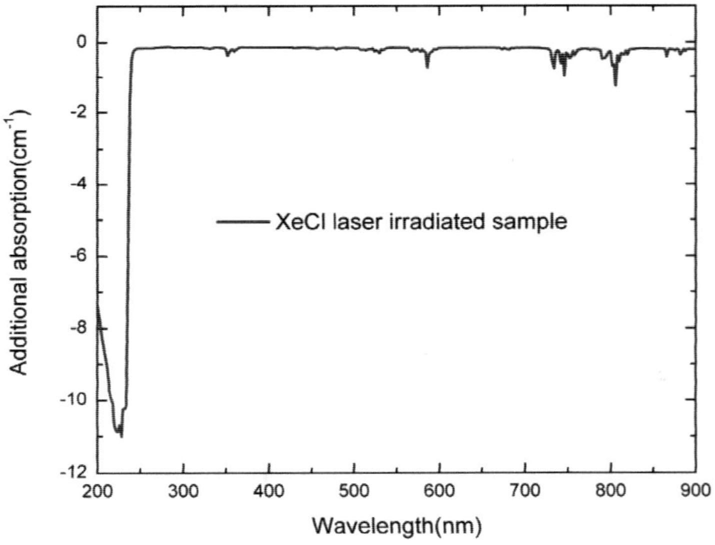

Figure 3. Additional absorption spectrum of the XeCl laser irradiated Nd:YAG sample (100 mJ/cm^2, 500 pulses, 25 Hz).

Figure 4 shows the AA absorption spectrum of the annealed Nd:YAG crystal with the broad absorption band centred at 256 nm, which has been observed in air annealed Nd:YAG and YAG crystals[4]. It has been attributed to charge transfer from oxygen ion to Fe^{3+} ion. Generally, some uncontrolled Fe^{3+} and Fe^{2+} ions often exist as impurities in Nd:YAG crystals. XRF analysis also detects about 400 ppm of iron ions in the Nd:YAG crystal samples which were used in the experiments. Some oxygen vacancies can be filled by oxidizing Fe^{2+} impurity during annealing in the air and then the number of Fe^{3+} ions localized near the oxygen ion will increase.

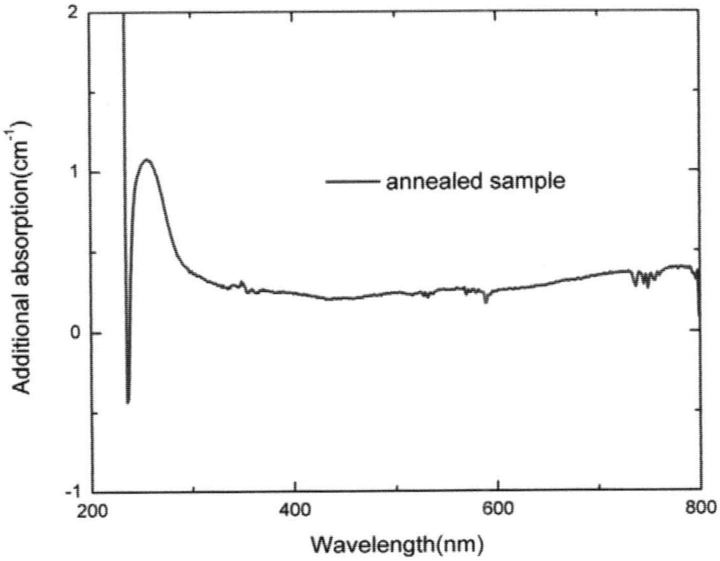

Figure 4. Additional absorption spectrum of the annealed sample (for 3 hours at 400°C and then 3 hours at1100°C).

In addition, Figure 4 shows a sharp negative absorption band at 236 nm that has not been reported earlier to the best of our knowledge. As mentioned above, annealing removes some oxygen vacancies in the Nd:YAG crystals and annihilation of the point defects creates negative bands in additional absorption spectrum. Then, there should be a negative band related to the annihilation of the oxygen vacancies in the additional absorption spectrum of the annealed sample, but it is observed no negative band in the additional absorption spectrum except the 236 nm negative absorption band. Thus, we can conclude that the 236 nm negative band is just related to the annihilation of the oxygen vacancy center.

Therefore, the 236 nm absorption band in additional absorption spectrum of the XeCl laser irradiated Nd:YAG sample is also related to the annihilation of the oxygen vacancy centers in Nd:YAG crystals.

To study the effect of the oxygen vacancy annihilation in LIDT of the Nd:YAG crystal, We measured the LIDT of the XeCl laser irradiated and un-irradiated Nd:YAG crystals at different lateral positions by gradually increasing the pulse energy at each location until optical breakdown occurred. Experiments were repeated several times and the results were averaged. The threshold fluence for un-irradiated crystal after dividing by $\sqrt{3}$ to scale for pulse duration of 10 ns was found to be 102±15 J/cm^2.

The LIDT of XeCl laser irradiated Nd:YAG crystal scaled for pulse duration was obtained to be 501±50 mJ/cm^2 that is nearly 5 times as high as that of un-irradiated Nd:YAG crystal.

According to the present results, we can conclude that the oxygen vacany annihilation increases the LIDT of the Nd:YAG crystal. Therefore, the oxygen vacancy is an important defect responsible for LIDT phenomenon in the Nd:YAG crystal. Indeed, oxygen vacancies can act as electron traps. When an oxygen vacancy captures one electron, it forms F^+ center. Also, F^+ center can capture another electron and change to natural vacancy named F center. These defects, which are generally called colour centers, have several absorption bands in the spectral range of UV-Vis-NIR spectral range.

Journal of Physics: Conference Series **497** (2014) 012012 doi:10.1088/1742-6596/497/1/012012

In high power laser interaction with Nd:YAG crystal, multi-photon absorption can ionize or excite electron to the conduction band, then many O^- hole centers produce and any trapping centers such as oxygen vacancy capture free electrons. Consequently, the defect structure in the crystal gradually changes and new defects appear with the new optical absorption bands. In this condition, the effective energy absorption of the crystal and also the coupling between the laser irradiation and the crystal increase till the breakdown happens in the crystal. If no trapping center such as oxygen vacancy exists in the crystal structure, the generated electrons and hole centers can recombine together and no new defects that can lead to damage would be generated in the crystal structure. Therefore, oxygen vacancies of Nd:YAG crystal structure help to onset of laser damage in the Nd:YAG crystal, so annihilation of the oxygen vacancies by low intensity XeCl laser pre-irradiation can increase the LIDT of the Nd:YAG crystal.

4. Conclusions

In conclusion, we presented five times higher laser induced damage threshold following pre-exposure to 100 mJ/cm^2 sub-threshold XeCl laser irradiation due to annihilation of the oxygen vacancy centers. The threshold fluences for un-irradiated and XeCl laser irradiated crystal after dividing by $\sqrt{3}$ to scale for pulse duration of 10 ns were obtained 102±15 and 501±50 J/cm^2, respectively.

We demonstrated the important role of oxygen vacancy centers, the most important Nd:YAG crystal defect, on laser induced damage of the Nd:YAG crystal for the first time to the best of our knowledge.

References

[1] Hu G, Zhao Y, Li D and Xiao Q 2012 *Opt. express* **20** 25169-80
[2] Ling X 2011 *Appl. Surf. Sci* **257** 5601-4
[3] Tan Q 2000 *J. cryst. growth* **209** 861-6
[4] Chen J, Lu T, Xu Y, Xu A and Chen D 2008 *J. Phy: Condens. Matter* **20** 325212
[5] Kaczmarek S, Domianiak-Dzik G, Ryba-Romanowski W, Kisielewski J and Wojtkowska J 1999 *Cryst. Res. Technol* **34** 1031-6
[6] Varney C, Mackay D, Reda S and Selim F 2012 *J. Phys. D: Appl. Phys* **45** 015103
[7] Akhmadullin I S, Migachev S and Mironov S 1992 *Nucl. Instrum. Meth B* **65** 270-4
[8] Kaczmarek S 1999 *Cryst. Res. Technol* **34** 1183-90

Thermal lens measurements in Yb-doped YAG, LuAG, Lu_2O_3, Sc_2O_3 ceramic lasers

A Pirri[1], G Toci[2,*], M. Ciofini[3], A. Lapucci[3], L. A. Gizzi[4], L. Labate[4], L. Esposito[5], J. Hostaša[5] and M. Vannini[2]

[1]Istituto di Fisica Applicata "N. Carrara", IFAC-CNR, Via Madonna del Piano 10 C, 50019 Sesto Fiorentino (FI), Italy.

[2]Istituto Nazionale di Ottica, INO-CNR, Via Madonna del Piano 10 C, 50019 Sesto Fiorentino (FI), Italy.

[3]Istituto Nazionale di Ottica, INO-CNR, Largo Enrico Fermi 6, 50125 Firenze, Italy.

[4]Istituto Nazionale di Ottica, INO-CNR, Via G. Moruzzi 1, 56124 Pisa, Italy.

[5]Istituto di Scienze e Tecnologie delle Ceramiche, ISTEC-CNR, Via Granarolo 64, 48018 Faenza, Italy.

E-mail: guido.toci@ino.it

Abstract: we report the last experimental results obtained measuring the thermal lens effect in Yb(1*at.%*):Lu_2O_3, Yb(1*at.%*):Sc_3O_3, Yb(10*at.%*):LuAG and Yb(10*at.%*):YAG ceramics. The experimental set-up apparatus using a Shack-Hartmann wavefront sensor allows the comparison of the thermal lens obtained under lasing and non lasing conditions, in order to highlight differences in the thermal loading.

1. Introduction

Diode pumped solid state lasers based on Yb-doped YAG [1-2] and LuAG ceramic [3-4] have shown some of the most promising results in the last few years, ranging from CW powers in the kW range to Kerr-lens mode-locked operation with fs pulses [2, 5-7].The use of materials having an appropriate thermal behaviour is extremely important because it can significantly improve the laser performance avoiding the possible decrease of the output power and the deterioration of the beam quality. The ceramic technology is playing a special role enabling the production of two important sesquioxides such as Lu_2O_3 [8-11] and Sc_2O_3 [12-14]. Basically, the thermal conductivity decreases with the doping level due to the different atomic weight of Yb^{3+} and the substituted host cation. LuAG and Lu_2O_3 play a special role because of their initial high thermal conductivity (8.0 W/mK [15] and 12.2 W/mK respectively [16]) is found to be insensitive to doping. Among the effects determining the thermo-mechanical behaviour such as the thermal lens, the re-absorption as well as the thermal conductibility,

* To whom any correspondence should be addressed

Journal of Physics: Conference Series **497** (2014) 012013 doi:10.1088/1742-6596/497/1/012013

the former is the most difficult to manage because it can dramatically change the resonator stability conditions and the diffraction losses.

In this work, we report the last experimental results obtained measuring the thermal lens effect in Yb-doped YAG, LuAG, Lu_2O_3, Sc_2O_3 ceramics. It was done by using an experimental set-up apparatus based on a Shack-Hartmann wavefront sensor, which allows the comparison of the thermal lens obtained in lasing and non lasing conditions in order to point out the differences in the thermal load.

2 Experimental apparatus

The experimental apparatus was designed to allow the measurement of the thermal lens in lasing and nonlasing samples, under longitudinal pumping. For this purpose the samples are inserted in a longitudinally pumped laser cavity, and a probe beam is used to measure the thermal lens effect. The pump source is a semiconductor laser emitting at 936 nm. The output of the laser is coupled in an optical fiber (200 μm diameter, 0.22 NA) whose output is refocused and magnified by a pair of achromatic doublet on a waist with 500 μm diameter @$1/e^2$ located in the center of the sample. A maximum incident pump power of 21.8 W CW was used in the experiments.

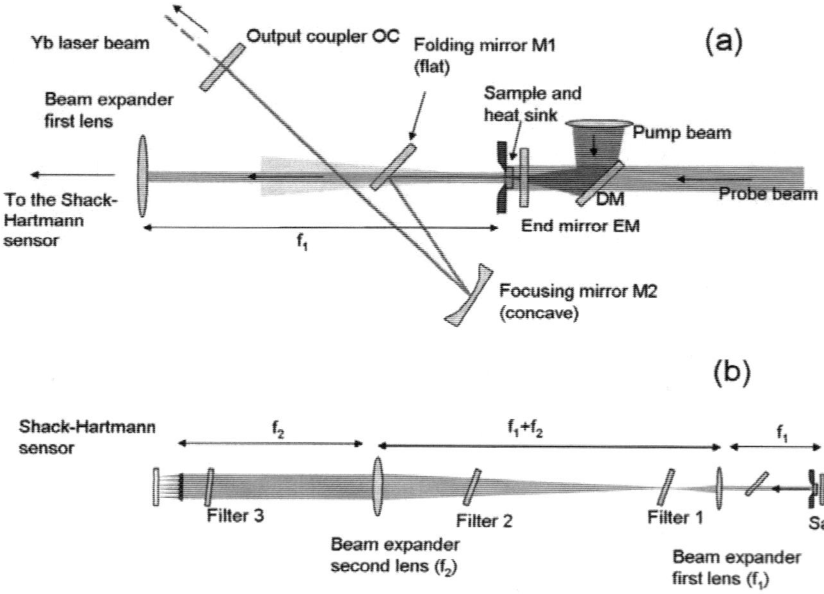

Figure 1: Experimental set-up for the measurement of the wavefront thermal distortion and of the thermal lens dioptric power. (a): Pump source: HeNe laser; Pump laser: CW fibre-coupled diode laser at 936 nm; DM: dichroic mirror; EM: End cavity mirror; M1: flat mirror; M2:spherical mirror (focal length 100 mm); OC: output coupler mirror. (b): Schematics of the probe beam path between the sample and the sensor.

The pump beam is injected into the laser cavity by the dichroic mirror DM and then through the end mirror EM (dichroic, high transmission at the pump wavelength, high reflection at the laser wavelength). The cavity is folded with the flat mirror M1 and with the spherical mirror M2 (focal length 100 mm), and it is closed by the output coupler OC. The overall distance between M2 and EM is 113 mm, and the distance between M2 and OC is 210 mm. The fundamental cavity mode radius on the waist located at the mirror EM was calculated as 100 μm for the "cold" cavity. The samples are brazed with Indium on a copper heat sink with a central circular aperture having a diameter of 2.5 mm, on the opposite side with respect to the incidence of the pump beam.

The thermal lens is probed by a beam emitted by a HeNe laser, see Fig. 1. The beam is expanded and collimated, sent into the sample collinearly with the pump beam through the mirrors DM and EM, and extracted from the cavity through the mirror M1. The probe beam illuminates the whole clear aperture of the sample holder. The wavefront distortion induced by thermal effects is measured by means of a Shack-Hartmann wavefront sensor (ML4010 by Metrolux GmbH, area 8.98 x 6.71 mm^2 and 200 μm pitch of the microlenses array). In order to measure the wavefront distortion occurring over an area of about 500 μm diameter, the probe beam was expanded by an afocal telescope formed by the two positive lenses with nominal focal lengths $f1$=155 mm and $f2$=750 mm placed at a distance $d=f1+f2$.

The sample and the sensor are placed in the conjugate planes of the telescope. The probe beam at the output of the sample is then reimaged on the sensor with a magnification $M=f2/f1$, without any further wavefront curvature introduced by the telescope. The magnification of the telescope was accurately measured and it resulted M=4.81. Several filters are used to reject the residual pump beam that would otherwise impinge on the sensor. The thermal deformation of the wavefront is measured at several levels of incident pump power. A circular portion of the wavefront with radius a corresponding to the pumped area of the sample is then selected and analyzed by fitting with Zernike polynomials up to the 6th order. The effective dioptric power D_{th} of the thermal lens is then calculated with the formula:

$$D_{th} = \frac{1}{f_{th}} = \frac{a^2}{4Z_{2,0}} M^2 \tag{1}$$

where $Z_{2,0}$ is the coefficient of the Zernike polynomial with radial index 2 and azimuthal index 0, corresponding to the wavefront defocus. In this analysis, we adopted a value of a=1.2 mm (on the sensor plane), corresponding to the pump beam radius at $1/e^2$ (*i.e.* 250 μm) on the sample plane. Due to the different absorption featured by the crystals, the dioptric power of the thermal lens was analyzed as a function of the absorbed pump power.

3. Experimental results and comments

Figures 2 (a) and (b) report the thermal lens results obtained with the Yb(10*at.* %):YAG and Yb(10*at.* %):LuAG respectively. Figures 3 (a) and 3 (b) show the measurements obtained with the Yb(1*at.* %):Sc_2O_3 and Yb(1*at.* %):Lu_2O_3 respectively. The lasing condition was obtained by an OC with a low transmission of T=2% in order to maximize the intracavity circulating power and, in turn, the difference in thermal load between lasing and non-lasing conditions. In non lasing conditions the dioptric power D of the thermal lens is expected to scale linearly with the absorbed pump power P_{abs}, according to the following relation [18]:

$$D_{off} = \Delta_{off} P_{abs} = A_{off}\left(1 - \eta_r \frac{\lambda_P}{\lambda_F}\right) P_{abs} \tag{2}$$

where P_{abs} is the absorbed pump power. The quantity in brackets express the fraction of the absorbed pump power that is converted in heat and it is influenced by the radiative quantum efficiency η_r; λ_P and λ_F are the pump wavelength and the average fluorescence wavelength respectively. The constant A_{off} incorporates the thermo-optical and the thermal properties of the material, as well as the geometric conditions [18].

In lasing conditions the overall probability of nonradiative decays is reduced by the stimulated emission probability that becomes dominant when the laser is pumped well above the threshold. In this latter condition the expression for the thermal lens can be modified as it follows:

$$D_{on} = \Delta_{on} P_{abs} = A_{on}\left(1 - \frac{\lambda_P}{\lambda_L}\right) P_{abs} \tag{3}$$

where λ_L is the lasing wavelength. It must be noticed that for the materials here considered is usually $\lambda_L > \lambda_F$; therefore, when the radiative quantum efficiency $\eta_r = 1$ the thermal load is higher under lasing than under non lasing conditions.

The value of the constant A_{on} is not necessarily the same as A_{off}, because the heat sources distribution in lasing conditions can be different from nonlasing conditions. Nevertheless if the laser is well above threshold we can make the approximation that the whole pumped region is lasing, so that the heat sources distribution becomes the same in the two cases and $A_{on} = A_{off}$.

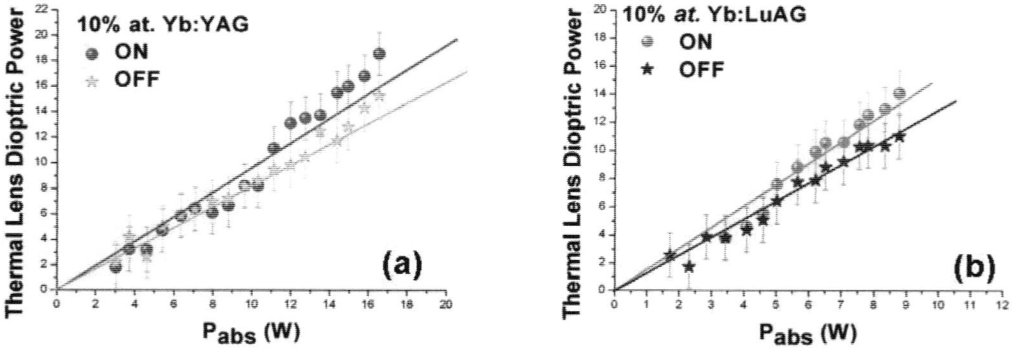

Figure 2. Thermal lens dioptric power (m^{-1}) in YAG (a) and in LuAG (b) as a function of the absorbed pump power. The solid curves are the best linear fits. ON and OFF stand for laser action on or off.

The thermal lens in the YAG and LuAG samples shows the expected linear dependence from the absorbed pump power, see Fig. 2 (a) and 2 (b); the slope in lasing and nonlasing conditions is slightly different for the reasons exposed above.

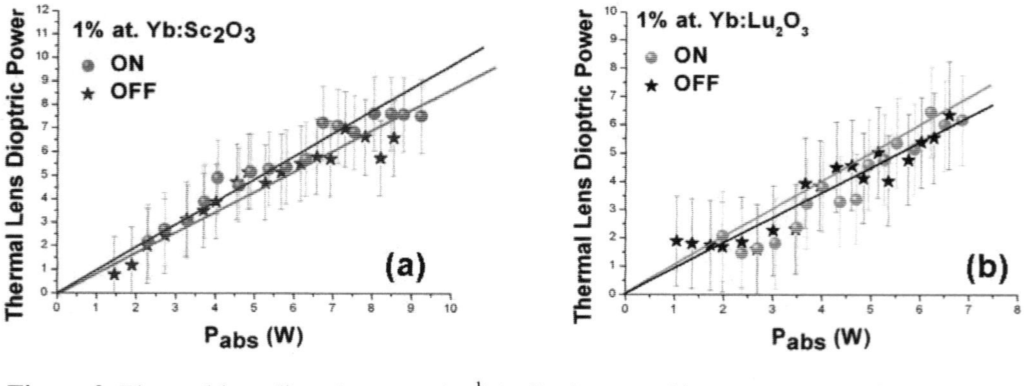

Figure 3. Thermal lens dioptric power (m^{-1}) in Sc_2O_3 (a) and in Lu_2O_3 (b) as a function of the absorbed pump power. The solid curves are the best linear fits. ON and OFF stand for laser action on or off.

From the analysis of the measurements shown in Figure 2 it appears that Yb:LuAG has a slightly smaller thermal lens effect than Yb:YAG: indeed, the slope of the thermal lens dioptic power with respect to the absorbed pump power is 0.77 $(m\ W)^{-1}$ for LuAG and 0.82 $(m\ W)^{-1}$ for YAG. In both cases the slope of the thermal lens dioptric in laser on conditions is higher than in the case of laser off. This indicates that the quantum efficiency of these two material is very close to unity (see Eq. 2).

Regarding the sesquioxides Yb:Sc_2O_3 and Yb:Lu_2O_3 (Figures 3(a) and 3(b)) the slopes of the thermal lens with respect to the absorbed pump power are respectively 1.0 and 1.1 $(m\ W)^{-1}$ (in laser off conditions), so Lu_2O_3 has an overall thermal lens effect slightly larger than Sc_2O_3. Moreover, the slopes in the laser on and laser off conditions are very similar, indicating that the quantum efficiency is slightly lower than in the case of the LuAG and YAG.

4. Conclusions

In conclusion, we have measured the thermal lens in several promising laser hosts doped with Yb, that is ceramic YAG, LuAG, and ceramic sesquioxides Sc_2O_3 and Lu_2O_3.

These measurements, carried out in uniform experimental conditions, allow to make a direct comparison among the different materials, showing that LuAG has a small advantage in terms of thermal lens. Moreover, the behavior of the thermal lens in laser on and laser off conditions can be analyzed in order to determine the quantum efficiency of the materials. This analysis will be the subject of forthcoming papers [18].

Acknowledgments

The research was supported by Regione Toscana, project "CTOTUS - Progetto integrato per lo sviluppo della Capacità Tecnologica e Operativa della Toscana per l'Utilizzo dello Spazio" (POR FESR 2007-2013 Attività 1.1 Linea d'intervento D); by the Consiglio Nazionale delle Ricerche, CNR-RSTL "Ricerca Spontanea a Tema libero", id. 959 and by the joint project by the joint project of ASCR and CNR and Czech GA AV project M100100910, 2012-2015.

References

[1] Takaichi K, Yagi H, Lu J, Shirakawa A, Ueda K, Yanagitani T, Kaminskii A A 2003 Phys. Stat. Sol. A **200** R5
[2] Ikesue A, Kinoshita T, Kamata K, Yoshida K, 1995 J. Am. Ceram. Soc. **78**, 1033
[3] Pirri A, Vannini M, Babin V, Nikl M, Toci G, 2013 Laser Phys. **23** 095002
[4] Luo D, Zhang J, Xu C, Yang H, Lin H, Zhu H, Tang D, 2012 Opt. Mat. Express **2** 1425
[5] Dong J, Ueda K, Shirakawa A, Yagi H, Yanagitani T, Kaminskii A A, 2007 Opt. Express **15** 14516
[6] Yoshioka H, Nakamura S, Ogawa T, Wada S, Yoshioka H, 2009 Opt. Express **17** 8919
[7] Griebner U, Petrov V, Petermann K, Peters V, 2004 Opt. Express **12** 3125
[8] Lu J, Takaichi K, Uematsu T, Shirakawa A, Musha M, Ueda K, Yagi H, Yanagitani T, Kaminskii A A, 2002 Appl. Phys. Lett. **81** 4324
[9] Tokurakawa M, Takaichi K, Shirakawa A, Ueda K, Yagi H, Hosokawa S, Yanagitani T, Kaminskii A A, 2006 Opt. Express **14** 12832
[10] Kaminskii A A, Bagayev S N, Ueda K, Takaichi K, Shirakawa A, Ivanov S N, Khazanov E N, Taranov A V, Yagi H, Yanagitani T, 2006 Laser Phys. Lett. **3** 375
[11] Kaminskii A A, Akchurin M Sh, Becker P, Ueda K, Bohatý L, Shirakawa A, Tokurakawa M, Takaichi K, Yagi H, Dong J, Yanagitani T, 2008 Laser Phys. Lett. **5** 300
[12] Sanghera J, Frantz J, Kim W, Villalobos G, Baker C, Shaw B, Sadowski B, Hunt M, Miklos F, Lutz A, Aggarwal I, 2011 Opt. Lett. **36** 576
[13] Pirri A, Toci G, Vannini M, 2011 Opt. Lett. 36 4284
[14] Lupei V, Lupei A, Ikesue A, 2005 Appl. Phys. Lett. **86** 111118
[15] Nakao H, Shirakawa A, Ueda K I, Yagi H, Yanagitani T, 2012 Opt. Express **20** 15385
[16] Petermann K, Fornasiero L, Mix E, Peters, V, 2002 Opt. Mat. **19** 67
[17] Chénais S, Balembois F, Druon F, Lucas-Leclin G, Georges P, 2004 IEEE J. Quantum Electron. **40** 1217
[18] Pirri A, Toci G, Nikl M, Babin V, Vannini M, 2013 Opt. Express, submitted Nov. 12, 2013

Molecular Attosecond Photoionization with Few Cycle XUV Laser Pulses

Kai-Jun Yuan and André D Bandrauk

Laboratoire de Chimie Theorique, Faculte des Sciences, Universite de Sherbrooke, Sherbrooke, Quebec, Canada, J1K 2R1

E-mail: andre.bandrauk@usherbrooke.ca

Abstract. We present molecular attosecond ionization with few cycle UV laser pulses from numerical solutions of time dependent Schrodinger equations. Simulations performed on aligned H_2^+ exhibit signature of red-shifts in photoelectron energy spectra. This is shown to be critically sensitive to the the pulse duration and wavelength and is attributed to the broad spectral width of the ionizing pulses and diminishing electronic Franck-Condon factors with short pulses. We analyze the laser parameter dependence of the energy spectra by a perturbative model.

1. Introduction

Investigation of electron dynamics in atoms, molecules and solids on its nature time scale of the attosecond (1 asec=10^{-18} s) nowadays has become a topical subject with the rapid developments of the generation and characterization of such ultrashort laser pulses [1, 2, 3]. In particular, the generation of attosecond laser pulses has led to imaging pump-probe techniques from laser Coulomb explosion imaging [4] of nuclear motion to laser induced electron diffraction (IED) [5, 6] of coupled electron-nuclear motion. To date the shortest single pulse width of 67 asec duration has been achieved from high-order harmonic generation (HHG) [7], thus offering a new tool to watch pure electronic quantum effects without interference from nuclear motion [8], and allowing in particular through attosecond imaging the creation of electron movies [9]. Attosecond pulses therefore allow one to separate electronic and nuclear effects due to the different time scales of attosecond for electron motion and femtosecond for nuclear motion.

Using extreme ultraviolet (XUV) light from high-order harmonics (HHG) [10] or free electron lasers (FE) [11], coherent electron wave packets (CEWPs) can be created inside atoms and molecules on the attosecond time scale and sub-nanometer dimension [12, 13, 14, 15]. Thus electron dynamics in CEWPs can be monitored by measuring ultrafast photoelectron angular and energy distributions. Many studies have focused on determining atomic and molecular orbital configurations [16, 17], multi-pathway quantum interference [18, 19, 20, 21, 22] and electron correlations [23, 24, 25] by XUV source or combined with ultrashort intense infrared (IR) laser pulses. An alternative efficient tool has been recently proposed to observe attosecond electron motion by measuring the generated radiation, molecular high-order harmonic generation (MHOHG), as a function of the pump-probe delay in a dissociating molecule [26]. XUV-pump-probe experiments have been adopted in tracing atomic coherences [27] and molecular nuclear wave packet dynamics [28]. Few cycle attosecond XUV pulses have recently been used also to control electronic motion [29, 30]. It has been shown that the symmetry of electron wave packet momentum and energy spectra is critically sensitive to the pulse carrier-envelope phase (CEP) and intensities [29]. Such sensitivity in the vicinity of two-electron doubly excited (autoionizing) states has also been examined [30]. We also have shown that intense attosecond magnetic field pulses and circular electronic currents can be induced [31] by means of few cycle circularly polarized attosecond XUV pulses which can be produced from circularly polarized MHOHG [32, 33].

In the present report, we investigate ultrafast molecular photoionization distributions in intense few cycle laser pulses from numerical solutions of the three dimensional (3D) time dependent Schrödinger equation (TDSE) for aligned H_2^+. It is found that an energy red shift occurs in the molecular photoelectron energy spectra relative to the classical above threshold ionization (ATI) kinetic energy $n\hbar\omega - I_p$, where I_p is the molecular ionization potential. The shift is shown to be sensitive to the pulse duration and is a function of ATI photoelectron kinetic energies. We adopt a perturbative theory model to describe such sensitivity. A fixed nuclei approach (Born-Oppenheimer approximation) is appropriate for the simulation due to the longer femtosecond (fs) time scale of nuclear motions. Throughout this paper, atomic units (a.u.) $e = \hbar = m_e = 1$ are used unless otherwise stated.

The paper is arranged as follows: In section 2, we briefly describe the computational methods for time-dependent quantum electron wave packet calculations from the corresponding TDSEs. Numerical results for intense ultrashort linearly polarized XUV laser pulses interacting with an aligned H_2^+ are shown in section 3. Effects of pulse duration and wavelength and photoelectron kinetic energies are reported. Finally, we summarize our findings in section 4.

2. Numerical methods

We here briefly describe the method for numerically solving the TDSE for an aligned molecular ion H_2^+ interacting with a linearly polarized laser pulse. The corresponding 3D TDSE with static nuclei in cylindrical coordinates (ρ, θ, z) is expressed as

$$i\frac{\partial}{\partial t}\Psi(\mathbf{t}, t) = H(\mathbf{r}, t)\Psi(\mathbf{r}, t), \tag{1}$$

where the field-mater Hamiltonian $H(\mathbf{r}, t) = T(\mathbf{r}) + V(\mathbf{r}) + \mathbf{r} \cdot \mathbf{E}(t)$, $\mathbf{r} = (\rho, \theta, z)$ with $x = \rho\cos\theta$ and $y = \rho\sin\theta$. (x, y) defines the molecular plane. The molecular kinetic energy (aplacian) term is given by

$$T(\rho, \theta, z) \quad = \quad -\frac{1}{2\rho}\frac{\partial}{\partial\rho}\left(\rho\frac{\partial}{\partial\rho}\right) - \frac{1}{2\rho^2}\frac{\partial^2}{\partial\theta^2} - \frac{1}{2}\frac{\partial^2}{\partial z^2}. \tag{2}$$

$V(\mathbf{r})$ is the two center Coulomb repulsion for protons separated by internuclear distance R. The field-molecule interaction is treated in the length gauge with a laser pulse of the form

$$E(t) = E_0\sin^2(\pi t/T)\cos(\omega t + \phi), \tag{3}$$

propagating along the z axis with polarization in the molecular (x, y) plane. $T = n\tau$ is the pulse duration where one optical cycle $1\tau = 2\pi/\omega$ and n is the number of cycles.

We solve numerically the 3D H_2^+ TDSE by a five-point finite difference method and fast Fourier transform technique combined with high-order split-operator methods [34, 35]. The time step is fixed at $\ t = 0.01$ a.u. (1 a.u.=24 asec) and the spatial discretization is $\ \rho = \ z = 0.25$ a.u. (1 a.u.=1 a_0, Bohr radius), ensuring a maximin momentum spread $\ p = \pi/\ \rho = 4\pi$ a.u., for radial grid sizes $0 \leq \rho \leq 128$ a.u., $|z| \leq 64$ a.u., and the angle grid size $\ \theta = 0.01$ radian. To prevent unphysical effects de to the reflection of the wave packet from the boundary, a "mask function" with the form $\cos^{1/8}[\pi(\rho - \rho_{\mathrm{a}})/2\rho_{\mathrm{abs}}]$ [17] is used at $\rho_{\mathrm{abs}} = \rho_{\mathrm{max}} - \rho_{\mathrm{a}} = 24$ a.u. with $\rho_{\mathrm{max}} = 128$ a.u..

For an intense laser pulse, a quite large grid range must be used to obtain via Fourier transform the high energy of the ejected electron in the ionization spectra. In the present work we use an efficient method by calculating a radial flux (electron current density) $S_e(t)$ to describe the ionization spectra, where high kinetic energy of ionized electron can be accurately calculated. The electron wave function $\psi(\rho, \theta, t)$ generates the radial flux $S_e(t)$ at an asymptotic point $\rho_{\mathrm{f}} = 100$ a.u. along the radial direction in the molecular (x, y) plane before the wave packet is absorbed. Then the electron angular distributions are given by [17],

$$S_e(\theta, t) = \frac{1}{2i}\left[\psi^*(t)\frac{\partial\psi(t)}{\partial\rho} - \frac{\partial\psi^*(t)}{\partial\rho}\psi(t)\right]_{\rho=\rho_{\mathrm{f}}}. \tag{4}$$

The corresponding energy-resolved angular differential yield (ionization spectra) is then obtained by a Fourier transform after integrating out z perpendicular to the (x, y) molecular plane, i.e., averaged over z:

$$S_e(\theta, E_e) \sim \int dz \int e^{-iE_e t} S_e(\theta, t) dt, \tag{5}$$

where $E_e = p_e^2/2$ is the kinetic energy of an ejected electron with electron wavelength $\lambda_e = 2\pi/p_e$ (in a.u.), and $p_e = (p_x^2 + p_y^2)^{1/2}$ is the total momentum of a photoelectron. With the transformation $p_x = p_e \cos\theta$ and $p_y = p_e \sin\theta$, we then obtain the 2D momentum distributions in the molecular (x, y) plane of photoelectrons from equation (5).

3. Numerical Results and Discussions

We first consider single photon ionization processes by XUV laser pulses with frequency $\omega > I_p$. For higher order ionization, their spectral intensity decrease rapidly as the photoelectron kinetic energy increases [22]. Figure 1 shows molecular photoelectron momentum (p_x, p_y) distributions for the x aligned H_2^+ initial ground $1s\sigma_g$ electronic state at equilibrium $R_e = 2$ a.u. by intense linearly polarized few cycle attosecond XUV laser pulses with intensity $I_0 = 5 \times 10^{15}$ W cm^2 ($E_0 = 0.38$ a.u.), wavelength $\lambda = 10$ nm ($\omega = 4.56$ a.u.), and phase $\phi = 0$. Four pulse durations $T = n\tau$ are used with numbers of cycles $n = 6$, 4, 3, and 2, respectively. The laser pulses are polarized along the x axis (the molecular internuclear axis R). Since the Keldysh parameter $\gamma = \sqrt{I_p/2U_p} = 18 \gg 1$, where $U_p = E_0^2/4\omega^2$ is the ponderomotive energy and $I_p = 1.1$ a.u., a multi-photon ionization regime defines the processes. Furthermore, since the wavelength of the continuum electron $\lambda_e = 2.36$ a.u. $> R_e$, no electron diffraction occurs [5].

Table 1. Photoelectron momenta p_e (kinetic energies E_e) and the corresponding momentum (energy) shift $p_e = |p_c - p_e|$ ($E_e = |E_c - E_e|$) relative to the classic values $p_c = \pm\sqrt{2(\omega - I_p)}$ ($E_c = \omega - I_p$) for the x-aligned equilibrium molecular ion H_2^+ with intensity $I_0 = 5 \times 10^{15}$ W cm^2, wavelengths $\lambda = 10$ nm (angular frequency $\omega = 4.56$ a.u.) and $\lambda = 20$ nm ($\omega = 2.28$ a.u.) few cycle attosecond XUV laser pulses at difference durations $n\tau$. Atomic units (a.u.) are used for the momentum p_e and energy E_e.

$n\tau$	$\lambda = 10$ nm, $p_c = \pm 2.63$ a.u.		$\lambda = 20$ nm, $p_c = \pm 1.54$ a.u.	
	p_e (E_e)	p_e (E_e)	p_e (E_e)	p_e (E_e)
6τ	±2.57 (3.3)	0.06 (0.16)	±1.51 (1.14)	0.02 (0.04)
4τ	±2.5 (3.13)	0.13 (0.33)	±1.49 (1.11)	0.04 (0.07)
3τ	±2.37 (2.81)	0.26 (0.65)	±1.45 (1.05)	0.08 (0.13)
2τ	±2.0 (2.0)	0.63 (1.46)	±1.38 (0.95)	0.15 (0.23)

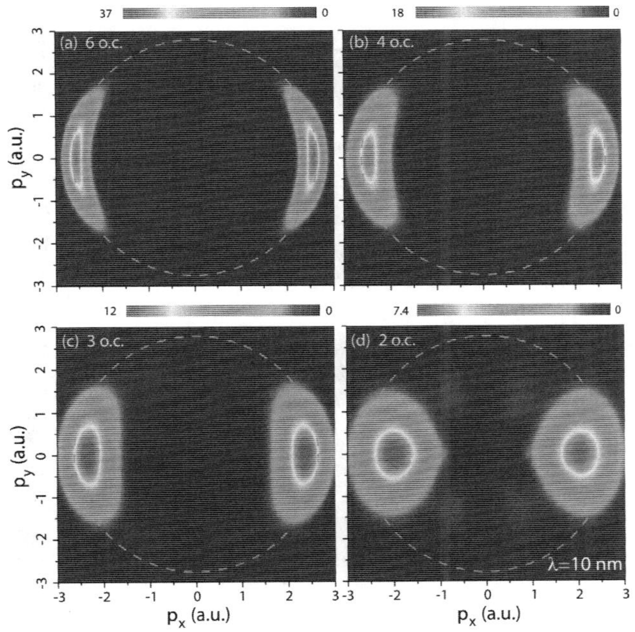

Figure 1. (Color online) Photoelectron momentum distributions for x aligned H_2^+ at equilibrium ($R_e = 2$ a.u.) by a single color linearly x polarized few cycle attosecond XUV laser pulse with intensity $I_0 = 5 \times 10^{15}$ W cm^2, wavelength $\lambda = 10$ nm (angular frequency $\omega = 4.56$ a.u.), and phase $\phi = 0$ for four different durations: (a) $T = 6$ o.c. (200 asec), (b) $T = 4$ o.c. (133 asec), (c) $T = 3$ o.c. (100 asec), and (d) $T = 2$ o.c. (66 asec). The classical momenta p_c corresponding to photoelectron energy $p_c^2/2 = \omega - I_p$ are shown as dash circles with radii $p_c = 2.63$ a.u. ($E_c = 3.46$ a.u.).

From figure 1 we see that in the ultrashort few cycle attosecond XUV laser pulses the momentum distributions are critically sensitive to the pulse duration $n\tau$. Comparing to the classical momentum predicted by $p_c = \sqrt{2(\omega - I_p)}$ (dash circles), a red shift of the photoelectron distribution occurs. As the duration or number n of pulse cycles decreases, the peak positions of the ionization distributions, i.e., the photoelectron energy, decreases as well. The corresponding photoelectron momenta and kinetic energies are listed in table 1. As shown in figure 1(a) for a duration of 6 cycles ($T = 200$ asec), the momenta of the photoelectron are $p_e = \pm 2.58$ a.u., which are nearly equal to the classical values $p_c = \pm\sqrt{2(\omega - I_p)} = \pm 2.63$ a.u. with a slight difference $p_e = 0.06$ a.u. ($E_e = 0.16$ a.u.). Decreasing the pulse duration $T = n\tau$ leads to an increase of the red shift of the photoelectron momentum p and energy E_e. At $T = 2\tau = 66$ asec, the largest red shift of the ionization distributions is obtained, where $p_e = 0.63$ a.u. ($E_e = 1.46$ a.u.), approximate 10 times the results at $T = 6\tau$. Moreover, as the pulse duration $n\tau$ decreases, the pulse spectral width broadens correspondingly.

Similar results are obtained in figure 2 with $\lambda = 20$ nm ($\omega = 2.28$ a.u.) laser pulses at different durations. The other parameters are the same as those in figure 1. It is found that the photoelectron distributions shift to low energy region, i.e., a red shift, with decrease of the number n of the pulse cycle as well. The values of the photoelectron momentum (kinetic energy) and corresponding energy shifts are also listed in table 1. Comparing with the ionization at shorter wavelength $\lambda = 10$ nm laser pulses, we note that these distributions exhibit smaller shifts p_e (E_e).

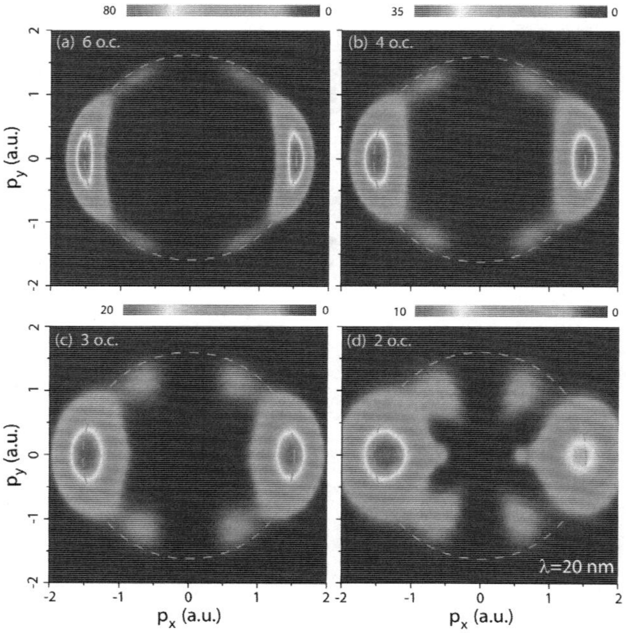

Figure 2. (Color online) Photoelectron momentum distributions in x aligned H_2^+ by a single color linearly x polarized few cycle attosecond XUV laser pulse with intensity $I_0 = 5 \times 10^{15}$ W cm^2, wavelength $\lambda = 20$ nm (angular frequency $\omega = 2.28$ a.u.), and phase $\phi = 0$ for four different durations: (a) $T = 6$ o.c. (400 asec), (b) $T = 4$ o.c. (266 asec), (c) $T = 3$ o.c. (200 asec), and (d) $T = 2$ o.c. (132 asec). The classical momenta p_e corresponding to photoelectron energy $E_c = p_c^2/2 = \omega - I_p$ are shown as dash circles with radii $p_c = 1.54$ a.u. ($E_c = 1.18$ a.u.).

To understand the occurrence of momentum and energy shifts of the photoelectron in intense few cycle attosecond XUV pulses, we adopt a perturbation ionization model to describe the effect of the pulse duration $n\tau$ on the distributions. Considering a direct one photon electrical dipole transition from the initial ground $1s\sigma_g$ electronic state to the continuum, the probability amplitude is given by

$$\frac{dP}{d\Omega} \propto |\langle \psi_c^{E_e} | r | \psi_{1s\sigma_g} \rangle|^2 \tilde{E}^2(\omega) \tag{6}$$

where $\psi_c^{E_e}$ is the continuum electron wave function with energy E_e and $\psi_{1s\sigma_g}$ is the initial state. The laser field $\tilde{E}(\omega)$ in the frequency domain obtained by a Fourier transform of the driving laser pulse, $\tilde{E}(\omega) = \int E(t)e^{-i\omega t}dt$, is a function of the pulse duration $n\tau$. Figure 3 illustrates the ionization amplitudes for the $\lambda = 10$ nm and 20 nm laser pulses with difference duration $n\tau$ obtained from equation (6) for a one dimensional (1D) model polarized along the molecular internuclear R axis (the x axis). We use a plane wave to describe the continuum wavefunction $\psi_c^{E_e} = A_0 \exp(ip_e x)$ and the initial $1s\sigma_g$ electronic state is a linear combination of the hydrogenic $1s$ orbitals $\psi_{1s\sigma_g} = [\psi_{1s}(-R/2) + \psi_{1s}(R/2)]/\sqrt{2}$. From equation (6) one sees that the photoelectron distributions are a product of an electronic Franck-Condon factor $|\langle \psi_c^{E_e} | r | \psi_{1s\sigma_g} \rangle|$ and the pulse spectral width $\tilde{E}(\omega)$. The first depends on the photoelectron kinetic energy E_e (p_e) and the second one is a function of the number n of cycle. The higher the photoelectron kinetic energy the smaller the ionization probability due to decreasing Franck-Condon factor. Of note is that we only consider the case of the photoelectron distributions along

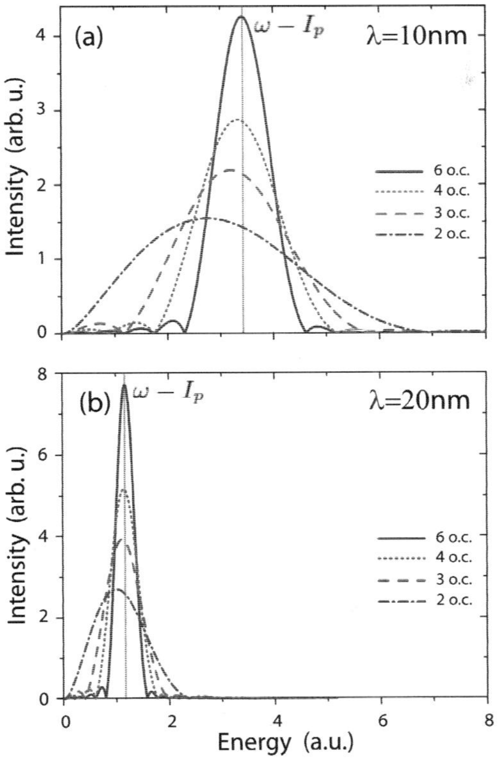

Figure 3. (Color online) Ionization yields obtained from the 1D perturbation model in equation (6) by $\lambda = 10$ nm and (b) $\lambda = 20$ nm few cycle laser pulses with different durations. Green line denotes central photon energy of $\omega - I_p$. Signal intensity units are arbitrary.

the laser polarization since no electron diffraction occurs.

For the ionizing $\lambda = 10$ nm and 20 nm laser pulses in figures 1 and 2, the spectral widths are limited. The ionization distribution is a resulting effect of $|\langle\psi_c^{E_e}|r|\psi_{1s\sigma_g}\rangle|$ and $\tilde{E}(\omega)$ in equation (6). As illustrated in figures 3(a) and 3(b), the photoelectron energy spectra are asymmetric with respect to the kinetic energy $E_c = \omega - I_p$, corresponding to the central frequency ω of the attosecond pulse. The ionization probability in the low energy region is slightly larger than that in the high energy region. As a results a red shift occurs in photoelectron energy spectra in figures 3(a) and 3(b). Decreasing the pulse duration T leads to broadening of the pulse field $\tilde{E}(\omega)$. The photoelectron energy spectra become broad as well. Then the asymmetry arising from the product of $|\langle\psi_c^{E_e}|r|\psi_{1s\sigma_g}\rangle|$ and $\tilde{E}(\omega)$ increases, i.e., the energy shift increases. Thus one gets in figures 1 and 2 the energy shifts in photoelectron energy spectra are sensitive to the pulse duration. For ionization with many cycle laser pulses the corresponding asymmetry and the energy shift in photoelectron energy spectra are very small and can be ignored due to narrow spectral widths. This can be seen in figure 3(b) for the 6τ pulses. Moreover, for the cases with fixed n, the pulse spectral width $\tilde{E}(\omega)$ at $\lambda = 10$ nm is broader than that at $\lambda = 20$ nm. Thus a larger momentum shift p_e (E_e) is obtained in figure 1 and table 1 by the $\lambda = 10$ nm laser pulse.

We next show the energy shift of photoelectron energy spectra in higher order molecular ATI spectra. Figure 3 shows ATI spectra of x-aligned H_2^+ for $\lambda = 20$ nm linearly polarized laser pulses at $I_0 = 5 \times 10^{15}$ W cm^2 and $\phi = 0$. Three ionization processes with pulse durations

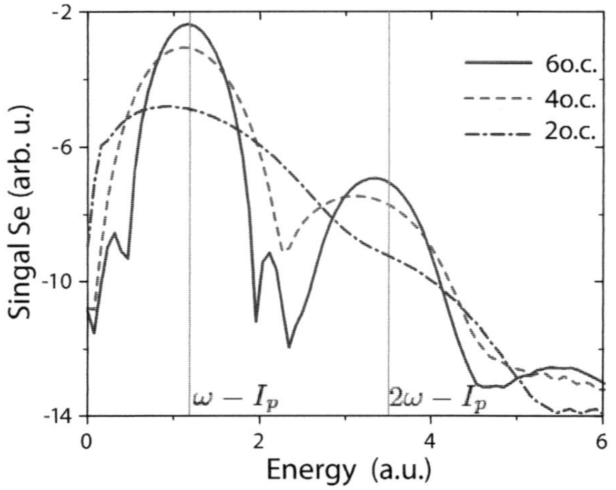

Figure 4. (Color online) Molecular ATI spectra in x aligned H_2^+ by a single color linearly x polarized few cycle attosecond XUV laser pulse with intensity $I_0 = 5 \times 10^{15}$ W cm^2, wavelength $\lambda = 20$ nm (angular frequency $\omega = 2.28$ a.u.), and phase $\phi = 0$ for different durations $T = 6$ o.c. (solid line) 4 o.c. (dash line), and 2 o.c. (dash-dot line). Green lines present the classical photoelectron energies $E_c = \omega - I_p = 1.18$ a.u. and $2\omega - I_p = 3.46$ a.u..

$T = 6\tau$, 4τ and 2τ are simulated. Form figure 4 one sees that the shift of photoelectron energy spectra is also sensitive to the photoelectron energy E_e. For both 6τ and 4τ pulses, small energy shifts E_e are obtained in the first ATI at energy $E_e = \omega - I_p$, as listed in table 1. For the second ATI peak at $E_e = 2\omega - I_p$, the corresponding energy shift E_e is larger. The higher the photoelectron energy E_e the smaller the ionization probability. The amplitude of the Franck-Condom factor $|\langle \psi_c^{E_e} | r | \psi_{1s\sigma_g} \rangle|$ in equation (6) decreases for higher order ATI spectra, therefore resulting in large energy shifts with same pulse durations. Due to effects of the pulse spectral width, the energy shift with $T = 4\tau$ is also larger than that with $T = 6\tau$ at $E_e = 2\omega - I_p$. Similar processes occur for higher order ATI spectra at energies $E_e > 2\omega - I_p$, illustrated at the end of the spectrum (figure 4). Furthermore, as illustrated in figure 4 at $T = 2\tau$ the ATI spectra become nearly a continuum and the corresponding red shift can not read any more.

4. Conclusions

In summary, we present attosecond photoelectron distributions of aligned H_2^+ by few cycle XUV laser pulses. Numerical results show pulse duration dependent red energy shifts in the spectra. This coincides with predictions from a perturbative model. The red energy shift can be interpreted as diminishing electronic Franck-Condon factors with shorter pulses durations which increase spectral band widths of the attosecond pulses around a central frequency ω. This is in contrast to high frequency, high intensity long pulses which induce ground state Stark shifts thus inducing blue shifts of ionized electron energies [36].

5. Acknowledgements

We thank RQCHP and Compute Canada for access to massively parallel computer clusters and NSERC for financial support in its ultrafast science program.

6. References

[1] Corkum P B and Krausz F 2007 Attosecond science *Nature Phys.* **3** 381

[2] Krausz F and Ivanov M 2009 Attosecond physics *Rev. Mod. Phys.* **81** 163
[3] Chang and Corkum P 2010 Attosecond photon sources: The rst decade and beyond *J. Opt. Soc. Am. B* **27** B9
[4] Chelkowski S, Corkum P B and Bandrauk A D 1999 Femtosecond Coulomb explosion imaging of vibrational wave functions *Phys. Rev. Lett.* **82** 3416
[5] uo T, Bandrauk A D and Corkum P B 1996 Laser-induced electron di raction: A new tool for probing ultrafast molecular dynamics *Chem. Phys. Lett.* **259** 313
[6] Meckel M, Comtois D, eidler D, Staudte A, Pavicic D, Bandulet H C, Pepin H, Kie er J C, Dorner R, Villeneuve D M and Corkum P B 2008 Laser-induced electron tunneling and di raction *Science* **320** 1478
[7] hao K, hang Q, Chini M, Wu Y, Wang and Chang 2012 Tailoring a 67 attosecond pulse through advantageous phase-mismatch *Opt. Lett.* **37** 3891
[8] Bandrauk A D, Chelkowski S, Diestler D J, Manz J and Yuan K J 2008 Quantum-mechanical models for photo-ionization: Uni-directional electron re-scattering by a laser pulse *Int. J. Mass Spectrom.* **277** 189
[9] Vrakking M 2009 Chemical physics: Electronic movies *Nature* **460** 960
[10] Popmintchev T, Chen M C, Arpin P, Murnane M M, Kapteyn H C 2010 The attosecond nonlinear optics of bright coherent -ray generation *Nature Photon.* **4** 822
[11] McNeil B W J and Thompson N R 2010 -ray free-electron lasers *Nature Photon.* **4** 814
[12] Chelkowski S, Yudin G L and Bandrauk A D 2006 Observing electron motion in molecules *J. Phys. B: At. Mol. Opt. Phys.* **39** S409
[13] Niikura H, Villeneuve D M and Corkum P B 2005 Mapping attosecond electron wave packet motion *Phys. Rev. Lett.* **94** 083003
[14] Marangos J P, Baker S, Kajumba N, Robinson J S, Tisch J W G and Torres R 2008 Dynamic imaging of molecules using high order harmonic generation *Phys. Chem. Chem. Phys.* **10** 35
[15] Shao H C and Starace A F 2010 Detecting electron motion in atoms and molecules *Phys. Rev. Lett.* **105** 263201
[16] Kelkensberg F, Rouzee A, Siu W, Gademann G, Johnsson P, Lucchini M, Lucchese R R and Vrakking M J J 2011 UV ionization of aligned molecules *Phys. Rev. A* **84** 051404(R)
[17] Yuan K J, Lu H and Bandrauk A D 2009 Laser-induced electron di raction in H_2 with linear and circular polarization ultrashort UV laser pulses *Phys. Rev. A* **80**, 061403(R)
Yuan K J, Lu H and Bandrauk A D 2011 Linear-and circular-polarization photoionization angular distributions in H_2 and H_2^+ by attosecond xuv laser pulses *Phys. Rev. A* **83** 043418
[18] Mauritsson J, Remetter T, Swoboda M, Klunder K, L'Huillier A, Schafer K J, Ghafur O, Kelkensberg F, Siu W, Johnsson P, Vrakking M J J, nakovskaya I, Uphues T, herebtsov S, Kling M F, Lepine F, Benedetti E, Ferrari F, Sansone G and Nisoli M 2010 Attosecond electron spectroscopy using a novel interferometric pump-probe technique *Phys. Rev. Lett.* **105** 053001
[19] Ranitovic P, Tong M, Hogle C W, hou , Liu Y, Toshima N, Murnane M M and Kapteyn H C 2011 Controlling the UV transparency of helium using two-pathway quantum interference *Phys. Rev. Lett.* **106** 193008
[20] Choi N N, Jiang T F, Morishita T, Lee M H and Lin C D 2010 Theory of probing attosecond electron wave packets via two-path interference of angle-resolved photoelectrons *Phys. Rev. A* **82** 013409
[21] Baggesen J C and Madsen L B 2011 Atomic and molecular phases through attosecond streaking *Phys. Rev. A.* **83** 021403(R)
[22] Yuan K J and Bandrauk A D 2012 Molecular above-threshold-ionization angular distributions with attosecond bichromatic intense UV laser pulses *Phys. Rev. A.* **85** 013413
[23] Feist J, Nagele S, Pazourek R, Persson E, Schneider B I, Collins L A and Burgdorfer J 2009 Probing electron correlation via attosecond xuv pulses in the two-photon double ionization of helium *Phys. Rev. Lett.* **103** 063002
[24] Moore L R, Lysaght M A, Parker J S, van der Hart H W and Taylor K T 2011 Time delay between photoemission from the 2p and 2s subshells of neon *Phys. Rev. A* **84** 061404(R)
[25] Sansone G, Pfeifer T, Simeonidis K and Kule A I 2012 Electron Correlation in Real Time *ChemPhysChem* **13** 661
[26] Bredtmann T, Chelkowski S and Bandrauk A D 2011 Monitoring attosecond dynamics of coherent electron-nuclear wave packets by molecular high-order-harmonic generation *Phys. Rev. A* **84** 021401
Chelkowski S, Bredtmann T and Bandrauk A D 2012 High-order-harmonic generation from coherent electron wave packets in atoms and molecules as a tool for monitoring attosecond electrons *Phys. Rev. A* **85** 033404
[27] Tzallas P, Skantzakis E, Nikolopoulos L A A, Tsakiris G D and Charalambidis D 2011 Extreme-ultraviolet pump-probe studies of one-femtosecond-scale electron dynamics *Nature Phys.* **7** 781
[28] Jiang Y H, Rudenko A, Herrwerth O, Foucar L, Kurka M, Kuhnel K U, Lezius M, Kling M F, van Tilborg

J, Belkacem A, Ueda K, Dusterer S, Treusch R, Schroter C D, Moshammer R and Ullrich J 2010 Ultrafast extreme ultraviolet induced isomerization of acetylene cations *Phys. Rev. Lett.* **105** 263002

[29] Peng L Y and Starace A F 2007 Attosecond pulse carrier-envelope phase e ects on ionized electron momentum and energy distributions *Phys. Rev. A* **76** 043401
Peng L Y, Pronin E A and Starace A F 2008 Attosecond pulse carrier-envelope phase e ects on ionized electron momentum and energy distributions: Roles of frequency, intensity and an additional IR pulse *New J. Phys.* **10** 025030

[30] Ngoko Djiokap J M, Hu S , Jiang W C, Peng L Y and Starace A F 2012 Enhanced asymmetry in few-cycle attosecond pulse ionization of He in the vicinity of autoionizing resonances *New J. Phys.* **14** 095010

[31] Yuan K J and Bandrauk A D 2013 Attosecond-magnetic- eld-pulse generation by intense few-cycle circularly polarized UV laser pulses *Phys. Rev. A* **88** 013417

[32] Yuan K J and Bandrauk A D 2011 Generation of circularly polarized attosecond pulses by intense ultrashort laser pulses from extended asymmetric molecular ions *Phys. Rev. A* **84** 023410
Yuan K J and Bandrauk A D 2012 Circularly polarized attosecond pulses from molecular high-order harmonic generation by ultrashort intense bichromatic circularly and linearly polarized laser pulses *J. Phys. B: At. Mol. Opt. Phys.* **45** 074001

[33] Yuan K J and Bandrauk A D 2013 Single circularly polarized attosecond pulse generation by intense few cycle elliptically polarized laser pulses and terahertz elds from molecular media *Phys. Rev. Lett.* **110** 023003

[34] Bandrauk A D and Lu H , in *High-Dimensional Partial Differential Equations in Science and Engineering*, edited by A. D. Bandrauk, M. Delfour, and C. LeBris, 2007 CRM Lecture Series Vol. 41, pp. 1-15 American Math. Soc., Philadelphia

[35] Bandrauk A D and Shen H 1993 Exponential split operator methods for solving coupled time-dependent Schrodinger equations *J. Chem. Phys.* **99** 1185

[36] Demekhin P V and Cederbaum L S 2012 Dynamic interference of photoelectrons produced by high-frequency laser pulses *Phys. Rev. Lett.* **108**, 253001

Novel aspects of radiation reaction in the classical and the quantum regime

Norman Neitz, Naveen Kumar, Felix Mackenroth, Karen Z. Hatsagortsyan, Christoph H. Keitel and Antonino Di Piazza

Max-Planck-Institut für Kernphysik, Saupfercheckweg 1, 69117 Heidelberg, Germany

E-mail: dipiazza@mpi-hd.mpg.de

Abstract.
This work is dedicated to the study of radiation reaction signatures in the framework of classical and quantum electrodynamics. Since there has been no distinct experimental validation of radiation reaction and its underlying equations so far and its impact is expected to be substantial for the construction of new experimental devices, e.g., quantum x-free electron lasers, a profound understanding of radiation reaction effects is of special interest. Here, we describe how the inclusion of quantum radiation reaction effects changes the dynamics of ultra-relativistic electron beams colliding with intense laser pulses significantly. Thereafter, the angular distribution of emitted radiation is demonstrated to be strongly altered in the quantum framework, if in addition to single photon emission also higher order photon emissions are considered. Furthermore, stimulated Raman scattering of an ultra-intense laser pulse in plasmas is examined and forward Raman scattering is found to be significantly increased by the inclusion of radiation reaction effects in the classical regime. The numerical simulations in this work show the feasibility of an experimental verification of the predicted effects with presently available lasers and electron accelerators.

1. Introduction

If a charged particle, an electron for definiteness, is exposed to an electromagnetic background field, it will be accelerated and subsequently emit radiation. Although this process is fundamental in electrodynamics, the usual classical treatment is insufficient, since it does not take into account radiation reaction (RR), i.e., the back reaction of the emitted radiation on the charged particle itself [1, 2]. In this work high intensity plane wave laser fields of electric field amplitude $\mathcal{E}_0$, central angular frequency ω_0, central wavelength λ_0 and propagation direction $\boldsymbol{n}_0$ are investigated, in order to theoretically probe the parameter regime relevant for RR. The study of electron-laser interactions in classical as well as quantum electrodynamics (QED) is of special interest, as they are expected to have impact on various fields like accelerators, quantum x-free electron lasers [3, 4, 5] or the production of multi-GeV photon beams [6]. However, RR effects are also of pure theoretical interest, as even in the framework of classical electrodynamics theoretical methods such as renormalization are necessary to describe the self-coupled dynamics [7].

Recently it has been reported, that the so-called Landau-Lifshitz (LL) equation is in theory the accurate equation of motion for an electron of mass m and charge $e < 0$ in the framework of classical electrodynamics [1, 2, 7, 8, 9, 10, 11]. (Units with $\hbar = c = 1$ are used.) In case of an electromagnetic plane wave the LL equation allows for an analytical solution [12] and it was demonstrated that if the parameter $R_c = \alpha\chi_0\xi_0$ is of order of unity, the dynamics of an electron with initial momentum p_i^μ colliding with a plane wave laser field is significantly altered by RR effects. Here, we introduced the classical and quantum nonlinearity parameters $\xi_0 = |e|\mathcal{E}_0/m\omega_0$ and $\chi_0 = ((n_0p_i)/m)\mathcal{E}_0/E_{cr}$, respectively, and defined $n_0^\mu = k_0^\mu/\omega_0 = (1, \boldsymbol{n}_0)$, where we made use of the abbreviation $(ab) = a_\mu b^\mu$ denoting the product of two four-vectors a^μ and b^μ. Furthermore, $\alpha = e^2$ is the fine structure constant and $E_{cr} = m^2/|e| = 1.3 \times 10^{16}\,\text{V/cm}$ is the critical field of QED. In addition, in a bichromatic laser pulse the RR force was shown to modify the trajectory of an electron also in the case of $R_c \ll 1$ [13] (see also [14]). The head-on collision of a laser field with an ultra-relativistic electron of initial energy ε_i yields $R_c = 3.2\,\varepsilon_i[\text{GeV}]I_0[10^{23}\ \text{W/cm}^2]/\omega_0[\text{eV}]$, where $I_0 = \mathcal{E}_0^2/4\pi$ is the laser peak intensity. Thus, the experimental challenges in observing RR effects can be understood, since this expression is usually very small. Nevertheless, an alternative method has been proposed to measure RR effects also at moderate laser intensities and larger pulse durations [15]. However, for upcoming ultra-high intensity laser facilities it was calculated that for laser intensities exceeding $I_0 > 2 \times 10^{23}\,\text{W/cm}^2$ RR effects have to be taken into account and for $I_0 > 4 \times 10^{24}\,\text{W/cm}^2$ quantum effects will become important [16] and lead to a strong alteration of the particle's dynamics [17].

In addition, the consistency of QED and the diverse classical approaches was examined [18]. In the quantum description a laser field is depicted as a stream of photons and the scattering process with an electron leads to the absorption of many photons from the field and to the subsequent emission of one or more photons. In fact, the emission of a single photon by an electron in strong laser pulses (nonlinear single Compton scattering (NSCS)) has been studied thoroughly [19, 20, 21, 22, 23] and the classical spectra without RR [24] were shown to coincide with the NSCS spectra for $\chi_0 \ll 1$. Hence, in the ultra-relativistic regime and for negligible pair creation the quantum analogue of RR can be understood as the emission of a higher number of photons [25]. Providing the first results for such higher order photon emission, the emission of two photons by an electron in a plane wave (nonlinear double Compton scattering (NDCS)) was considered recently [26, 27, 28, 29].

The influence of the RR force on the collective particle dynamics of a system can be considerably different from the single particle dynamics as the collective energy loss of the particles due to RR can give rise to unexpected physical phenomena in a medium. In the classical electrodynamics regime, an analysis of the influence of the Landau-Lifshitz RR force [7] on the collective plasma dynamics of the particles has been carried out recently, where it was found that the inclusion of RR counterintuitively strongly enhances the forward Raman scattering (FRS) of the laser radiation in plasmas [30]. This growth enhancement is attributed to the nonlinear mixing of the two Raman sidebands mediated by the RR force.

In the sections concerning quantum RR (sections 2 and 3), we will employ light-cone coordinates, which for a four-vector $a^\mu = (a^0, \boldsymbol{a})$ are defined as $a^\mu = (a^+, a^-, \boldsymbol{a}^\perp)$, where $a^\pm = a^0 \pm a^\parallel$, with $a^\parallel = \boldsymbol{k}_0 \cdot \boldsymbol{a}/\omega_0$ and $\boldsymbol{a}^\perp = \boldsymbol{a} - a^\parallel \boldsymbol{k}_0/\omega_0$.

2. Kinetic approach to quantum radiation reaction

In this section, we investigate RR effects in the collision of an intense laser pulse with an ultra-relativistic electron beam. In classical electrodynamics RR was shown to reduce the energy width of electron [31] and ion [32, 33, 34, 35] bunches. However, in the quantum regime we find that RR has the opposite tendency and leads to a broadening in the width of the energy distribution describing the electron beam. The difference between the classical and the quantum regime can be explained by the increasing importance of the stochastic nature of photon emission in the quantum regime. The classical LL equation ignores the stochasticity of photon emission and even for small χ_0 a correct treatment of RR requires an additional stochastic term. In fact, a Langevin-like equation can be employed for not too large χ_0's, though in the full quantum regime at $\chi_0 \sim 1$ this approximated description is not valid anymore. The broadening in the energy distribution of the electron beam displayed by our numerical simulations is expected to be detectable with nowadays available electron accelerators and strong laser fields.

An exact treatment of RR in the realm of strong-field QED would in principal result in the determination of the full S-matrix, taking into account multiple photon emission, radiative corrections and pair creation following photon emission [1, 25]. However, RR mainly stems from incoherent multi-photon emission in the so-called "nonlinear moderately-quantum" regime $\xi_0 \gg 1$, $\chi_0 \lesssim 1$ [25], where nonlinear effects in the laser field are considered to be large and quantum effects are already important but pair production can still be neglected. In order to investigate RR in this regime we apply a kinetic approach [36, 37, 38] and in turn characterize electrons and photons by distribution functions. Due to the neglect of pair production the distribution function of positrons is considered to vanish and the kinetic equation of the electrons is decoupled from that of the photons [36, 37, 38]. If the average energy of the electron beam ε^* fulfills the constraint $\varepsilon^* \gg m\xi_0$, the transverse momentum of the electrons can be disregarded, since throughout the whole interaction it will remain much smaller than the longitudinal momentum [7]. Assuming the collision of presently available optical ($\omega_0 = 1.55$ eV) laser fields of intensity 10^{22} W/cm^2 [39] with electron bunches with typical energies of $\varepsilon^* = 1$ GeV yields $m\xi_0 = 25$ MeV and allows us to treat the present problem as an one-dimensional one. Considering a linearly polarized plane wave propagating along the positive y direction, we introduce the electric field of the laser field by $\boldsymbol{\mathcal{E}}(\eta) = \mathcal{E}_0 g(\eta)\hat{\boldsymbol{z}}$ depending on the laser phase $\eta = \omega_0(t-y)$ via a pulse-shape function with $|g(\eta)|_{\text{max}} \leq 1$. Since the ultra-relativistic regime $\xi_0 \gg 1$ is investigated, the emission in a plane wave field of a photon with four-momentum $k^\mu = (\omega, \boldsymbol{k})$ by an electron with initial four-momentum $p^\mu = (\varepsilon, \boldsymbol{p})$ can be characterized by adopting the well-known single photon differential emission probability dP_{p^-} per unit phase and per unit $u = k^-/(p^- - k^-)$ [40]

$$\frac{dP_{p^-}}{d\eta du} = \frac{\alpha}{\sqrt{3}\pi}\frac{m^2}{\omega_0 p^-}\frac{1}{(1+u)^2}\left[\left(1+u+\frac{1}{1+u}\right)\mathrm{K}_{\frac{2}{3}}\left(\frac{2u}{3\chi}\right) - \int_{\frac{2u}{3\chi}}^{\infty} dx\,\mathrm{K}_{\frac{1}{3}}(x)\right], \tag{1}$$

where $\mathrm{K}_\nu(\cdot)$ is the modified Bessel function of νth order. The quantum nonlinearity parameter is given by $\chi \equiv \chi(\eta, p^-) = (p^-/m)|\mathcal{E}(\eta)|/E_{\text{cr}}$ and is now depending on the laser phase due to the oscillating electric field $\mathcal{E}(\eta) = \mathcal{E}_0 g(\eta)$. The fact that Eq. (1) only depends on the variables η and p^- allows us to employ an electron distribution $f_e(\eta, p^-)$ for the description of an electron beam and its phase evolution is determined by the kinetic equation (see Ref. [36])

$$\frac{\partial f_e}{\partial \eta} = \int_{p^-}^{\infty} dp'^- \frac{dP_{p'^-}}{d\eta dp^-} f_e(\eta, p'^-) - f_e(\eta, p^-)\int_0^{p^-} dk^- \frac{dP_{p^-}}{d\eta dk^-} \tag{2}$$

with

$$\frac{dP_{p'^-}}{d\eta dp^-} = \frac{p'^-}{(p^-)^2}\frac{dP_{p'^-}}{d\eta du}\bigg|_{u=\frac{p'^- - p^-}{p^-}}, \tag{3}$$

$$\frac{dP_{p^-}}{d\eta dk^-} = \frac{p^-}{(p^- - k^-)^2}\frac{dP_{p^-}}{d\eta du}\bigg|_{u=\frac{k^-}{p^- - k^-}}. \tag{4}$$

Eq. (2) is non-local in the momentum p^- since it is an integro-differential equation. Hence, an electron with initial momentum p_0^- is coupled to the electron with momentum $p_0^- - k^-$, where the momentum of the emitted photon k^- varies between 0 and p_0^-. In turn, the evolution of $f_e(\eta, p^-)$ is not only influenced by the neighborhood of p^- but by all possible values of p'^-. In order to study the classical limit of RR, we expand Eq. (2) up to the order of $\chi^3(\eta, p^-)$

$$\frac{\partial f_e}{\partial \eta} = -\frac{\partial}{\partial p^-}[A(\eta, p^-) f_e] + \frac{1}{2}\frac{\partial^2}{\partial (p^-)^2}[B(\eta, p^-) f_e], \tag{5}$$

which is a Fokker-Planck-like equation [38, 41, 42]. Here, we introduced the "drift" coefficient $A(\eta, p^-) = -\frac{2\alpha m^2}{3\omega_0}\chi^2(\eta, p^-)[1 - \frac{55\sqrt{3}}{16}\chi(\eta, p^-)]$ and the "diffusion" coefficient $B(\eta, p^-) = \frac{\alpha m^2}{3\omega_0}\frac{55}{8\sqrt{3}}p^-\chi^3(\eta, p^-)$. Eq. (5) is no longer an integro-differential equation and the evolution of $f_e(\eta, p^-)$

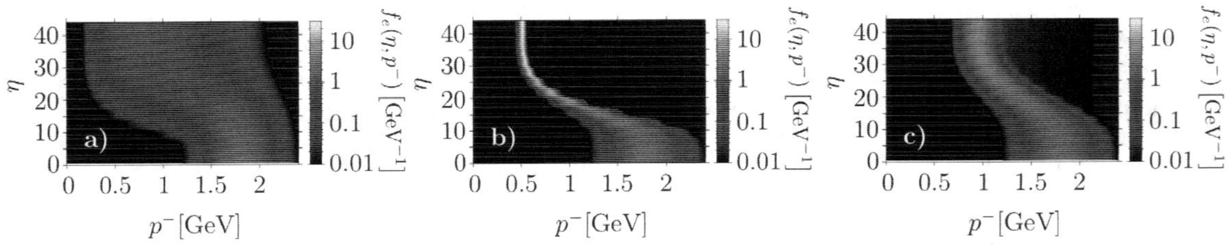

Figure 1. (color online) Phase evolution of the electron distribution for a 7-cycle $\sin^2$-like laser pulse according to Eq. (2) (part a)), to the classical kinetic equation without (part b)) and with the replacement $I_{cl}(\eta, p^-) \to I_q(\eta, p^-)$ (part c)). The laser and the initial electron distribution parameters are given in the text.

is determined by the momenta p'^- close to p^-, due to the locality in p^-. If also higher-order corrections in $\chi(\eta, p^-)$ are considered, the expansion of Eq. (2) leads to terms with higher derivatives of $f_e(\eta, p^-)$ with respect to p^-.

By expanding the kinetic equation (2) we obtain two quantum corrections to the classical kinetic equation $\partial f_e/\partial\eta = -\partial/\partial p^- \, (f_e dp^-/d\eta)$, with $dp^-/d\eta = -I_{cl}(\eta, p^-)/\omega_0$ and the classical radiation intensity $I_{cl}(\eta, p^-) = (2/3)\alpha m^2 \chi^2(\eta, p^-)$. The first correction modifies the drift coefficient $A(\eta, p^-)$, but does not affect the analytical structure of the classical equation. Therefore the phase evolution is purely deterministic [43] and this correction coincides with the well-known leading-order correction to the total intensity of radiation [36, 40]. As the correction term is negative, we expect the reduction of the energy width to be smaller than in classical electrodynamics. However, if the classical radiation intensity $I_{cl}(\eta, p^-)$ is substituted by the corresponding quantum one $I_q(\eta, p^-)$ (see, e.g., [40]), the corresponding Liouville equation still predicts a decrease in the energy width, due to the fact that electrons with higher energy on average will emit more radiation. On the other hand, the second quantum correction introduces the diffusion coefficient $B(\eta, p^-)$ and transforms the classical kinetic equation into a Fokker-Planck-like equation, which is equivalent to the single-particle stochastic equation $dp^- = -A(\eta, p^-)d\eta + \sqrt{B(\eta, p^-)}dW$, with dW an infinitesimal stochastic function [41]. Reflecting the stochasticity of photon emission, this equation is no longer deterministic and the stochastic evolution of the system causes the broadening of the energy distribution [43].

We solved Eq. (2) numerically by employing a finite difference method and we ensured that for $\chi^* = (p^{*,-}/m)\mathcal{E}_0/E_{\rm cr} \ll 1$, where $p^{*,-}$ is the average momentum, our numerical simulations coincide with classical results, as well as with the results in the quantum regime [25]. We now consider a laser pulse with shape function $g(\eta) = \sin^2(\eta/2N_L)\sin(\eta)$, where N_L is the number of laser cycles and with $\omega_0 = 1.55$ eV, to collide head-on with an initial Gaussian electron distribution that is normalized to unity. Further, we assume the laser peak intensity $I_0 = 2.5 \times 10^{22}$ W/cm^2, the average momentum $p^{*,-} = 1.8$ GeV ($\varepsilon^* \approx 900$ MeV) and the initial width of the electron distribution $\sigma_{p^-} = 0.18$ GeV corresponding to $\chi^* = 0.8$, and $N_L = 7$ corresponding to about 21 fs. The results of our numerical simulations are shown in Fig. 1. As expected, Fig. 1a) displays a broadening of the electron distribution in the quantum regime. However, if the full kinetic Eq. (2) is not applied but the quantum intensity $I_q(\eta, p^-)$ (see, e.g., [36, 40]) is set into the classical kinetic equation, the photon emission still reduces the energy spread of the electron distribution (Fig. 1c)) as in the classical case (Fig. 1b)). This clearly indicates that the broadening of the electron distribution is induced by the importance of the stochasticity of photon emission in the full quantum regime.

3. Nonlinear double Compton scattering

In this section, we are going to demonstrate how in the ultra-relativistic quantum regime, distinguished by the conditions $\xi_0 \gg 1$, $\chi_0 \gtrsim 1$, the NDCS signal can be spatially separated from the dominant NSCS signal. The investigation of the given parameter regime is timely as it is about to come into experimental reach. To clearly interpret the results of this analysis and to establish an intuitive understanding of the underlying physics we are going to work out a semi-classical picture of two smoothly joined classical electron trajectories from which the two separate photons are consecutively emitted. These trajectories are found to feature a discontinuity only in the electron's energy, which can be attributed to the emission of a photon of finite energy. We are going to connect this picture of a discontinuous change of the trajectory to the classical account of RR which leads to a smooth change of the electron's trajectory [25]. Finally, we are going to demonstrate that the discussed effect is likely to be observable with already available lasers, featuring intensities beyond 10^{22} W/cm^2 [39], and electron acceleration technology either from conventional accelerators [44] or modern plasma-based laser accelerators [45, 46].

As in the regime $m\xi_0 \ll \varepsilon_i$, with the electron's initial energy ε_i, the radius of the laser's focal volume routinely exceeds the electron's perpendicular excursion, which is on the order of $\lambda_0(m\xi_0/\varepsilon_i)$ it is justified to approximate the laser field by a plane wave. In the present study we are thus going to model the laser pulse by a plane wave field of arbitrary temporal shape $A_0^\mu(\eta) = (\mathcal{E}_0/\omega_0)\epsilon_0^\mu\psi(\eta)$ depending on the space-time coordinates only via the invariant phase $\eta = k_0^\mu x_\mu$. Here ϵ_0^μ is the wave's polarization four-vector and the shape function $\psi(\eta)$ describes the laser pulse's arbitrary temporal shape. Since in the regime $\xi_0 \gg 1$ nonlinear effects have to be taken into account exactly we will perform our calculations in the so-called Furry picture of quantum dynamics [47, 48]. The essence of this procedure is to attribute terms in the QED Lagrangian describing the strong background field, i.e. the laser pulse in this case, to the free Lagrangian and to subsequently quantize the charged fermionic fields in the presence of this strong background. Technically this amounts to a replacement of the vacuum wave function of an electron with momentum p^μ and spin quantum number σ by a solution of the Dirac equation in the presence of the strong plane wave background, known as Volkov wave function $\Psi_{p,\sigma}(x)$ [49]. It is then straightforward to obtain an expression for the scattering matrix element S_{fi} of an electron with initial (final) four-momentum $p_i^\mu = (\varepsilon_i, \boldsymbol{p}_i)$ ($p_f^\mu = (\varepsilon_f, \boldsymbol{p}_f)$) and spin quantum number σ_i (σ_f) emitting two photons with wave vectors k_1^μ and k_2^μ and polarization four-vectors $\epsilon_{k_1,\lambda_1}^\mu$ and $\epsilon_{k_2,\lambda_2}^\mu$, respectively. The resulting expression can be written as $S_{fi}^{(1)} + S_{fi}^{(2)}$ with

$$S_{fi}^{(1)} = -e^2 \int d^4x d^4y\, \overline{\Psi}_{p_f,\sigma_f}(y)\, \not{\epsilon}^*_{k_2,\lambda_2} e^{ik_2y}\, G(y,x)\, \not{\epsilon}^*_{k_1,\lambda_1} e^{ik_1x}\, \Psi_{p_i,\sigma_i}(x) \tag{6}$$

and $S_{fi}^{(2)} = S_{fi}^{(1)}(1 \leftrightarrow 2)$. Here $\not{a} = \gamma^\mu a_\mu$ is the common Feynman slash notation with the four-vector of the Dirac matrices γ^μ. Furthermore, we introduced the Dirac conjugate wave function $\overline{\Psi}_{p,\sigma}(x) = \Psi_{p,\sigma}^\dagger(x)\gamma^0$ and made use of the laser dressed electron propagator $G(y,x)$ [40]. Due to symmetry reasons it is sufficient to only compute the quantity $S_{fi}^{(1)}$, whence the cross-channel amplitude $S_{fi}^{(2)}$ can be obtained by the exchange of indices $(1 \leftrightarrow 2)$ in the final expression. According to earlier work [28, 29] this partial amplitude naturally splits up into two contributions in the following way

$$S_{fi}^{(1)} = (2\pi)^3 \sum_{r,s=0}^{2} (a_r f_r \delta_{r,s} + b_{r,s} f_{r,s})\delta(p_i^- - k_1^- - k_2^- - p_f^-)\delta^{(2)}(\boldsymbol{p}_i^\perp - \boldsymbol{k}_1^\perp - \boldsymbol{k}_2^\perp - \boldsymbol{p}_f^\perp). \tag{7}$$

The matrix coefficients a_r and $b_{r,s}$ are rather involved but not needed here and thus not given explicitly. Anyway, the important information on the dynamics of the two-photon emission process is encoded in the dynamic integrals [28, 29]

$$f_r = \int d\eta \psi^r(\eta) \exp\{-i\,[S_x(\eta) + S_y(\eta)]\}, \tag{8a}$$

$$f_{r,s} = \int d\eta_x d\eta_y \Theta(\eta_y - \eta_x)\, \psi^s(\eta_x)\psi^r(\eta_y) \exp\{-i\,[S_x(\eta_x) + S_y(\eta_y)]\}, \tag{8b}$$

where $S_{x/y}(\eta) = \int_0^\eta d\eta'[\alpha_{x/y}\,\psi(\eta') + \beta_{x/y}\,\psi^2(\eta') + \gamma_{x/y}]$, with $\alpha_x = -m\xi[(p_i\epsilon_0)/(k_0p_i) - (p_t\epsilon_0)/(k_0p_t)]$, $\beta_x = -m^2\xi^2(k_0k_1)/2(k_0p_t)(k_0p_i)$, $\gamma_x = -(k_1p_i)/(k_0p_t)$. The parameters α_y, β_y

and γ_y are obtained from α_x, β_x and γ_x, respectively, substituting $p_t^\mu \to p_f^\mu$, $p_i^\mu \to p_t^\mu$ and $k_1^\mu \to k_2^\mu$. In these expressions we defined the transitional electron momentum p_t^μ, distinguished by the four conditions $p_t^- = p_i^- - k_1^-$, $\boldsymbol{p}_t^\perp = \boldsymbol{p}_i^\perp - \boldsymbol{k}_1^\perp$ and $p_t^2 = m^2$. From the above expressions one can read off that the integrals f_0, $f_{0,s}$ and $f_{r,0}$ are divergent. These divergences, however, can be analytically regularized by an integration by parts technique [28, 29]. Carrying out the resulting replacements in the scattering matrix element, taking the modulus square and summing over the discrete and continuous phase spaces of all particles involved in the scattering the final result for the differential energy spectrum is found to be

$$\mathrm{d}E = \frac{\omega_1+\omega_2}{2}\frac{d^3\boldsymbol{p}_f}{(2\pi)^3}\prod_{i=1}^{2}\frac{d^3\boldsymbol{k}_i}{(2\pi)^3}\sum_{\{\sigma,\lambda\}}\left|S_{fi}^{(1)}+S_{fi}^{(2)}\right|^2, \tag{9}$$

where $\{\sigma,\lambda\} \equiv \sigma_i, \sigma_f, \lambda_1, \lambda_2$. The δ-functions contained in $S_{fi}^{(1,2)}$ serve to fix the final electron momentum p_f^μ whence we only have to integrate over the phase spaces of the emitted photons.

As a next step we wish to give an exemplary numerical study of Eq. (9) to demonstrate its tractability as well as to highlight the aforementioned spatial separation of the NSCS and the NDCS emission signals. For the numerical evaluation we assume the laser's shape function to be $\psi(\eta) = \sin^4(\eta/4)\sin(\eta)$ for $\eta \in [0, 4\pi]$ and zero elsewhere with a central frequency $\omega_0 = 1.55$ eV, corresponding to an optical laser pulse of approximately 5 fs duration. We consider the case of an ultra-intense laser pulse of peak intensity $I_0 = 10^{22}\mathrm{W/cm}^2$ corresponding to $\xi_0 \approx 48$, as is already available at nowadays working laser facilities such as the Hercules laser [39]. Furthermore, we study the case of an ultra-relativistic electron of initial energy $\varepsilon_i = 5$ GeV colliding head on with the specified laser pulse, resulting in a quantum nonlinearity parameter $\chi_0 = 2.8$, indicating the importance of quantum effects. Next, we wish to recall several properties of the NSCS signal, that are to be expected for the given experimental setup [50]: At $\xi_0 \gg 1$ the single photon signal is limited to a polar angular range $\pi - \theta \leq \psi_0\,\theta_0$, where the typical opening angle of the single photon emission cone is given by $\theta_0 = m\xi_0/\varepsilon_i$. The maximum value of the shape function specified above is $\psi_0 = |\max(\psi(\eta))| = 0.8$. Any emission predicted towards polar angles outside this emission cone will be clearly separated from the dominant NSCS signal.

As our reference frame we choose the coordinate frame in which the laser pulse is polarized along the x-direction and propagates towards the positive z-axis. This axis we also choose as polar axis, i.e. $\theta = \pi$ corresponds to the electron's initial propagation direction opposite to the laser's propagation direction at $\theta = 0$. To stress that two-photon emission will be important for the electron's radiation pattern given the specified experimental parameters we estimate the average number of emitted photons to be $\mathcal{N}_{NSCS} = 1.6$ [40].

In the given reference frame we consider an experiment where one photon detector observes any photon emitted towards the direction ($\theta_1 = \pi - \theta_0/2, \phi_1 = \pi$) and a second detector to trace photons emitted towards the two different directions ($\theta_2 = \pi - \theta_0/2, \phi_2 = 0$) (see fig. 2a)) and ($\theta_2 = \pi - \theta_0, \phi_2 = 0$) (see fig. 2b)). While the first detector is positioned inside the NSCS emission cone the choice for the second observation direction corresponds to photon detection inside (see fig. 2a)) and outside (see fig. 2b)) this cone, respectively. The azimuthal angles $\phi_{1,2} = 0, \pi$ are chosen for observation of the emitted photons within the laser's plane of polarization, where most radiation is emitted.

In fig. 2a) we wish to highlight that the cutoff for the emitted photon's frequencies (white line in fig. 2a)), set by energy-momentum-conservation, is closely approached, whence we infer that quantum effects indeed are non-negligible. Comparing now this figure to Fig. 2b), which is plotted in the same color-scale, we immediately conclude that there is a considerable amount of radiation emitted to directions outside the NSCS emission cone (recall that this latter figure corresponds to the detection of a photon outside this cone). We can thus conclude a clear spatial separation between the NSCS signal, confined exclusively to the angular range $\pi - \theta \leq 0.8\,\theta_0$, and the NDCS signal, which is also detectable under $\pi - \theta = \theta_0$. The classical analog of this quantum result is an effect attributed to RR changing the angular distribution of the radiation emitted by an electron [14]. For a qualitative interpretation of the presented separation of the single and two photon signals we take advantage of a

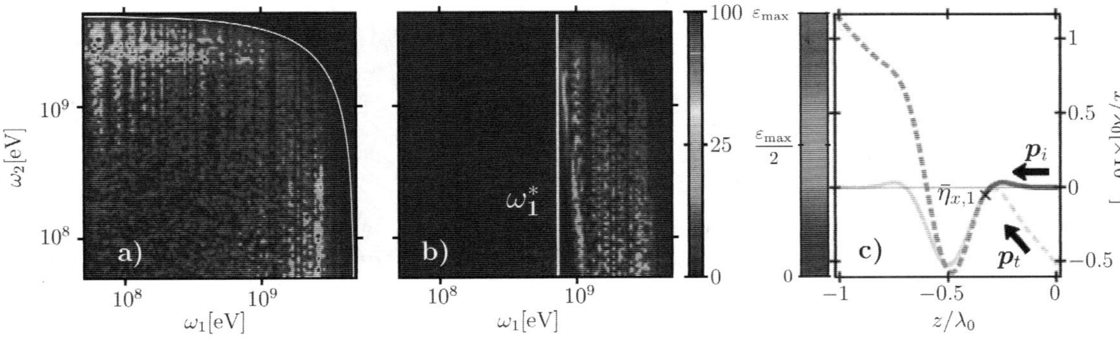

Figure 2. (color online) Two-photon energy emission spectra $\mathrm{d}E/\Pi_{i=1}^{2}\mathrm{d}\omega_i\mathrm{d}\Omega_i[\mathrm{eV}^{-1}\ \mathrm{sr}^{-2}]$ at $\chi_0 \approx 2.8$ observed at $\theta_1 = \pi - \theta_0/2$, and at $\theta_2 = \theta_1$ (part a)) and at $\theta_2 = \pi - \theta_0$ (part b)), with $\theta_0 = 5 \times 10^{-3}$ rad. Other numerical parameters are given in the text. The solid white lines correspond to the cutoff-energy $\omega_1 + \omega_2 = \varepsilon_i$ (part a)) and the threshold frequency ω_1^* (part b)), respectively. Part c): Two classical electron trajectories with initial electron momentum $\boldsymbol{p}_i$ (solid line) and $\boldsymbol{p}_t$ (dashed line). In color-code: Actual electron trajectory for a photon with energy $\omega_1 = 2$ GeV emitted at $\bar{\eta}_{x,1}$ towards (θ_1, ϕ_1).

stationary-phase analysis that was recently developed for the analysis of NDCS spectra [29]. It was shown that it was only necessary to analyze the part of the scattering matrix element proportional to the bivariate dynamic integrals $f_{r,s}$ as this partial amplitude largely dominates the scattering amplitude. The corresponding dynamic integrals were approximated as [29] $f_{r,s} \approx \sum_{l,n} \Theta(\bar{\eta}_{y,n} - \bar{\eta}_{x,l}) f_r^y(\bar{\eta}_{y,n}) f_s^x(\bar{\eta}_{x,l})$ with $f_r^{x/y} = \int \mathrm{d}\eta \psi^r(\eta) \exp[-\mathrm{i}S_{x/y}(\eta)]$ with the exponential phases $S_{x/y}(\eta)$ defined above. The sum over the indices l and n is a sum over all stationary points in whose vicinities the only non-negligible contributions to the dynamic integrals are formed. These stationary points are found as solutions of the equations $\psi(\bar{\eta}_{x,l}) = -1/2$ and $\psi(\bar{\eta}_{y,n}) = \Delta\vartheta_2 - \omega_1/\varepsilon_i(1/2 + \Delta\vartheta_2)$ with $\Delta\vartheta_2 = (\pi - \theta_2)/\theta_0$. In these expressions the fixed values of θ_1, ϕ_1 and ϕ_2 were already inserted and only θ_2 was left variable. The given equations formally correspond to the stationary point conditions of two separate single photon emissions, where first a photon with wave vector k_1^μ is emitted by an electron with initial momentum p_i^μ, whereas the second photon with wave vector k_2^μ is emitted by an electron with initial momentum p_t^μ. We can consequently interpret two-photon emission in the regime $\xi_0 \gg 1$ as the sequential emission of the first and second photon emitted from the classical trajectory an electron would take if it entered the laser field with an initial four-momentum p_i^μ and p_t^μ, respectively. For the sake of continuity these trajectories naturally have to be joined at the emission point of the first photon k_1^μ. However, the energies a classical electron will have when following these two separate trajectories will be discontinuous at the connection point, reflecting the finite loss of energy and momentum due to the quantum photon emission. We note that, of course, there are several solutions for the stationary points $\bar{\eta}_{x,l}$ and hence several combinations of classical trajectories for each choice of observation directions for the emitted photons. One exemplary combination of two such classical trajectories computed for an electron scattered from a laser pulse of the given parameters is shown in fig. 2c). An electron with initial momentum p_i^μ would follow the solid trajectory, whereas, on the other hand, an electron of initial electron momentum p_t^μ would take the dashed trajectory. In the case shown here the transitional momentum p_t^μ is computed from p_i^μ by assuming the emission of a photon of energy $\omega_1 = 2$ GeV into the direction (θ_1, ϕ_1). Since this corresponds to the loss of a significant portion of its initial energy by the electron the momentum p_t^μ significantly differs from p_i^μ and thus the two corresponding trajectories are clearly distinguishable. This feature then also explains the considerable spatial separation of the NSCS and the NDCS signal in the regime $\chi_0 \sim 1$. We note, however, that the tangent vectors of the two shown trajectories are parallel at the junction, as they have to be for an ultra-relativistic electron emits photons almost exclusively into its instantaneous direction

Journal of Physics: Conference Series **497** (2014) 012015 doi:10.1088/1742-6596/497/1/012015

of propagation and thus can lose momentum only in this direction. The aforementioned loss of energy, however, can be read off from the color-coding of the two trajectories and is clearly discontinuous at the point of emission of the first photon. It is also this significantly decreased energy of the electron that leads to a stronger deflection of the second trajectory inside the laser field. To find an analytical prediction of the energy threshold the first photon has to carry away to render this deflection strong enough to facilitate emission towards the chosen observation direction $\theta_2 = \pi - \theta_0$ we solve the defining equation of the stationary point $\bar{\eta}_{y,n}$ for the first emitted photon's frequency ω_1. From this procedure we find the threshold $\omega_1^* = \varepsilon_i(1-0.8)/(1+1/2) \approx 660$ MeV, which is well confirmed in fig. 2b). Furthermore, we have to stress that in the overall scattering amplitude we also have to include the cross-channel term $S_{fi}^{(2)}$ where the photons' wave vectors k_1^μ and k_2^μ are exchanged. The interpretation for this cross-channel is in terms of classical trajectories is analogous to the above given arguments, however, with the order of the emission of the two photons exchanged as well.

4. Influence of RR on the parametric instabilities in plasmas

Parametric instabilities of a laser pulse in a plasma are important due to their applications in the area of laser-driven fusion, laser wakefield acceleration, and have been investigated for decades [51, 52, 53, 54, 55, 56, 57, 58, 59, 60, 61, 62]. The FRS—a scattering process belonging to the general Stimulated Raman scattering (SRS) processes in plasmas—is one of the prominent examples of parametric instabilities in plasmas. In the FRS, the incident pump laser decays into two forward moving daughter electromagnetic waves, and a plasma wave. The daughter waves have their frequencies upshifted (anti-Stokes waves) and downshifted (Stokes wave) from the pump laser by the magnitude which equals the excited plasma wave frequency.

Though, at high laser intensities $I_0 \geq 10^{19}$ W/cm^2, the growth rate of the parametric instabilities becomes smaller due to the relativistic Lorentz factor [57], the role of RR force becomes also important especially at ultra-high laser intensities, $I_0 \geq 10^{22}$ W/cm^2[1, 33, 34, 14, 63, 64, 65]. Such ultra-intense laser systems are expected to be available in a near future after the commissioning of the Extreme Light Infrastructure (ELI) project in Europe [66]. Due to the RR force, the laser pulse suffers damping while propagating in a plasma. As the laser loses energy due to RR force it facilitates, apart from the usual parametric decay processes, the availability of an additional source of free energy for perturbations to grow in the plasma. Its effective intensity also decreases which lowers the relativistic Lorentz factor. Moreover, the phase shift, caused by the RR force, in the nonlinear current densities causes polarization rotations of the scattered daughter electromagnetic waves. This necessitates to include the effect of RR force in the theoretical formalism of the parametric instabilities in the plasma.

We study the FRS of an ultra-intense laser pulse in a plasma including the RR force effects in the classical electrodynamics regime where quantum effects arising due to photon recoil and spin are negligible [1]. This approach is valid if the wavelength and magnitude of the external electromagnetic field in the instantaneous rest frame of the electron satisfy $\lambda \gg \lambda_C$, $\mathcal{E} \ll E_{\rm cr}$, where $\lambda_C = 3.9 \times 10^{-11}$ cm is the Compton wavelength and E_{cr} is the critical field of the quantum electrodynamics [1]. For the laser intensities planned in the ELI project $I_0 \sim 10^{22-23}$ W/cm^2 [66], these two criteria can be fulfilled. In the classical electrodynamics regime, the Landau-Lifshitz RR force [7] correctly accounts for the radiation emitted by a relativistic charged particle [1]. Consider the propagation of a circularly polarized (CP) pump laser along the $\hat{\boldsymbol{z}}$ direction in an underdense plasma with uniform plasma electron density n_e. Ions are assumed to be at rest. Eq. of motion for an electron including the leading order term of the Landau-Lifshitz RR force in the laser field is

$$\frac{\partial \boldsymbol{p}}{\partial t} + \boldsymbol{v}\cdot\nabla\boldsymbol{p} = -e(\boldsymbol{E}+\boldsymbol{v}\times\boldsymbol{B}) - \frac{2e^4}{3m^2}\gamma^2\boldsymbol{v}\Big[(\boldsymbol{E}+\boldsymbol{v}\times\boldsymbol{B})^2 - (\boldsymbol{v}\cdot\boldsymbol{E})^2\Big], \tag{10}$$

where $\gamma = 1/\sqrt{1-v^2}$ and the velocity of the light in vacuum $c = 1$. The other terms of the Landau-Lifshitz RR force are ignored as they are $1/\gamma$ times smaller than the leading order term [7]. We first ignore the RR term and express the electric and magnetic fields in potentials as $\boldsymbol{E} =$

$-\nabla\phi - \partial \boldsymbol{A}/\partial t$, $\boldsymbol{B} = \nabla \times \boldsymbol{A}$. In a 1D approximation valid when $r_0 \gg \lambda_0$ (where r_0 is the spot-size and λ_0 is the wavelength of the pump laser pulse), the transverse momentum and z-component of motion are $\boldsymbol{p}_\perp = e\boldsymbol{A}$, and $\partial v_z/\partial t = e\nabla\phi/(m\gamma_0) - e^2\nabla|A|^2/(2m^2\gamma_0^2)$, where $\boldsymbol{A} = \boldsymbol{A}_0 e^{i\eta_0}/2 + c.c$, $\boldsymbol{A}_0 = \boldsymbol{\sigma} A_0, \boldsymbol{\sigma} = (\hat{\boldsymbol{x}} + i\hat{\boldsymbol{y}})/\sqrt{2}, \gamma_0 = (1+\xi_0^2/2)^{1/2}, \xi_0 = eA_0/m, \eta_0 = k_0 z - \omega_0 t$, and ω_0, k_0 are the carrier frequency and wavevector of the pump laser respectively [57, 67]. A plane monochromatic CP light doesn't cause any charge separation since for it $\nabla|A_0|^2 = 0$ and consequently there is no component of velocity in the $\hat{z}$ direction. This solution is known as the Akhiezer-Polovin solution for a purely transverse monochromatic CP light in plasmas [57, 68, 69]. The scattering of the laser pulse results into the total vector potential of the form $\boldsymbol{A} = [\boldsymbol{A}_0 e^{i\eta_0} + \boldsymbol{\delta A}_+ e^{i\boldsymbol{k}_\perp\cdot\boldsymbol{x}_\perp} e^{i\eta_+} + \boldsymbol{\delta A}_-^* e^{-i\boldsymbol{k}_\perp\cdot\boldsymbol{x}_\perp} e^{-i\eta_-^*}]/2 + c.c.$, where $\boldsymbol{\delta A}_+ = \boldsymbol{\sigma}\delta A_+$, $\boldsymbol{\delta A}_-^* = \boldsymbol{\sigma}\delta A_-^*$, $\boldsymbol{\delta A}_+$ and $\boldsymbol{\delta A}_-$ represent the anti-Stokes and the Stokes waves respectively, $\eta_+ = (k_z + k_0)z - (\omega + \omega_0)t$, $\eta_-^* = (k_z - k_0)z - (\omega^* - \omega_0)t$ [57, 67]. Beating of the Stokes and the anti-Stokes waves with the pump laser leads to the plasma wave excitation $\delta n/n_e$, which can be estimated from the equation of continuity, Poisson equation, and the z-component of equation of motion, yielding $\delta\tilde{n} = (e^2 k_z^2/2m^2\gamma_0^2 D_e)(A_0^*\delta A_+ + A_0\delta A_-)$, where $D_e = \omega^2 - \omega_p'^2$, $\omega_p'^2 = \omega_p^2/\gamma_0$, $\omega_p^2 = 4\pi n_e e^2/m$, $\delta n/n_e = \delta\tilde{n} e^{i\eta} e^{i\boldsymbol{k}_\perp\cdot\boldsymbol{x}_\perp}/2 + c.c$, and $\eta \equiv \eta_+ - \eta_0 \equiv \eta_- + \eta_0 = k_z z - \omega t$ [57, 67]. Plasma wave oscillation causes an axial component of velocity and momentum $v_z \ll 1, p_z \ll p_\perp$.

On using the above solutions for transverse and longitudinal components of momenta to simplify the RR term in Eq.(10), the full equation of motion after expressing the CP laser pulse as $\boldsymbol{A} = \boldsymbol{A}_\perp(\boldsymbol{x}_\perp, z, t)e^{i\eta_0}/2 + c.c.$, with its amplitude varying slowly i.e. $|\partial\boldsymbol{A}_\perp/\partial t| \ll |\omega_0\boldsymbol{A}_\perp|$, $|\partial\boldsymbol{A}_\perp/\partial z| \ll |k_0\boldsymbol{A}_\perp|$, and $|\phi| \ll |\boldsymbol{A}|, \omega_p^2/\gamma\omega_0^2 \ll 1$, and $\gamma = (1 + e^2|\boldsymbol{A}|^2/m^2)^{1/2}$, yields,

$$\frac{\partial}{\partial t}(\boldsymbol{p}_\perp - e\boldsymbol{A}) = -e\mu\omega_0\boldsymbol{A}\gamma|\boldsymbol{A}|^2(1 - 2v_z), \tag{11}$$

where $\mu = 2e^4\omega_0/3m^3$, $v_z = (\omega/k_z)\,\delta\tilde{n}\, e^{i\boldsymbol{k}_\perp\cdot\boldsymbol{x}_\perp} e^{i\eta}/2 + c.c.$, and we have assumed $|\mu\gamma|\boldsymbol{A}|^2| \ll 1$, valid for laser intensity $I_0 \le 10^{23}$ W/cm^2. Since $|\phi| \ll |A|$ and the RR effects in the case of the collinear movement of plasma electrons and the plasma wave are negligible, we don't consider the effect of RR on plasma oscillations. One can solve Eq.(11) by expressing the transverse component of the quiver momentum as *e.g.* $\boldsymbol{p}_\perp = [\boldsymbol{p}_0 e^{i\eta_0} + \boldsymbol{p}_+ e^{i\boldsymbol{k}_\perp\cdot\boldsymbol{x}_\perp} e^{i\eta_+} + \boldsymbol{p}_-^* e^{-i\boldsymbol{k}_\perp\cdot\boldsymbol{x}_\perp} e^{-i\eta_-^*}]/2 + c.c.$, where $\boldsymbol{p}_+$ and $\boldsymbol{p}_-$ have similar polarizations as the anti-Stokes and the Stokes modes. The wave equation for the vector potential after the density perturbation $n = n_e + \delta n$ reads as

$$\nabla^2\boldsymbol{A} - \frac{\partial^2\boldsymbol{A}}{\partial t^2} = \frac{\omega_p^2}{\gamma}\left(1 + \frac{\delta n}{n_e}\right)\frac{\boldsymbol{p}_\perp}{e}. \tag{12}$$

The dispersion relation for the equilibrium vector potential can be obtained after collecting the terms containing $e^{i\eta_0}$, and it reads as $\omega_0^2 = k_0^2 + \omega_p'^2(1 - i\mu|A_0|^2\gamma_0/2)$, implying that the RR term causes damping of the pump laser field. We incorporate this damping by defining a frequency or a wavenumber shift in the pump laser [1]. On writing $\omega_0 = \omega_{0r} - i\delta\omega_0$, $\delta\omega_0 \ll \omega_{0r}$ we get the frequency shift $\delta\omega_0$ as $\delta\omega_0 = \omega_p'^2\varepsilon_{\rm rad}\gamma_0\xi_0^2/2\omega_{0r}$, where $\varepsilon_{\rm rad} = r_e\omega_{0r}/3$, $r_e = e^2/m$ is the classical radius of the electron and without the loss of generality we have assumed $\xi_0 = \xi_0^*$. For the SRS growth to occur, this frequency shift must be less than the growth rate. On collecting the terms containing $e^{i\eta_\pm}e^{i\boldsymbol{k}_\perp\cdot\boldsymbol{x}_\perp}$ in Eq.(12), we get $D_+\delta A_+ = R_+(\delta A_+ + \delta A_-)$ and $D_-\delta A_- = R_-(\delta A_+ + \delta A_-)$ which yields the dispersion relation

$$\left(\frac{R_+}{D_+} + \frac{R_-}{D_-}\right) = 1, \tag{13}$$

[1] One can also incorporate the RR term by appropriately modifying the plasma frequency, which essentially implies change in the laser pump wavevector arising due to the it's dispersion in the plasma.

where

$$D_{\pm} = (\omega \pm \omega_0)^2 - \omega_p^{'2}\left(1 - \frac{i\varepsilon_{\text{rad}}\xi_0^2\gamma_0\omega_0}{\omega \pm \omega_0}\right) - [(k_z \pm k_0)^2 + k_{\perp}^2],$$

$$R_{\pm} = \frac{\omega_p^2\xi_0^2}{4\gamma_0^3}\left[\frac{k_z^2}{D_e}\left(1 \mp i\varepsilon_{\text{rad}}\xi_0^2\gamma_0 + \frac{2i\varepsilon_{\text{rad}}\xi_0^2\gamma_0}{k_z}\frac{\omega\omega_0}{\omega \pm \omega_0}\right) - \left(1 \mp i\varepsilon_{\text{rad}}\xi_0^2\gamma_0\frac{\omega}{\omega \pm \omega_0} + 4i\varepsilon_{\text{rad}}\gamma_0^3\frac{\omega_0}{\omega \pm \omega_0}\right)\right].$$

The RR term modifies the coupling between the Stokes and the anti-Stokes modes ($R_+ \neq R_-$), and the form of dispersion relation from the dispersion relation derived before [51, 53, 57, 67].

For the estimation of the FRS growth rate in a low-density plasma, $\omega_p^{'} \ll \omega_{0r}$, both the Stokes and the anti-Stokes modes have to be taken into account [30]. Substituting the pump laser frequency shift $\delta\omega_0$ and ignoring the finite $k_\perp$ gives $D_{\pm} = (\omega \pm \omega_{0r})^2 - \omega_p^{'2} - (k_z \pm k_0)^2$. On expressing $\omega = \omega_p^{'} + i\Gamma_{\text{frs}}$, where Γ_{frs} is the growth rate of the FRS instability, yields $D_{\pm} \approx 2i\Gamma_{\text{frs}}(\omega_p^{'} \pm \omega_{0r})$, $D_e \approx 2i\omega_p^{'}\Gamma_{\text{frs}}$. On assuming $k_z^2 \approx \omega_p^{'2}$, $\omega_p^{'2} - \omega_{0r}^2 \approx -\omega_{0r}^2$, we get, in the weakly-coupled regime $\Gamma_{\text{frs}} \ll \omega_p^{'}$, the growth rate of the FRS as

$$\Gamma_{\text{frs}} = -\frac{\omega_p^2\varepsilon_{\text{rad}}\xi_0^2}{2\omega_{0r}} \pm \frac{\omega_p^2\xi_0\cos(\theta/2)}{\sqrt{8}\gamma_0^2\omega_{0r}}\sqrt[4]{(1 + 2\varepsilon_{\text{rad}}^2\xi_0^2\gamma_0^4)^2 + \left(\frac{\varepsilon_{\text{rad}}\xi_0^2\gamma_0\omega_{0r}}{\omega_p^{'}}\right)^2},$$
$$\tan\theta = \left(\frac{-\varepsilon_{\text{rad}}\xi_0^2\gamma_0(\omega_{0r}/\omega_p^{'})}{(1 + 2\varepsilon_{\text{rad}}^2\xi_0^2\gamma_0^4)}\right). \tag{14}$$

Without the RR force $\varepsilon_{\text{rad}} = 0$, one recovers the relativistic growth rate of the FRS instability as derived before [57, 67]. Fig.3 shows the growth rate ratio of the FRS with ($\Gamma_{\text{frs}} - \delta\omega_0$) and without ($\Gamma_0$) RR force. It is evident that the RR force strongly enhances the growth rate of the FRS at lower plasma densities $\omega_p^{'}/\omega_{0r} \ll 1$ and higher laser amplitude $\xi_0 \gg 1$, which is also apparent from Eq.(14). The strong growth enhancement due to the RR force is counterintuitive as the later is generally considered as a damping force similar to collisions in plasmas. One can attribute this enhancement in the growth rate of the FRS due to the mixing between the Stokes and the anti-Stokes modes mediated by the RR force. Without the RR force, nonlinear currents driving the Stokes and the anti-Stokes modes have opposite polarizations. Since the phase shift induced by the RR force is polarization dependent, it is opposite for the Stokes and the anti-Stokes modes. The opposite phase shifts, consequently, lead to the interaction between the nonlinear current terms and phase shifts accumulation in Eq.(13). This phase shift accumulation is termed as the manifestation of the nonlinear mixing of the two modes, and it is responsible for the enhanced growth rate of the FRS instability. Intuitively this growth enhancement can be imagined to occur due to the availability of an additional channel of the laser energy decay due to the RR force induced damping and its subsequent utilization by both the Stokes and the anti-Stokes modes.

Since, the resonant excitation of both the Stokes and the anti-Stokes modes is the essential condition for the growth enhancement of the FRS, let us estimate the conditions under which both the modes are excited. Resonant excitation of the Stokes modes ($D_- = 0$) is always possible due to kinematical considerations. However, the simultaneous resonant excitation of both the modes is only possible in a tenuous plasma ($\omega_p^{'} \ll \omega_{0r}$). The resonant excitation of the Stokes mode leads to frequency mismatch for the anti-Stokes mode defined as $\Delta\omega_m = \omega_p^{'} + \omega_{0r} - [\omega_p^{'2} + (k_p^{'} + k_0)^2 + D_+]^{1/2}$, and it reads as $\Delta\omega_m = -\omega_p^{'3}/\omega_{0r}^2 + 9\omega_p^{'4}/4\omega_{0r}^3$. As shown in Ref. [30], this frequency mismatch is smaller than the actual growth rate $\Gamma_{\text{frs}} - \delta\omega_0$ of the FRS instability. This makes the inclusion of both the modes important while deriving the growth rate of the FRS. The RR force only marginally enhances the growth rate of the FRS, if only the Stokes mode is resonantly excited in the plasma. This can be understood easily as the nonlinear mixing of the two Raman sidebands is absent in this case. The phase shift caused by the RR force maintains the laser energy transfer to the Stokes mode for a longer time causing minor enhancement in the growth rate of the FRS.

Journal of Physics: Conference Series **497** (2014) 012015 doi:10.1088/1742-6596/497/1/012015

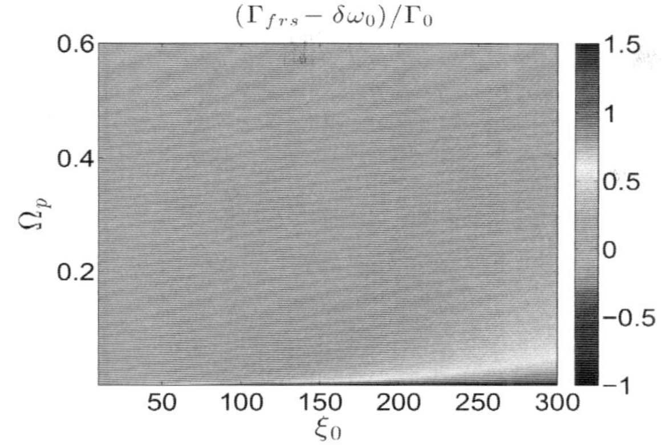

Figure 3. (Color online) Growth rate ratio $(\Gamma_{\rm frs} - \delta\omega_0)/\Gamma_0$ of the FRS in the presence $(\Gamma_{\rm frs} - \delta\omega_0)$, and in the absence of the RR force Γ_0 as a function of the normalized plasma density $\Omega_p \equiv \omega_p/\omega_{0r}$ and normalized pump laser amplitude $\xi_0 = eA_0/m$. The normalized growth rate is plotted on $\log_{10}$ scale.

These results are important for the ultra-intense laser-plasma interaction as the onset of parametric instabilities appears again, changing the frequency spectra and shapes of extremely intense short laser pulses. Moreover, enhanced FRS of the laser pulse provides an alternative way to detect the RR effects on the spectra of low-energy optical photons. This is in contrast to the scheme of the nonlinear Compton scattering of a counter-propagating relativistic electron in a strong laser field, which aims to discern the signatures of the RR force on the spectra of high-energy gamma-ray photons [1].

5. Conclusions

To summarize, we have investigated RR effects for the collision of an ultra-intense laser pulse with ultra-relativistic electrons in the quantum regime as well as with a plasma in the classical regime. In contrast to the classically predicted narrowing of the energy width of particle beams, we have shown by employing a kinetic approach that the stochastic nature of photon emission spreads up the energy distribution of the electrons, if quantum effects are substantial. Further, the quantum computation including two-photon emission indicates an extensive broadening of the predicted angular range of the emitted radiation. In the quantum regime this can be associated with the discontinuous energy loss of the electron and the following modification of its trajectory, whereas in classical electrodynamics RR effects alter an electron's trajectory smoothly. Moreover, the forward Raman scattering in plasmas is shown to be significantly amplified by the inclusion of classical RR, due to the induced nonlinear mixing of the anti-Stokes and the Stokes modes. Finally, our numerical examples indicate that all the discussed effects should be detectable with laser-based electron accelerators with presently available and next-generation lasers.

[1] Di Piazza A, Müller C, Hatsagortsyan K Z and Keitel C H 2012 Rev. Mod. Phys. **84** 1177
[2] Rohrlich F 2002 Phys. Lett. A **303** 307
[3] Bonifacio R and Casagrande F 1984 Opt. Comm. **50** 251
[4] Bonifacio R and Casagrande F 1985 Nucl. Instr. Meth. Phys. Res. A **237** 168
[5] Moshammer R and Ullrich J 2009 J. Phys. B: At. Mol. Opt. Phys. **42** 130201
[6] Apyan A et al. 2005 Nucl. Instr. and Meth. B, **234** 128
[7] Landau L D and Lifshitz E M 1975 *The classical theory of fields* (Oxford: Butterworth-Heinemann)
[8] Spohn H 2000 Europhys. Lett. **50** 287
[9] Spohn H 2004 *Dynamics of Charged Particles and Their Radiation Field* (Cambridge: Cambridge University Press)
[10] Rohrlich F 2007 *Classical Charged Particles* (Singapore: World Scientific)
[11] Bulanov S V, Esirkepov T Zh, Kando M, Koga J K and Bulanov S S 2011 Phys. Rev. E **84** 056605
[12] Di Piazza A 2008 Lett. Math. Phys. **83** 305

[13] Tamburini M, Keitel C H and Di Piazza A 2013 Electron dynamics controlled via self-interaction *Preprint* arXiv:1306.3328
[14] Di Piazza A, Hatsagortsyan K Z and Keitel C H 2009 Phys. Rev. Lett. **102** 254802
[15] Heinzl T, Harvey C, Ilderton A, Marklund M, Bulanov S S, Rykovanov S, Schroeder C B, Esarey E and Leemans W P 2013 Detecting radiation reaction at moderate laser intensities *Preprint* arXiv:1310.0352
[16] Bulanov S V et al. 2011 Nucl. Instr. Meth. Phys. Res. A **660** 31
[17] Zhidkov A, Masuda S, Bulanov S S, Hosokai T, Koga J and R. Kodama 2013 Radiation Reaction Effects in Cascade Scattering of Intense, Tightly Focused Laser Pulses by Relativistic Electrons *Preprint* arXiv:1308.1608
[18] Ilderton A and Torgrimsson G 2013 Radiation reaction from QED: lightfront perturbation theory in a plane wave background *Preprint* arXiv:1304.6842
[19] Boca M and Florescu V 2009 Phys. Rev. A **80** 053403
[20] Mackenroth F and Di Piazza A 2011 Phys. Rev. A **83** 032106
[21] Seipt D and Kämpfer B 2011 Phys. Rev. A **83** 022101
[22] Krajewska K and Kamiński J Z 2012 Phys. Rev. A **85** 062102
[23] Harvey C, Heinzl T, Ilderton A and Marklund M 2012 Phys. Rev. Lett. **109** 100402
[24] Jackson J D 1975 *Classical Electrodynamics* (New York: Wiley)
[25] Di Piazza A, Hatsagortsyan K Z and Keitel C H 2010 Phys. Rev. Lett. **105** 220403
[26] Lötstedt E and Jentschura U D 2009 Phys. Rev. A **80** 053419
[27] Lötstedt E and Jentschura U D 2009 Phys. Rev. Lett. **103** 110404
[28] Seipt D and Kämpfer B 2012 Phys. Rev. D **85** 101701
[29] Mackenroth F and Di Piazza A 2013 Phys. Rev. Lett **110** 070402
[30] Kumar N, Hatsagortsyan K Z and Keitel C H 2013 Phys. Rev. Lett. **111** 105001
[31] Zhidkov A, Koga J, Sasaki A and Uesaka M 2002 Phys. Rev. Lett. **88** 185002
[32] Naumova N M, Schlegel T, Tikhonchuk V T, Labaune C, Sokolov I V and Mourou G A 2009 Phys. Rev. Lett. **102** 025002
[33] Chen M, Pukhov A, Yu T P and Sheng Z M 2011 Plasma Phys. Controlled Fusion **53** 014004
[34] Tamburini M, Pegoraro F, Di Piazza A, Keitel C H and Macchi A 2010 New J. Phys. **12** 123005
[35] Tamburini M, Pegoraro F, Di Piazza A, Keitel C H, Liseykina T V and Macchi A 2011, Nucl. Instr. Meth. Phys. Res. **653** 181
[36] Baier V N, Katkov V M and Strakhovenko V M 1998, *Electromagnetic processes at high energies in oriented single crystals* (Singapore: World Scientific)
[37] Khokonov M Kh 2004 Soviet Phys. JETP **99** 690
[38] Sokolov I V, Naumova N M, Nees J A and Mourou G A 2010 Phys. Rev. Lett. **105** 195005
[39] Yanovsky V *et al.* 2008 Opt. Express **16** 2109
[40] Ritus V I 1985 J. Sov. Laser Res. **6** 497
[41] Gardiner C 2009 *Stochastic Methods: A Handbook for the Natural and Social Sciences* (Berlin: Springer)
[42] Lifshitz E M and Pitaevskii L P 1981 *Physical Kinetics* (Oxford: Pergamon Press)
[43] Neitz N and Di Piazza A 2013 Phys. Rev. Lett. **111** 054802
[44] Bazarov I V, Bilderback D H, Gruner S M, Padamsee H S, Talman R, Tigner M, Krafft G A, Merminga L and Sinclair C K 2001 *Proc. Particle Accelerator Conference 2001* 230
[45] Leemans W P, Nagler B, Gonsalves A J, Tóth Cs, Nakamura K, Geddes C G R, Esarey E, Schroeder C B and Hooker S M 2006 Nature Phys. **2** 696
[46] Clayton C E et al. 2010 Phys. Rev. Lett. **105** 105003
[47] Furry W H 1951 Phys. Rev. **81** 115
[48] Fradkin E S, Gitman D M and Shvartsman S M 1991 *Quantum Electrodynamics* (Berlin: Springer)
[49] Berstetskii V B, Lifshitz E M and Pitaevskii L P 1982 *Quantum Electrodynamics* (Oxford: Elsevier)
[50] Mackenroth F, Di Piazza A and Keitel C H 2010 Phys. Rev. Lett. **105** 063903
[51] Kruer W 2003, *The Physics of Laser Plasma Interactions*, Frontiers in Physics (Boulder, CO: Westview)
[52] Brueckner K A and Jorna S 1974 Rev. Mod. Phys. **46** 325
[53] Drake J F, Kaw P K, Lee Y C, Schmid G, Liu C S and Rosenbluth M N 1974 Phys. Fluids **17** 778
[54] Tripathi V K and Liu C S 1991 Phys. Fluids. B **3** 468
[55] McKinstrie C J and Bingham R 1992 Phys. Fluids. B **4** 2626
[56] Sakharov A S and Kirsanov V I 1994 Phys. Rev. E **49** 3274
[57] Decker C D, Mori W B, Tzeng K C and Katsouleas T 1996 Phys. Plasmas **3** 2047
[58] Barr H C, Mason P and Parr D M 1999 Phys. Rev. Lett. **83** 1606
[59] Antonsen J T M and Mora P 1993 Phys. Fluids. B **5** 1440
[60] Guerin S, Laval G, Mora P, Adam J C, Heron A and Bendib A 1995 Phys. Plasmas **2** 2807
[61] Quensel B, Mora P, Adam J C, Guerin S, Heron A and Laval G 1997 Phys. Rev. Lett. **78** 2132
[62] Esarey E, Schroeder C B and Leemans W P 2009 Rev. Mod. Phys. **81** 1229 and references therein

[63] Schlegel T and Tikhonchuk V T 2012 New. J. Phys. **14** 073034
[64] Sokolov I V, Nees J A, Yanovsky V P, Naumova N M and Mourou G A 2010 Phys. Rev. E **81** 036412
[65] Keitel C H, Szymanowski C, Knight P L and Maquet A 1998 J. Phys. B **31** L75
[66] http://www.extreme-light-infrastructure.eu
[67] Gibbon P 2005, *Short Pulse Laser Matter Interaction with Matter:An Introduction* (Singapore: World Scientific)
[68] Akhiezer A I and Polovin R 1956 Sov. Phys. JETP **3** 696
[69] Bulanov S V *et al.* 2001 *Reviews of Modern Physics* Vol.22 edited by Shafranov V D (New York: Kluwer/Plenum)

22nd International Laser Physics Worksop (LPHYS 13) IOP Publishing
Journal of Physics: Conference Series **497** (2014) 012016 doi:10.1088/1742-6596/497/1/012016

Breit Wheeler pair creation by nite laser pulses

K. Krajewska and J. Z. Kamiński

Institute of Theoretical Physics, Faculty of Physics, University of Warsaw,
Hoża 69, 00-681 Warszawa, Poland

E-mail: `katarzyna.krajewska@fuw.edu.pl`

Abstract. Probability distributions of electron-positron pair creation in collisions of a laser beam and a nonlaser photon (Breit-Wheeler process) are calculated in the framework of strong- eld quantum electrodynamics. The driving laser eld is modeled as a nite pulse, similar to the formulation introduced in Phys. Rev. A **86**, 052104 (2012). The sensitivity of pair production to the driving pulse peak intensity and duration, in the context of spin e ects, are investigated.

1. Introduction

With the rapid development of high-power laser technology, the laser radiation in the near-visible spectrum with intensities as high as 10^{22} W cm^2 can be produced in the laboratory [1]. These new capabilities have led to a renaissance of theoretical interest in strong-field quantum electrodynamics (QED), in which electron dynamics in a powerful laser field must be treated relativistically and in which nonlinear QED processes become important. One of such processes is nonlinear electron-positron (e^-e^+) pair creation which occurs in collisions of an intense laser beam with a nonlaser photon. This process, known as the *nonlinear Breit-Wheeler process*, is the topic of our paper (for recent reviews, see, Refs. [2, 3, 4]).

In the original paper of Breit and Wheeler [5], a collision of two light photons which combined their energies to produce the e^-e^+ pair was considered. This idea was further developed for the case when one of the photons is replaced by a strong laser field [6, 7, 8]. In these studies, the laser field was treated as a monochromatic plane wave. The process in a weakly nonmonochromatic laser field was investigated by Narozhny and Fofanov [9]. It is crucial to realize that the process was realized in the series of experiments performed at S AC [10, 11], which aimed at testing the hypothesis of light-light scattering.

Since the synthesis of laser pulses with well-controlled properties is currently feasible, it is important to investigate their impact on the nonlinear Breit-Wheeler process. This was realized in recent works where pair creation stimulated by a pulsed laser field was analyzed [12, 13, 14, 15, 16]. While different aspects were studied in these papers, including the dependence of probability distributions of created pairs on intensity, shape, duration, and carrier-envelope phase of the driving pulse, neither of them was analyzed in the context of spin effects. In this paper, we demonstrate that generation of lepton pairs in well-defined spin states can be effectively altered by varying the parameters of the driving pulse. This is in contrast to previous investigations of spin effects in strong-field QED processes which rely on a monochromatic plane wave field approximation (see, for instance, Refs. [17, 18, 19, 20, 21, 22, 23, 24, 25, 26]).

This paper is organized as follows. In Section 2, we present the theoretical formulation of the nonlinear Breit-Wheeler process which is along the lines presented in our previous paper [15].

In Section 3, we specify the laser pulse shape for which the actual numerical calculations are performed. The latter are presented in Section 4, showing a strong dependence of spin effects on parameters of a driving laser pulse such as pulse duration and peak intensity. A brief summary of our results is given in Section 5.

In the theoretical formulas below we keep $\hbar = 1$, however the numerical results are presented in relativistic units such that $\hbar = c = m_{\rm e} = 1$ where $m_{\rm e}$ is the electron mass.

2. Theory

The probability amplitude for the Breit-Wheeler pair creation process, $\gamma_{\boldsymbol{K}\sigma} \to e^-_{\boldsymbol{p}_{\rm e^-}\lambda_{\rm e^-}} + e^+_{\boldsymbol{p}_{\rm e^+}\lambda_{\rm e^+}}$, with electron and positron momenta and spin polarizations $\boldsymbol{p}_{\rm e^-}\lambda_{\rm e^-}$ and $\boldsymbol{p}_{\rm e^+}\lambda_{\rm e^+}$, respectively, equals

$$\mathcal{A}(\gamma_{\boldsymbol{K}\sigma} \to e^-_{\boldsymbol{p}_{\rm e^-}\lambda_{\rm e^-}} + e^+_{\boldsymbol{p}_{\rm e^+}\lambda_{\rm e^+}}) = -{\rm i}e \int {\rm d}^4x\, j^{(+-)}_{\boldsymbol{p}_{\rm e^-}\lambda_{\rm e^-},\boldsymbol{p}_{\rm e^+}\lambda_{\rm e^+}}(x) \cdot A^{(+)}_{\boldsymbol{K}\sigma}(x), \tag{1}$$

where $\boldsymbol{K}\sigma$ denotes the initial nonlaser photon momentum and polarization. Here,

$$A^{(+)}_{\boldsymbol{K}\sigma}(x) = \sqrt{\frac{1}{2\varepsilon_0\omega_{\boldsymbol{K}}V}}\,\varepsilon_{\boldsymbol{K}\sigma}{\rm e}^{-{\rm i}K\cdot x}, \tag{2}$$

where V is the quantization volume, ε_0 is the vacuum electric permittivity, $\omega_{\boldsymbol{K}} = cK^0 = c|\boldsymbol{K}|$ $(K\cdot K = 0)$, and $\varepsilon_{\boldsymbol{K}\sigma} = (0, \boldsymbol{\varepsilon}_{\boldsymbol{K}\sigma})$ is the linear polarization four-vector satisfying the conditions,

$$K\cdot\varepsilon_{\boldsymbol{K}\sigma} = 0, \quad \varepsilon_{\boldsymbol{K}\sigma}\cdot\varepsilon_{\boldsymbol{K}\sigma'} = -\delta_{\sigma\sigma'}, \tag{3}$$

for $\sigma, \sigma' = 1, 2$. The matrix element of the pair current operator is defined as

$$[j^{(+-)}_{\boldsymbol{p}_{\rm e^-}\lambda_{\rm e^-},\boldsymbol{p}_{\rm e^+}\lambda_{\rm e^+}}(x)]^\nu = \bar{\psi}^{(+)}_{\boldsymbol{p}_{\rm e^-}\lambda_{\rm e^-}}(x)\gamma^\nu\psi^{(-)}_{\boldsymbol{p}_{\rm e^+}\lambda_{\rm e^+}}(x). \tag{4}$$

Moreover, $\psi^{(\beta)}_{\boldsymbol{p}\lambda}(x)$ (with $\beta = +1$ for electron and $\beta = -1$ for positron) is the Volkov solution of the Dirac equation coupled to the electromagnetic field [27, 28]

$$\psi^{(\beta)}_{\boldsymbol{p}\lambda}(x) = \sqrt{\frac{m_{\rm e}c^2}{VE_{\boldsymbol{p}}}}\Big(1 - \beta\frac{e}{2k\cdot p}\not{A}\not{k}\Big)u^{(\beta)}_{\boldsymbol{p}\lambda}{\rm e}^{-{\rm i}\beta S^{(\beta)}_p(x)}, \tag{5}$$

with the phase:

$$S^{(\beta)}_p(x) = p\cdot x + \int_{-\infty}^{k\cdot x}\Big[\beta\frac{eA(\phi)\cdot p}{k\cdot p} - \frac{e^2A^2(\phi)}{2k\cdot p}\Big]{\rm d}\phi. \tag{6}$$

Here, $E_{\boldsymbol{p}} = cp^0 \geqslant m_{\rm e}c^2$, $p = (p^0, \boldsymbol{p})$, $p\cdot p = (m_{\rm e}c)^2$, and $u^{(\beta)}_{\boldsymbol{p}\lambda}$ are the free-electron (positron) bispinors normalized such that

$$\bar{u}^{(\beta)}_{\boldsymbol{p}\lambda}u^{(\beta')}_{\boldsymbol{p}\lambda'} = \beta\delta_{\beta\beta'}\delta_{\lambda\lambda'}. \tag{7}$$

The four-vector potential $A(k\cdot x)$ in Eq. (5) represents an external electromagnetic radiation generated by lasers such that $k\cdot A(k\cdot x) = 0$ and $k\cdot k = 0$.

We consider a linearly polarized laser pulse that propagates in the direction determined by the unit vector $\boldsymbol{n}$ and that lasts for time $T_{\rm p}$. Its fundamental frequency is $\omega = 2\pi/T_{\rm p}$ whereas its wave four-vector is $k = k^0(1, \boldsymbol{n})$ with $k^0 = \omega/c$. The laser pulse is characterized by the following four-vector potential,

$$A(k\cdot x) = A_0\varepsilon f(k\cdot x), \tag{8}$$

where A_0 relates to the peak laser field intensity, and where a shape function $f(k\cdot x)$ vanishes for $k\cdot x < 0$ and $k\cdot x > 2\pi$. Here, a linear polarization vector $\varepsilon = (0, \boldsymbol{\varepsilon})$ is such that $\varepsilon^2 = -\boldsymbol{\varepsilon}^2 = -1$

and $k \cdot \varepsilon = 0$. As explained in Ref. [15], such a choice of the polarization four-vector ε as a spacelike vector (the same concerns also $\varepsilon_{\boldsymbol{K}\sigma}$) is justified by gauge invariance of the probability amplitude (1). Also in Ref. [15], we showed that this amplitude can be represented such that

$$\mathcal{A}(\gamma_{\boldsymbol{K}\sigma} \to e^-_{\boldsymbol{p}_{e^-}\lambda_{e^-}} + e^+_{\boldsymbol{p}_{e^+}\lambda_{e^+}}) = \mathrm{i}\sqrt{\frac{2\pi\alpha c(m_e c^2)^2}{E_{\boldsymbol{p}_{e^-}} E_{\boldsymbol{p}_{e^+}} \omega_{\boldsymbol{K}} V^3}}\, \mathcal{A}_{\mathrm{BW}}, \tag{9}$$

where $\alpha = e^2/(4\pi\varepsilon_0 c)$ is the fine-structure constant, V is the quntization volume, and

$$\mathcal{A}_{\mathrm{BW}} = \sum_N (2\pi)^3 \delta^{(1)}(P_N^-)\delta^{(2)}(\boldsymbol{P}_N^\perp) D_N \frac{1 - \mathrm{e}^{-2\pi\mathrm{i}P_N^+/k^0}}{\mathrm{i}P_N^+}. \tag{10}$$

Here, a four-vector

$$P_N = K - \bar{p}_{e^-} - \bar{p}_{e^+} + Nk, \tag{11}$$

is expressed in terms of the laser-dressed momenta of electron and positron,

$$\bar{p}_{e\mp} = p_{e\mp} \mp \mu m_e c \frac{\varepsilon \cdot p_{e\mp}}{k \cdot p_{e\mp}} \langle f \rangle k + \frac{1}{2}(\mu m_e c)^2 \frac{\langle f^2 \rangle}{k \cdot p_{e\mp}} k, \tag{12}$$

where the averaged values $\langle f^i \rangle$ for $i = 1, 2$ are defined as

$$\langle f^i \rangle = \frac{1}{2\pi} \int_0^{2\pi} \mathrm{d}(k \cdot x) [f(k \cdot x)]^i. \tag{13}$$

In Eq. (10), the light cone coordinates have been used $P_N = (P_N^-, P_N^+, \boldsymbol{P}_N^\perp)$ in accordance with Ref. [29]. In addition, the Fourier coefficients, D_N, are defined as

$$\begin{aligned} D_N \;=\; & \frac{1}{2}\mu m_e c \Big[\frac{2k^0}{Q^0} \Big(\frac{\varepsilon \cdot p_{e^+}}{k \cdot p_{e^+}} - \frac{\varepsilon \cdot p_{e^-}}{k \cdot p_{e^-}} \Big) \bar{u}^{(+)}_{\boldsymbol{p}_{e^-}\lambda_{e^-}} \not{\varepsilon}_{\boldsymbol{K}\sigma} u^{(-)}_{\boldsymbol{p}_{e^+}\lambda_{e^+}} - \frac{1}{k \cdot p_{e^+}} \bar{u}^{(+)}_{\boldsymbol{p}_{e^-}\lambda_{e^-}} \not{\varepsilon}_{\boldsymbol{K}\sigma} \not{\varepsilon} \not{k} u^{(-)}_{\boldsymbol{p}_{e^+}\lambda_{e^+}} \\ & - \frac{1}{k \cdot p_{e^-}} \bar{u}^{(+)}_{\boldsymbol{p}_{e^-}\lambda_{e^-}} \not{\varepsilon} \not{k} \not{\varepsilon}_{\boldsymbol{K}\sigma} u^{(-)}_{\boldsymbol{p}_{e^+}\lambda_{e^+}} \Big] G_N^{(1)} + \frac{(\mu m_e c)^2}{4(k \cdot p_{e^+})(k \cdot p_{e^-})} \Big[\frac{2k^0}{Q^0} \Big(k \cdot p_{e^+} + k \cdot p_{e^-} \Big) \\ & \times\; \bar{u}^{(+)}_{\boldsymbol{p}_{e^-}\lambda_{e^-}} \not{\varepsilon}_{\boldsymbol{K}\sigma} u^{(-)}_{\boldsymbol{p}_{e^+}\lambda_{e^+}} + \bar{u}^{(+)}_{\boldsymbol{p}_{e^-}\lambda_{e^-}} \not{\varepsilon} \not{k} \not{\varepsilon}_{\boldsymbol{K}\sigma} \not{\varepsilon} \not{k} u^{(-)}_{\boldsymbol{p}_{e^+}\lambda_{e^+}} \Big] G_N^{(2)}, \end{aligned} \tag{14}$$

where $Q^0 = k^0 - p^0_{e^-} - p^0_{e^+} \neq 0$, and where we have introduced a relativistically invariant parameter, $\mu = |eA_0|/(m_e c)$. The functions $G_N^{(i)}$, with $i = 1, 2$, are expressed in terms of the generalized Bessel functions (for more details, see [15, 29]).

Now, it is possible to derive the probability distribution for positrons produced in the Breit-Wheeler process by a finite laser pulse [15]

$$\frac{\mathrm{d}^3\mathsf{P}^{(\mathrm{p})}}{\mathrm{d}E_{\boldsymbol{p}_{e^+}} \mathrm{d}^2\Omega_{\boldsymbol{p}_{e^+}}} = \frac{\alpha(m_e c)^2 k^0 |\boldsymbol{p}_{e^+}|}{(2\pi)^2 \omega_{\boldsymbol{K}} (k \cdot p_{e^-})} \Big| \sum_N D_N \frac{1 - \mathrm{e}^{-2\pi\mathrm{i}P_N^0/k^0}}{P_N^0} \Big|^2, \tag{15}$$

where the electron four-momentum p_{e^-} equals

$$\boldsymbol{p}^\perp_{e^-} = \boldsymbol{w}^\perp, \tag{16}$$

$$p^\parallel_{e^-} = \frac{(m_e c)^2 + (\boldsymbol{w}^\perp)^2 - (w^-)^2}{2w^-}, \tag{17}$$

$$p^0_{e^-} = \frac{(m_e c)^2 + (\boldsymbol{w}^\perp)^2 + (w^-)^2}{2w^-}, \tag{18}$$

with $w = K - p_{e^+}$. The last formulas follow from solving the momentum conservation conditions, $P_N^- = 0$ and $\boldsymbol{P}_N^{\perp} = \boldsymbol{0}$, which are defined by the delta functions in Eq. (10). One can check that these solutions satisfy the on-mass shell relation, $p_{e^-} \cdot p_{e^-} = (m_e c)^2$. Note also that the corresponding probability distribution for electrons can be obtained from (15) by interchanging e^+ and e^-.

3. Laser pulse shape

For numerical illustrations, we choose the four-vector potential of the form

$$A(k \cdot x) = A_0 B \varepsilon f(k \cdot x), \tag{19}$$

with the following shape function:

$$f(k \cdot x) = N_A \sin^2\Big(\frac{k \cdot x}{2}\Big) \sin(N_{\rm osc} k \cdot x). \tag{20}$$

Here, A_0 is related to the parameter μ whereas $N_{\rm osc}$ denotes the number of field oscillations. As one can understand, the pulse lasts for time $T_{\rm p} = 2\pi N_{\rm osc}/\omega$. In addition, the parameter B is introduced in Eq. (19). If we choose $B = N_{\rm osc}$ the mean intensity contained in the pulse is fixed, irrespective of its duration. For this to happen, we have to keep

$$\frac{1}{2\pi} \int_0^{2\pi} \mathrm{d}(k \cdot x) [f'(k \cdot x)]^2 = \frac{1}{2}, \tag{21}$$

where the derivative is with respect to $k \cdot x$ (for more details, see [30]). In the following, we assume that the laser pulse propagates in the z-direction ($\boldsymbol{n} = \boldsymbol{e}_z$) and its polarization vector is $\boldsymbol{\varepsilon} = \boldsymbol{e}_x$. In the considered configuration, the nonlaser photon is counterpropagating ($\boldsymbol{n}_{\boldsymbol{K}} = -\boldsymbol{e}_z$) and its polarization vector is either parallel, $\boldsymbol{\varepsilon}_{\boldsymbol{K}} = \boldsymbol{e}_x$, or perpendicular, $\boldsymbol{\varepsilon}_{\boldsymbol{K}} = -\boldsymbol{e}_y$, to the polarization direction of the laser pulse. For these two configurations, in the next section we compare the angle-fixed energy spectra of created positrons.

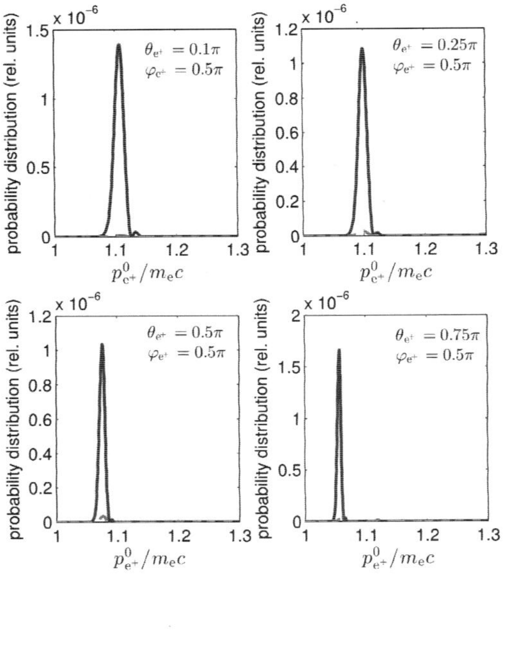

Figure 1. The angle-fixed energy spectra of positrons [defined by Eq. (15)] which are produced in a head-on collision of a laser pulse with a nonlaser photon. The parameters of the colliding pulse are such that $\mu = 1$, $\omega_{\rm L} = 0.3 m_e c^2$, and $N_{\rm osc} = 32$, while for the nonlaser photon we have $\omega_{\boldsymbol{K}} = m_e c^2$. Both polarization directions are parallel, i.e., $\boldsymbol{\varepsilon} = \boldsymbol{\varepsilon}_{\boldsymbol{K}} = \boldsymbol{e}_x$. While the solid blue line corresponds to the creation of a positron and an electron with the same helicity ($\lambda_{e^+}\lambda_{e^-} = 1$), the dashed red line (which is hardly visible on the scale of this figure) corresponds to the opposite helicity configuration for the produced particles ($\lambda_{e^+}\lambda_{e^-} = -1$).

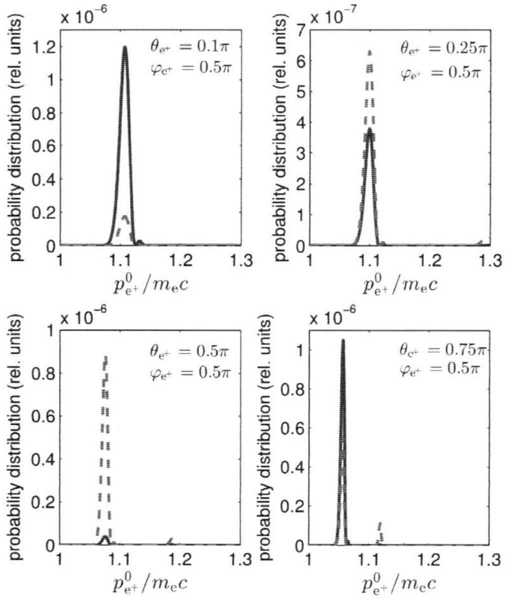

Figure 2. The same as in Fig. 1 but for $\boldsymbol{\varepsilon_K} = -\boldsymbol{e}_y$.

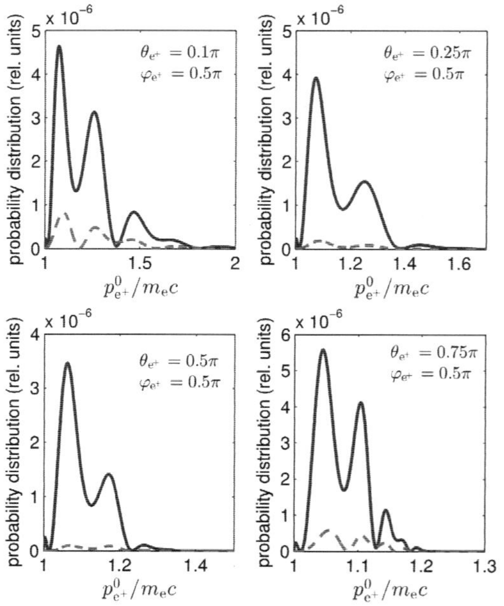

Figure 3. The same as in Fig. 1 but for $N_{\rm osc} = 4$.

4. Numerical illustrations

The distribution (15) is defined for well-defined spin states of produced electrons and positrons. Based on this formula, we shall study now the spin effects in the nonlinear Breit-Wheeler process.

In Figs. 1 and 2 we show the positron spectra (15) calculated at fixed detection angles in the (yz)-plane, as denoted in each panel. We present the results for the case when linear polarizations of the laser pulse and the colliding photon are parallel (Fig. 1), $\boldsymbol{\varepsilon} = \boldsymbol{\varepsilon_K} = \boldsymbol{e}_x$, or perpendicular (Fig. 2) to each other, $\boldsymbol{\varepsilon} = \boldsymbol{e}_x$ and $\boldsymbol{\varepsilon_K} = -\boldsymbol{e}_y$. The remaining parameters are given in the caption to Fig. 1. Here, we focus on spin effects of the created pairs. Therefore, we compare the energy distributions for the case when the positron and electron are created with the same (solid lines) or the opposite (dashed lines) spin projections. As one can see in Fig. 1, the former configuration dominates. This tendency, however, can be manipulated by changing the polarization direction of the colliding photon or that of the laser pulse, as shown in Fig. 2.

The results presented in Figs. 1 and 2 relate to a rather long driving pulse, which contains 32 field oscillations. In this case, we observe separate multiphoton peaks in the energy spectra of created particles, with dominant peaks as shown in the figures and smaller peaks which are hardly visible on the scale of these figures (see, also Ref. [15]). The question arises whether the pulse duration has any observable impact on the spin effects. To answer this question we present the results for a much shorter pulse, which contains only four oscillations of the electromagnetic field, $N_{\rm osc} = 4$. Both configurations, when the polarization vectors of the pulse and the nonlaser photon are parallel or perpendicular to each other, are considered (Fig. 3 and 4, respectively).

When comparing Figs. 1 and 3, we observe the same tendency which is that electrons and positrons are mostly created with the same spin projection. It turns out that for perpendicular polarizations, the creation of particles with the opposite spin projections is more significant for shorter pulses (compare Figs. 2 and 4). The latter depends also on the peak intensity of the driving pulse. This can be seen when comparing Figs. 4 and 5. Both figures relate to the same physical geometry, but the peak field strength is different in these two cases (it is either $\mu = 1$

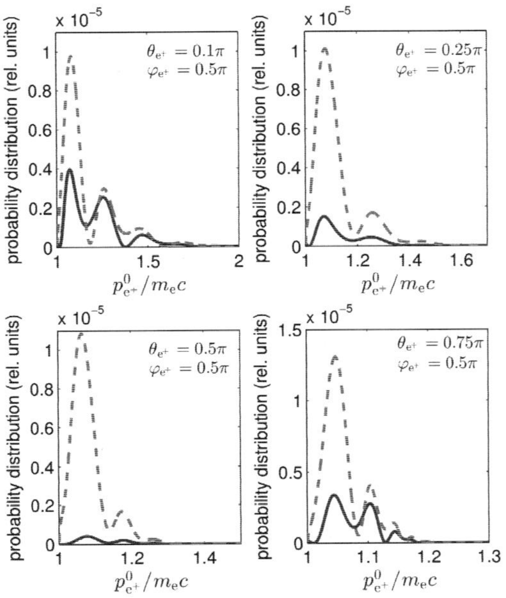

Figure 4. The same as in Fig. 2 but for $N_{\rm osc} = 4$.

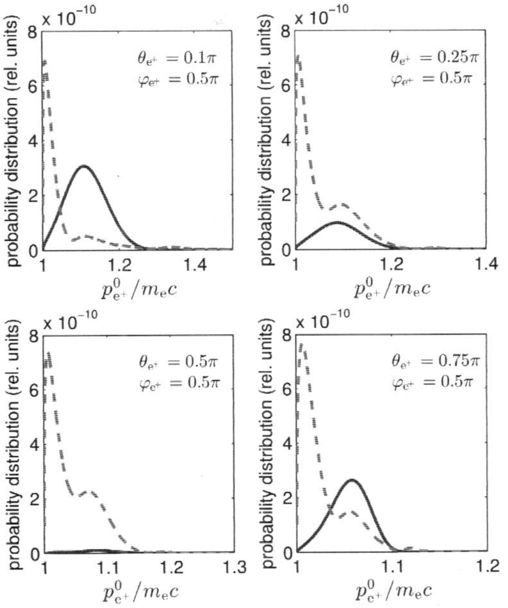

Figure 5. The same as in Fig. 4 but for $\mu = 0.1$.

in Fig. 4, or $\mu = 0.1$ in Fig. 5). Note that for a weaker field (Fig. 5), we observe quite broad regions of positron energy where both product particles are created in the same spin states [this specifically concerns the case when the particles are detected in the propagation direction of the pulse, i.e., for small θ_{e+} (see, the upper left panel for $\theta_{e+} = 0.1\pi$)]. However, if we increase the peak field strength (Fig. 4), this energy region significantly deteriorates.

Here, we have looked at the energy spectra of created positrons in the (yz)-plane, i.e., in the plane normal to the polarization direction of the pulse. While for the colinear polarizations of the pulse and the nonlaser photon the electron and the positron are predominantly created in the same spin states (Figs. 1 and 3), the situation is changed for the perpendicular polarizations of the pulse and the nonlaser photon (Figs. 2, 4, and 5). Although we do not present here the corresponding results we have also looked at the positron spectra in the (xz)-plane, i.e., in the plane defined by the polarization and the propagation direction of the driving field. In this case, we observe that the configuration with the opposite spin projections becomes dominant for a colinear configuration of $\boldsymbol{\varepsilon}$ and $\boldsymbol{\varepsilon_K}$ unless the particles are detected in the propagation direction of the pulse (at small angles θ_{e+}). The electrons and the positrons produced at small angles θ_{e+} are in the same spin states. On the other hand, when the laser pulse and the nonlaser photon polarizations are perpendicular, the leptons which are created at small angles are in the opposite spin states. This stays true for different values of μ and $N_{\rm osc}$.

In closing, we note that the stronger the peak field intensity the more effective is pair creation, as illustrated by Figs. 4 and 5. This explains also why the signal of pair creation becomes larger when, for a fixed μ, we decrease the incident pulse duration. Once we keep the average intensity contained in a pulse fixed (see, the discussion in Sec. 3), the maximum value of the electric and magnetic fields must be effectively increased when decreasing the pulse duration. Consequently, an enhancement of the probability of pair production induced by shorter laser pulses is observed. Specifically, if one compares Figs. 1 and 3 the spectra are enhanced roughly four times; if one compares Figs. 2 and 4 the spectra are enhanced by one order of magnitude.

5. Conclusions

In this paper, we have focused on spin effects in the electron-positron pair creation which happens via collisions of a strong laser pulse with a nonlaser photon; both being linearly polarized. We have compared two cases of the colinear and perpendicular polarizations of the laser pulse and the nonlaser photon. We have observed that while for the colinear polarizations, both leptons (when detected at small angles) tend to be in the same spin state thus for the perpendicular polarizations this tendency is reversed. This cannot be generalized for particles detected at bigger angles. We have demonstrated that the spin effects in the nonlinear Breit-Wheeler process are sensitive to both the driving pulse peak strength and its duration. Therefore, they can be used as control parameters when producing highly polarized positron (electron) beams.

Acknowledgments

This work is supported by the Polish National Science Center (NCN) under Grant No. 2011 01 B ST2 00381.

References

[1] Yanovsky V, Chvykov V, Kalinchenko G, Rousseau P, Planchon T, Matsuoka T, Maksimchuk A, Nees J, Cheriaux G and Krushelnick K 2008 *Opt. Exp.* **16** 2109
[2] Ehlotzky F, Krajewska K and Kaminski J 2009 *Rep. Prog. Phys.* **72** 046401
[3] Ruffini R, Vereshchagin G and ue S S 2010 *Phys. Rep.* **487** 1
[4] Di Piazza A, Muller C, Hatsagortsyan K and Keitel C H 2012 *Rev. Mod. Phys.* **84** 1177
[5] Breit G and Wheeler J A 1934 *Phys. Rev.* **46** 1087
[6] Reiss H R 1962 *J. Math. Phys.* **3** 59
[7] Nikishov A I and Ritus V I 1964 *Sov. Phys. JETP* **19** 529
[8] Narozhny N B, Nikishov A I and Ritus V I 1965 *Sov. Phys. JETP* **20** 622
[9] Narozhny N B and Fofanov M S 1997 *Laser Phys.* **7** 141
[10] Bula C *et al* 1996 *Phys. Rev. Lett.* **76** 3116
[11] Bamber C *et al* 1999 *Phys. Rev. D* **60** 092004
[12] Heinzl T, Ilderton A and Marklund M 2010 *Phys. Lett. B* **692** 250
[13] Titov A I, Takabe H, Kampfer B and Hosaka A 2012 *Phys. Rev. Lett.* **108** 240406
[14] Nousch T, Seipt D, Kampfer B and Titov A I 2012 *Phys. Lett. B* **715** 246
[15] Krajewska K and Kaminski J 2012 *Phys. Rev. A* **86** 052104
[16] Titov A I, Kampfer B, Takabe H and Hosaka A 2013 *Phys. Rev. A* **87** 042106
[17] Bunkin F, Kazakov A and Fedorov M V 1973 *Sov. Phys. Usp.* **15** 416
[18] Szymanowski C, Veniard V, Taıeb, Maquet A and Keitel C H 1997 *Phys. Rev. A* **56** 3846
[19] Walser M V, Szymanowski C and Keitel C H 1999 *Europhys. Lett.* **48** 533
[20] Panek P, Kaminski J and Ehlotzky F 2002 *Phys. Rev. A* **65** 033408
[21] Panek P, Kaminski J and Ehlotzky F 2004 *Phys. Rev. A* **69** 013404
[22] Di Piazza A, Milstein A I and Muller C 2010 *Phys. Rev. A* **82** 062110
[23] Muller T O and Muller C 2011 *Phys. Lett. B* **696** 201
[24] Muller T O and Muller C 2012 *Phys. Rev. A* **86** 022109
[25] Ahrens S, Bauke H, Keitel C H and Muller C 2012 *Phys. Rev. Lett.* **109** 043601
[26] Ahrens S, Muller T O, Villalba-Chavez S, Bauke H and Muller C 2013 *J. Phys.: Conf. Ser.* **414** 012012
[27] Volkov D M 1935 *Z. Phys.* **94** 250
[28] Krajewska K and Kaminski J 2010 *Phys. Rev. A* **82** 013420
[29] Krajewska K and Kaminski J 2012 *Phys. Rev. A* **85** 062102
[30] Krajewska K, Muller C and Kaminski J 2013 *Phys. Rev. A* **87** 062107

Journal of Physics: Conference Series **497** (2014) 012017 doi:10.1088/1742-6596/497/1/012017

Interaction of a classical laser field with a model Rydberg atom in a mixed state prepared by entanglement with few-photon quantum light

P.R.Sharapova and O.V.Tikhonova

Skobeltsyn Institute of Nuclear Physics, Lomonosov Moscow State University, 119234, Moscow,Russia

Physics Department, Lomonosov Moscow State University, 119991, Moscow, Russia

E-mail: Sharapova-Polina@yandex.ru

Abstract: The dynamics a model Rydberg atom prepared initially in a mixed state is investigated in a classical laser field. The possibility of interference stabilization regime is studied and the influence of parameters of initial mixed state on the ionization process is analyzed. In the case when initial atomic mixed state is formed due to entanglement with few-photon non-classical light the possibility to extract the information about initial atomic density matrix from the atomic behavior in the further applied classical laser field is discussed.

Introduction

Today one of the most interesting directions of research in laser physics is the interaction of an atom or molecule with strong electromagnetic field. In the case of strong field new phenomena appear and one of them is stabilization against ionization. Two different mechanisms of stabilization are known: the interference stabilization of Rydberg atoms and the Kramers-Henneberger stabilization. The mechanism of interference stabilization was firstly predicted theoretically for Rydberg atoms interacting with classical laser field [1-5].

However today it is possible to generate different types of non-classical field: squeezed vacuum field [6, 7] and few photon states [8, 9]. That's why the question about interference stabilization in quantum electromagnetic field becomes relevant. The possibility of interference stabilization of a model Rydberg atom in quantum field was firstly discussed in [10, 11] and the suppression of ionization was established. In addition, an efficient entanglement between atomic and field subsystems was found which provided the atomic system to occupy a certain mixed state due to the interaction with quantum field. Therefore any further dynamics of such an atom should be described in the density matrix formalism. In connection with, two interesting questions appear. At first, what is further dynamics of such specislly prepared in a mixed state atom in classical laser field? This is an interesting problem, since it is known that interference stabilization is very sensitive to the phase difference between Rydberg states. But in situation when atom is described by density matrix, such phases can hardly be introduced. The second question concerns the possibility to extract information about the initial density matrix of the atom from its dynamics in a further applied laser field. Such measurements allow to calculate the degree of entanglement [12, 13] induced during the interaction with quantum field [11].

So, in this work the dynamics of a model Rydberg atom prepared initially in a mixed state is considered in classical laser field. The atom is studied in a model which includes several discrete levels and continuum and allows one to investigate both: the field-induced ionization and resonance transitions between low-lying and high exciting atomic states. The possibility of interference stabilization regime in the case of interaction of such atom prepared initially in a mixed state with laser field and the influence of non-diagonal elements of initial density matrix on this dynamics is studied. The possibility to extract the information about initial atomic density matrix from the measurements of ionization and population trapping induced by the classical laser field is discussed.

Analytical model

The interaction between an atom and classical field in the density matrix formalism is described by non-stationary equation:

$$i\hbar\frac{\partial\hat{\rho}}{\partial t} = [\hat{H}_{at}+\hat{V}_{int},\hat{\rho}] \qquad (1)$$

where $\hat{H}_{at}$ is the atomic Hamiltonian, $\hat{V}_{int} = -\vec{d}\vec{\varepsilon}_0\cos(\omega t)$ is the atomic-laser field interaction in the dipole approximation , $\vec{d}$ stands for the dipole moment of atomic electron, and $\vec{\varepsilon}_0$ represents the strength of classical laser field.

In our case the Rydberg atom is considered in a model of "3 bound levels and continuum" and is characterized by 2 highly excited Rydberg levels repopulated by the field-induced Λ-type transitions via the continuum and the resonantly coupled third low-lying bound atomic state. In this case V-type transitions become also possible. The atomic eigenstates obey the stationary Schroedinger equation:

$$\hat{H}_{at}\varphi_n = E_n\varphi_n \qquad n = 0,1,2$$

$$\hat{H}_{at}\varphi_E = E\varphi_E$$

with φ_n, E_n and φ_E, E representing the wavefunctions and eigenvalues of atomic bound and continuum eigenstates respectively.

The frequency of laser field is chosen to provide the exact resonance condition between the lower state and the middle between the two Rydberg levels. We assume that transitions between low-lying atomic state and continuum are negligible small. The considered atomic system is discussed in details in [11]. We considered the hydrogen – like ion with nuclear charge Z, in which fine doublet $n\,P_{\frac{1}{2},\frac{3}{2}}\,(\varphi_1,\varphi_2)$ was chosen for highly excited neighboring Rydberg levels and the state $n_0S\,(\varphi_0)$ - for the lower lying resonant level. The laser frequency is much larger than the splitting energy $\Delta = E_2 - E_1$ between neighboring Rydberg levels and was chosen to correspond to $\hbar\omega \approx 2eV$ for $= 6, n_0 = 4, Z = 2$.

Initially the atom is supposed to occupy a certain mixed state which is formed during primary interaction with non-classical few photons field [11]. However, in general, an arbitrary mixed state can be considered as an initial one. The relaxation processes are not taken into account in our case since their characteristic time is of the order of $10^{-9}\,s$ or even more that is much larger than the largest time characterizing the atomic dynamics in a laser field.

Using the mentioned above basis of discrete and continuum atomic states and carrying out the procedure of adiabatic elimination of the continuum [2] we obtain the system of differential equations for density matrix elements between bound atomic states:

$$\begin{aligned}
i\hbar\frac{d\rho_{01}}{dt} &= \frac{\Delta}{2}\rho_{01} + \Omega(\rho_{11}+\rho_{21}-\rho_{00}) - i\frac{\Gamma}{2}(\rho_{01}+\rho_{02})\\
i\hbar\frac{d\rho_{02}}{dt} &= -\frac{\Delta}{2}\rho_{02} + \Omega(\rho_{12}+\rho_{22}-\rho_{00}) - i\frac{\Gamma}{2}(\rho_{01}+\rho_{02})\\
i\hbar\frac{d\rho_{00}}{dt} &= \Omega(\rho_{10}+\rho_{20}-\rho_{01}-\rho_{02})\\
i\hbar\frac{d\rho_{11}}{dt} &= \Omega(\rho_{01}-\rho_{10}) - i\Gamma\left(\frac{\rho_{12}+\rho_{21}}{2}+\rho_{11}\right)\\
i\hbar\frac{d\rho_{22}}{dt} &= \Omega(\rho_{02}-\rho_{20}) - i\Gamma\left(\frac{\rho_{12}+\rho_{21}}{2}+\rho_{22}\right)\\
i\hbar\frac{d\rho_{12}}{dt} &= -\Delta\rho_{12} + \Omega(\rho_{02}-\rho_{10}) - i\Gamma\left(\frac{\rho_{11}+\rho_{22}}{2}+\rho_{12}\right)
\end{aligned} \qquad (2)$$

where $\hbar\omega = \frac{E_1+E_2}{2} - E_0$, $\Delta = E_2 - E_1$, E_0, E_1, E_2 are the energies of atomic bound states, $\Omega = \frac{d_{0n}\varepsilon_0}{2\hbar}$ is the Rabi frequency of resonant transition and $\Gamma_{mn} = \frac{2\pi\left(\frac{\varepsilon_0}{2}\right)^2 d_{mE}d^*_{En}}{\hbar}$ are the elements of ionization rate tensor, calculated by using the Fermi golden rule and supposed to be the same and equal to Γ.

Because of cumbrous analytical solution the system (2) is solved numerically. The initial density matrix is chosen as that formed during atom- quantum field interaction [11] or in an arbitrary form. The developed approach is rather general and includes the cases of both pure and mixed initial state of the atom. The solution of the system (2) gives the time-dependent population of bound Rydberg states:

$$W_1(t) = \rho_{11}(t),\quad W_2(t) = \rho_{22}(t),\quad W_0(t) = \rho_{00}(t) \quad (3)$$

and the probability of ionization of the system:

$$W_i = 1 - (\rho_{00}+\rho_{11}+\rho_{22}) \quad (4)$$

To analyze the change of initially mixed atomic state during the interaction with laser field we introduce the parameter "purity" P which is quantitative degree of purity of atomic state and is equal to the trace of atomic density matrix squared.

$$P = Tr[\rho^2] \quad (5)$$

Since it is well known that for pure state $\rho = \rho^2$ and for any density matrix ρ $Tr[\rho] = 1$, the case $P = 1$ corresponds to the pure atomic state which can be characterized by a wavefunction and $P < 1$ – to a mixed state of the atom. Thus, the closer is P to unity, the closer is the atomic state to the pure state. Since the Eqs. (2) assume possible decay of density matrix elements with time due to atomic ionization the procedure of renormalization is used to obtain the value of parameter P in a correct way. Finally, it is rewritten as follows:

$$P(t) = \frac{{\rho_{00}}^2 + {\rho_{11}}^2 + {\rho_{22}}^2 + 2\rho_{12}\rho_{21} + 2\rho_{10}\rho_{01} + 2\rho_{20}\rho_{02}}{(\rho_{00}+\rho_{11}+\rho_{22})^2} \quad (6)$$

It should be mentioned, that the worked out approach based on the density matrix formalism is a general description of the interaction of a model Rydberg atom with a classical laser field and is valid for both the pure and mixed initial state of the atom. In the case of initial density matrix corresponding to a pure atomic state we immediately obtain from (2) the same results as have been obtained earlier in [3] using the wave function formalism.

Results and discussions

First we investigate the interaction of a model Rydberg atom with a classical laser field in the case when atom is initially in an arbitrary mixed state. Similarly to [3] we consider the regime of rather strong field when the Rabi frequency of resonant transition Ω is the largest energy parameter in the system and the relations $\Omega \gg \Delta$, $\Omega \gg \Gamma$ are fulfilled. The main feature of the field-induced atomic dynamics is found to consist in the transformation of initial mixed state of the atom into the pure state due to the atom-field interaction during rather short time interval. The typical dependencies of the purity parameter and ionization probability on time are presented on Fig. 1 a, b respectively. The calculations are performed for the following initial density matrix chosen with arbitrary non-zero matrix elements:

$$\rho|_{t=0} = \begin{pmatrix} 0.5 & 0.2-0.1i & 0.1+0.3i \\ 0.2+0.1i & 0.2 & 0.1-0.2\,i \\ 0.1-0.3i & 0.1+0.2\,i & 0.3 \end{pmatrix} \quad (7)$$

It is clearly seen that during the atom-field interaction the value of "purity" reaches rather fast its maximal value equal to unity. Thus, the coherent laser field converts the initial atomic mixed state into a pure state inducing certain phases between wave functions in the coherent superposition of bound Rydberg states formed during the interaction. The observed "passage" from initial mixed state to the pure one is accompanied by rather efficient atomic ionization, which, however, saturates at the level significantly less than unity. This means that the field-induced ionization process leads to the "purification" of the bound atomic fraction in such a way that the formed coherent wave packet is characterized by a high resistance to the further ionization. That's why the system is not fully ionized even in the limit of large time and the probability of ionization becomes constant and doesn't reach unity. Thus the regime of interference stabilization takes place. It should be emphasized that interference stabilization is known to result from the certain phase relations between bound Rydberg states, while in the case of mixed states the phase information is lost. It is the coherent laser field that brings coherence into incoherent state and induces proper phase relations.

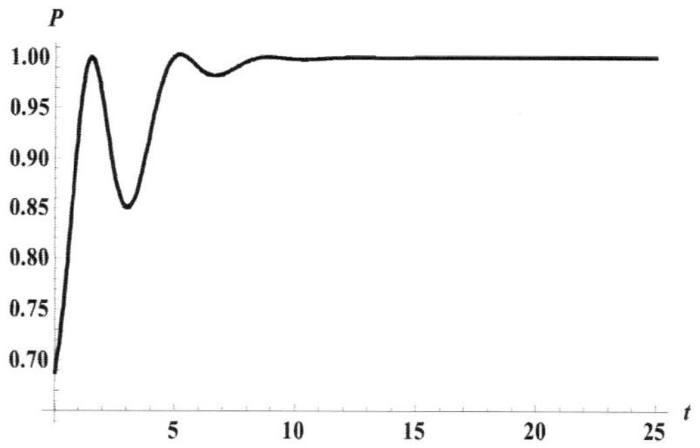

Fig 1a

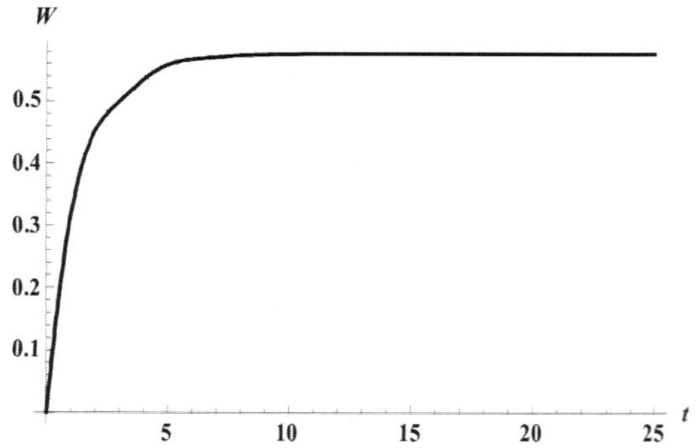

Fig 1b

Fig. 1. Parameter "purity" (a) and probability of ionization of the system (b) versus time in optical cycles.

It is rather easy to find that in the steady state regime the residual bound atomic fraction is characterized by the following stable quasienergy wave function:

$$\psi_{\gamma_0}^{QES} = \frac{1}{\sqrt{\left(1+\frac{\Delta^2}{8\Omega^2}\right)}}\left(\frac{\varphi_1-\varphi_2}{\sqrt{2}}+\frac{\Delta}{\sqrt{8}\Omega}\varphi_0\right) \quad (8)$$

Journal of Physics: Conference Series **497** (2014) 012017 doi:10.1088/1742-6596/497/1/012017

The expression (8) coincides exactly with the quasienergy state which can be found for the interaction of a laser field with pure state of the atom and also can be obtained from the Schroedinger formalism [3] under proper resonant condition and only one low-lying Rydberg state taken into consideration.

Thus the general feature of the atomic dynamics in a classical laser field is the formation of a pure atomic state which is stable quasienergy state of the system. This effect is general and is observed for arbitrary initial mixed states including different initial pure states as well as thermodynamic mixture with all non-diagonal matrix elements equal to zero. The latest are interesting since all the phase information is absent in such states.

From Eqs. (2) the connection between the ionization probability and the matrix elements of initial density matrix can be found:

$$W_{ion} = 1 - W_b = 1 - \rho_{00}|_{t\to\infty} - \rho_{11}|_{t\to\infty} - \rho_{22}|_{t\to\infty} =$$

$$= 1 - \frac{1}{1+\frac{\Delta^2}{8\Omega^2}}\left[\frac{\Delta^2}{8\Omega^2}\rho_{00}|_{t=0} + \frac{\rho_{11}|_{t=0} + \rho_{22}|_{t=0}}{2} - Re\rho_{12}|_{t=0} + \frac{\Delta}{2\Omega}(Re\rho_{10}|_{t=0} - Re\rho_{20}|_{t=0})\right] \quad (9)$$

For example, in the case of the data presented on Fig 1 $\frac{\Delta^2}{8\Omega^2} = 1$ and from (9) using the initial condition (7) we obtain $W_{ion} = 0.6043$. However the inverse problem – to retrieve the information of initial mixed state from ionization dynamics – doesn't seem to be easy solved, since all elements of initial density matrix should be found while, in fact, only one value - the ionization probability - can be measured.

However in some special cases this problem can be solved. We will show that in the case when the mixed state of the atom is prepared due to its interaction with few-photon quantum light the further applying classical field gives the possibility to measure experimentally the degree of entanglement, which was induced between the atom and quantum light. For this purpose it is necessary to recover the formed density matrix of the atom from its dynamics in classical laser field, because a quantitative degree of entanglement, for example, the value of Schmidt parameter, depends on the components of atomic density matrix. The Schmidt number cannot be obtained by direct experimentally measurement since it depends on both the non-diagonal and diagonal elements of density matrix. The latest are nothing more than a probability of population of Rydberg states and can be in principal measured experimentally.

Further we will demonstrate the possibility to extract the information about the initial atomic density matrix from atomic dynamics in a classical laser field. First, we specify the initial condition in the form of rather general atomic density matrix formed during the interaction of the considered model atom with few-photon quantum field in a steady state regime [11]:

$$\rho|_{t=0} = \begin{pmatrix} \rho_{00} & 0 & 0 \\ 0 & \rho_{11} & \rho_{21} \\ 0 & \rho_{12} & \rho_{22} \end{pmatrix} \quad (11)$$

This density matrix can be found from the evolution of the system "atom+ quantum field" which is shown to transfer the system to the following bipartite quasienergy state [11]:

$$\psi^{QES} = \left(\alpha\frac{\varphi_1-\varphi_2}{\sqrt{2}}\Phi_{k-1} + \beta\varphi_0\Phi_k\right) \quad (12),$$

where Φ_k is k-photon Fock state and the amplitudes α, β are determined by the atomic and quantum field parameters and obey the normalization condition: $|\alpha|^2 + |\beta|^2 = 1$.

The corresponding reduced density matrix describing the atom in a quantum field in a steady state regime is given by:

$$\rho_r^{at} = \frac{|\alpha|^2}{2}\begin{pmatrix} \frac{|\beta|^2}{|\alpha|^2} & 0 & 0 \\ 0 & 1 & -1 \\ 0 & -1 & 1 \end{pmatrix} \quad (13)$$

It is important that such matrix has zero elements: $\rho_{10} = \rho_{20} = \rho_{01} = \rho_{02} = 0$ due to the resonant transitions $\varphi_0 \leftrightarrow \varphi_{1,2}$ providing these states always associated with different number of photons.

It is convenient to move from initial basis of bound atomic states to the new basis which provides the initial density matrix (11) to be diagonal. This basis exactly coincides with Schmidt modes found for atomic subsystem in the case of the steady state (12) describing the entanglement between the atom and few-photon quantum field. By solving the eigenproblem for such reduced atomic density matrix (11) one can obtain the following expression for Schmidt modes or new basis atomic states:

$$\tilde{\varphi}_0 = \varphi_0, \tilde{\varphi}_1 = \frac{\sqrt{\frac{\rho_{21}}{\rho_{12}}}\varphi_1 + \varphi_2}{\sqrt{2}}, \tilde{\varphi}_2 = \frac{\varphi_1 - \sqrt{\frac{\rho_{12}}{\rho_{21}}}\varphi_2}{\sqrt{2}} \quad (14)$$

The basis of Schmidt modes is seen to depend strongly on the ratio $\frac{\rho_{21}}{\rho_{12}}$ and the initial density matrix becomes diagonal in terms of these states instead of being non-diagonal in the previous basis. However, the diagonal elements correspond now to the population of the Schmidt states that hardly can be measured experimentally. What is also very important is that for the quasienergy state (12) $\rho_{12} = \rho_{21}$ due to the stabilization regime pound for the Rydberg atom in few photon quantum field. Under this condition the Schmidt modes do not depend on the amplitudes α, β and are given by:

$$\tilde{\varphi}_0 = \varphi_0, \tilde{\varphi}_1 = \frac{\varphi_1 + \varphi_2}{\sqrt{2}} \equiv 0, \tilde{\varphi}_2 = \frac{\varphi_1 - \varphi_2}{\sqrt{2}} \quad (15)$$

And the initial atomic mixed state is characterized by the diagonal matrix written as follows:

$$\tilde{\rho}|_{t=0} = \begin{pmatrix} \tilde{\rho}_{00}|_{t=0} & 0 & 0 \\ 0 & 0 & 0 \\ 0 & 0 & \tilde{\rho}_{22}|_{t=0} \end{pmatrix} = \begin{pmatrix} |\beta|^2 & 0 & 0 \\ 0 & 0 & 0 \\ 0 & 0 & |\alpha|^2 \end{pmatrix} \quad (16)$$

Our goal now is to retrieve only one of the two diagonal matrix elements (due to the normalization condition) performing one measurement only. The system (2) can be rewritten in the Schmidt basis (15) by following way:

$$i\hbar\frac{d\tilde{\rho}_{10}}{dt} = -\frac{\Delta}{2}\tilde{\rho}_{20} - \sqrt{2}\Omega(\tilde{\rho}_{11} - \tilde{\rho}_{00}) - i\Gamma\tilde{\rho}_{10}$$

$$i\hbar\frac{d\tilde{\rho}_{20}}{dt} = -\frac{\Delta}{2}\tilde{\rho}_{10} - \sqrt{2}\Omega\tilde{\rho}_{21}$$

$$i\hbar\frac{d\tilde{\rho}_{00}}{dt} = \sqrt{2}\Omega(\tilde{\rho}_{10} - \tilde{\rho}_{01})$$

$$i\hbar\frac{d\tilde{\rho}_{11}}{dt} = -\sqrt{2}\Omega(\tilde{\rho}_{10} - \tilde{\rho}_{01}) - i\Gamma\tilde{\rho}_{11} - \frac{\Delta}{2}(\tilde{\rho}_{21} - \tilde{\rho}_{12}) \quad (17)$$

$$i\hbar\frac{d\tilde{\rho}_{22}}{dt} = \frac{\Delta}{2}(\tilde{\rho}_{21} - \tilde{\rho}_{12})$$

$$i\hbar\frac{d\tilde{\rho}_{21}}{dt} = -\sqrt{2}\Omega(\tilde{\rho}_{20} + \tilde{\rho}_{02}) - i\Gamma\tilde{\rho}_{21} - \frac{\Delta}{2}(\tilde{\rho}_{11} - \tilde{\rho}_{22})$$

It's easy to see from (17) that some initially zero matrix elements ($Re[\tilde{\rho}_{10}], Im[\tilde{\rho}_{20}], Re[\tilde{\rho}_{21}]$) remain equal to zero during the dynamics of the system. However some of them being initially zero ($\tilde{\rho}_{11}, Im[\tilde{\rho}_{10}], Im[\tilde{\rho}_{21}]$) vary during the dynamics of the system but turn to zero again in the steady regime of the interaction. This feature is a direct consequence of the interference stabilization phenomenon and the formation of a stable atomic pure state (8) (from the initial mixed one) in a strong classical laser field. Fig. 2 represents the growth of the population of initially non-populated Schmidt state $\tilde{\varphi}_1$ due to the interaction with laser field and its decrease up to zero in the steady state regime when the bound atom is characterized by a pure quasienergy state (8) resistant to the further ionization.

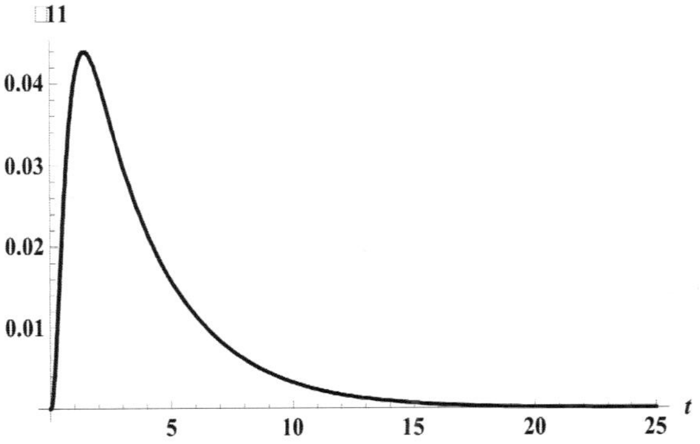

Fig 2. Dynamics of population of initially non-populated Schmidt state $\tilde{\varphi}_1$ versus time in optical cycles.

In terms of Schmidt basis the steady density matrix corresponding to the stable atomic state (8) is given by:

$$\tilde{\rho}|_{t\to\infty} = \frac{W_b}{1+\frac{\Delta^2}{8\Omega^2}} \begin{pmatrix} \frac{\Delta^2}{8\Omega^2} & 0 & \frac{\Delta}{\sqrt{8}\Omega} \\ 0 & 0 & 0 \\ \frac{\Delta}{\sqrt{8}\Omega} & 0 & 1 \end{pmatrix} \quad (18),$$

where W_b denotes the residual bound probability of the atom. Notice, the fact that $Tr[\tilde{\rho}|_{t\to\infty}] = W_b < 1$ is a consequence of the decay of the atomic bound fraction due to the ionization process. Thus the evolution of atomic system in a laser field starting from a mixed state (16) comes to the pure state (18).

It should be emphasized that from the system of differential equations in Schmidt basis (17) a certain combination of density matrix elements S is found to be conserved during the interaction with the laser field and its derivative is identically zero at any instants of time:

$$i\hbar\frac{d}{dt}\left[\frac{\Delta^2}{8\Omega^2}\tilde{\rho}_{00} + \tilde{\rho}_{22} + \frac{\Delta}{\sqrt{8}\Omega}\, 2\, Re[\tilde{\rho}_{20}]\right] \equiv i\hbar\frac{d}{dt}S = 0 \qquad (19)$$

It means that such value is directly proportional to the residual bound probability surviving in a classical laser field. Indeed, from (18) we immediately obtain:

$$S = W_b\left(1 + \frac{\Delta^2}{8\Omega^2}\right) \qquad (20)$$

From the other hand, the value of S can be obtained from the initially density matrix (16). Equating both values we obtain:

$$\frac{\Delta^2}{8\Omega^2}\tilde{\rho}_{00}|_{t=0} + \tilde{\rho}_{22}|_{t=0} + \frac{\Delta}{\sqrt{8}\Omega}\, 2\, Re[\tilde{\rho}_{20}|_{t=0}] = W_b\left(1 + \frac{\Delta^2}{8\Omega^2}\right) \qquad (21)$$

In account of the zero values of the non-diagonal element $\tilde{\rho}_{20}|_{t=0}$ we obtain the connection between the non-zero matrix elements of initial density matrix (16) describing the initial atomic mixed state and the final probability of ionization of the atom in a laser field:

$$W_{ion} = 1 - \rho_{00}|_{t\to\infty} - \rho_{11}|_{t\to\infty} - \rho_{22}|_{t\to\infty} = 1 - \frac{1}{1+\frac{\Delta^2}{8\Omega^2}}\left[\frac{\Delta^2}{8\Omega^2}\tilde{\rho}_{00}|_{t=0} + \tilde{\rho}_{22}|_{t=0}\right] \qquad (22)$$

where the diagonal elements are connected with each other by the condition of the unity trace.

The obtained expression (22) means that by measuring the probability of ionization in the steady state regime during the interaction of a Rydberg atom prepared initially in a mixed state with classical electromagnetic field we can retrieve the initial atomic density matrix in the Schmidt basis and therefore calculate the degree of entanglement (the value of the Schmidt parameter K) between the atom and few-photon non-classical light induced during their interaction before the classical field is applied:

$$K^{-1} = \tilde{\rho}_{00}|_{t=0}{}^2 + \tilde{\rho}_{22}|_{t=0}{}^2 = P|_{t=0} \qquad (23)$$

Thus it is possible to measure the degree of entanglement between the atom and few-photon quantum light by further applying of classical laser field.

Conclusion

The interaction of a model Rydberg atom prepared initially in a mixed state with classical electromagnetic laser field is studied. A general feature of atomic dynamics is found to consist in the transformation of the initial mixed state of the atom to the pure state due to the interaction with laser field. In other words, classical field brings coherence into initially incoherent state. The "purification" is accompanied by significant ionization of the atom. However the laser field induces certain phases between bound Rydberg states composing the formed coherent wave packet in such a way that the residual atomic population becomes resistant to further ionization. In the steady state regime the regime of interference stabilization takes place and the bound atomic fraction is described by the stable quasienergy state of the system. The observed "purification" is a common feature of atomic dynamics in a classical laser field and can be used as a tool to retrieve the information about initial density matrix of the atom. In the case when the initial mixed state of the atom is formed during its primary interaction with few-photon quantum field it is possible to extract the information about the mixed atomic state and therefore about the entanglement reached between the atom and few-photon field by performing only one measurement of the probability of ionization induced by applying classical laser light.

Acknowledgement This work was supported by Russian Fund for Basic Research (grant N14-02-31232-mol_a, N14-02-31084-mol_a, N14-02-00389-a).

References

1. M.V.Fedorov, A.M.Movsesian, J.Phys.B **21**, L155 (1988)
2. M.V.Fedorov, World Scientific, 1997, 452pp, Singapore
3. M.V.Fedorov, N.P.Poluektov, Laser.Phys. **11**, 255 (2001)
4. M.Gavrila, J.Phys.B. **35**, R147 (2002)
5. A.M.Popov, O.V.Tikhonova and E.A.Volkova, J.Phys.B. **36**, R125-R165 (2003)
6. Slusher et al, Phys.Rev.Lett. **55**, 2409-2412 (1985).
7. L.Shangqing, C.Yansong, J. Opt. Soc. Am. B **12**, 829 (1995)
8. C.Santori, M.Petlon, G.Solomon, et.al. Phys. Rev. Lett. **86**, 1502 (2001).
9. M.Sondermann, R.Maiwald, H.Konermann et. al. Appl. Phys. B **89**, 489 (2007)
10. P R Sharapova, O V Tikhonova, Quantum Electronics, 2012, **42** (3), 199–207, (2012)
11. P R Sharapova, O V Tikhonova, Laser.Phys.Lett. 10, 075204, (2013)
12. R.Grobe, K.Rzazewski and J.H.Eberly J.Phys.B **27**, 2503 (1994)
13. M.V.Fedorov, M.A.Efremov, P.A.Volkov et al J.Phys.B **39**, 467 (2006)

Ionization and entanglement of two interacting Rydberg atoms in a strong laser field

Ivan A. Burenkov
Skobeltsyn Institute of Nuclear Physics, Lomonosov Moscow State University, Leninskie Gory 1, 119234, Moscow, Russia
E-mail: ivan.burenkov@gmail.com

Olga V. Tikhonova
Skobeltsyn Institute of Nuclear Physics, Lomonosov Moscow State University, Leninskie Gory 1, 119234, Moscow, Russia
Department of physics, Lomonosov Moscow State University, Leninskie Gory 1, 119991, Moscow, Russia

Abstract. Dynamics of ionization and entanglement of two Rydberg atoms interacting with each other and with external strong laser field is investigated. The phenomenon of interference stabilization of the bipartite atomic system is established in a strong field limit. The production of highly-entangled state of the studied atoms is shown to be realized. The possibility to control the evolution and time-dependent entanglement of the examined multilevel q-dit system is demonstrated.

PACS numbers: 32.80.Ee, 32.80.Fb, 32.80.Qk, 32.80.Wr, 03.67.Bg

1. Introduction

Coherent control of multi-partical systems is one of the most interesting and important problems of the modern quantum optics. There are many types of the q-bit systems and many different ways to control them [1, 2, 3, 4]. Many papers are aimed on the creation of highly entangled atomic q-bit systems and on their applications in different problems of quantum information [5, 6, 7, 8, 9, 10]. Usually coupling between atomic q-bits is supposed to arise due to the interaction with a common reservoir (atom trap, microcavity, resonator, etc.) and quantum control is performed by applying an external field. Now it is possible to trap Rydberg atoms and ions together in pairs or chains during rather long time [11] with different constructions of traps being used for these purposes [12, 13]. According to recent investigations a pair of tightly trapped atoms excited into the Rydberg states appears to be very attractive as a simple q-bit system and can be controlled either by optical or microwave field [5, 6, 14]. Coupling of these highly-excited atoms results from their dipole interaction caused by rather broad electron density distribution of both Rydberg atoms. This interaction is known to influence strongly on the eigenstate energies of individual atoms and leads to

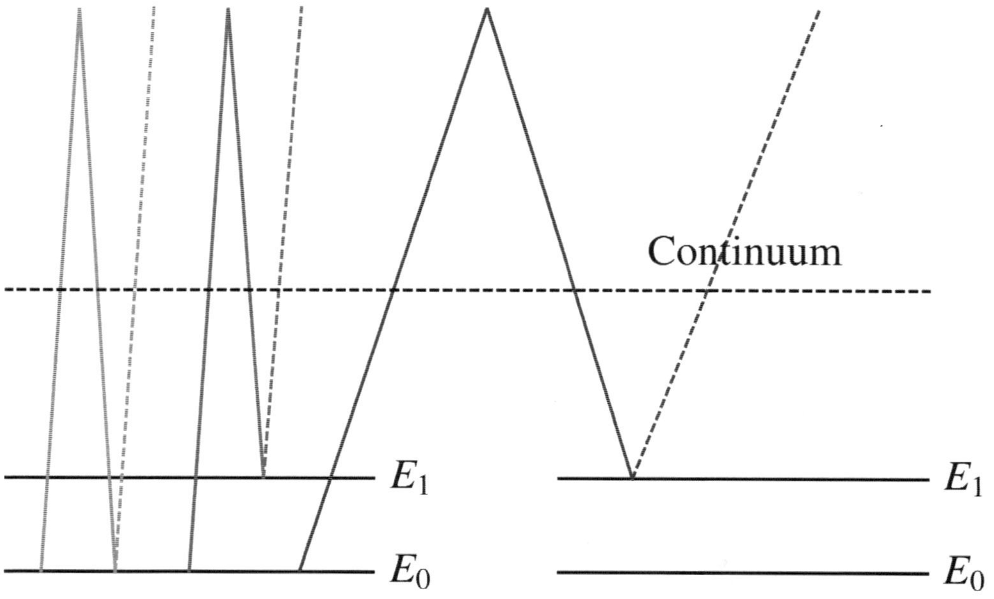

Figure 1. Different types of Raman Λ-type transitions taken into consideration for the system of two interacting Rydberg atoms.

so-called "dipole blockade" effect [6, 14, 15, 16]. Nevertheless tuning of the laser frequency to the shifted by interaction excitation energy results in "anti-blockade". In such situation it is very important to find and analyze the quasienergy states of the bipartite system in a laser field, including the interaction between subsystems and a laser field action. The wave functions and their features are strongly dependent on the interatomic separation R. This problem does not seem to be solved up to now. In addition to study the influence of each type of interaction on the atomic dynamics and the possibility to obtain and to control the entanglement in such bipartite system remains still an open question.

Another interesting aspect of the interaction of a Rydberg q-bit with a laser field can lead to the ionization of a system which is usually not taken into account in theoretical considerations. Often a large number of neighbouring Rydberg levels existing in a real atom can be simultaneously involved both in the excitation/ionization process. The ionization process can be partially suppressed in a strong laser field due to the well known interference stabilization phenomenon [17, 18]. In such case again the influence of the strong external field and the role of the interatomic interaction should be correctly taken into account to analyze the dynamics of the system and to perform the coherent control of atomic q-bits.

In this paper we investigate the behavior of two coupled Rydberg atoms interacting with each other in a presence of external non-resonant laser field and analyze the possibility to control the dynamics of the system including the suppression of the ionization process and the production highly entangled bipartite states. Each atom is considered in a model of "n bound Rydberg states + continuum" [17] and the non resonant laser field is supposed to provide multi-photon transitions to the continuum being accompanied by the repopulation of

Journal of Physics: Conference Series **497** (2014) 012018 doi:10.1088/1742-6596/497/1/012018

Rydberg states via the Raman Λ-type transitions. The scheme of the considered field induced transitions is presented on Fig. 1 and different multi-photon processes of odd order are seen to be taken into account. Direct field-induced transitions between different Rydberg levels of each atom are supposed to be small. In approach used the quasi-energies of the system are found in a presence of both the external laser field and interatomic interaction and the dynamics of the system is analyzed. The strong laser field will be shown to result in ionization suppression and coherent control of the two atomic system, that can be used in many Rydberg q-bit applications. Production of the highly entangled bipartite states is demonstrated.

2. Analytical Model

The evolution of considered bipartite system is described by the time dependent Schroedinger equation (TDSE):

$$i\hbar\partial_t\psi(\vec{r}_1,\vec{r}_2,t) = \left[\hat{H}_1 + \hat{H}_2 + \hat{V}_1 + \hat{V}_2 + \hat{W}_{12}\right]\psi, \tag{1}$$

Here $\psi(\vec{r}_1,\vec{r}_2,t)$ is the wave function (WF) of electronic subsystem of two atoms depending on time, indexes "1" and "2" correspond to electrons of different atoms, the Hamiltonian $\hat{H}_i$ includes the kinetic energy and interaction of "i" electron with the own and other nucleus:

$$\hat{H}_i = \hat{T}_i - \frac{ze^2}{|\vec{r}_i - \vec{r}_A|} - \frac{ze^2}{|\vec{r}_i - \vec{r}_B|}, \tag{2}$$

where ze and $\vec{r}_{A,B}$ are the charge and the radius-vector of nucleus "A" and "B" correspondingly. The operator $\hat{W} = e^2/|\vec{r}_1 - \vec{r}_2|$ stands for the interaction between electrons and the interaction with the laser field is considered in the dipole approximation $\hat{V}_i = -(\hat{\vec{d}}_i \cdot \vec{\varepsilon})$ where $\vec{d}_i = e\vec{r}_i$ is the dipole moment corresponding to "i"-th electron and $\vec{\varepsilon}$ is the electric field of the laser pulse with the polarization direction chosen along the direction from one nucleus to another. The laser field carrier envelope can be constant or depend slowly on time $\vec{\varepsilon}(t) = \vec{\varepsilon}_0(t)\cos\omega t$ and the laser field wavelength is assumed to exceed the interatomic separation R to keep us in the frame of the dipole approximation.

To describe the field induced evolution of the system correctly an appropriate basis of the field-free two electron WF should be used, which obey the proper symmetry conditions in relation to exchange of both electron and nucleus positions. Such states can be found as a product of the single electron eigenfunctions $\varphi_i(\vec{r}_1,R)$ and $\varphi_k(\vec{r}_2,R)$ (including the continuum states), depending on the relative position between nuclei $R = |\vec{r}_B - \vec{r}_A|$ and obeying the following stationary problems:

$$\begin{aligned} \hat{H}_1\varphi_i(\vec{r}_1,R) &= E_i\varphi_i(\vec{r}_1,R), \\ \hat{H}_2\varphi_k(\vec{r}_2,R) &= E_k\varphi_k(\vec{r}_2,R), \end{aligned} \tag{3}$$

The solution of (3) can be found using the elliptical coordinates [19] and obey the symmetry conditions in relation to the exchange between nuclear positions. The two-electron WF $|ik\rangle$ are the solutions of the eigenproblem:

$$\begin{aligned} (\hat{H}_1 + \hat{H}_2)|ik\rangle &= E_{ik}|ik\rangle \\ |ik\rangle &= \varphi_i(\vec{r}_1,R)\varphi_k(\vec{r}_2,R) \end{aligned} \tag{4}$$

and are characterized by the energies $E_{ik} = E_i + E_k$. To avoid the autoionization process we suppose internuclear separation R or the charge z of the ion in (2) are large enough provide electron states to be bound.

To provide the symmetry conditions in relation to the electron positions we use the following states:

$$\begin{aligned} &|ii\rangle \text{ for } i = k, \\ \Psi^{\pm}_{ik} &= \tfrac{|ik\rangle \pm |ki\rangle}{\sqrt{2}} \text{ for } i \neq k, \end{aligned} \tag{5}$$

where $\Psi^{\pm}_{ik}$ can be considered as generalized Bell functions [20].

It should be noticed that many of states (5) should be taken into account to describe correctly the behavior of the two-atomic system in a laser field. However we would like to demonstrate that some general features of the system dynamics - the ionization suppression and time dependent entanglement - can be found in the frame of a rather simple model when only two bound states (3) for each electron are taken into account, and these features can be also seen in the case of larger number of the basis functions (5).

Let us consider firstly the continuum and only two bound states (3) for each electron keeping in mind that they do not coincide with the eigenfunctions of the individual atom since each Hamiltonian $\hat{H}_i$ includes the interaction of the "i"-th electron with the own and and the other nuclei (2). Under these condition using the procedure of adiabatic elimination of the continuum [17, 18] we obtain the differential equations for probability amplitudes of bound bipartite states separately for spatially symmetric (singlet) $\Psi^{bound}_{S=0}$ and anti-symmetric (triplet) $\Psi^{bound}_{S=1}$ states, that can be represented by following way:

$$\begin{aligned} \Psi^{bound}_{S=0} &= C_{00}|00\rangle + C_{11}|11\rangle + \\ &+ C^{+}_{01}\Psi^{+}_{01} \\ \Psi^{bound}_{S=1} &= C^{-}_{01}\Psi^{-}_{01} \end{aligned} \tag{6}$$

Then the system of equations for spatially symmetric and antisymmetric WF are independent in accordance with the intercombination prohibition principle and are given by:

$$\begin{cases} i\dot{C}_{00} = (2E_0 - i\Gamma + W)C_{00} + WC_{11} + \frac{2W - i\Gamma}{\sqrt{2}}C^{+}_{01} \\ i\dot{C}_{11} = WC_{00} + (2E_1 - i\Gamma + W)C_{11} + \frac{2W - i\Gamma}{\sqrt{2}}C^{+}_{01} \\ i\dot{C}^{+}_{01} = \frac{2W - i\Gamma}{\sqrt{2}}(C_{00} + C_{11} + \sqrt{2}C^{+}_{01}) + (E_0 + E_1)C^{+}_{01} \end{cases} \tag{7}$$

$$i\dot{C}^{-}_{01} = (E_0 + E_1 - i\Gamma)C^{-}_{01}. \tag{8}$$

Equations (7-8) are written with following significations:

$$\begin{aligned} \Gamma = \Gamma_{i,k} &= 2\pi \frac{V_{i,E}}{2}\frac{V_{E,i}}{2} \\ V_{\alpha,\beta} &= -\vec{\varepsilon}\langle\alpha|\vec{d}|\beta\rangle \\ W = \langle ik|\hat{W}|i'k'\rangle &= W^{i'k'}_{ik}, \end{aligned} \tag{9}$$

where for simplicity we have supposed that all elements of tensor $\Gamma_{i,k}$ are the same and all matrix elements $\langle ik|\hat{W}|i'k'\rangle$ are equal to each other. It should be noticed that the value of W depends strongly on the interatomic separation R and the further will analyze different cases for wide range of values $W(R)$ in relation to the values of ionization width Γ and energy separation between atomic levels $\Delta = E_2 - E_1$.

Journal of Physics: Conference Series **497** (2014) 012018 doi:10.1088/1742-6596/497/1/012018

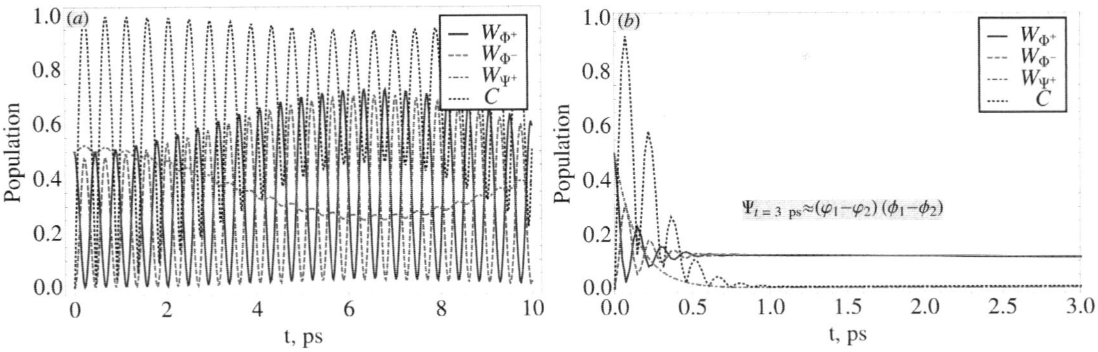

Figure 2. Population of different Bell states (W_{Φ^+}, W_{Ψ^+}, W_{Φ^-}) atomic states in (6) and concurrence C calculated in dependence on time for two interacting Rydberg atoms (with 2 bound Rydberg levels) in the absence of laser field **(a)** and in a strong field regime **(b)** for $W = 30\Delta$, $\Gamma = 10\Delta$ and $\Delta = \frac{2Ry}{50^3}$. The state $\varphi_0(\vec{r}_1)\varphi_0(\vec{r}_2)$ is populated initially.

3. Results and discussion

According to equation (8) in the case of only two bound levels (3) taken into account in each atom there is the only spatially antisymmetric bipartite state which decays fast in a strong laser field. However as we will show later in the case of larger number of bound levels (3) the dynamics is more rich and the stabilization can take place in the case of spatially antisymmetric states too.

Before we'll analyze the solution of equation (7) let us discuss a fundamental and very interesting feature of composite systems referred to as the entanglement. The physical meaning of entanglement consists in the fact that it is impossible to characterize each subsystem of a multi-particle system by its own wave function. In other words, the total WF of the system can not be factorized into the product of two WF separately describing each of the subsystems.

To characterize th entanglement of the system different quantitative degrees can be used: the Schmidt number [21, 22], concurrence [23, 24], negativity, etc. According to [21, 22] the Schmidt number defined as following:

$$K^{-1} = Tr\{\rho_r^2\}, \tag{10}$$

where ρ_r is the reduced density matrix of the system. And for our spatially symmetric state $\Psi^{bound}_{S=0}$ the expression (10) can be calculated as:

$$K_S^{-1} = 1 - 2\left|C_{00}C_{11} - (C_{01}^{+})^2/2\right|^2. \tag{11}$$

This value appears to coincide exactly with that obtained for the biphoton q-trite state considered in [24]. This fact is a direct evidence that in our case the considered two atomic system is an example of two atomic q-bit, which can be analyzed in terms of different q-bit states without special attention on the continuous variable dependence of the single atom WF. In addition, using the well known connection between Schmidt number K and concurrence

C [23, 24] one can obtain for the case of the symmetric singlet state $\Psi^{bound}_{S=0}$:

$$C_S = 2\left|C_{00}C_{11} - C_{01}^2\right|. \tag{12}$$

Taking into account the antisymmetric spin part of the total WF in this case we found the total Schmidt number to be twice larger:

$$K_{SA} = 2K_S. \tag{13}$$

Since the spin variable is not included in the Hamiltonian The factor "2" is not changed during the evolution and results in a renormalization of the concurrence from the range $[0, 1]$ to $[1, \sqrt{3/2}]$ in a case of (12).

The change of the range of possible values of the concurrence (or of the Schmidt parameter) means that now (including the spin variables) the system is always entangled. However in our work we will be interested firstly in the possible entanglement over spatial variables (excluding the independent spin degree of freedom). So only K_S (11) and C_S (12) are calculated and presented in this paper.

Let us consider firstly the field-free dynamics of our system. Fig. 2a represents the time-dependent population of Bell states:

$$\begin{aligned}\Phi^{\pm} &= \frac{\varphi_0(\vec{r}_1)\varphi_0(\vec{r}_2)\pm\varphi_1(\vec{r}_1)\varphi_1(\vec{r}_2)}{\sqrt{2}},\\ \Psi^{\pm} &= \frac{\varphi_0(\vec{r}_1)\varphi_1(\vec{r}_2)\pm\varphi_1(\vec{r}_1)\varphi_0(\vec{r}_2)}{\sqrt{2}}\end{aligned} \tag{14}$$

in order to demonstrate that maximally entangled states are populated periodically during the evolution of the system, even though the dis-entangled ground state of two atomic system was populated initially. Fast oscillations of the populations are caused by the electron-electron (interatomic) interaction. Period of the oscillations is determined by the interaction energy W and slow modulation is proportional to the energy difference Δ between bound atomic levels of each atom.

In a presence of a rather strong field these fast oscillations of atomic population are damped out at times of about $\sim \hbar/\Gamma$ (see figure 2b) due to the dominant role of the laser field in the dynamics of the system. In a strong field regime ($\Gamma \gg \Delta$) the ionization process is found to be partly suppressed and there is a non-zero probability to find the system bound even at large time, so the stabilization phenomenon is observed. The mechanism of stabilization is similar to that predicted for a single Ry atom in a strong field [17, 18] and is reffered to as "the interference stabilization". The state which describes the bound fraction of population of two atomic q-bits is found to be given by:

$$\Psi^{--} = \frac{(\varphi_0(\vec{r}_1) - \varphi_1(\vec{r}_1))(\varphi_0(\vec{r}_2) - \varphi_1(\vec{r}_2))}{2}. \tag{15}$$

and appears to be factorized so as the two atomic subsistems are totally dis-entangled. The state (15) is found to be the quasienergy state (QES) of the system in a strong field limit and is seen to be a product of stable in a strong field QES of each independent atom discussed in [17, 18].

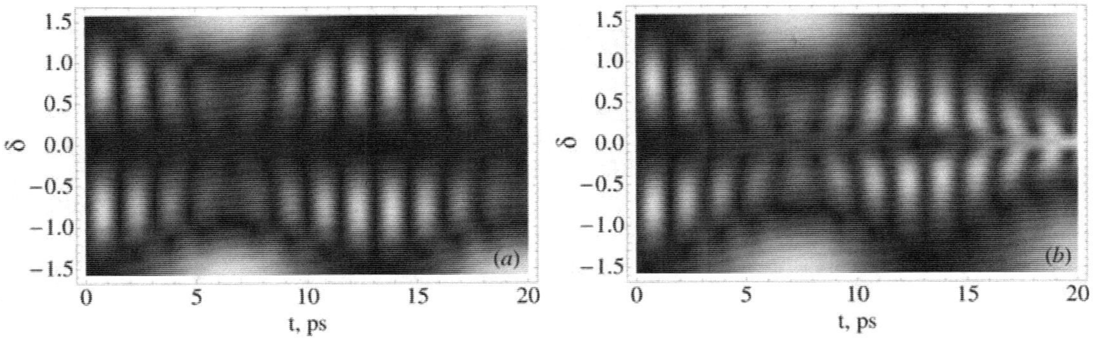

Figure 3. Concurrence calculated for two interacting atoms ($W = 3\Delta$) in dependence on time and relative phase of initial state (16) for the field free regime **(a)** and in a presence of a weak field with $\Gamma = \Delta/4$ **(b)**.

In the case of the intermediate field strength ($\Gamma \sim \Delta$) it is possible to provide both ionization suppression and entanglement in the system. Fig. 3 represents entanglement map calculated without laser field and in a presence of a rather week field at different phases of initial state chosen for the two atomic q-bit system:

$$\Psi|_{t=0} = \frac{(\varphi_0(\vec{r}_1) + \varphi_1(\vec{r}_1)e^{2i\delta})(\varphi_0(\vec{r}_2) + \varphi_1(\vec{r}_2)e^{2i\delta})}{2}. \tag{16}$$

What is interesting here is that the presence of a classical field leads to higher degree of entanglement at some initial conditions (see the case of $\delta = 0$ of fig. 3b) and such entanglement can not be achieved in a field free dynamics of even strongly interacting atoms.

It should be emphasized that the factorized state Ψ^{--} which describes two atomic q-bits remaining bound in a strong laser field can be used to produce a maximally entangled Bell state. If we prepare the system in a stable Ψ^{--} state, then turn off the laser field and look at the field-free dynamics of the system, the population of this state will be changed in time due to the coupling between atoms caused by electron (interatomic) interaction. It is found that two of four QES characterizing the field-free dynamics of two coupled atoms are the superpositions of the state Ψ^{--} and the Bell state Φ^{-} (14):

$$\begin{aligned} \Psi_-^{QE} e^{-\frac{i}{\hbar}\gamma_- t} &= \tfrac{\Psi^{--} - \Phi^-}{\sqrt{2}} e^{-\frac{i}{\hbar}\gamma_- t}, \\ \Psi_+^{QE} e^{-\frac{i}{\hbar}\gamma_+ t} &= \tfrac{\Psi^{--} + \Phi^-}{\sqrt{2}} e^{-\frac{i}{\hbar}\gamma_+ t}, \end{aligned} \tag{17}$$

with the corresponding approximate quasienergies:

$$\begin{aligned} \gamma_+ &\approx E_0 + E_1 + \Delta/\sqrt{2} + o(\Delta^2), \\ \gamma_- &\approx E_0 + E_1 - \Delta/\sqrt{2} + o(\Delta^2). \end{aligned} \tag{18}$$

Thus the population of state Ψ^{--} corresponds to the population (with equal probability amplitudes) of both QES (17) and the time dependent probability to find system in Ψ^{--} state is given by:

$$\begin{aligned} W_{\Psi^{--}} = \tfrac{1}{2} \left| \langle \Psi^{--} | \Psi_-^{QE} e^{-\frac{i}{\hbar}\gamma_- t} + \Psi_+^{QE} e^{-\frac{i}{\hbar}\gamma_+ t} \rangle \right|^2 = \\ = \cos^2\left(\tfrac{t\Delta}{\hbar\sqrt{2}}\right). \end{aligned} \tag{19}$$

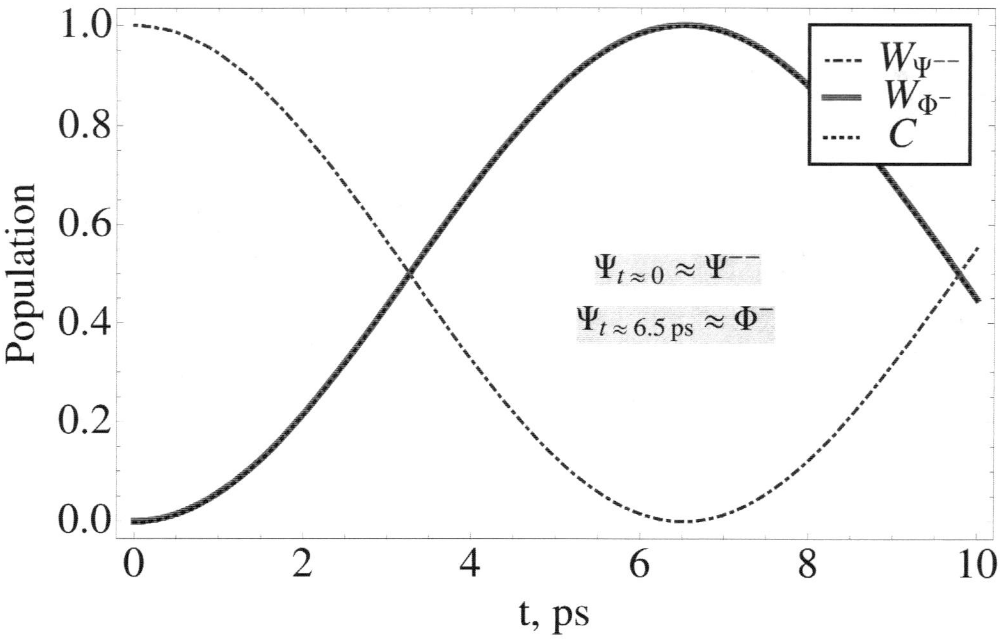

Figure 4. The time dependent population of the state Ψ^{--} ($W_{\Psi^{--}}$), the Bell state Φ^- (W_{Φ^-}) and the concurrence C of the two-atomic q-bit in the after pulse regime.

The population (19) oscillates in time with the frequency $\Delta\sqrt{2}/\hbar$ and there is a population exchange between only two states: Ψ^{--} and the Bell state Φ^- (see fig. 4), so as $W_{\Psi^{--}} + W_{\Phi^-} = 1$. In other words it is possible to create maximally entangled electron spatial Bell state of the two atomic system existing for relatively long time. Since the system reveals periodically maximal entanglement or the highest resistance to the the ionization by a second strong laser pulse, it seems to be rather easy to measure experimentally the population of the stable state Ψ^{--} or the degree of entanglement (concurrence) which coincides exactly with population $W_{\Phi^-} = 1 - W_{\Psi^{--}}$. For this purpose the second ("probe") strong laser pulse should be applied to the system with the delay τ in relation to the first pulse, which prepared Ψ^{--} state. Since $\Psi^{--}$is almost stable against ionization in a strong laser field the ionization signal W_i obtained in the second ("probe") pulse in dependence on the time delay τ gives the population W_{Φ^-} or concurrence of the system in dependence on time ($W_{\Phi^-} = W_i = 1 - W_{\Psi^{--}} = C$). As a result the dynamics of the population of the Bell state Φ^- and the entanglement of the system can be measured experimentally and can be used to extract information about the QES (17) in their dynamics as well as about the average number of the populated Rydberg states because the oscillation period is proportional to Δ and in the case of large number of highly excited states (3) corresponds to the modified Kepler period of the system.

Now let us consider the case of large number n of bound states (3) for each electron. Now we obtain again the system of independent equation for spatially symmetric and antisymmetric states. However for $n > 2$ we do observe the interference stabilization regime for states of both symmetries. The stabilization assumes here that several of the QES are characterized by much more slower decay in a strong field in comparison to the ionization rate calculated according to the Fermy Golden Rule, similarly to the case of individual Rydberg atom with large number of Rydberg bound states discussed in [17, 18]. The initially fast and further very slow decay to the continuum is observed for two coupled Rydberg atoms interacting with a strong laser field (see fig. 5) in the case of $n = 4$.

During this interaction the system reveals high degree of stability due to significant population of several QES that are very resistant to the ionization process. Since these QES are characterized by different quasienergy values (taking into account their real parts), the population of different spatially symmetric states is found to be changed significantly during the laser field action, as it can be seen on fig. 5. Because of the simple initial condition chosen, the stable QES appear to be populated in such a way that the system remains almost totally dis-entangled during all the time of interaction with a laser field. This fact is confirmed by the time dependent behavior of the Schmidt parameter K presented on fig. 5 and the value of K in the case of large number of bound levels is calculated using the expression:

$$K = \frac{\sum_{i,k} |C_{ik}|^2}{\sum_{i,k,l,n} C_{il} C_{in}^* C_{kn} C_{kl}^*} \tag{20}$$

However in the after pulse regime the interatomic interaction $\hat{W}$ plays significant role and appears to provide efficient entanglement between atoms and the increase of the Schmidt parameter possibly to its maximum value $K_{max} = n$. Thus for spatially symmetric states the regime of interference stabilization is found to be accompanied by almost independent dynamics of two atoms in a strong field while the the significant entanglement can be achieved due to the interaction between the atoms $\hat{W}$ in the after pulse regime only.

Another situation is observed for spatially antisymmetric initial state. In this case similarly to the case of two bound level ($n = 2$) for each electron the interatomic interaction term doesn't influence the dynamics of the system (under the assumption of equal matrix elements $W_{ik}^{i'k'}$) and the antisymmetric Bell states ψ_{ik}^- appear to be the eigenstates of the field-free two atomic system. In a strong laser field the QES of the system can be found as a antisymmetrized product of the QES of two non-interacting atoms in a strong field. All the antisymmetric QES can be represented then as a superposition of different Bell states ψ_{ik}^- (5):

$$\psi_l^{QES} = \sum_{i,k} \alpha_{ik}^l \Psi_{ik}^- \tag{21}$$

and some of ψ_l^{QES} appear to be almost stable in a strong field limit if the number of considered Rydberg levels n for each atom is greater than 2. Thus for $n > 2$ the interference stabilization is observed for antisymmetric states too.

At fig. 6 the dynamics of the total bound probability and the populations of different Bell states are presented during the strong laser pulse action and in the after pulse regime for the case when one of the stable antisymmetric QES is initially populated. During the laser pulse the populations of different antisymmetric Bell states as well as the total bound probability are seen to be characterized by the very slow exponential decay with the same small ionization

Journal of Physics: Conference Series **497** (2014) 012018 doi:10.1088/1742-6596/497/1/012018

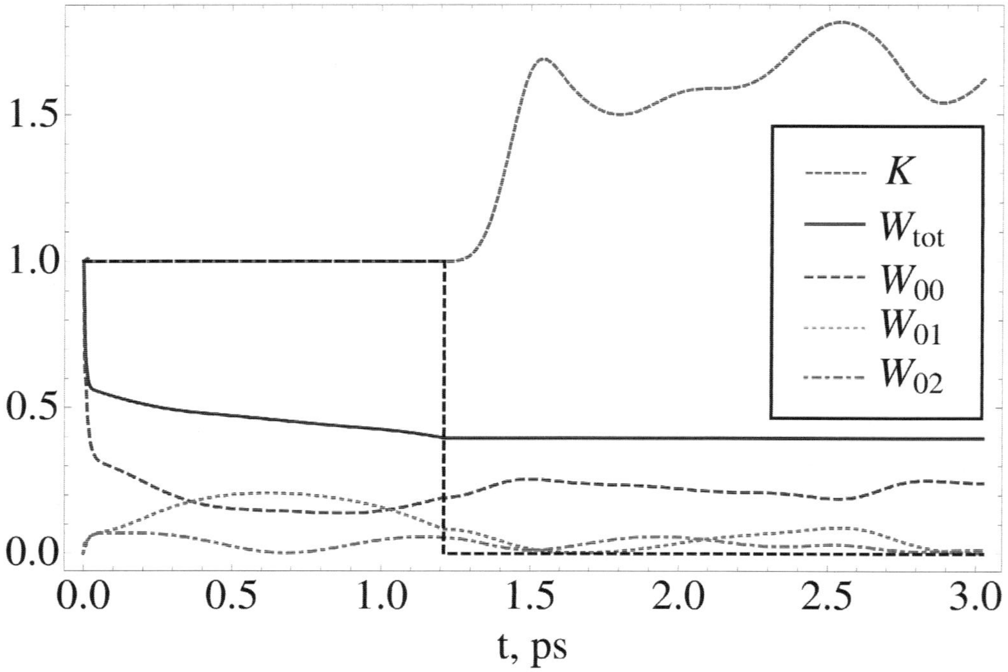

Figure 5. The population of several bipartite states (W_{00}, W_{01}, W_{02}) and Schmidt parameter K calculated in dependence on time for two interacting Rydberg atoms (with $n = 4$ electron bound excited levels) during the laser pulse (dashed line) and in the after pulse regime, for $W = 10\Delta$ and $\Gamma = 10\Delta$. The state $\varphi_0(\vec{r}_1)\varphi_0(\vec{r}_2)$ is populated initially.

rate Γ_{stable} for all states since the dynamics of initially populated QES with quasienery γ is given by:

$$\psi^{QES}(t) = (\sum_{i,k} \alpha_{ik}\Psi^-_{ik})e^{-i\gamma t/\hbar} \tag{22}$$

where α_{ik} are the probability amplitudes of the Bell states ψ^-_{ik} at $t = 0$. Thus the total bound probability is not changed much during many periods of the laser field and the system reveals high stability against ionization in a strong laser field. In the after pulse regime the presented probabilities are conserved. As for entanglement, it is shown that for any superposition of antisymmetric Bell states the value of Schmidt parameter (20) equals 2 ($K = 2$), so as the system is found to be always entangled. As a result, if the system is initially prepared in any arbitrary spatially antisymmetric state, its dynamics is characterized by surviving part of bound population in a strong field due to trapping in the stable QES and by conservation of the degree of entanglement with $K = 2$ both during the laser pulse and in the after pulse regime. Due to different energies of the Bell states ψ^-_{ik} the projection of the total WF on the QES (22) existing in the laser field is changed significantly in the after pulse regime (fig. 6). However full reconstruction of the wave packet is seen to take place periodically in time so as the system periodically reveals a very high resistance against ionization if the second strong laser pulse will be applied.

It should be emphasized that the stability of the two atomic system in a strong laser field

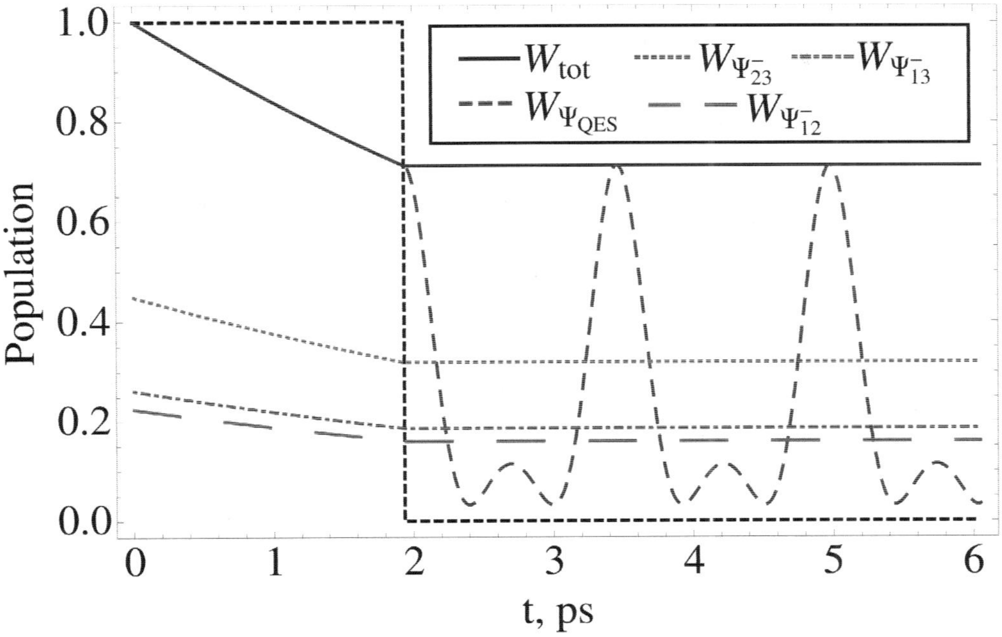

Figure 6. The total bound probability W_{tot}, the population of antisymmetric initially populated QES $W_{\Psi_{QES}}$ and population of several antisymmetric Bell states during the laser pulse and in the after pulse regime for two atomic system (with $n = 4$ bound Rydberg levels for aech atom) for $\Gamma = 10\Delta$

increases with the number of bound levels (3) taken into account for each electron. According to [17, 18] the individual Rydberg atom (with n Rydberg levels) in a strong field regime has only one "unstable" QES, which decays fast in a presence of the laser field. Since in a presence of a strong laser field the influence of the electron-electron interaction on the system dynamics appears to be relatively small the field induced QES of the system can be approximated simply as a product of single electrone QES with a performed symmetrization if needed. As a result for considered two atomic system it can be shown that there are n "unstable" and $n(n-1)/2$ "stable" spatially symmetric states and $(n-1)$ "unstable" and $(n-2)(n-1)/2$ "stable" antisymmetric states. Thus the number of "stable" states grows much faster than the number of states which decays rapidly in a strong laser field. So the interference stabilization expected to be well pronounced for the system of two real Rydberg atoms.

4. Conclusions

The dynamics of two interacting Rydberg atoms in external laser field is studied and interference stabilization against ionization is found in a strong field regime. The mechanism of stabilization is similar to that proposed in [18] for isolated Rydberg atom in a strong field. In the case of more then 2 bound levels taken into account for each electron the interference stabilization is found to occur for both spatially symmetric and antisymmetric states and appears to be more pronounced for larger number of electron bound levels. The surviving of the stable QES under the spatially symmetric initial condition leads to the full dis-entanglement between two electrons in the system, while in the case of antisymmetric initial state the entanglement is found to be conserved both for stable QES in the strong laser field and in the after pulse regime. The methods to obtain strongly entangled bipartite states and to measure the degree of entanglement experimentally are suggested. Thus we show the possibility to control the studied two atom q-bit system by external strong non-resonant laser field and create highly entangled and dis-entangled bipartite states providing significant stabilization against the field induced ionization process. The main approximation made in the worked out theoretical approach consists in the assumption of equal matrix elements (9). Since the wave functions for each electron (3) depend on the interatomic separation and can be a superposition of the large number of Rydberg states of individual atom, the matrix elements $\langle ik|\hat{W}|i'k'\rangle$ can be different for different states and depend dramatically on R. For this reason we analyzed the dynamics of the system in a wide range of the this matrix element values. The predominant role of the strong laser field was found in relation to the electron-electron interaction even under condition of large value of W. In a real system the difference of matrix elements $W_{ik}^{i'k'}$ can lead to the influence of the electron-electron interaction on the dynamics of the antisymmetric electron states. This can reduce the degree of entanglement quantitatively. However the main features of observed dynamics of the studied two atomic system - the strong field stabilization and the possibility of entanglement - seem to be general. Thus the general analytical approach to describe the dynamics of two interacting Rydberg atoms in a strong laser field is developed, providing simple way to calculate for wide range of quantum system parameters, based on the scaling parameter Δ - separation between atomic levels (inverse Kepler period).

4.1. Acknowledgments

This work was supported by Russian Fund for Basic Research grant N14-02-00389-a and N12-02-31232-mol_a.

The calculations were performed on the supercomputer SKIF-MSU "Chebyshev".

References

[1] J Gruska 1999 *Quantum computing* (London: McGrow-Hill)
[2] Sackett C A, Kielpinski D, King B E, Langer C, Meyer V, Myatt C J, Rowe M, Turchette Q A, Itano W M, Wineland D J and Monroe C 2000 *Nature* **404** 256–259 ISSN 0028-0836 URL `http://dx.doi.org/10.1038/35005011`
[3] Calarco T, Briegel H J, Jaksch D, Cirac J I and Zoller P 2000 *Journal of Modern Optics* **47** 2137–2149 (*Preprint* `http://www.tandfonline.com/doi/pdf/10.1080/09500340008235137`) URL `http://www.tandfonline.com/doi/abs/10.1080/09500340008235137`
[4] Saffman M, Walker T G and Mølmer K 2010 *Rev. Mod. Phys.* **82**(3) 2313–2363 URL `http://link.aps.org/doi/10.1103/RevModPhys.82.2313`
[5] Jaksch D, Cirac J I, Zoller P, Rolston S L, Côté R and Lukin M D 2000 *Phys. Rev. Lett.* **85**(10) 2208–2211 URL `http://link.aps.org/doi/10.1103/PhysRevLett.85.2208`
[6] Lukin M D, Fleischhauer M, Cote R, Duan L M, Jaksch D, Cirac J I and Zoller P 2001 *Phys. Rev. Lett.* **87**(3) 037901 URL `http://link.aps.org/doi/10.1103/PhysRevLett.87.037901`

[7] Barrett M D, Chiaverini J, Schaetz T, Britton J, Itano W M, Jost J D, Knill E, Langer C, Leibfried D, Ozeri R and Wineland D J 2004 *Nature* **429** 737–739 ISSN 0028-0836 URL `http://dx.doi.org/10.1038/nature02608`

[8] Cho J and Lee H W 2004 *Phys. Rev. A* **70**(3) 034305 URL `http://link.aps.org/doi/10.1103/PhysRevA.70.034305`

[9] Lloyd S, Shahriar M S, Shapiro J H and Hemmer P R 2001 *Phys. Rev. Lett.* **87**(16) 167903 URL `http://link.aps.org/doi/10.1103/PhysRevLett.87.167903`

[10] Wang Z y, Feng X l, Gong S q and Xu Z z 2001 *Phys. Rev. A* **63**(6) 062308 URL `http://link.aps.org/doi/10.1103/PhysRevA.63.062308`

[11] Thompson R C, Donnellan S, Crick D R and Segal D M 2009 *Journal of Physics B: Atomic, Molecular and Optical Physics* **42** 154003 URL `http://stacks.iop.org/0953-4075/42/i=15/a=154003`

[12] Casterjon-Pita J R, Ohadi H, Crick D R, Winters D F A, Segal D M and Thompson R C 2007 *Journal of Modern Optics* **54** 1581–1594 (*Preprint* `http://www.tandfonline.com/doi/pdf/10.1080/09500340600736793`) URL `http://www.tandfonline.com/doi/abs/10.1080/09500340600736793`

[13] Crick D R, Ohadi H, Bhatti I, Thompson R C and Segal D M 2008 *Opt. Express* **16** 2351–2362 URL `http://www.opticsexpress.org/abstract.cfm?URI=oe-16-4-2351`

[14] Ryabtsev I I, Tretyakov D B and Beterov I I 2005 *Journal of Physics B: Atomic, Molecular and Optical Physics* **38** S421 URL `http://stacks.iop.org/0953-4075/38/i=2/a=032`

[15] Ryabtsev I I, Tretyakov D B, Beterov I I and Entin V M 2007 *Phys. Rev. A* **76**(1) 012722 URL `http://link.aps.org/doi/10.1103/PhysRevA.76.012722`

[16] Comparat D and Pillet P 2010 *J. Opt. Soc. Am. B* **27** A208–A232 URL `http://josab.osa.org/abstract.cfm?URI=josab-27-6-A208`

[17] Fedorov M 1997 *Atomic and free electrons in a strong light field* (World Scientific Publishing Co. Pte. Ltd.)

[18] Fedorov M V and Movsesian A M 1988 *Journal of Physics B: Atomic, Molecular and Optical Physics* **21** L155 URL `http://stacks.iop.org/0953-4075/21/i=7/a=003`

[19] Landau L D and Lifshitz E M 1969 *Mechanics* (Pergamon Press)

[20] Scully M O and Zubairy M S 1997 *Quantum Optics* (Cambridge University Press)

[21] Grobe R, Rzazewski K and Eberly J H 1994 *Journal of Physics B: Atomic, Molecular and Optical Physics* **27** L503 URL `http://stacks.iop.org/0953-4075/27/i=16/a=001`

[22] Fedorov M V, Efremov M A, Volkov P A and Eberly J H 2006 *Journal of Physics B: Atomic, Molecular and Optical Physics* **39** S467 URL `http://stacks.iop.org/0953-4075/39/i=13/a=S20`

[23] Wootters W K 1998 *Phys. Rev. Lett.* **80**(10) 2245–2248 URL `http://link.aps.org/doi/10.1103/PhysRevLett.80.2245`

[24] Fedorov M V, Volkov P A, Mikhailova J M, Straupe S S and Kulik S P 2011 *New Journal of Physics* **13** 083004 URL `http://stacks.iop.org/1367-2630/13/i=8/a=083004`

Journal of Physics: Conference Series **497** (2014) 012019 doi:10.1088/1742-6596/497/1/012019

Shape effects in nonlinear Thomson and Compton processes

M Twardy[1], **K Krajewska**[2] **and J Z Kamiński**[2]

[1] Faculty of Electrical Engineering, Warsaw University of Technology, Pl. Politechniki 1, 00-661 Warszawa, Poland.

[2] Institute of Theoretical Physics, Faculty of Physics, University of Warsaw, Hoża 69, 00-681 Warszawa, Poland.

E-mail: `katarzyna.krajewska@fuw.edu.pl`

Abstract. Spectra of Thomson and Compton radiation, emitted during electron scattering o an intense laser beam, are calculated using the frameworks of classical and strong- eld quantum electrodynamics, respectively. Both approaches use a plane-wave-fronted pulse approximation regarding the driving laser beam. Within this approximation, a very good agreement between Thomson and Compton frequency distributions is observed provided that frequencies of the emitted radiation is relatively low. The dependence of frequency spectra on the laser pulse envelope is analysed. This becomes important in the context of ultra-short pulse generation, as illustrated by numerical examples.

1. Introduction

Compton scattering occurs when an electron scattered against a laser beam emits electromagnetic radiation [1, 2]. A complete description of this phenomenon is given within the framework of strong-field quantum electrodynamics (QED) using the Furry picture [3]. The classical counterpart of the Compton process is called Thomson scattering [4, 5]. In the Thomson approach the emitted radiation spectrum is calculated from the orentz-Maxwell equations with the use of the iénard-Wiechert potentials [6, 7]. In this paper the incident laser beam will be modeled as a plane-wave-fronted pulse [8] and both the Compton and Thomson approaches will be studied.

In many works devoted to nonlinear Compton and Thomson scattering the driving laser beam is treated as a monochromatic plane wave field [9, 10, 11, 12, 13, 14, 15, 16, 17, 18, 19, 20, 21, 22, 23, 24, 25, 26]. In fact, only a few works on Compton scattering, which go beyond this approximation, can be found in literature. This includes the case when the slowly-varying envelope approximation [27] (see, also Refs. [28, 29, 30]) and, more recently, the plane-wave-fronted pulse approximation [31, 32, 33, 34, 35, 36, 37, 38] is used with regard to the driving laser field. The latter is applicable when highly energetic electrons move in a laser pulse, as the action of the ponderomotive force pushing these electrons aside with respect to the pulse propagation direction is negligible [39]. In this case it is assumed that the laser pulse has infinite extension in the transverse direction. In the classical limit, on the other hand, a more accurate description of the scattering process is available. This indicates the importance of studies which underline the relation between quantum and classical approaches. These are of particular interest in light of various applications of Compton and Thomson scattering, e.g., the production of ultra-short

Journal of Physics: Conference Series **497** (2014) 012019 doi:10.1088/1742-6596/497/1/012019

laser pulses in the x-ray domain [16], determining the carrier envelope phase of intense ultra-short pulses [32], measuring the electron beam parameters [40], and generating coherent comb structures in strong-field QED for radiation and matter waves [41].

Note that a comparison of Compton and Thomson scattering, based on a plane-wave-fronted pulse approximation, was carried out in Refs. [32, 33, 34, 35]. In this context, we compare the respective spectra for pulse envelopes which consists of subpulses. We investigate the possibility of generating coherent frequency combs. Specifically, we look at the sensitivity of these structures to a time delay between the incident subpulses. As we demonstrate, these frequency combs can be synthesized into ultra-short pulses with a repetition rate depending on the time delay between the subpulses.

In this paper we use units such that $\hbar = 1$. Numerical results are given in relativistic units where also $m_{\rm e} = c = 1$ (here, $m_{\rm e}$ is the electron rest mass).

The paper is organized as follows. In Sec. 2 we introduce the main results for Thomson scattering based on classical electrodynamics. In Sec. 3 we introduce the Compton scattering theory arising from strong-field QED and present numerical illustrations comparing Thomson and Compton spectra. In Sec. 4 we discuss the frequency comb generation and synthesis of short pulses. The main results are summarized in Sec. 5.

2. Nonlinear Thomson scattering

By Thomson scattering we mean the process consisting in scattering of electrons by a laser beam, described entirely within the framework of classical mechanics and classical electrodynamics. The two most important results relevant to our considerations are the Newton- orentz equation and the frequency-angular distribution of energy radiated by an accelerating electron. The Newton- orentz equation [7, 42]

$$\ddot{\boldsymbol{r}} = \frac{e}{m_{\rm e}}\sqrt{1-\boldsymbol{\beta}^2}\Big[\boldsymbol{\mathcal{E}}(k\cdot x) - \boldsymbol{\beta}(\boldsymbol{\beta}\cdot\boldsymbol{\mathcal{E}}(k\cdot x)) + c\,\boldsymbol{\beta}\times\boldsymbol{\mathcal{B}}(k\cdot x)\Big], \tag{1}$$

describes the acceleration of the electron moving at the reduced velocity $\boldsymbol{\beta} \equiv \dot{\boldsymbol{r}}/c$ when placed in the electromagnetic field generated by a laser. Distribution of energy radiated by the electron is given by [6]

$$\frac{{\rm d}^3 E_{\rm Th}}{{\rm d}\omega_{\boldsymbol{K}}{\rm d}^2\Omega_{\boldsymbol{K}}} = \alpha|\mathcal{A}_{\rm Th}|^2, \tag{2}$$

where $\alpha = e^2/(4\pi\varepsilon_0 c)$ is the fine-structure constant (ε_0 denotes the vacuum electric permittivity) and

$$\mathcal{A}_{\rm Th} = \frac{1}{2\pi}\int \frac{\boldsymbol{n}_{\boldsymbol{K}}\times[(\boldsymbol{n}_{\boldsymbol{K}}-\boldsymbol{\beta})\times\dot{\boldsymbol{\beta}}]}{(1-\boldsymbol{\beta}\cdot\boldsymbol{n}_{\boldsymbol{K}})^2}\exp[{\rm i}\omega_{\boldsymbol{K}}(t-\boldsymbol{n}_{\boldsymbol{K}}\cdot\boldsymbol{r}/c)]{\rm d}t. \tag{3}$$

In order to make use of this formula we must compute a specific trajectory of the scattered electron, according to Eq. (1), i.e., we must know the electric and magnetic fields of the laser pulse. The laser pulse is specified by a shape function that we choose as follows. et us assume that the total pulse consists of $N_{\rm rep}$ identical subpulses that are separated by the time interval $T_{\rm d}$ and each of them lasts for $T_{\rm sub}$ and contains $N_{\rm osc}$ laser field oscillations of the frequency $\omega_{\rm L}$. This means that $\omega_{\rm L}T_{\rm sub} = 2\pi N_{\rm osc}$ and for the envelope function we choose the sine-squared function. For the time interval $0 \leqslant t \leqslant T_{\rm d} + T_{\rm sub}$ we define the function

$$F(t) = \sin^2\Big(\pi\frac{t-T_{\rm d}/2}{T_{\rm sub}}\Big)\sin(\omega_{\rm L}(t-T_{\rm d}/2)+\chi), \tag{4}$$

for $T_{\rm d}/2 \leqslant t \leqslant T_{\rm d}/2 + T_{\rm sub}$ and 0 otherwise, and repeat it $N_{\rm rep}$ times. In this equation the real parameter, χ, denotes a carrier envelope phase. Now, we introduce the frequency $\omega = 2\pi/T_{\rm p}$

with $T_\mathrm{p} = N_\mathrm{rep}(T_\mathrm{d} + T_\mathrm{sub})$, and define the shape function $f(\phi)$ for $0 \leqslant \phi \leqslant 2\pi$ such that its derivative over the phase ϕ equals

$$f'(\phi) = N_\mathrm{A} F(\phi/\omega). \tag{5}$$

From now on, we use the Coulomb gauge for the radiation field, in which case the electric and magnetic field components are equal to

$$\boldsymbol{\mathcal{E}}(k \cdot x) = -\partial_t \boldsymbol{A}(k \cdot x) = -ck^0 \boldsymbol{A}'(k \cdot x), \tag{6}$$
$$\boldsymbol{\mathcal{B}}(k \cdot x) = \boldsymbol{\nabla} \times \boldsymbol{A}(k \cdot x) = -\boldsymbol{k} \times \boldsymbol{A}'(k \cdot x). \tag{7}$$

Because the electric field generated by lasers has to fulfill the following condition,

$$\int_{-\infty}^{\infty} \boldsymbol{\mathcal{E}}(ck^0 t - \boldsymbol{k} \cdot \boldsymbol{r}) \mathrm{d}t = 0, \tag{8}$$

we have also that

$$\lim_{t \to -\infty} \boldsymbol{A}(ck^0 t - \boldsymbol{k} \cdot \boldsymbol{r}) = \lim_{t \to \infty} \boldsymbol{A}(ck^0 t - \boldsymbol{k} \cdot \boldsymbol{r}), \tag{9}$$

and, hence, we can assume that in the remote past and far future the vector potential vanishes. Therefore, for the electromagnetic potential we choose

$$A(k \cdot x) = A_0 N_\mathrm{osc} \varepsilon f(k \cdot x), \tag{10}$$

with the shape function $f(k \cdot x)$ such that $f(k \cdot x) = 0$ for $k \cdot x < 0$ and for $k \cdot x > 2\pi$. In addition, ε is the linear polarization vector of the laser field such that $\varepsilon^2 = -1$ and $k \cdot \varepsilon = 0$. Moreover, we define the dimensionless and relativistically invariant parameter

$$\mu = \frac{|eA_0|}{m_\mathrm{e} c}, \tag{11}$$

which determines the intensity of the laser pulse.

Note that the shape function (5) determines the electric and magnetic fields of the laser pulse, Eqs. (6) and (7). Thus, the shape function for the electromagnetic potential equals

$$f(k \cdot x) = \int_0^{k \cdot x} \mathrm{d}\phi f'(\phi), \tag{12}$$

and, as desired, vanishes for $k \cdot x < 0$ and $k \cdot x > 2\pi$. In Eq. (5), the normalization constant N_A is defined such that

$$\frac{1}{2\pi} \int_0^{2\pi} \mathrm{d}\phi \, [f'(\phi)]^2 = \frac{1}{2}. \tag{13}$$

In our investigations we use the plane-wave-fronted pulse approximation. An applicability of this approximation is based on the assumption that the transverse variation of the electron trajectory in the laser field is negligible when compared to the size of the laser focus, as illustrated in Fig. 1. In Fig. 1(a), we present the solution of the relativistic Newton- orentz equation (1) for the electron initially at rest and in the presence of the laser pulse. The carrier frequency of the laser field ω_L is equal to $10^{-4} m_\mathrm{e} c^2$ (which corresponds to the laser photon energy of about 50 eV), $N_\mathrm{osc} = 32$, $N_\mathrm{rep} = 1$, $T_\mathrm{d} = 0$, and $\mu = 1$. The conditions are set such that initially the electron is at rest in the center of the coordinate system, so its velocity is zero before the arrival of the laser pulse. Fig. 1(b) depicts the x- and z-coordinates of the electron's position, Fig. 1(c) – the reduced velocity, and Fig. 1(d) – the acceleration of the electron. The colors of

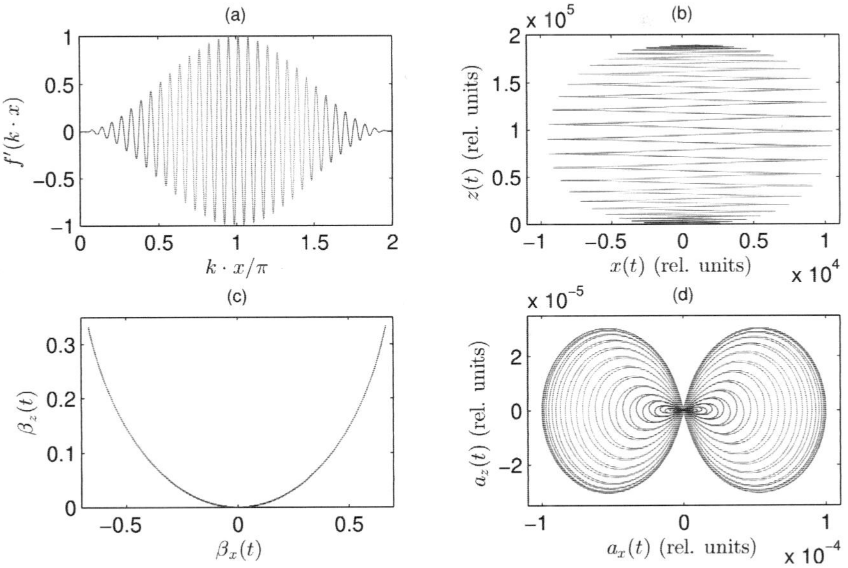

Figure 1. In panel (a), the shape function $f'(k \cdot x)$ defining the electric and magnetic field components of the 32-cycle laser pulse is shown for $T_\mathrm{d} = 0$ and $N_\mathrm{rep} = 1$. In the remaining panels, the solution of the relativistic Newton- orentz equation (1) for the electron moving in the depicted laser pulse is presented (i.e., electron position, speed, and acceleration). The parameters of the pulse are such that $\mu = 1$, $\omega_\mathrm{L} = 10^{-4} m_\mathrm{e}c^2$, and $N_\mathrm{osc} = 32$, where the scaled amplitude of vector potential, μ, measures the peak intensity of the laser field. Initially, the electron is at rest in the center of the coordinate system.

these distributions correspond to the colors of the laser pulse [Fig. 1(a)]. It turns out, that for higher amplitudes of the field depicted in Fig. 1(a), the trajectory of the electron preserves its shape form [Fig. 1(b)]; however, the values of the speed components $\beta_x(t)$ and $\beta_z(t)$ [Fig. 1(c)] attain higher values.

As one can see from Fig. 1, the classical electron placed in a linearly polarized laser field exhibits an oscillatory motion along the direction of the electric field component (i.e., along the x-axis) together with a drift motion in the propagation direction of the laser pulse (i.e., in the z-direction). It may also be observed that for a given laser field frequency the displacement of the electron is larger for a stronger laser field than for a weaker laser field. A useful measure of the relativistic length unit is the reduced electron Compton wavelength

$$\frac{1}{2\pi}\lambda_\mathrm{C} = \frac{\hbar}{m_\mathrm{e}c} = 3.8616 \times 10^{-13}\,\mathrm{m}\,, \tag{14}$$

which equals 1 in relativistic units. Note that for the chosen laser field parameters the electron displacement along the electric field vector is of the order of 10^{-9}m (Fig. 1). Taking this into account, we find that the electron displacement along the electric field direction is of the order of 0.001μm. For lasers available today, a typical linear dimension of their focus is a few μm. Therefore, the electron displacement in the transverse direction (even for very powerful laser fields) can be neglected on the scale of the focus, provided that the laser frequency is sufficiently large in the reference frame in which initially electrons are at rest. We conclude that the plane-wave-fronted pulse approximation is perfectly suitable for describing the nonlinear Thomson scattering processes generated by currently available laser sources.

3. Nonlinear Compton scattering

Using the S-matrix formalism of strong-field QED, we derive that the probability amplitude for the Compton process, $e^-_{\boldsymbol{p}_\mathrm{i}\lambda_\mathrm{i}} \to e^-_{\boldsymbol{p}_\mathrm{f}\lambda_\mathrm{f}} + \gamma_{\boldsymbol{K}\sigma}$, with the initial and final electron momenta and spin polarizations $\boldsymbol{p}_\mathrm{i}\lambda_\mathrm{i}$ and $\boldsymbol{p}_\mathrm{f}\lambda_\mathrm{f}$, respectively, equals

$$\mathcal{A}(e^-_{\boldsymbol{p}_\mathrm{i}\lambda_\mathrm{i}} \to e^-_{\boldsymbol{p}_\mathrm{f}\lambda_\mathrm{f}} + \gamma_{\boldsymbol{K}\sigma}) = -\mathrm{i}e\int \mathrm{d}^4x\, j^{(++)}_{\boldsymbol{p}_\mathrm{f}\lambda_\mathrm{f},\boldsymbol{p}_\mathrm{i}\lambda_\mathrm{i}}(x)\cdot A^{(-)}_{\boldsymbol{K}\sigma}(x), \tag{15}$$

where $\boldsymbol{K}\sigma$ denotes the Compton photon momentum and polarization. Here, we consider the case when both the laser pulse and the Compton photon are linearly polarized. In Eq. (15),

$$A^{(-)}_{\boldsymbol{K}\sigma}(x) = \sqrt{\frac{1}{2\varepsilon_0\omega_{\boldsymbol{K}}V}}\,\varepsilon_{\boldsymbol{K}\sigma}\mathrm{e}^{\mathrm{i}K\cdot x}, \tag{16}$$

where V is the quantization volume, $\omega_{\boldsymbol{K}} = cK^0 = c|\boldsymbol{K}|$ $(K\cdot K = 0)$, and $\varepsilon_{\boldsymbol{K}\sigma} = (0, \boldsymbol{\varepsilon}_{\boldsymbol{K}\sigma})$ are the polarization four-vectors satisfying the conditions

$$K\cdot\varepsilon_{\boldsymbol{K}\sigma} = 0, \quad \varepsilon_{\boldsymbol{K}\sigma}\cdot\varepsilon_{\boldsymbol{K}\sigma'} = -\delta_{\sigma\sigma'}, \tag{17}$$

for $\sigma, \sigma' = 1, 2$. Moreover, $j^{(++)}_{\boldsymbol{p}_\mathrm{f}\lambda_\mathrm{f},\boldsymbol{p}_\mathrm{i}\lambda_\mathrm{i}}(x)$ is the matrix element of the electron current operator with its ν-component equal to

$$[j^{(++)}_{\boldsymbol{p}_\mathrm{f}\lambda_\mathrm{f},\boldsymbol{p}_\mathrm{i}\lambda_\mathrm{i}}(x)]^\nu = \bar{\psi}^{(+)}_{\boldsymbol{p}_\mathrm{f}\lambda_\mathrm{f}}(x)\gamma^\nu\psi^{(+)}_{\boldsymbol{p}_\mathrm{i}\lambda_\mathrm{i}}(x). \tag{18}$$

Here, $\psi^{(+)}_{\boldsymbol{p}\lambda}(x)$ is the so-called Volkov solution of the Dirac equation coupled to the electromagnetic field [43, 44] (see, also Refs. [45, 46, 47] for possible generalizations)

$$\psi^{(+)}_{\boldsymbol{p}\lambda}(x) = \sqrt{\frac{m_\mathrm{e}c^2}{VE_{\boldsymbol{p}}}}\Big(1 - \frac{e}{2k\cdot p}\not{A}\not{k}\Big)u^{(+)}_{\boldsymbol{p}\lambda}\mathrm{e}^{-\mathrm{i}S^{(+)}_p(x)}, \tag{19}$$

with

$$S^{(+)}_p(x) = p\cdot x + \int^{k\cdot x}\Big[\frac{eA(\phi)\cdot p}{k\cdot p} - \frac{e^2A^2(\phi)}{2k\cdot p}\Big]\mathrm{d}\phi. \tag{20}$$

Moreover, $E_{\boldsymbol{p}} = cp^0$, $p = (p^0, \boldsymbol{p})$, $p\cdot p = m_\mathrm{e}^2c^2$, and $u^{(+)}_{\boldsymbol{p}\lambda}$ is the free-electron bispinor normalized such that

$$\bar{u}^{(+)}_{\boldsymbol{p}\lambda}u^{(+)}_{\boldsymbol{p}\lambda'} = \delta_{\lambda\lambda'}. \tag{21}$$

The four-vector potential $A(k\cdot x)$ in Eq. (19) represents an external electromagnetic radiation generated by lasers, in the case when a transverse variation of the laser field in a focus is negligible. In other words, $A(k\cdot x)$ represents the plane-wave-fronted pulse. In this case, $k\cdot A(k\cdot x) = 0$ and $k\cdot k = 0$, which allows one to exactly solve the Dirac equation for such electromagnetic fields.

The probability amplitude for the Compton process (15) becomes

$$\mathcal{A}(e^-_{\boldsymbol{p}_\mathrm{i}\lambda_\mathrm{i}} \longrightarrow e^-_{\boldsymbol{p}_\mathrm{f}\lambda_\mathrm{f}} + \gamma_{\boldsymbol{K}\sigma}) = \mathrm{i}\sqrt{\frac{2\pi\alpha c(m_\mathrm{e}c^2)^2}{E_{\boldsymbol{p}_\mathrm{f}}E_{\boldsymbol{p}_\mathrm{i}}\omega_{\boldsymbol{K}}V^3}}\,\mathcal{A}, \tag{22}$$

where

$$\begin{aligned}\mathcal{A} &= \int \mathrm{d}^4x\,\bar{u}^{(+)}_{\boldsymbol{p}_\mathrm{f}\lambda_\mathrm{f}}\Big(1 - \mu\tfrac{m_\mathrm{e}c}{2p_\mathrm{f}\cdot k}f(k\cdot x)\not{\varepsilon}\not{k}\Big)\not{\varepsilon}_{\boldsymbol{K}\sigma} \\ &\times\Big(1 + \mu\tfrac{m_\mathrm{e}c}{2p_\mathrm{i}\cdot k}f(k\cdot x)\not{\varepsilon}\not{k}\Big)u^{(+)}_{\boldsymbol{p}_\mathrm{i}\lambda_\mathrm{i}}\,\mathrm{e}^{-\mathrm{i}S(x)},\end{aligned} \tag{23}$$

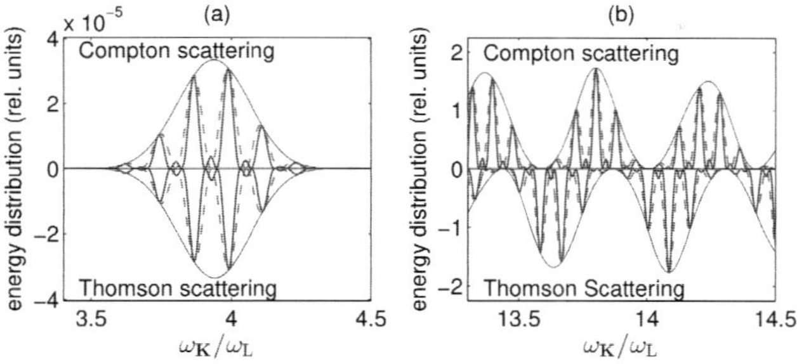

Figure 2. The energy distribution of radiation emitted by the scattered electron. The black solid line corresponds to $N_{\rm rep} = 1$, the red dashed line to $N_{\rm rep} = 2$, whereas the blue solid line is for $N_{\rm rep} = 3$. Each subpulse contains eight field oscillations ($N_{\rm osc} = 8$). The detection angles are $\varphi_{\boldsymbol{K}} = 0$ and $\theta_{\boldsymbol{K}} = \pi/10$. While the results presented in panel (a) relate to $\mu = 1$ and $\omega_{\rm L} = 3 \times 10^{-3} m_{\rm e}c^2$, the results presented in panel (b) are for $\mu = 10$ and $\omega_{\rm L} = 3 \times 10^{-2} m_{\rm e}c^2$.

with

$$S(x) = S_{p_{\rm i}}^{(+)}(x) - S_{p_{\rm f}}^{(+)}(x) - K \cdot x \,. \tag{24}$$

While moving in a laser pulse, the electron acquires an additional momentum shift [36, 48], this leads to a notion of the laser-dressed momentum:

$$\bar{p} = p - \mu m_{\rm e} c \frac{p \cdot \varepsilon}{p \cdot k} \langle f \rangle k + \frac{1}{2}(\mu m_{\rm e} c)^2 \frac{1}{p \cdot k} \langle f^2 \rangle k \,. \tag{25}$$

Having this in mind we can define

$$N_{\rm eff} = \frac{K^0 + \bar{p}_{\rm f}^0 - \bar{p}_{\rm i}^0}{k^0} = cT_{\rm p} \frac{K^0 + \bar{p}_{\rm f}^0 - \bar{p}_{\rm i}^0}{2\pi}, \tag{26}$$

which is both gauge- and relativistically invariant [36].

The frequency-angular distribution of energy of the emitted photons for an unpolarized electron beam is given by

$$\frac{{\rm d}^3 E_{\rm C}}{{\rm d}\omega_{\boldsymbol{K}} {\rm d}^2\Omega_{\boldsymbol{K}}} = \frac{1}{2} \sum_{\sigma=1,2} \sum_{\lambda_{\rm i}=\pm} \sum_{\lambda_{\rm f}=\pm} \frac{{\rm d}^3 E_{{\rm C},\sigma}(\lambda_{\rm i}, \lambda_{\rm f})}{{\rm d}\omega_{\boldsymbol{K}} {\rm d}^2\Omega_{\boldsymbol{K}}}, \tag{27}$$

where

$$\frac{{\rm d}^3 E_{{\rm C},\sigma}(\lambda_{\rm i}, \lambda_{\rm f})}{{\rm d}\omega_{\boldsymbol{K}} {\rm d}^2\Omega_{\boldsymbol{K}}} = \alpha \, |\mathcal{A}_{{\rm C},\sigma}(\omega_{\boldsymbol{K}}, \lambda_{\rm i}, \lambda_{\rm f})|^2 \,, \tag{28}$$

and the scattering amplitude equals

$$\mathcal{A}_{{\rm C},\sigma}(\omega_{\boldsymbol{K}}, \lambda_{\rm i}, \lambda_{\rm f}) = \frac{m_{\rm e} c K^0}{2\pi\sqrt{p_{\rm i}^0 k^0 (k \cdot p_{\rm f})}} \sum_N D_N \frac{1 - {\rm e}^{-2\pi {\rm i}(N - N_{\rm eff})}}{{\rm i}(N - N_{\rm eff})}, \tag{29}$$

with the functions D_N defined in [36].

In Fig. 2, we compare the Thomson and Compton energy distributions for a single laser pulse in the reference frame in which initially electrons are at rest. As expected, for $\omega_{\boldsymbol{K}} \ll m_{\rm e}c^2$ both theories give the same results, as depicted in the left panel. However, for larger laser carrier frequency $\omega_{\rm L}$ and energies of generated photons $\omega_{\boldsymbol{K}}$ the Compton distribution is red-shifted with

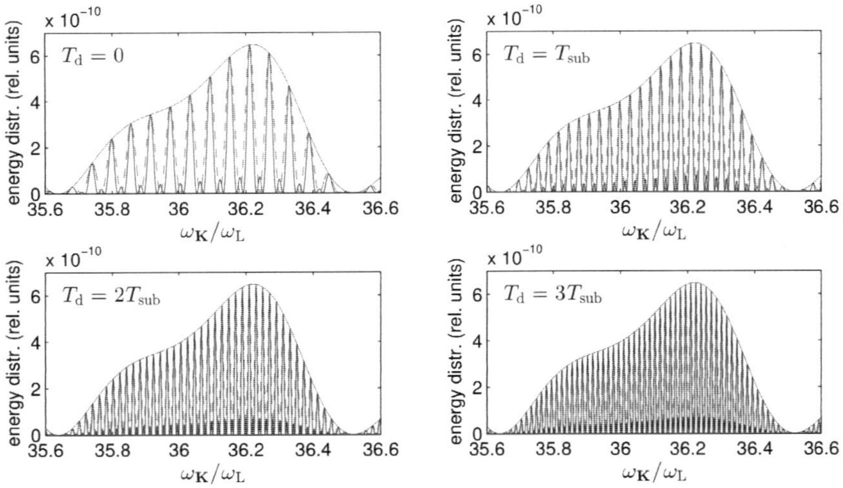

Figure 3. The Compton energy distribution for the laser pulse parameters: $\omega_L = 4.15 \times 10^{-4} m_e c^2$, $\mu = 1$, $N_{osc} = 16$, $\theta_{\boldsymbol{K}} = 0.2\pi$, $\varphi_{\boldsymbol{K}} = 0$, and $\chi = \pi$. Different panels correspond to four different delay times T_d. The solid black lines (envelopes) present the energy distributions for a single pulse ($N_{rep} = 1$), the dashed red lines are for $N_{rep} = 2$, whereas the solid blue lines are for $N_{rep} = 3$. The corresponding energy distributions are divided by N_{rep}^2, which proves the coherent properties of the generated high-order harmonics comb structures.

respect to the Thomson one. Such a shift has been analyzed in [33] for long laser pulses. This analysis has been extended to arbitrarily short laser pulses in [49], together with the discussion of the significant role played by the polarization of emitted radiation and the spin degrees of freedom of the electron initial and final states.

4. Compton high-order harmonics and generation of ultra-short pulses

High-order harmonics generated via non-relativistic interaction of intense laser pulses with atoms allow to synthesize attosecond pulses of coherent radiation [50]. Currently, the energy bandwidth of the harmonic plateau can reach a few keV [51]. In order to extend this spectrum up to MeV domain a relativistic treatment is necessary. The Compton process offers such a possibility, as it is presented in Fig. 3. For a single laser pulse ($N_{rep} = 1$), we observe a broad and smooth energy distribution from which we choose a part of the bandwidth approximately equal to the carrier frequency ω_L. However, if we apply the sequence of N_{rep} pulses the energy distribution shows an equally spaced peaks with maxima which scale as N_{rep}^2. This scaling law indicates that the generated comb structure is temporarily coherent. Moreover, the distance between the peaks can be controlled by a delay of the laser subpulses. Therefore, one can interpret the emergence of such a structure as the result of the interference of Compton photons emitted from different subpulses [41].

In order to further investigate the coherent properties of such a high-order harmonic spectrum let us consider the temporal power distribution of emitted radiation. This power distribution is related to the Compton amplitude (29) by the formula

$$\frac{d^2 P_{C,\sigma}(\phi_r, \lambda_i, \lambda_f)}{d^2\Omega_{\boldsymbol{K}}} = \frac{\alpha}{\pi}(\Re \tilde{\mathcal{A}}^{(+)}_{C,\sigma}(\phi_r, \lambda_i, \lambda_f))^2. \tag{30}$$

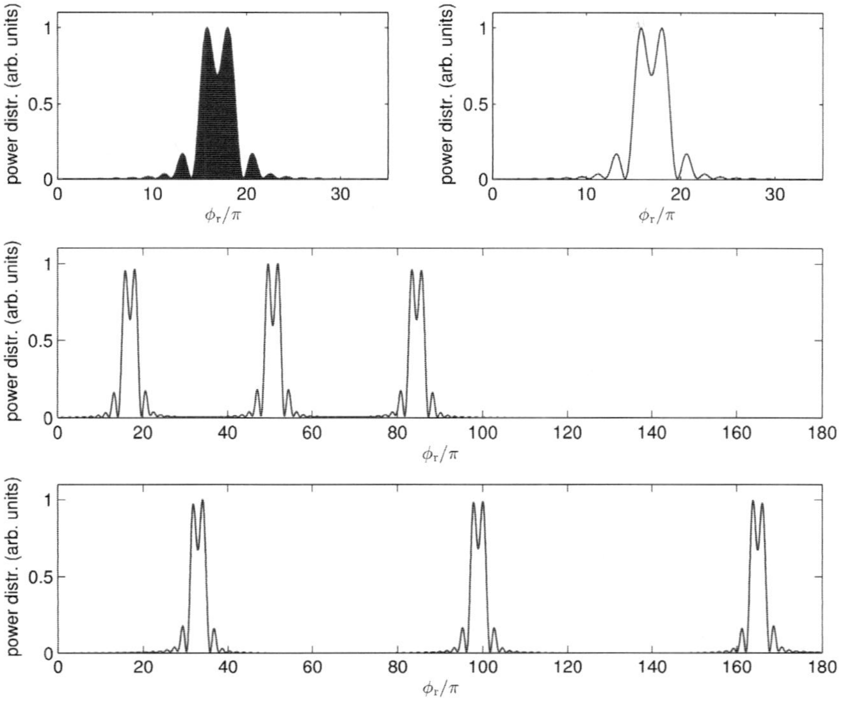

Figure 4. Temporal power distributions of generated radiation, synthesized from the Compton amplitude (29) for $\lambda_i\lambda_f = 1$ and for the laser field parameters specified in Fig. 3. The upper two panels show the power distributions defined by Eqs. (30) and (32) for a single pulse ($N_{rep} = 1$), after being normalized to the maximum value. The middle and the bottom panels show the temporal power distributions (32) composed from the energy distributions represented in Fig. 3 by the blue lines ($N_{rep} = 3$) for $T_d = 0$ and $T_d = T_{sub}$, respectively.

where

$$\tilde{\mathcal{A}}^{(+)}_{C,\sigma}(\phi_r, \lambda_i, \lambda_f) = \int_0^\infty d\omega \mathcal{A}_{C,\sigma}(\omega, \lambda_i, \lambda_f) e^{-i\omega\phi_r/\omega_L}. \tag{31}$$

Here, $\Re$ denotes the real part and $\phi_r = \omega_L(t - R/c)$, with R being a distance from the scattering region to the observation point. In general, the power distribution (30) is a very rapidly oscillating function of time. One can define the temporal power distribution of generated radiation, avaraged over these oscillations,

$$\frac{d^2\langle P_{C,\sigma}\rangle(\phi_r, \lambda_i, \lambda_f)}{d^2\Omega_{\boldsymbol{K}}} = \frac{\alpha}{2\pi}|\tilde{\mathcal{A}}^{(+)}_{C,\sigma}(\phi_r, \lambda_i, \lambda_f)|^2. \tag{32}$$

Fig. 4 depicts the temporal power distributions, (30) or (32), as functions of the dimensionless retarded phase $\phi_r = \omega_L(t - R/c)$, instead of the observation time t. The power distributions have been synthesized from the energy distributions presented in Fig. 3. As we see, the radiation is emitted in the form of short pulses. Similar conclusions can be drawn from the classical Thomson scattering, although some significant discrepancies between the classical and quantum theories can be observed.

5. Conclusions

In this paper, we have presented the theory of Thomson and Compton processes in intense laser pulses. We have shown that, by applying the sequence of laser pulses, it is possible to create high-order harmonics structures in the emitted radiation. We have investigated this problem in the electron beam reference frame. It appears, however, that in the laboratory frame, when electrons have the energy of the order of GeV, the bandwidth of the generated harmonic spectrum is of the order of a few MeV. This can be further synthesized into zepto- or even yoctosecond pulses, as it is presented elsewhere [52].

Acknowledgments

This work is supported by the Polish National Science Center (NCN) under Grant No. 2012 05 B ST2 02547.

References

[1] Ehlotzky F, Krajewska K and Kaminski J 2009 *Rep. Prog. Phys.* **72** 046401
[2] Di Piazza A, Muller C, Hatsagortsyan K and Keitel C H 2012 *Rev. Mod. Phys.* **84** 1177
[3] Furry W H 1951 *Phys. Rev.* **81** 115
[4] Lau L L, He F, Umstadter D P and Kowalczyk R 2003 *Phys. Plasmas* **10** 2155
[5] Umstadter D P 2003 *J. Phys.* D **36** R151
[6] Jackson J D 1998 *Classical Electrodynamics* (New York: John Wiley & Sons)
[7] Landau L D and Lifshitz E M 1980 *The Classical Theory of Fields* (Portsmouth, NH: Heinemann)
[8] Neville R A and Rohrlich F 1971 *Phys. Rev.* D **3** 1692
[9] Brown L S and Kibble T W B 1964 *Phys. Rev.* **133** A705
[10] Goldman I 1964 *Sov. Phys. JETP* **46** 1412
[11] Nikishov A I and Ritus V I 1964 *Sov. Phys. JETP* **19** 1191
[12] Sengupta N D 1949 *Bull. Math. Soc.* **41** 187
[13] Vaschaspati 1962 *Phys. Rev.* **128** 664
[14] Vaschaspati 1963 *Phys. Rev.* **130** E2598
[15] Sarachik E S and Schappert G T 1970 *Phys. Rev.* D **1** 2738
[16] Esarey E, Ride S K and Sprangle P 1993 *Phys. Rev.* E **48** 3003
[17] Ride S K, Esarey E and Baine M 1995 *Phys. Rev.* E **52** 5425
[18] Salamin Y I and Faisal F H M 1996 *Phys. Rev.* A **54** 4383
[19] Salamin Y I and Faisal F H M 1997 *Phys. Rev.* A **55** 3964
[20] Salamin Y I and Faisal F H M 1998 *J. Phys.* A **31** 1319
[21] Goreslavskii S P, Popruzhenko S V and Shcherbachev O V 1999 *Laser Phys.* **9** 1039
[22] Panek P, Kaminski J and Ehlotzky F 2002 *Phys. Rev.* A **65** 022712
[23] Ivanov D Y, Kotkin G L and Serbo V G 2004 *Eur. Phys. J.* C **36** 127
[24] Hartin A and Moortgat-Pick G 2011 *Eur. Phys. J.* C **71** 1729
[25] Popa A 2011 *Phys. Rev.* A **84** 023824
[26] Popa A 2012 *Laser Part. Beams* **30** 591
[27] Narozhny N B and Fofanov M S 1996 *Sov. Phys. JETP* **83** 14
[28] Voroshilo A I, Roshchupkin S P and Nedoreshta V N 2011 *Laser Phys.* **21** 1675
[29] Roshchupkin S P, Lebed' A A, Padusenko E A and Voroshilo A I 2012 *Laser Phys.* **22** 1113
[30] Nedoreshta V N, Voroshilo A I and Roshchupkin S P 2013 *Phys. Rev.* A **88** 052109
[31] Boca M and Florescu V 2009 *Phys. Rev.* A **80** 053403
[32] Mackenroth F, Di Piazza A and Keitel C H 2010 *Phys. Rev. Lett.* **105** 063903
[33] Seipt D and Kampfer B 2011 *Phys. Rev.* A **83** 022101
[34] Mackenroth F and Di Piazza A 2011 *Phys. Rev.* A **83** 032106
[35] Boca M and Florescu V 2011 *Eur. Phys. J.* D **61** 449
[36] Krajewska K and Kaminski J 2012 *Phys. Rev.* A **85** 062102
[37] Boca M, Dinu V and Florescu V 2012 *Phys. Rev.* A **86** 013414
[38] Krajewska K and Kaminski J 2013 *Laser Part. Beams* **31** 503
[39] Bulanov S V et al. 2011 *Nucl. Instrum. Methods Phys. Res.* A **600** 31
[40] Leemans W P et al. 1996 *Phys. Rev. Lett.* **77** 4182
[41] Krajewska K and Kaminski J 2013 Universality of comb structures in strong- eld QED (*Preprint* ar iv:1307.5433)

[42] Griffiths D J 1999 *Introduction to Electrodynamics* (Englewood Cli s, NJ: Prentice-Hall)
[43] Volkov D M 1935 *Z. Phys.* **94** 250
[44] Krajewska K and Kaminski J 2010 *Phys. Rev.* A **82** 013420
[45] Varro S 2013 *Laser Phys. Lett.* **10** 095301
[46] Varro S 2013 *Nuclear Instrum. Methods Phys. Res.* A (http://dx.doi.org/10.1016/j.nima.2013.11.091)
[47] Raicher E and Eliezer S 2013 *Phys. Rev.* A **88** 022113
[48] Krajewska K and Kaminski J 2012 *Phys. Rev.* A **86** 052104
[49] Krajewska K and Kaminski J 2013 Frequency scaling law for nonlinear Compton and Thomson scattering: Relevance of spin and polarization e ects (*Preprint* ar iv:1308.1663)
[50] Farkas G and Toth C 1992 *Phys. Lett.* A **168** 447
[51] Popmintchev T *et al* 2012 *Science* **336** 1287
[52] Krajewska K, Twardy M and Kaminski J 2013 Supercontinuum and ultra-short pulse generation from nonlinear Thomson and Compton scattering (*Preprint* ar iv:1311.4872)

Nonlinear Bethe–Heitler Pair Creation in an Intense Two-Mode Laser Field

Sven Augustin[1,2] **and Carsten Müller**[1,2]

[1] Institut für Theoretische Physik I, Heinrich-Heine-Universität, 40225 Düsseldorf, Germany
[2] Max-Planck-Institut für Kernphysik, Saupfercheckweg 1, 69117 Heidelberg, Germany

E-mail: sven.augustin@mpi-hd.mpg.de

Abstract. We investigate electron-positron pair creation in the interaction of a nuclear Coulomb field and a highly intense two-mode laser field. For bichromatic laser fields, we examine the differences arising for commensurable and incommensurable frequencies in a continuous variation of the laser frequency ratio and the quantum interference effects, which may occur in the commensurable case. We show that the interference manifests in the angular distributions and the total pair-production rates of the created particles. Additionally, by varying the amplitudes of the two modes we study pair creation in a monochromatic laser wave of arbitrarily elliptical polarization.

1. Introduction

The creation of matter from laser light has already been theoretically investigated [1–4] shortly after the realization of the first laser itself [5]. Recently this interest has experienced a revival [6, 7], due to the large and still ongoing progress in high-intensity laser technology. Additionally, the experimental feasibility of electron-positron (e^-e^+) pair production by multiphoton absorption was demonstrated by a pioneering experiment at the Stanford Linear Accelerator Center (SLAC) [8, 9], applying the nonlinear Breit–Wheeler process [1–3]. In a similar manner, the nonlinear Bethe–Heitler process

$$Z + n\omega \to Z + e^- + e^+ \tag{1}$$

is in principle accessible by modern experimental techniques, e.g., by using the highly relativistic nuclear beam from the Large Hadron Collider (LHC) at CERN in conjunction with a counterpropagating highly intense laser beam. In the nuclear rest frame, the laser frequency and intensity are largely amplified by a relativistic Doppler shift, reaching the levels required for pair production.

The prospect of an experimental test has led to further theoretical investigations dedicated to the nonlinear Bethe–Heitler effect. On the one hand the various field parameter regimes have been studied by calculating total and differential pair-production rates (e.g., [10–15]), while on the other hand more specialized features were examined, such as the effects of the electron spin [16] and the nuclear recoil [17, 18]. In all these studies the laser field was assumed to be a monochromatic plane wave with either linear or circular polarization.

The subject of this contribution is Bethe–Heitler pair creation in a two-mode laser field. In particular, the modes may oscillate with different frequencies, leading to a bichromatic laser field:

$$Z + n_1\omega_1 + n_2\omega_2 \to Z + e^- + e^+. \tag{2}$$

Journal of Physics: Conference Series **497** (2014) 012020 doi:10.1088/1742-6596/497/1/012020

While in general arbitrary combinations of the laser frequencies ω_i are allowed, so far mainly two special cases have been investigated: largely differing frequencies with $\omega_1/\omega_2 \gtrsim 10^2$ [19,20] and, as only there quantum path interference may occur, commensurable frequencies, i.e., frequencies of rational ratio [21–24]. In [22–24] it could be concluded that the relative phase between the two laser modes distinctly influences the angular spectrum of the created pairs. In [22,23] both field modes were assumed to be linearly polarized along the same direction. For other examples of interference effects in field-induced pair creation, we refer the interested reader to [25–31].

In the present contribution, we will extend our previous study [24] to incommensurable frequencies, again assuming both field modes to be linearly polarized with mutually orthogonal polarization vectors and propagating in the same direction. Additionally, we will examine further aspects of the interference in the commensurable case by comparing a laser pair combination from our previous study to a new one with a higher number of photons, emphasizing on the features arising from this increase. Finally, we investigate the special case of a monochromatic laser wave, by a variation of the individual intensities of the two laser modes. This allows a transition from a single linearly polarized to a circularly polarized laser wave, covering elliptical polarizations in between.

The present article is organized as follows. First, we will outline our calculational approach in Sec. 2. Into the $\mathscr{S}$-matrix describing the nonlinear Bethe–Heitler process we insert the Volkov solutions of the Dirac equation for a bichromatic laser field with linearly polarized modes of orthogonal field vectors. From this, an expression for the total pair-production rate is derived, containing a six-fold integral over the momenta of the created particles and a four-fold sum over photon numbers, which both can effectively be reduced by one due to constraints from energy conservation. The remaining integrations are performed numerically, leading to the results presented in Sec. 3, where we show total pair-production rates or pair-production rates differential in the polar emission angle. Section 3.1 is devoted to the case of commensurable laser frequencies for which quantum interference effects may occur. We will compare rates obtained for two different photon orders but identical total photon energy and discuss the influence of the relative phase between the laser modes. In Sec. 3.2 the difference between commensurable and incommensurable field frequencies for the nonlinear Bethe–Heitler process will be analyzed by showing pair-creation rates in a variation of the frequency ratio of the two laser modes. The special case of pair creation in a monochromatic laser field with elliptical polarization will be discussed in Sec. 3.3. The dependence of the total pair-creation rate on the ellipticity will be examined by varying the field polarization continuously from circular to linear in two different ways. The conclusions that can be drawn from our study are summarized in Sec. 4.

2. Theoretical Framework

2.1. Pair-Production Amplitude and Rate for a Two-Mode Field

We model pair creation in the superposition of a nuclear field and a laser field as a transition from a negative continuum state $\Psi^{(+)}$ to one of the positive continuum $\Psi^{(-)}$, induced by the nuclear Coulomb potential

$$A_{\mathrm{N}} = \frac{Ze}{|\mathbf{r}|}. \tag{3}$$

Treating the nuclear Coulomb field in the lowest order of perturbation theory leads to the pair-creation amplitude [4, 10, 12–14]

$$\mathscr{S} = \frac{\mathrm{i}Ze^2}{\hbar c}\int \bar{\Psi}^{(-)}\gamma_0\Psi^{(+)}\frac{\mathrm{d}^4x}{|\mathbf{r}|}, \tag{4}$$

in the rest frame of the nucleus. As continuum wave functions, in order to fully account for the interaction of the leptons with the laser field, we use the Volkov solutions [32] for electrons and positrons, labelled by the sign of their charge $-$ and $+$, respectively,

$$\Psi^{(\pm)}_{p_\pm,s_\pm} = N_\pm\left(1 \pm \frac{e\not{\kappa}\not{A}}{2c\kappa_\mu p^\mu_\pm}\right)\exp\left(\frac{\mathrm{i}}{\hbar}S^{(\pm)}\right)u^{(\pm)}_{p_\pm,s_\pm}, \tag{5}$$

containing the normalizer N, the action S, and the free Dirac spinor u with the respective particle's momentum p and spin s. Herein the positive elementary charge e and the unit vector in wave propagation direction κ are used, and Feynman slash notation $\not{A} = \gamma_\mu A^\mu$ is applied. The Volkov wave functions are gained as exact solutions of the Dirac equation for an electron moving in the field of an electromagnetic plane wave in vacuum, where the vector potential A, given in Lorenz gauge $\partial_\mu A^\mu = \kappa_\mu A^\mu = 0$, depends only on a phase variable η.

The Volkov wave functions can be inserted into the pair-creation amplitude from Eq. (4), leading to

$$\begin{aligned} \mathscr{S} &= N_- N_+ \frac{\mathrm{i}Ze^2}{\hbar c} \int \frac{\mathrm{d}^4x}{|\mathbf{r}|} G \exp\left(\frac{\mathrm{i}}{\hbar}\left(-S^{(-)} + S^{(+)}\right)\right), \\ G &= \bar{u}^{(-)}_{p_-,s_-}\left(1 - \frac{e\not{A}\not{\kappa}}{2c\kappa_\mu p_-^\mu}\right)\gamma_0\left(1 + \frac{e\not{\kappa}\not{A}}{2c\kappa_\mu p_+^\mu}\right)u^{(+)}_{p_+,s_+}, \end{aligned} \tag{6}$$

where the abbreviation G contains all γ-matrices.

The laser field A is defined as superposition of two plane waves,

$$A = A_1 + A_2 \quad \text{with} \quad A_i = a_i \cos(\eta_i + \varphi_i) \quad (i = 1 \text{ or } 2), \tag{7}$$

with relative phases φ_i, phase coordinates $\eta_i = (\omega_i/c)\,\kappa_\mu x^\mu$, where the direction of propagation $\kappa = (1,0,0,1)$ is shared among the wave vectors $k_i = (\omega_i/c)\kappa$, and perpendicular field vectors a_i, given by $a_1 = (0,1,0,0)\,|\mathbf{a}_1|$ and $a_2 = (0,0,1,0)\,|\mathbf{a}_2|$, measuring their absolute amplitudes using the dimensionless intensity parameters

$$\xi_i = \frac{e}{mc^2}\frac{|\mathbf{a}_i|}{\sqrt{2}}, \tag{8}$$

with the electronic rest mass m and the speed of light in vacuum c.

Due to the chosen field geometry, functions of the two phase coordinates are still separable when the laser amplitude is squared. We can thus give the action as sum over the laser modes:

$$S^{(\pm)} = \pm p_\mu x^\mu + \sum_{i=1}^{2} \frac{e}{cp_\mu k_i^\mu}\left[p_\mu a_i^\mu \sin(\eta_i + \varphi_i) \mp \frac{e}{4c} a_i^2 \left(\frac{\sin(2[\eta_i + \varphi_i])}{2} + (\eta_i + \varphi_i)\right)\right]. \tag{9}$$

Upon insertion of the action from Eq. (9) and the laser fields from Eq. (7) into the pair-creation amplitude from Eq. (6) we obtain for each laser mode a set of three functions periodic in the respective phase coordinate η_i. These functions can be expanded in Fourier series and the resulting coefficients are built from generalized Bessel functions [33]. With this expansion we can write the amplitude as a summation over two indices, which can be interpreted as counts for the number of photons taken from each of the two modes:

$$\mathscr{S} = \frac{\mathrm{i}Ze^2 mc}{\hbar c\sqrt{q_-^0 q_+^0}} \sum_{n_1,n_2} M^{(n_1,n_2)}_{p_-p_+} \int \frac{\mathrm{d}^4x}{|\mathbf{r}|} \exp\left(\frac{\mathrm{i}}{\hbar} x_\mu Q^\mu_{(n_1,n_2)}\right). \tag{10}$$

Here we have introduced the matrix element $M^{(n_1,n_2)}_{p_-p_+}$, which consists of all slashed quantities and the aforementioned Fourier coefficients, the normalizers $N_\pm = \sqrt{mc/q_\pm^0}$, and the momentum transfer to the nucleus $Q_{(n_1,n_2)} = q_+ + q_- - n_1\hbar k_1 - n_2\hbar k_2$, where the latter two are defined using the effective momentum [34]

$$q_\pm = p_\pm + \frac{e^2\overline{\mathbf{A}^2}}{2c^2\kappa_\mu p_\pm^\mu}\kappa, \tag{11}$$

with the averaged squared laser amplitude $\overline{\mathbf{A}^2} = \frac{1}{2}\left(|\mathbf{a}_1|^2 + |\mathbf{a}_2|^2\right) = \frac{m^2c^4}{e^2}\left(\xi_1^2 + \xi_2^2\right)$. The four-dimensional integral in Eq. (10) can be solved by using the Fourier transform of the Coulomb potential and a representation of the δ-function for the integral in space and time, respectively [35]. Note that by definition of $Q^0_{(n_1,n_2)}$ the newly introduced δ-function ensures energy conservation.

Squaring the amplitude leads to a sum over four indices:

$$|\mathscr{S}|^2 = \sum_{\substack{n_1',n_2' \\ n_1,n_2}} \mathscr{P}_{[n_1,n_1',n_2,n_2']}, \tag{12}$$

with the addends being the thereby defined partial contributions

$$\mathscr{P}_{[n_1,n_1',n_2,n_2']} = \frac{Z^2 e^4 m^2}{q_+^0 q_-^0} 32\pi^3\hbar^3 \bar{M}^{(n_1,n_2)}_{p_-p_+} M^{(n_1',n_2')}_{p_-p_+} \frac{cT}{Q^4_{(n_1,n_2)}} \delta\left(Q^0_{(n_1,n_2)}\right). \tag{13}$$

Here we have used $Q_{(n_1,n_2)} = Q_{(n_1',n_2')}$ and $n_1k_1 + n_2k_2 = n_1'k_1 + n_2'k_2$ as enforced by the δ-function, and introduced the time T from the squared δ-function [35]. The product of the two matrix elements $\bar{M}^{(n_1,n_2)}_{p_-p_+}$ and $M^{(n_1',n_2')}_{p_-p_+}$ is a rather cumbersome summation of products of Dirac γ-matrices and thus shall not be shown here.

Finally, the partial contributions $\mathscr{P}$ enter the differential partial rates

$$\mathrm{d}^6 R_{[n_1,n_1',n_2,n_2']} = \frac{1}{T} \sum_{s_+,s_-} \mathscr{P}_{[n_1,n_1',n_2,n_2']} \frac{\mathrm{d}^3 q_-}{(2\pi\hbar)^3} \frac{\mathrm{d}^3 q_+}{(2\pi\hbar)^3}, \tag{14}$$

where we also summed over the final spin states and divided by the time T. Using the δ-function in Eq. (13), we can perform one integration analytically. The remaining integrals are calculated numerically to obtain angular differential and fully integrated partial rates. Additionally, the summation over photon numbers from Eq. (12) is performed to find differential and total rates. Results from these computations are presented in Sec. 3.

We emphasize that the partial rates $R_{[n_1,n_1',n_2,n_2']}$ introduced above are, in general, no experimental observables. It is especially important to note that, for commensurable laser frequencies, they may be negative and accordingly decrease the total rate

$$R = \sum_{\substack{n_1',n_2' \\ n_1,n_2}} R_{[n_1,n_1',n_2,n_2']}. \tag{15}$$

The latter, on the other hand, is an experimentally measurable quantity and always positive. A negative contribution to the four-index sum may only arise for certain index combinations and will subsequently be interpreted as destructive interference between the two laser modes, just as a positive contribution from these particular index combinations will be understood as constructive interference.

2.2. *Terminology*

In the following, the terminology applied in the presentation of our results is briefly discussed. The minimal number of photons from a single mode i with frequency ω_i needed to create an electron-positron pair will be denoted by $\tilde{n}_i$. In the case of commensurable frequencies, a pair of two laser modes $(\tilde{n}_1, \tilde{n}_2)$ is then given by their respective minimal photon numbers if their total photon energies are identical, $E_{\mathrm{tot}} = \tilde{n}_1\hbar\omega_1 = \tilde{n}_2\hbar\omega_2$, and thus it is indistinguishable whether $\tilde{n}_1$ photons were absorbed from the first mode or $\tilde{n}_2$ from the second mode.

It is useful to introduce three categories for the index combinations $[n_1, n_1', n_2, n_2']$ from Eq. (12): *direct* terms, *symmetrically mixed* terms and *asymmetrically mixed* terms. In Tab. 1 the exact conditions and some examples are summarized. The *direct* terms stem solely from one of the two laser modes and are thus the only terms that could be measured individually in an experiment by turning off the respective

Table 1. Types, conditions, and examples of the terms in the summation from Eq. (12).

Type of term	Condition	Example: $(4,8)$
Direct	$n_1 = n_1', n_2 = n_2'$ and only one is not 0	$[4,4,0,0]$ or $[0,0,8,8]$
Symmetrically mixed	$n_1 = n_1' \neq 0$ and $n_2 = n_2' \neq 0$	$[2,2,4,4]$
Interference (Asymmetrically mixed)	$n_1 \neq n_1'$ and $n_2 \neq n_2'$	$[0,4,8,0]$ or $[0,2,8,4]$

other mode. For the *symmetrically mixed* terms the energy to overcome the pair-creation threshold is gained by taking a certain number of photons from the first mode and another number of photons from the second mode. In contrast, for the *asymmetrically mixed* terms this intuitive explanation in terms of photon numbers does not hold, instead they can be understood as stemming from interference of the two laser modes, which may only occur in the commensurable case discussed in Sec. 3.1. As we will also see later, only these *interference terms* are sensitive to a variation of the relative phases φ_i from Eq. (7).

Note that, due to symmetry reasons we always find pairs of interference terms with interchanged n_i and n_i', which give identical contributions to the four-index sum. In figures where individual terms are shown, they thus overlay each other and are only depicted by a single line. If the sum over all terms is performed these lines need to be counted twice. Besides, any term that would be allowed by energy conservation (enforced by the δ-function in Eq. (13)) but is several orders of magnitude smaller than the strongest terms will not be depicted.

Finally, we would like to note that for all results in Sec. 3 a proton beam target is assumed by setting $Z = 1$. As long as our first-order treatment of the nuclear field does not need Coulomb corrections, pair-creation rates for higher nuclear charges can be inferred from the proton target results by multiplying with an overall scaling factor of Z^2. The rates in Sec. 3 will be given in *atomic units* where 1 au $= 4.13\times10^{16}$ s^{-1}.

2.3. Intensity parameters

To study the effects of combining two modes in a laser field, we have to ensure both modes have a sizeable influence on the results. In order to prevent one mode from dominating over the other, which would correspond to an effectively monochromatic laser wave, it is favourable to choose the intensity parameters so that the direct terms for the two modes contribute equally. Consequently the contribution strength of all mixed terms will be maximized. Provided that $\xi_i \ll 1$, the ξ_i-scaling of the direct term of mode i can be given as

$$R_i \sim \xi_i^{2\tilde{n}_i}, \tag{16}$$

where the abbreviations $R_1 = R_{[\tilde{n}_1,\tilde{n}_1,0,0]}$ and $R_2 = R_{[0,0,\tilde{n}_2,\tilde{n}_2]}$ are used. To achieve $R_1 \approx R_2$ one can choose a common parameter ζ so that the two laser intensity parameters ξ_i (compare Eq. (7)) are connected to the other wave's minimal photon number $\tilde{n}_i$ by

$$\xi_1 \approx \zeta^{\tilde{n}_2} \quad \text{and} \quad \xi_2 \approx \zeta^{\tilde{n}_1}, \tag{17}$$

leading to

$$R_1 \approx R_2 \sim \zeta^{2\tilde{n}_1\tilde{n}_2}. \tag{18}$$

This definition only takes the general scaling into account and due to differences in the respective proportionality factors a small adjustment has to be made to gain fully equal direct terms in our results. This is achieved by calculating the fully-integrated partial rates of the two direct terms once and inferring the corrected ξ_i from their ratio.

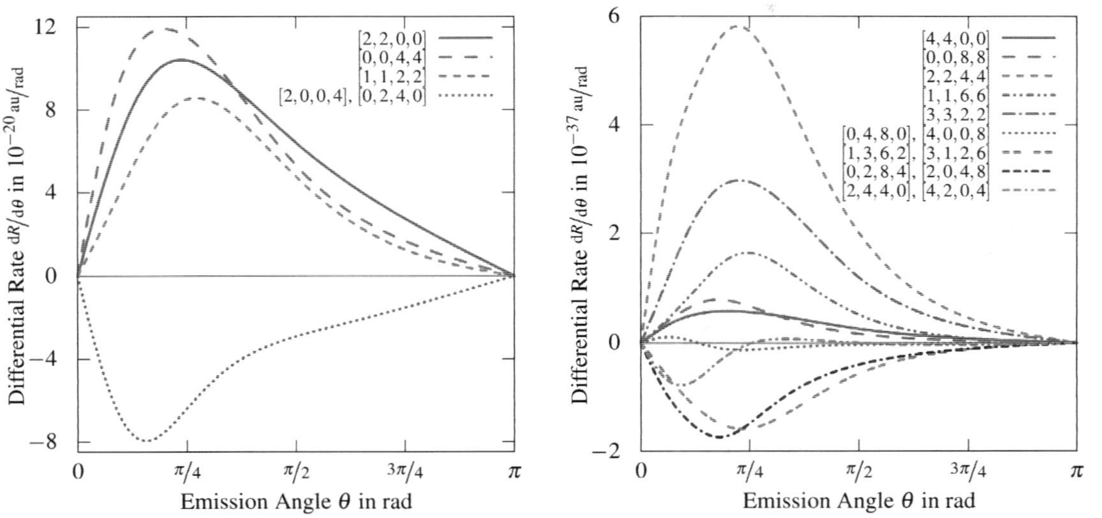

(a) Laser pair $(2,4)$ with $\xi_1 = 10^{-4}$ and $\xi_2 = 1.46\times10^{-2}$ (b) Laser pair $(4,8)$ with $\xi_1 = 10^{-4}$ and $\xi_2 = 1.31\times10^{-2}$

Figure 1. (Color online) Angular-differential partial rates for laser pairs $(2,4)$ and $(4,8)$ with total photon energy 1.15 MeV in the nuclear rest frame, corresponding to (a) $\omega_1 = 575$ keV, $\omega_2 = 287.5$ keV and (b) $\omega_1 = 287.5$ keV, $\omega_2 = 143.75$ keV. The frequencies are chosen such that the total photon energy is reached for $\tilde{n}_1$ photons from the first mode and $\tilde{n}_2$ photons from the second mode. The emission angle θ is measured with respect to the laser propagation direction. The relative phases are set as $\varphi_1 = \varphi_2 = 0$.

For a general term $[n_1, n'_1, n_2, n'_2]$ a scaling corresponding to Eq. (16) can be given and the individual ξ_i can be replaced by the newly introduced ζ:

$$R_{[n_1,n'_1,n_2,n'_2]} \sim \xi_1^{(n_1+n'_1)} \xi_2^{(n_2+n'_2)} \approx \zeta^{\tilde{n}_2(n_1+n'_1)+\tilde{n}_1(n_2+n'_2)}. \tag{19}$$

3. Results and Discussion

3.1. Interference and Phase Effects

At first we consider the case of commensurable frequencies. In [24] we used laser pair $(2,4)$ as prime example as it showed the strongest contribution from interference terms. There we compared it explicitly to laser pair $(1,2)$, which has a lower photon number but identical frequency ratio. It is interesting to also study laser pair $(4,8)$ where the frequency ratio is again identical but the total number of photons is higher.

In Fig. 1 a comparison between the laser pairs $(2,4)$ and $(4,8)$ is shown. Due to the higher photon number in the latter example, a larger set of contributing terms needs to be discussed. For $(2,4)$, there is one direct term for each mode ($[2,2,0,0]$ and $[0,0,4,4]$), one symmetrically mixed term $[1,1,2,2]$, using half of the needed energy from each mode, and two interference terms ($[2,0,0,4]$ and $[0,2,4,0]$). All five contributing terms are of identical order in the common parameter ζ and, consequently, they give similar absolute contributions.

For laser pair $(4,8)$, the corresponding set of terms is again clearly visible: The two direct terms $[4,4,0,0]$ and $[0,0,8,8]$, the symmetrically mixed term $[2,2,4,4]$, and the two interference terms $[4,0,0,8]$ and $[0,4,8,0]$. In contrast to the former case, the symmetrically mixed term $[2,2,4,4]$ is now the strongest contribution. Additionally, we find two more interference terms, $[1,3,6,2]$ and $[3,1,2,6]$, with $n_1 + n'_1 = 4$ and $n_2 + n'_2 = 8$. Finally, two symmetrically mixed terms with different $n_i + n'_i$ are

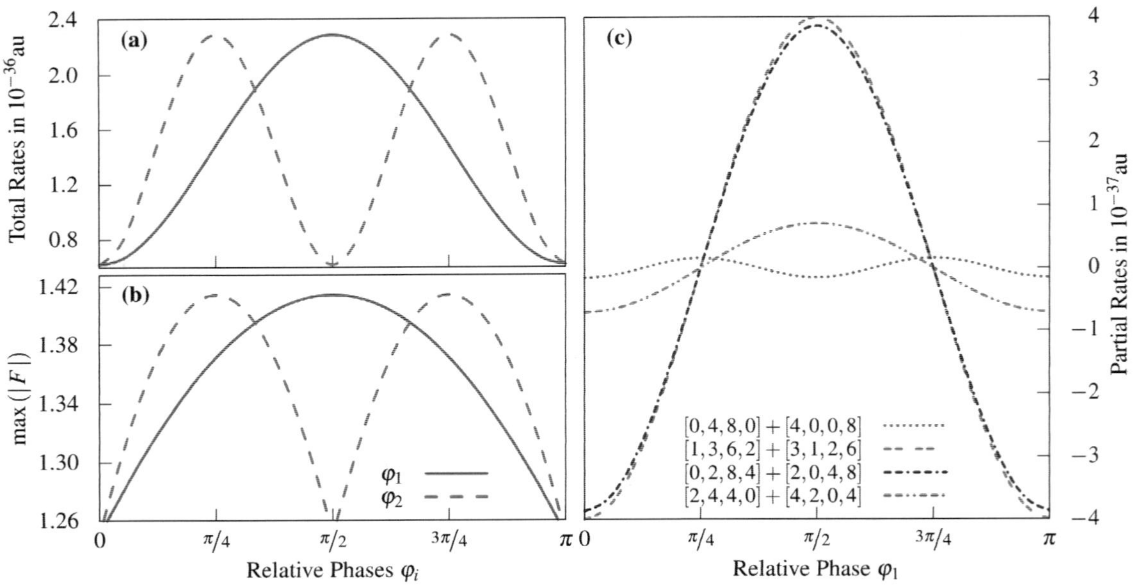

Figure 2. (Color online) Variation of the relative phases φ_1 and φ_2 for laser pair $(4,8)$ as in Fig. 1 (b) – Comparison of (a) the total pair-production rate and (b) the measure defined via the absolute maximum of the sum of the normalized electric fields from Eq. (21). Additionally, in (c) the relevant partial rates of the interference terms are shown for the variation of φ_1. Note the different periodicity of the terms $[4,0,0,8]$ and $[0,4,8,0]$.

visible: $[1,1,6,6]$ and $[3,3,2,2]$. Both have corresponding interference terms: $[0,2,8,4]$, $[2,0,4,8]$ and $[2,4,4,0]$, $[4,2,0,4]$, respectively. It is interesting to note, that all non-negligible contributions are again of the same order in ζ, despite the differing $n_i + n_i'$ in the latter. Small differences in the contribution strengths remain due to the different prefactors contained in each term (cf. Sec. 2.3).

We find that for laser pair $(4,8)$ the two direct terms are only weak contributions and all symmetrically mixed terms are stronger. Of the interference terms, the straightforward ones $[4,0,0,8]$ and $[0,4,8,0]$ are weakest, while the more complex ones, with only one or none index equal to zero, are quite strong. This is especially interesting in the comparison of $[1,1,6,6]$, which is the weakest of the three symmetrically mixed terms, and the interference terms with identical $n_i + n_i'$, $[0,2,8,4]$ and $[2,0,4,8]$, which are the strongest of that type.

Of all terms shown in Fig. 1, only the interference terms are affected by a variation of the relative phase between the two laser modes. For laser pair $(2,4)$, their periodicity Φ_i in the relative phases φ_i can be given explicitly as [24]

$$\Phi_i = {}^{2\pi}/_{\tilde{n}_i}. \tag{20}$$

The phase dependence can be related to the modulus of the electric field of the combined laser modes. The measure $\max(|F|)$, derived from the electric fields E_i using an appropriate scaling of the mode intensities, is defined via the squared quantity

$$F^2(ct-z) = \sum_{i=1}^{2} \frac{c^2}{\mathbf{a}_i^2 \omega_i^2} E_i^2 = \sum_{i=1}^{2} \sin^2(\eta_i + \varphi_i), \tag{21}$$

and exhibits qualitatively the same phase dependence as the total pair-production rate.

The connection between the phase dependences of $\max(|F|)$ and the total pair-production rate also holds for laser pair $(4,8)$, as Figs. 2 (a) and (b) show. However, Eq. (20) does not apply to this case but requires a generalization. From the analytical calculations, we find that a more general expression for the periodicity in φ_i can be given as

$$\Phi_i = {}^{2\pi}/_{\Delta n_i} \tag{22}$$

using the difference of the indices of the respective laser mode i: $\Delta n_i = |n_i - n_i'|$. An illustration of Eq. (22) is given in Fig. 2 (c) where the phase dependences of the interference terms of laser pair $(4,8)$ are shown. Interestingly, not all of them share the same behaviour. The six stronger interference terms share $\Delta n_1 = 2$ and accordingly they all exhibit the same periodicity $\Phi_1 = \pi$, which is also the periodicity of $\max(|F|)$. In contrast, the weakest terms $[4,0,0,8]$ and $[0,4,8,0]$ show a different periodicity of $\Phi_1 = \pi/2$, in accordance with $\Delta n_1 = 4$.

Thus we can conclude that, for laser pair $(4,8)$, those interference terms which exhibit a different phase dependence than $\max(|F|)$ are contributing only marginally, while all strong interference terms share the periodicity of this parameter. It is thus retained by the sum of all terms and, eventually, by the total rate. For laser pair $(2,4)$ the situation is simpler because all interference terms that give relevant contributions to the total pair-production rate satisfy $\Delta n_i = \tilde{n}_i$, in agreement with Eq. (20).

3.2. Commensurable vs. Incommensurable Frequencies

In Sec. 3.1 we have discussed examples of two laser modes with commensurable frequencies ω_1 and ω_2, where $\tilde{n}_1\omega_1 = \tilde{n}_2\omega_2$. As, in this case, it is indistinguishable whether $\tilde{n}_1$ or $\tilde{n}_2$ photons were taken from mode one or two, respectively, quantum paths may interfere, giving rise to the effects discussed in said section.

These examples should principally be distinguished from those with incommensurable frequencies, as there interference cannot occur. However, due to two arguments, one physical and the other mathematical, this strict distinction may not be an ideal choice (a similar line of argument can be found in [36]): On the one hand, laser fields used in a real experimental situation have a finite pulse length and, thus, comprise a continuous range of frequencies; physical properties should possess a smooth dependence under small variations therein. On the other hand, any irrational number can be approximated by a ratio of integers with arbitrary precision. Combining these arguments it becomes obvious that commensurability can be understood as always partially fulfilled and thus not allowing the intended distinction.

A physically meaningful replacement is whether the frequency ratio is comprised of two *small* integers or not. For a small-integer frequency ratio the maximally achievable field strength in the superposition of the two laser modes will occur often and repeatedly. For example, it may happen that the maxima of the mode with the slower oscillation always coincide with a maximum of the other one. In contrast, for a large-integer frequency ratio, two coinciding maxima are very rare. Furthermore, the latter case also means that a high number of photons is needed for an interference term, making it less likely. The rational approximation of an irrational number will certainly consist of two large integers (thus implying minor contributions from interference) and consequently, incommensurable frequencies are treated properly by this distinction.

Applying this idea, we can study a continuous variation of the frequency ratio ${}^{\omega_1}/_{\omega_2}$, passing by the commensurable laser pairs $(2,4)$ and $(2,6)$ discussed in the earlier study [24], and thus gain further insights into the processes leading to the interferences. In Fig. 3 this variation is plotted by keeping ω_1 fixed and varying ω_2 for the total pair-production rate and several summed-up partial rates. The spectrum is comprised of the constant direct term of the first mode, $[2,2,0,0]$, and slowly rising and falling peaks for the direct terms of the second mode, $[0,0,n,n]$ for $n = (2,3,4,5,6)$. Note that, the increase of these direct terms around their respective $\tilde{n}_2$ value can be explained by the applied scaling of the intensity as given in the figure caption. On top of these peaks, for those commensurable frequencies where both $\tilde{n}_i$ are even integers, constructive ($\tilde{n}_2 = 2$ or 6) and destructive ($\tilde{n}_2 = 4$) interference terms lead to prominent

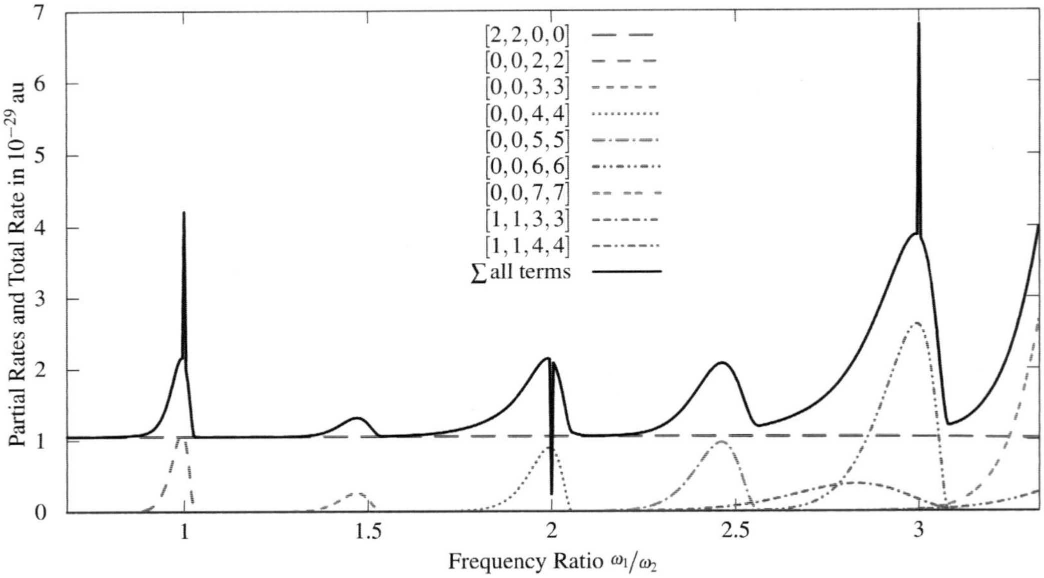

Figure 3. (Color online) Variation of the frequency ratio ω_1/ω_2 of the two laser modes – Total rate and integrated partial rates for the total photon energy $E_{\text{tot}} = 1.05$ MeV in the nuclear rest frame with $\omega_1 = E_{\text{tot}}/2$ fixed and ω_2 varied from $E_{\text{tot}}/1.8$ to $E_{\text{tot}}/6.2$. The relative phases are set as $\varphi_1 = \varphi_2 = 0$. The intensity parameter $\xi_2 = \sqrt{\omega_1/\omega_2}\zeta^{\omega_2/\omega_1}$ is scaled similar to Eq. (17), with $\xi_1 = \zeta = 10^{-6}$ and an empirically found prefactor of $\sqrt{\omega_1/\omega_2}$ taking the small adjustment explained in Sec. 2.3 into account. Note that the partial contributions of the interference terms are not shown separately as they are clearly visible as the δ-spikes for the integer frequency ratios.

δ-spikes, which are – in our treatment – infinitely narrow but of finite height. Note that the interference terms are not shown separately, but the spike they contribute can be clearly seen in the sum of all terms.

We note that in an actual experimental realisation of the effects studied here, the laser would not be an infinite plane wave but a finite pulse. This would lead to the δ-spikes being smeared out and the resulting spectrum being continuous. Nevertheless, the strong enhancement or depletion at the even-numbered commensurable frequencies can be expected to remain.

A variation of the relative phases φ_i will only affect the spikes, sinusoidally changing their height and sign with the periodicity as discussed in Sec. 3.1. This has been examined extensively in the aforementioned earlier study, particularly for the laser pairs $(2,4)$ and $(2,6)$. In the next section, we would like to concentrate on the special case $(2,2)$ instead, representing a single linearly polarized or a circularly polarized laser wave for the difference of the relative phases $|\varphi_1 - \varphi_2|$ set to 0 or $\pi/2$, respectively. Here, a variation of the phase corresponds to a variation of the ellipticity of the combined laser wave.

3.3. Monochromatic Laser Wave of Elliptical Polarization

A vector potential of the form given in Eq. (7) can also describe a monochromatic laser wave if we set $\omega_1 = \omega_2$. Particularly, the setup denoted as laser pair $(2,2)$ in the Sec. 3.2 corresponds to a single linearly polarized laser wave with the polarization axis lying diagonally in the x-y-plane for vanishing relative phases $\varphi_1 = \varphi_2 = 0$, and to a circularly polarized laser wave for one φ_i set to $\pi/2$ instead. By a continuous variation between these two extreme conditions a laser wave of arbitrary ellipticity can be studied. The

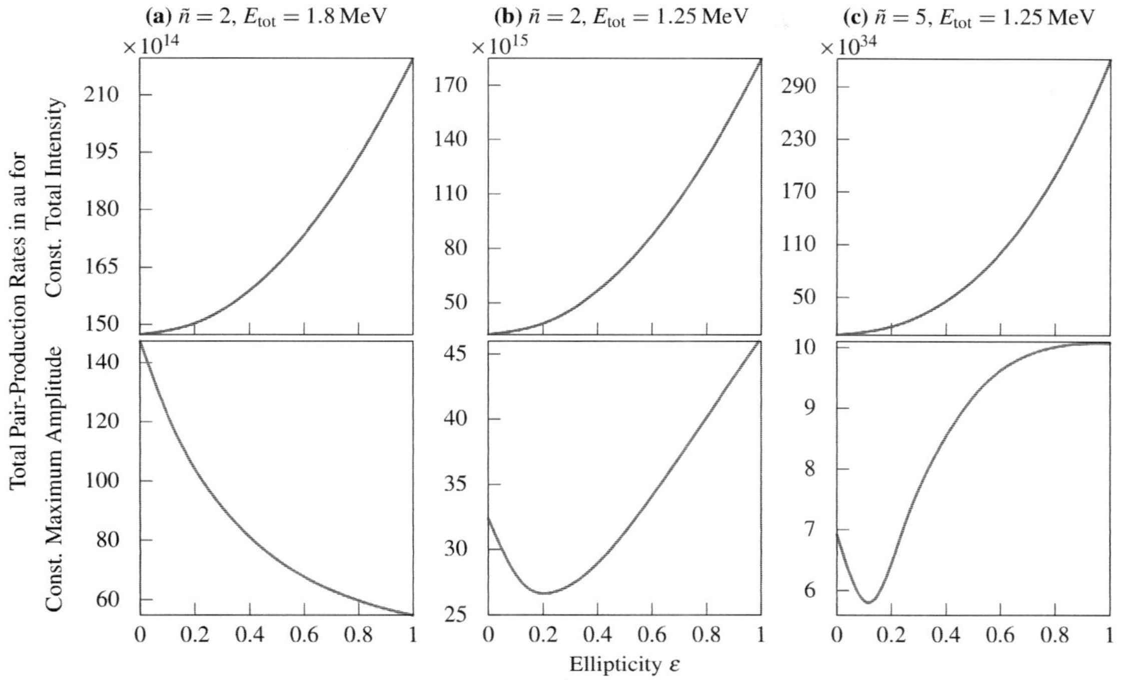

Figure 4. (Color online) Variation of the ellipticity ε – Total rates in the nuclear rest frame for laser pair $(2,2)$ with total photon energies (a) 1.8 MeV and (b) 1.25 MeV, and for (c) laser pair $(5,5)$ with total photon energy 1.25 MeV. Comparisons are shown between (top panels) fixed total intensity using $\xi_1^2+\xi_2^2=2\times10^{-6}$, and (bottom panels) fixed maximum laser wave amplitude using $\xi_1^2=10^{-6}$ and $0\leq\xi_2\leq\xi_1$. For the two-photon cases, where it is applicable, the fully analytical graphs obtained using the formulas derived in [37] coincide perfectly with our results.

variation may be carried out in two different ways, both starting from a circularly polarized wave: On the one hand one can fix the maximum amplitude of the laser field and decrease one intensity half-axis until it disappears, while keeping the other half-axis constant. On the other hand one can fix the total intensity supplied to the system, which means decreasing one half-axis while increasing the other, so that $\xi_1^2+\xi_2^2=\text{const.}$[1] In both cases the ellipticity will be measured using the parameter

$$\varepsilon=\frac{|\,\xi_1^2-\xi_2^2\,|}{\xi_1^2+\xi_2^2}=\begin{cases}0 & \text{for circular polarization}\\ 0<\varepsilon<1 & \text{for elliptical polarization}\\ 1 & \text{for linear polarization.}\end{cases}\tag{23}$$

In Fig. 4 (a) the two pathways are shown in comparison for a total photon energy of 1.8 MeV and $\tilde{n}=2$ photons absorbed from the wave. For the first case of a constant maximum laser wave amplitude (bottom panel), we see a decreasing total pair-production rate, when going from circular to linear polarization. This can be intuitively explained by the decrease of the supplied total laser intensity as we gradually turn off the second laser mode. For the second case of constant total intensity (top panel), we see an increase

[1] It can be shown that the latter pathway corresponds directly to the variation of the relative phase mentioned at the end of Sec. 3.2 for fixed intensity parameters, which is interesting as φ_i and ξ_i are independent parameters. For a variation of φ_i the ellipticity is then given as $\varepsilon=|\cos\varphi_i|$.

from left to right instead. Here the peak amplitude of the combined laser waves allows an explanation. In the circularly polarized case the field amplitude is constant, while only the field vector direction changes:

$$\mathbf{A}_{\rm circ} = \mathbf{a}_1 \sin\eta + \mathbf{a}_2 \sin(\eta + \pi/2) = a\,(\mathbf{e}_1 \sin\eta + \mathbf{e}_2 \cos\eta)\,, \qquad |\mathbf{A}_{\rm circ}| = a. \tag{24}$$

In the linearly polarized case the field amplitude is subject to the sinusoidal variation of the wave, but the peak amplitude is increased by a factor of $\sqrt{2}$ due to the superposition of the two waves:

$$\mathbf{A}_{\rm lin} = \mathbf{a}_1 \sin\eta + \mathbf{a}_2 \sin\eta \qquad = a\,(\mathbf{e}_1 + \mathbf{e}_2)\sin\eta\,, \qquad |\mathbf{A}_{\rm lin}| = a\sqrt{2}\sin\eta. \tag{25}$$

The explanation given for the ellipticity dependence for a fixed total field intensity also applies to lower total photon energies and higher photon orders. Corresponding examples are shown in the top panels of Figs. 4 (b) and (c) for a total photon energy of 1.25 MeV and $\tilde{n} = 2$ or 5 absorbed photons, respectively. However, remarkably, the argument provided before for the ellipticity dependence for a fixed maximum field amplitude is not applicable to these parameters. As the bottom panels of Figs. 4 (b) and (c) illustrate, the rate exhibits a non-monotonous dependence on the ellipticity here. After passing through a minimum at about $\varepsilon \approx 0.2$ and 0.12, respectively, the rate starts growing again and reaches its maximum value for a linearly polarized wave. The reason for this striking difference seems to be related to the excess energy above the pair creation threshold $\Delta E = \tilde{n}\omega - 2mc^2$, which is much smaller here than in Fig. 4 (a). We note that a non-monotonous dependence on the applied field ellipticity has also been obtained recently for the rate of pair production by the nonlinear Breit–Wheeler process [38].

Bethe–Heitler pair creation by an elliptically polarized, monochromatic laser wave has also been studied in [37] using a polarization-operator approach. In particular, analytical expressions for the total pair-production rate by two-photon absorption were obtained in Eqs. (15) and (26) therein. We stress that the ellipticity dependences following from these expressions coincide perfectly with our numerically calculated results for the two-photon case in Figs. 4 (a) and (b) (top and bottom panels). Moreover, our present approach allows us to extend these results straightforwardly to higher photon orders. The five-photon case shown in Fig. 4 (c) serves as an example. It exhibits features qualitatively similar to the two-photon case.

4. Summary and Conclusion

We have studied electron-positron pair creation on a nucleus by multiphoton absorption from an intense two-mode laser field. If the laser field contains two different frequencies of commensurable ratio we found quantum interference effects to be visible in the angular distribution of the produced particles as well as in the total production rate. The latter can be explained within an intuitive picture based on the phase dependence of the maximum electric field amplitude. Additionally, the case of two incommensurable frequencies was examined by continuously varying their ratio. Finally, for identical frequencies of both field modes, the dependence of the pair production rate on the ellipticity of the laser wave was studied. While this dependence shows the expected behaviour when the field intensity is held constant, interesting features were found when the maximum field amplitude is kept fixed instead.

Our results could be tested experimentally, for instance, by combining a relativistic proton beam of Lorentz factor $\gamma \sim 50$ with a counterpropagating intense X-ray laser beam comprising two field modes [39].

Acknowledgments

Funding for this project by the German Research Foundation (DFG) under Grant No. MU 3149/1-1 is gratefully acknowledged. S. A. also wishes to thank the Heidelberg Graduate School of Fundamental Physics (HGSFP) for the generous travel support.

References

[1] Reiss H R 1962 *J. Math. Phys.* **3** 59–67
[2] Nikishov A I and Ritus V I 1964 *Sov. Phys. JETP* **19** 529–541
[3] Narozhny N B, Nikishov A I and Ritus V I 1965 *Sov. Phys. JETP* **20** 622
[4] Yakovlev V P 1966 *Sov. Phys. JETP* **22** 223
[5] Maiman T H 1960 *Nature* **187** 493–494
[6] Ehlotzky F, Krajewska K and Kamiński J Z 2009 *Rep. Prog. Phys.* **72** 046401
[7] Di Piazza A, Müller C, Hatsagortsyan K Z and Keitel C H 2012 *Rev. Mod. Phys.* **84**(3) 1177–1228
[8] Burke D L, Field R C, Horton-Smith G, Spencer J E, Walz D, Berridge S C, Bugg W M, Shmakov K, Weidemann A W, Bula C, McDonald K T, Prebys E J, Bamber C, Boege S J, Koffas T, Kotseroglou T, Melissinos A C, Meyerhofer D D, Reis D A and Ragg W 1997 *Phys. Rev. Lett.* **79**(9) 1626–1629
[9] Bamber C, Boege S J, Koffas T, Kotseroglou T, Melissinos A C, Meyerhofer D D, Reis D A, Ragg W, Bula C, McDonald K T, Prebys E J, Burke D L, Field R C, Horton-Smith G, Spencer J E, Walz D, Berridge S C, Bugg W M, Shmakov K and Weidemann A W 1999 *Phys. Rev. D* **60**(9) 092004
[10] Müller C, Voitkiv A B and Grün N 2003 *Phys. Rev. A* **67**(6) 063407
[11] Avetissian H K, Avetissian A K, Mkrtchian G F and Sedrakian K V 2003 *Nucl. Instrum. Meth. Phys. Res. A* **507** 582–586
[12] Müller C, Voitkiv A B and Grün N 2004 *Phys. Rev. A* **70**(2) 023412
[13] Sieczka P, Krajewska K, Kamiński J Z, Panek P and Ehlotzky F 2006 *Phys. Rev. A* **73**(5) 053409
[14] Kuchiev M Y and Robinson D J 2007 *Phys. Rev. A* **76**(1) 012107
[15] Fillion-Gourdeau F, Lorin E and Bandrauk A D 2013 *Phys. Rev. Lett.* **110**(1) 013002
[16] Müller T O and Müller C 2012 *Phys. Rev. A* **86**(2) 022109
[17] Müller S J and Müller C 2009 *Phys. Rev. D* **80**(5) 053014
[18] Krajewska K and Kamiński J Z 2011 *Phys. Rev. A* **84**(3) 033416
[19] Lötstedt E, Jentschura U D and Keitel C H 2009 *New J. Phys.* **11** 013054
[20] Di Piazza A, Lötstedt E, Milstein A I and Keitel C H 2010 *Phys. Rev. A* **81**(6) 062122
[21] Roshchupkin S P 2001 *Phys. At. Nucl.* **64** 243–252
[22] Krajewska K and Kamiński J Z 2012 *Phys. Rev. A* **85**(4) 043404
[23] Krajewska K and Kamiński J Z 2012 *Phys. Rev. A* **86**(2) 021402
[24] Augustin S and Müller C 2013 *Phys. Rev. A* **88**(2) 022109
[25] Narozhny N B and Fofanov M S 2000 *JETP* **90**(3) 415–427
[26] Fedorov M V, Efremov M A and Volkov P A 2006 *Opt. Commun.* **264** 413–418
[27] Voitkiv A B and Najjari B 2011 *Phys. Rev. A* **84**(4) 042708
[28] Cheng T, Chott N I, Su Q and Grobe R 2008 *Laser Phys.* **18** 190–196
[29] Jiang M, Su W, Lv Z Q, Lu X, Li Y J, Grobe R and Su Q 2012 *Phys. Rev. A* **85**(3) 033408
[30] Dumlu C K and Dunne G V 2011 *Phys. Rev. D* **83**(6) 065028
[31] Akkermans E and Dunne G V 2012 *Phys. Rev. Lett.* **108**(3) 030401
[32] Wolkow D M 1935 *Z. Phys.* **94**(3-4) 250–260
[33] Reiss H R 1980 *Phys. Rev. A* **22**(5) 1786–1813
[34] Berestetskii V B, Lifshitz E M and Pitaevskii L P 1971 *Relativistic Quantum Theory* 1st ed (*Course of Theoretical Physics* vol 4) (Oxford: Pergamon Press) ISBN 0080171753 in Sec. §40
[35] Bjorken J D and Drell S D 1964 *Relativistic Quantum Mechanics* (New York: McGraw-Hill) ISBN 0070054932 in Sec. 7.1
[36] Potvliege R M and Smith P H G 1992 *J. Phys. B* **25** 2501
[37] Milstein A I, Müller C, Hatsagortsyan K Z, Jentschura U D and Keitel C H 2006 *Phys. Rev. A* **73**(6) 062106
[38] He L Y, Xie B S, Guo X H and Wang H Y 2012 *Commun. Theor. Phys.* **58** 863
[39] Lutman A A, Coffee R, Ding Y, Huang Z, Krzywinski J, Maxwell T, Messerschmidt M and Nuhn H D 2013 *Phys. Rev. Lett.* **110**(13) 134801

Journal of Physics: Conference Series **497** (2014) 012021 doi:10.1088/1742-6596/497/1/012021

Silver doped metal layers for medical applications

T Kocourek[1,2], M Jelínek[1,2], J Mikšovský[1,2], K Jurek[1], M Weiserová[3]
[1] *Institute of Physics AS CR, Na Slovance 2, 182 21 Praha 8, Czech Republic*
[2] *Czech Technical University in Prague, Faculty of Biomedical Engineering, nam. Sitna 3105, 27201 Kladno, Czech Republic*
[3] *Institute of Microbiology ASCR, Videnska cesta 1083, Praha 4, Czech Republic*

E-mail: kocourek@fzu.cz

Abstract. Biological, physical and mechanical properties of silver-doped layers of titanium alloy Ti6Al4V and 316L steel prepared by pulsed laser deposition were studied. Metallic silver-doped coatings could be a new route for antibacterial protection in medicine. Thin films of silver and silver-doped materials were synthesized using KrF excimer laser deposition. The materials were ablated from two targets, which were composed either from titanium alloy with silver segments or from steel with silver segments. The concentration of silver ranged from 1.54 at% to 4.32 at% for steel and from 3.04 at% to 13.05 at% for titanium alloy. The layer properties such as silver content, structure, adhesion, surface wettability, and antibacterial efficacy (evaluated by *Escherichia coli* and *Bacillus subtilis* bacteria) were measured. Film adhesion was studied using scratch test. The antibacterial efficacy changed with silver doping up to 99.9 %. Our investigation was focused on minimum Ag concentration needed to reach high antibacterial efficiency, high film adhesion, and hardness.

1. Introduction

Some of the most serious complications of surgical treatment of fractures are infections [1-3]. The infectious complications prolong healing, prevent fractures healing, and deplete the body's immune system [4-7]. The solution usually requires repeated surgeries and multiplies healing costs and inconvenience for patients [4, 8, 9]. The aim of this work is to find suitable antibacterial materials that will reduce the possibility of an infection or severity of infectious complications in patients after surgical treatment of fractures. The use of fixation screws for temporary immobilization of broken bones entails considerable risk of infection due to the possibility of bacteria spreading along the outer surface of the fixing screws into the body, see Fig. 1. Since silver is known for its excellent antibacterial properties, its use as a suitable dopant seemed a viable road to take [10-17].

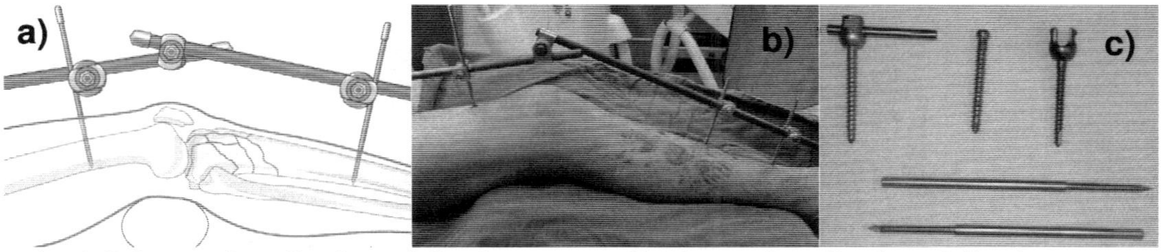

Fig. 1. a) Scheme of application of fixation screws. b) Photo of application of fixation screws. c) Fixation screws.

In this contribution the silver doped metallic materials (titanium alloy Ti6Al4V and 316L steel) of various concentrations was created by pulsed laser deposition (PLD) method. Silver concentration for various deposition conditions, films crystalline structure, adhesion, hardness, surface morphology, antibacterial properties (evaluated by Gram-positive strain *Bacilus subtilis* and Gram-negative strain *Escherichia coli*) and wettability were also studied.

Our primary goal was to prepare layers with sufficient Ag content for antibacterial effect with maximum adhesion and hardness as close as possible to the substrate.

2. Experimental

Deposition - Silver-doped layers of titanium alloy Ti6Al4V and 316L steel were prepared by PLD using a KrF excimer laser (λ = 248 nm, τ = 20 ns, rep. rate of 10 Hz). The laser beam was focused on a silver target with energy density of 2 Jcm^{-2}, silver with Ti6Al4V or silver with 316L steel targets with energy density of 5 Jcm^{-2}. The material was ablated from one target composed from silver and titanium alloy or steel segments. The substrate (Ti6Al4V, 316L steel or Si (100)) was placed 35 mm away from the target and held at room temperature. Films were grown in argon atmosphere at 0.25 Pa. The substrates were cleaned by RF discharge before the deposition process.

Thickness and roughness was measured by Alpha-step IQ mechanical profilometer (KLA TENCOR Co.).

Structure of layers was determined by XRD in parallel beam geometry and detector scan with stationary sample and grazing angle of incidence (GAOI) were used.

Concentration of silver was determined using WDX measurement (WDX – wavelength dependence X-ray analysis) was analysed with EDAX Jeol Supersprobe 733.

Adhesion - For determination of adhesion we used macroscratch tester REVETEST (CSM Instruments co.). The REVETEST system is a standard tester which is compatible with DIN EN 1071-3 and ASTM C1624. We used acoustic emission to estimate the delamination point and we visually confirmed it from the photography of the scratch. Rockwell diamond with tip radius of 200 μm was used as indenter.

Micro adhesion and micro hardness - For determination of micro adhesion and micro hardness the nanosclerometric head was used. The head is an extension of AFM microscope Solver Next (NT-MDT co.). This system provided us with graphic information of the surfaces by SPM technique with high resolution. The probe with diamond tip was of Berkovich type (radius about eighty nanometres). Calibration measurement on fused silica was performed and the calibration curve was obtained before measurement.

Antibacterial properties – The antibacterial properties of Ag-doped films were studied using the gram-negative strain *Escherichia coli* K12 C 600 [18] and the gram-positive strain *Bacillus subtilis* 168, commonly found in soil [19]. Bacteria were grown in Lysogeny Broth (LB) [20] medium and plated on LA plates (LA; LB with addition of 1.5 % of agar). The tested layers were placed into sterile glass chambers and covered with 0.5 ml of overnight cultures diluted in the sterile physiological saline solution at a concentration of cells 10^6 CFU/ml (Colony Forming Units). Glass chambers were cultivated at 37°C and after 24 hours aliquots of 100 μl were withdrawn and serially diluted up to 10^{-6}

in physiological saline solution. 100 µl of each diluted sample were spread over an agar plate. The plates were incubated overnight and the number of growing colonies was counted. The bactericidal efficacy expressing the antibacterial properties of different silver doped films was estimated according the formula [10, 11, 21]:

$$ABE\ (\%) = \frac{AN_{REF} - AN_{EXP}}{AN_{REF}} \times 100\%,$$

where ABE is the antibacterial efficiency, AN_{REF} is the alive number in reference group, and AN_{EXP} is the alive number in experiment group.
Wettability - The wettability of the films was determined by means of contact angle method measured by DSA100 (Krüss co.). The liquid used was distilled water.

3. Results and discussion

Thickness of created silver layers was 100 nm and 350 nm while that of created silver-doped layers was from 94 nm to 398 nm, depending on target, see Table 1.
Roughness of silver-doped 316L steel films was from 12 nm to 29 nm and the roughness of silver-doped titanium alloy Ti6Al4V films was from 5 nm to 28 nm, similarly to silver layer, see Table 1.
Composition - WDX measurement confirmed the increasing concentration of silver with increasing segment of the silver piece on the target - during deposition process. The concentrations of silver ranged from 1.54 at% to 4.32 at% for steel and from 3.04 at% to 13.05 at% for titanium alloy, see Table 1.

Table 1. The deposition condition for fabrication of silver-doped thin films by PLD process: the thickness, roughness and concentration of silver in the layers.

Sample	*Substrate*	*Roughness Ra [nm]*	*Thickness [nm]*	*Target (Area)*	*Ag [at%]*
Ag-1	Ti6Al4V	10	100	Ag	**100**
Ag-2	Ti6Al4V	11	350	Ag	**100**
S-1	316L steel	13	161	316L steel : Ag (40:1)	**1.54**
S-2	316L steel	29	260	316L steel : Ag (20:1)	**3.53**
S-3	316L steel	12	94	316L steel : Ag (10:1)	**4.32**
T-1	Ti6Al4V	5	398	Ti6Al4V : Ag (40:1)	**3.04**
T-2	Ti6Al4V	28	294	Ti6Al4V : Ag (20:1)	**5.05**
T-3	Ti6Al4V	25	198	Ti6Al4V : Ag (10:1)	**13.05**

Crystallinity - In the XRD spectrum of silver-doped 316L steel films the peak of intermetallic compounds with Ag (compound Ag with Fe, Cr, Ni, and Mo) was identified. In the XRD spectrum of the titanium alloy doped with silver there are no new peaks, see Fig. 2.

Journal of Physics: Conference Series **497** (2014) 012021 doi:10.1088/1742-6596/497/1/012021

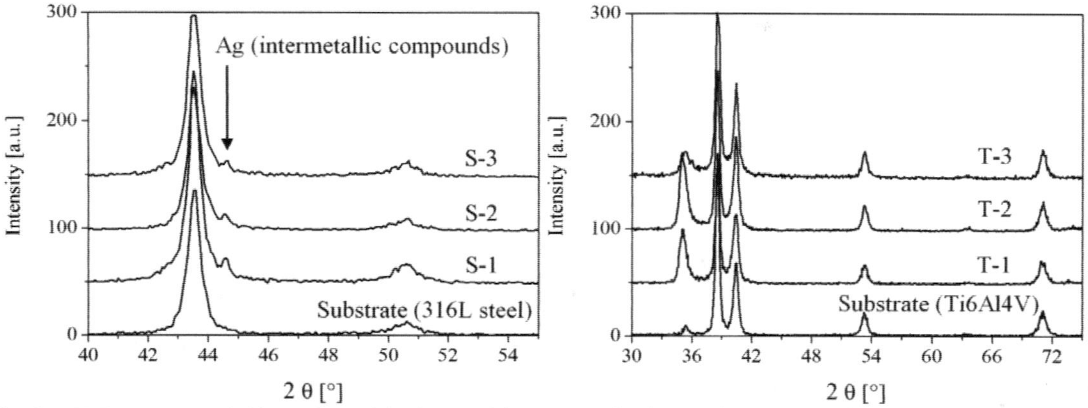

Fig. 2. XRD spectra of silver-doped 316L steel layers on 316L steel substrate (left) and silver-doped titanium alloy Ti6Al4V layers on titanium alloy Ti6Al4V substrate (right).

Surface morphology was studied using scanning electron microscopy (SEM). The surface of the steel layers was smooth and covered with droplets with diameter up to several micrometres. No changes were observed with increasing amount of silver in the layers, see Fig. 3a. Titanium alloy layers were also smooth and their surfaces were also covered by droplets of similar size. Nevertheless, in contrast with steel layers there was an increase in number of droplets on the layers with increased silver amount, see Fig. 3b.

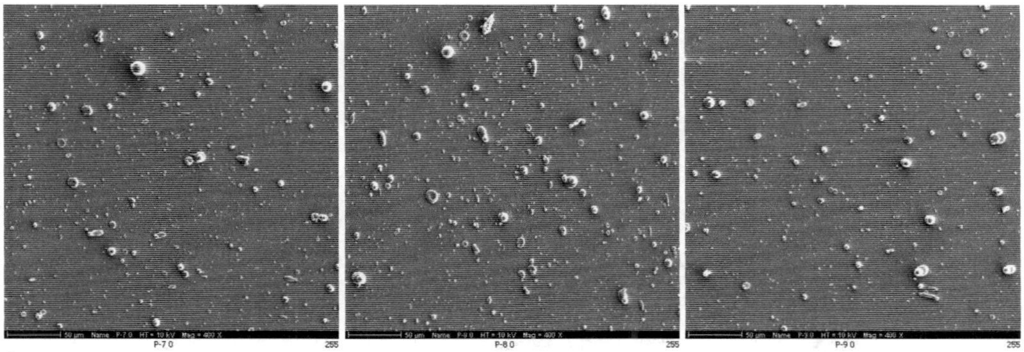

Fig. 3a. Surface morphology measured by SEM (mag = 400x) of layers of steel 316L with Ag. From left: S-1, S-2, S-3.

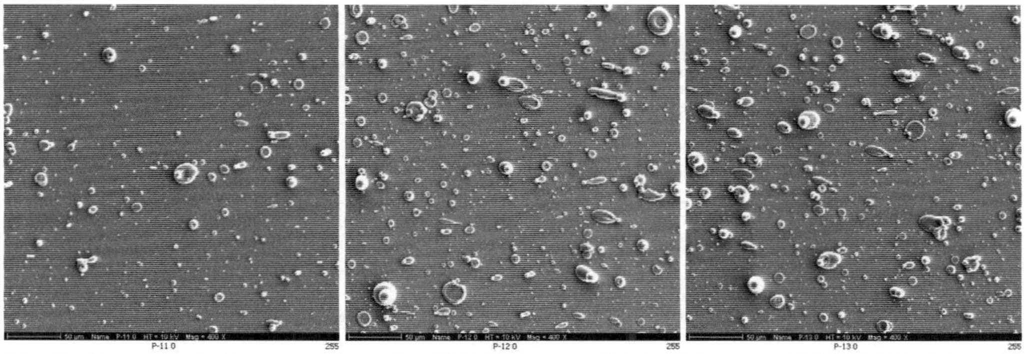

Fig. 3b. Surface morphology measured by SEM (mag = 400x) of layers of Ti6Al4V with Ag. From left: T-1, T-2, T-3.

Journal of Physics: Conference Series **497** (2014) 012021 doi:10.1088/1742-6596/497/1/012021

Adhesion – A macro scratch tester was used for adhesion measurements. For samples Ag-1 and Ag-2 of pure silver layers we used linear progressive scratch with initial load 1 N and the end load 5 N. Loading rate was 4 N/min. The lengths of the scratches were 8 mm. Two scratches were performed on each sample. The layers were very soft and the tip penetrated at the start by initial load. For Ag-1 sample there was no delamination observed. In case of Ag-2 sample we observed delamination for critical force approximately 1.75 N, see Fig. 4. Pure silver layers exhibit low adhesion.

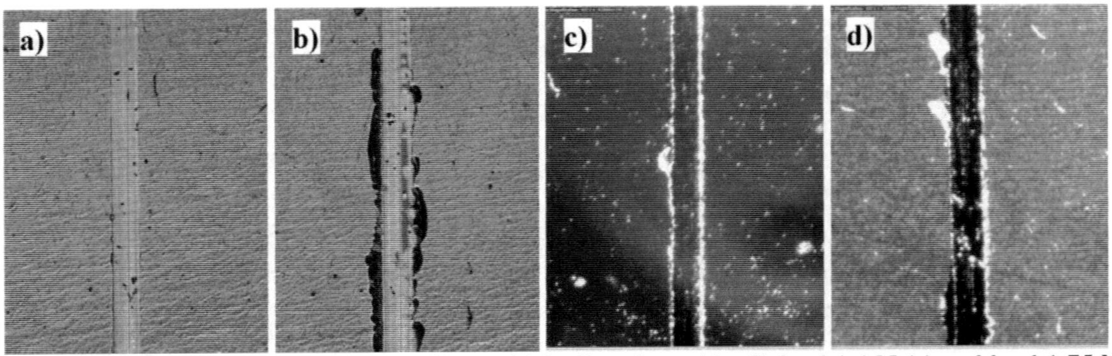

Fig. 4. Example of delamination of a 350 nm thick silver layer (Ag-2) load 1.4 N (a) and load 1.75 N (b). Comparison of adhesion for 100 nm (Ag-1) (c) and 350 nm (Ag-2) (d) thick silver layers for the force of 4 N.

For samples S-1, S-2, S-3, T-1, T-2, and T-3 we used linear progressive scratch with initial load 1 N and the end load 30 N. Loading rate was 15 N/min. The length of the scratches was 5 mm. Two scratches were performed on each sample. Samples were tested for two different substrate roughnesses, polished and lathed. Both substrates had similar behaviour. We did not observe any penetration through layer or delamination. The behaviour of the samples was similar to bulk material, see Fig. 5a., 5b.

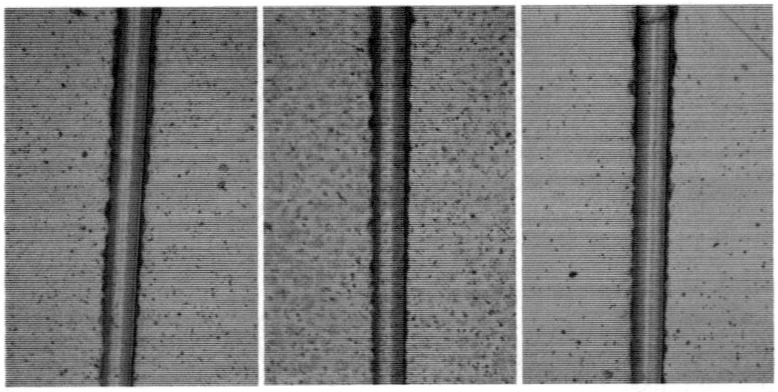

Fig. 5a. Example of macro scratch on the layers of Ag doped steel 316L. From left to right: sample S-1, S-2 and S-3 (load for all samples 24 N).

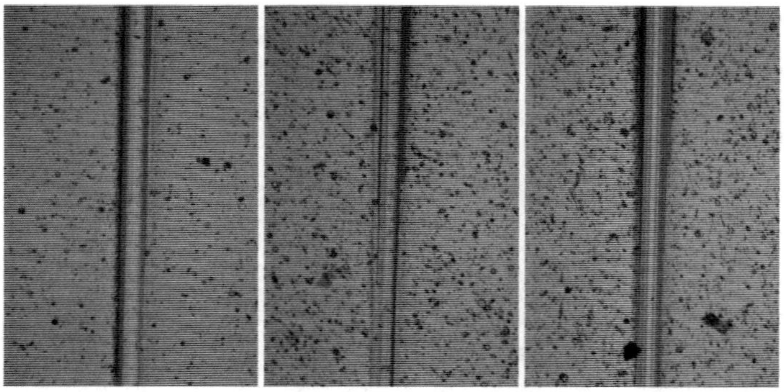

Fig. 5b. Example of macro scratch on the layers of Ag doped Ti6Al4V alloy. From left to right: sample T-1, T-2 and T-3 (load for all samples 24 N).

Micro adhesion and micro hardness - The adhesion behaviour was verified by micro adhesion measurement, too. Two scratches with different loads were made on each sample. Lengths of the scratches were approximately 10 μm and loads were (0.5 ÷ 3) mN and (2.5 ÷ 5) mN. Scratching speed was 100 nm/s. For the micro scratch test we observed no adhesion failure. The hardness was evaluated from the scratch test too, see Table 2. Tabular values of hardness are approximately 0.25 GPa for pure silver, 5.0 GPa for 316L steel, and 3.4 GPa for titanium alloy Ti6Al4V. For the steel 316L there is obvious decrease in hardness with increasing silver amount in the layers. There is no such decrease in hardness for the titanium layers and the hardness values lie in a narrow band of 0.1 GPa between 4.12 and 4.22 GPa.

Table 2. The micro hardness of silver-doped 316L steel layers on 316L steel substrate and silver-doped titanium alloy Ti6Al4V layers on titanium alloy Ti6Al4V substrate.

Sample	*Hardness H [GPa]*
S-1	5.22 ± 0.79
S-2	4.02 ± 0.66
S-3	3.74 ± 0.54
T-1	4.12 ± 0.43
T-2	4.22 ± 0.50
T-3	4.15 ± 0.15

Antibacterial properties - Layers of steel and titanium alloy doped with various concentration of silver (1.54 at%, 3.53 at%, and 4.32 at% for steel; and 3.04 at%, 5.05 at%, and 13.05 at% for titanium alloy) were tested for their antibacterial effect against the *Escherichia coli* (gram-negative) and the *Bacillus subtilis* (gram-positive) strains. Antibacterial efficiency increased with increasing silver content for Ti6Al4V layers, and was good for both of bacteria strains (reached 95% for *Bacillus subtilis* and 99.9% for *Escherichia coli*). For 316L steel we did not observed clear dependence on silver content, but the efficiency was high, too (reached 93% for *Bacillus subtilis* and 99.9% for *Escherichia coli*). Antibacterial efficiency significantly increased over time, see Fig. 6. Pure silver has excellent antibacterial properties (exceeds 99.9 %), but is very soft and has a low adhesion, inappropriate for practical use. After 24 h cultivation the silver doped metal layers with more than 4 at% of Ag exhibit good anti-bacterial ability. This is consistent with the results of silver-doped ceramic layers [10, 11].

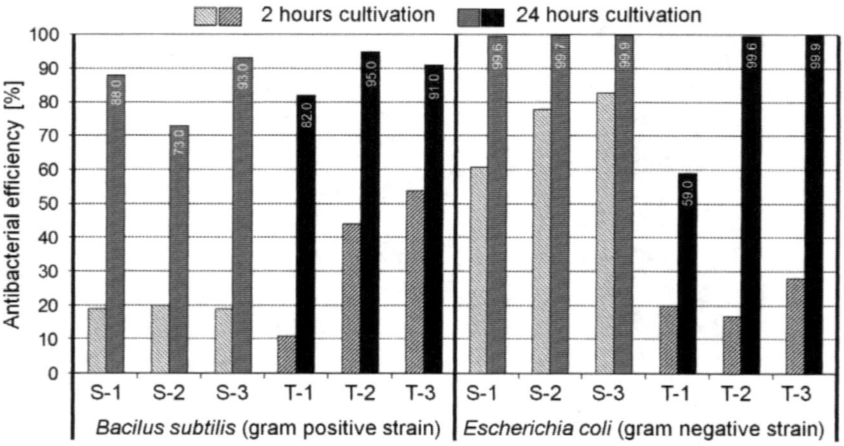

Fig. 6. Antibacterial efficiency of silver-doped metal layers against the *Bacillus subtilis* (gram-positive) and the *Escherichia coli* (gram-negative) strains, after 2 and 24 hours.

Wettability - The wettability of the films was determined by using contact angle method. Distilled water was used as the liquid. The highest values of the contact angle were observed for pure silver layers. For layers of Ti6Al4V doped with silver, increasing contact angle values with increasing content of silver were observed. For layers of steel 316L there was no significant change in contact angle with silver content. Each type of layer was prepared on substrates with two different roughnesses, polished (P) and lathe turned (T). Contact angles on lathe turned substrates were slightly higher, see Fig. 7.

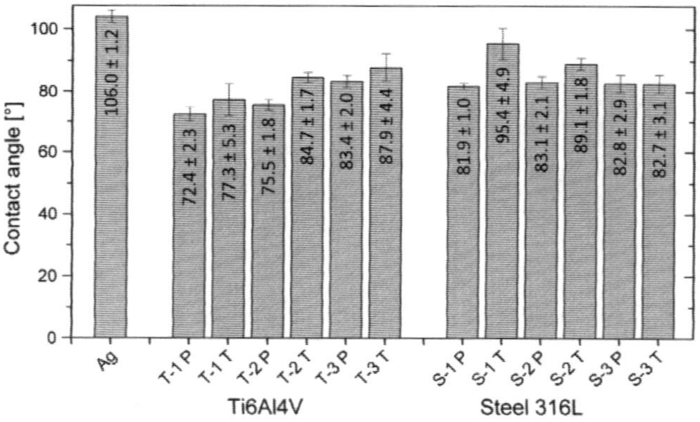

Fig. 7. The wettability of silver-doped 316L steel layers on 316L steel substrate (P – polished surface and T – turning surface) and silver-doped titanium alloy Ti6Al4V layers on titanium alloy Ti6Al4V substrate (liquid was distilled water).

4. Conclusions

The metallic (titanium alloy Ti6Al4V and 316L steel) layers with various concentration of silver were prepared by PLD. Composition was determined by WDX. The amount of silver in Ti6Al4V layers was from 3.04 to 13.05 at%. The amount of silver in 316L steel layers was from 1.54 to 4.32 at%. Minor changes were observed in crystalline structure of the doped steel, which can be assigned to intermetallic compounds with silver. We did not observe new peaks in XRD spectrum of the titanium alloy doped with silver. No silver crystalline phase was found. All surfaces of the layers were covered with droplets. In case of the titanium layers the number of droplets increases with the increasing silver concentration. The adhesion of the silver-doped 316L steel and Ti6Al4V was outstanding. We did not observe any delamination of layers. The transition between the layer and substrate was not observed. The hardness of 316L steel layers decreased with increasing silver content. There is no such decrease in hardness for the titanium layers. Antibacterial efficiency increased with increasing silver content for Ti6Al4V layers, and was good for both bacteria strains. For 316L steel we did not observe any clear dependence on silver content, but the efficiency was high, too. Antibacterial efficiency significantly increased over time. The silver doped metal layers with over 4 at% of Ag exhibit good anti-bacterial ability. Wettability measurement showed that the contact angle of Ti6Al4V layers doped with silver increases with increasing silver content. For layers of 316L steel there was no significant change in the contact angle with varying silver content. The prepared layers were verified as a new possibility to prepare effective protection against incidence and propagation of infections during treatment of complicated fractures.

Acknowledgements

The authors are grateful for support from the grant projects SGS 12/167/OHK4/2T/17, SGS13/092/OHK4/1T/17 of the Czech Technical University in Prague, and from the grant FR- TI 3/088.

References

[1] Džupa V, Ryantová V, Skála-Rosenbaum J, Vyhnálek F, Fric M, Grill R, Horák L and Pavelka T 2008 *Acta chirurgiae orthopaedicae et traumatologiae Cechoslovaca* **75** 293
[2] Weber G B 2004 *Asepsis and the risk of infection* In: ed. Weber G B *Minimax fracture fixation (Davos Platz, AO Publishing)* 1–18
[3] Wininger D A and Fass R F 2003 *Bone and joint infections* In: ed. Finch R G, Greenwood D, Norrby S R and Whitley R J *Antibiotic and chemotherapy (8th ed. Edinburgh, Churchill Livingstone)* 733–739
[4] Hanssen A D and Rand J A 1998 *J. Bone Jt Surg.* **80-A** 910
[5] Jahoda D, Nyč O, Imša J, Kučera E, Hanek P, Chrz P, Pokorný D, Tawa N, Landor I and Sosna A 2007 *Acta Chir. orthop., Traum. čech.*, **74** 397
[6] Stuhldreier G, Gaebel G, Kramer W and Neugebauer W 1989 *Akt. Traumatol.* **19** 28
[7] Šťastník M 2004 *Klin. Mikrobiol. Inf. Lek.* **10** 73
[8] Gallo J, Landor I and Vavřík P 2006 *Acta Chir. orthop., Traum. čech.* **73** 229
[9] Hanssen A D 2005 *Clin. Orthop.* **437** 91
[10] Jelinek M, Weiserová M, Kocourek T, Jurek K and Strnad J 2010 *Laser Physics* **20** 562
[11] Jelinek M, Weiserová M, Kocourek T, Zezulová M and Strnad J 2011 *Laser Physics* **21** 1265
[12] Chen W, Liu Y, Courtney H S, Bettenga M, Agrawal C M, Bumgardner J D and Ong J L 2006 *Biomaterials* **27** 5512
[13] Chung R J, Hsieh M R, Huang K C, Perng L H, Chou F I and Chin T S 2005 *Journal of Sol-Gel Science and Technology* **33** 229
[14] Lansdown A B G, Samson B, Laupattarakasem P and Vuttiviorjan A 1997 *Br. J. Dermatos* **137** 728
[15] Gosheger G, Hardes J, Ahrens H, Streitburger A, Buerger H, Erren M, Gunsel A, Kemper F H, Winkemann W and Eiff C 2004 *Biomaterials* **25** 5547
[16] Chen Y, Zheng X, Xie Y, Ding C, Ruan H and Fan C 2008 *J. Mater Sci: Mater Med* **19** 3603
[17] Choi J W, Cho H M, Kwak E K, Kwon T G, Ryoo H M, Jeong Y K, Oh K S and Shin H I 2004 *Key Engineering Materials* **254-256** 47
[18] Appleyard R K 1954 *Genetics* **39** 440

[19] Madigan M, Martinko J 2005 *Brock Biology of Microorganisms (11th edition), Prentice Hall*
[20] Luria S E 1953 *Cold Spring Harb Symp Quant Biol* **18** 237
[21] Huang L, Li D Q, Lin Y J, Wei M, Evans D G and Duan X 2005 *J Inorg Biochem* **99** 986

Journal of Physics: Conference Series **497** (2014) 012022 doi:10.1088/1742-6596/497/1/012022

Assessment of the Suitability of Excimer Lasers in Treating Onychomycosis

Jaroslava Kymplová[1], Miroslav Jelínek[2,3], Jana Urzová[2] , Jan Mikšovský[2,3] , Karel Dušek[1] ,Lenka Bauerová [3]

1 Institute of Biophysics and Informatics of 1st Faculty of Medicine of Charles University in Prague,
Salmovská 1, 120 00 Praha 2, Czech Republic
2 Faculty of Biomedical Engineering, Czech Technical University in Prague, nám. Sítná 3105, 272 01 Kladno, Czech Republic
3 Institute of Physics, ASCR, Na Slovance 1999/2, 18221 Praha 8, Czech Republic
4 Institute of Pathology, First Faculty of Medicine, Charles University in Prague and General University Hospital in Prague

jkymplova@volny.cz, jelinmi4@fbmi.cvut.cz, jana.urzova.ga@seznam.cz, jan.miksovsky@centrum.cz

Abstract

Since it is known that UV-C radiation kills fungus, we wanted to verify the hypothesis that the use of excimer laser could be an alternative method for treating onychomycosis - nail fungus. The aim of the first stage of this work was to determine the transmission, reflection and absorption of nails. In the following stage we focused on irradiation of fungi. Our final task is to assess whether it is possible to determine the parameters of radiation (a total dose,a dose per pulse frequency, a repetition rate, a number of pulses) for which the elimination of fungi would be the most effective but without damaging the nail and soft tissue underneath it. The results so far have showed that UV-C radiation does not pass through a fingernail to such an extent that it could damage the soft tissue beneath it. Fungi are destroyed by the application of only small doses of radiation using the excimer laser. Additional measurements will be required to determine the modulation parameters of the excimer laser radiation for the treatment of onychomycosis.

Keywords: Excimer laser, UV-C radiation, nails, onychomycosis

Journal of Physics: Conference Series **497** (2014) 012022 doi:10.1088/1742-6596/497/1/012022

Introduction

Onychomycosis is caused by various kinds of parasitic fungi, mold or yeast. Our research is focused on the fungus Trichophyton rubrum. This type of fungus was cultured from samples from patients treated for onychomycosis. Such cultivation is time consuming and is carried out in Petri dishes or in glass tubes on glucose agar. Trichophyton rubrum is the anamorphic fungus from the genus of fungi in the family Gymnoascaceae (Teleomorfa Arthroderma). T. rubrum has minor antigenic properties and the human immune system responds to it only mildly. Patients, who are free from any difficulties caused by T. Rubrum, serve as a long-term source of infection. Trichophyton rubrum induces pathological changes in the keratin structures - in skin, hair and nails.The most common cutaneous mycosis caused by T. Rubrum is Tinea Pedis, mostoften localized on the skin between toes. Onychomycosis is a medical condition where nails are infected by fungus that feeds on keratin (a tissue forming the nail). It gradually grows through and decomposes it.
It is said that this disease affects 8% of population. The main causes of the disease include: wearing closed and airtight shoes made of inappropriate materials, use of public swimming pools, trauma affecting legs during longer exercise, certain diseases - especially diabetes and vascular diseases, but also obesity, increased sweating of feet - especially in the summer, wider use of antibiotics and corticosteroids.

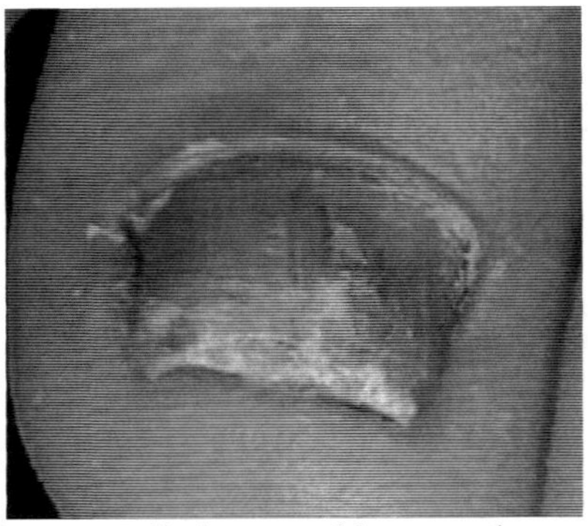

Fig.1 - Nail damaged by onychomycosis

Treatment is topical, in more serious cases systemic (oral), or combined, but always long and problematic. Its choice depends on the seriousness of each clinical condition. Local therapy using antifungal (AF) solutions have a chance only if there is a maximum of 30% nail plate damage. AF varnish can be applied in conditions with up to 50% of nail plate damage. When changes affect 50-90% of a nail plate, the topical treatment usually fails. If changes are visible all the way to the rear ball of a nail, it means that the fungal infection (which in fact extends 3-5 mm further beyond the visible changes) has hit the nail matrix. This is a clear indication that the use of an oral, or combined, treatment is needed. The local treatment does not significantly affect the patient's body; however, it puts high demand on patient's cooperation. Before the administration of medication, it is important to remove all visibly damaged parts of the nail (by cutting out, sanding or chemical ablation) and then the area has to be thoroughly treated until its full clinical recovery.
The treatment can take many months. Throughout all this time, the patient must apply AF solution twice a day or AF varnish twice a week. AF creams are not suitable for nail

Journal of Physics: Conference Series **497** (2014) 012022 doi:10.1088/1742-6596/497/1/012022

treatment, as cream can be quickly wiped off from the nail surface, so that the active substance does not manage to permeate the keratin. AF solution, on contrary, seeps into crevices of the damaged nail, the nail plate and the nail balls. In addition, AF varnish creates after drying an occlusion (enclosed film) that further supports the diffusion of the active ingredient. In the case of a systemic (overall) treatment, the AF ingredient permeates after oral administration from the blood to the nail bed. If the AF level is sufficient, the originator of the problem is unable to multiply and destroy the other parts of the nail. A new healthy nail starts to grow and continuously displaces the affected parts. The overall AF cure must be taken daily or in monthly cycles. However, systemic therapy has a number of side effects as well as contraindications.

The first trials in treating fungal nail disease using laser radiation appeared in 1980. Initially, however, it was more of a thinning or ablation of the afflicted nail, or digging holes in the nail using high power lasers with the subsequent application of ointment. It was not until 2009, when the use of lasers that cause no destruction of the nail began. This type of lasers should act upon the originator of the mold.

For the treatment of onychomycosis, there are currently used so called Footlasers. As of 15 October 2010, the Pinpointe FootLaser (Nd: YAG pulse laser operating near 1064 nm) has been granted FDA (Food and Drug Administration) clearance for use on onychomycosis.

Nd: YAG laser (1064 nm) is a fixed-phase device with active environment consisting of yttrium aluminum garnet. The laser is driven by a burning krypton discharge tube and has a power output of 100-200 W. Some modern workplaces use a Q-switched Nd: YAG laser, which uses very short pulses with high energy.

The CoolBreeze laser uses 1320nm mid-infrared wavelength.

The next type of such a laser is the Noveon laser that uses 870nm and 930nm concurrently. Results have been published in the Journal of the American Podiatric Medical Association. In 2010 physicians in 12 podiatric offices conducted a study of 199 toenail patients with a total of 687 infected toes. Patients were evaluated at 90 day intervals. Following treatment, 85% of the treated toenails showed clear linear nail improvement at day 180 and 35% of these showed further improvement at day 270.

A problem is that there are no studies that would explain how these lasers affect onychomycosis from the physical point of view. Considering their wavelength and high energy, it is assumed that it is the heat effect that acts on the mold. Infrared radiation, especially in the area of IR-A (760-1400 nm), penetrates into the tissue most deeply. It can penetrate skin to a depth of several millimeters.

Similarly, there is a lack of extensive clinical studies that would in the longer term evaluate and compare this method with pharmacological treatments.

No study provides information about the long term safety of this method, since there is a risk of damage to the nail bed and, therefore, temporary or even permanent loss of a nail. Published experience indicates great effectiveness of this treatment, but these are files of only several dozen patients, and the evaluation criterion is only a photograph and evidence of nail growth without any signs of onychomycosis.

Last year an article about laser systems for the new treatment area of onychomycosis was published in the Journal of the American Podiatric Medical Association. As of January 2012, the US FDA has approved four laser systems for the "temporary increase of clear nail in onychomycosis." The FDA has approved these devices on the basis of "substantial equivalence" to predicate devices with similar technical specifications and applications. Laser therapy appears to be a promising alternative to traditional pharmacotherapy, but these systems have been tested in only limited clinical trials.

Radiation with a wavelength less than 300 nm is extremely effective in destroying micro-organisms. Most sterilizing is the UVC light at a wavelength of 253.7 nm. This area is called the germicidal. UVC has a very small penetration capability and does not penetrate beyond the dead skin cells. Germicidal UV radiation does not kill cells directly. It damages their DNA, and the cells then become sterile.
We assume therefore that the use of a laser emitting a laser beam in the UVC area should damage the DNA mold even by lower density of the energy radiation, and should not result in the transmission of radiation into the soft tissue beneath the nail.
Excimer laser active area consists of a specific type of molecules called dimers. Most excimer lasers radiate in the UV spectrum. They have minimal absorption depth in the tissue, allowing the removal of microscopic layers of tissue with minimal damage to the surrounding area. In our experiment, we used an excimer laser with a wavelength of 248 nm (with krypton - fluorine content).
Excimer is an unstable molecule that is produced only on a temporary basis as a result of the interaction of excited atoms (or excited molecules) with the atom (or molecule) in the basic state, and that cannot be created if both atoms (molecules) are in the basic state. Excimer lasers use electrical discharge excitation and electron beams. Excitation by electron beam is utilized in the construction of large lasers because these exhibit greater efficiency (about 10%).
Lasers excited by an electrical discharge are of a simpler structure than lasers excited by an electron beam. The mixture produces ions and excited atoms of rare earth elements. Ions, then, in a triple collision with atoms of the same type create molecular ions. These are then, in a subsequent process of recombination, converted into excimers.

Methods

Measuring transmission, reflection and absorption of nails.

A fiber spectrophotometer from Ocean Optics, model DH2000, was used as a source of UV light. This device emits electromagnetic radiation at wavelengths of 220-1050 nm. It consists of two sources. Deuterium, radiating at wavelengths between 220-400 nm, and halogen, radiating at wavelengths of 300-1050 nm. It was used to determine transmission, absorption and reflection properties of nail tissue. For the actual measuring of nail parameters, we used four samples. The first two were taken from a big toe, the other two from a thumb. Each of them had different thicknesses, which was necessary to take into account when analyzing the results. The nails were cleaned before taking the measurements mechanically as well as in an ultrasound bath. After assembling the apparatus for measuring transmission (see Figure 2), we performed reference measurements without the sample.

Fig.2 – Set for testing nail transmission and absorption

For observed reflections we used the same equipment as in the previous experiment. The only difference was in the optical cable, which was adapted to collect the reflected radiation (see Figure 3). The nail transmissions were measured for an area of 248 nm using high-performance KrF laser in pulse mode.

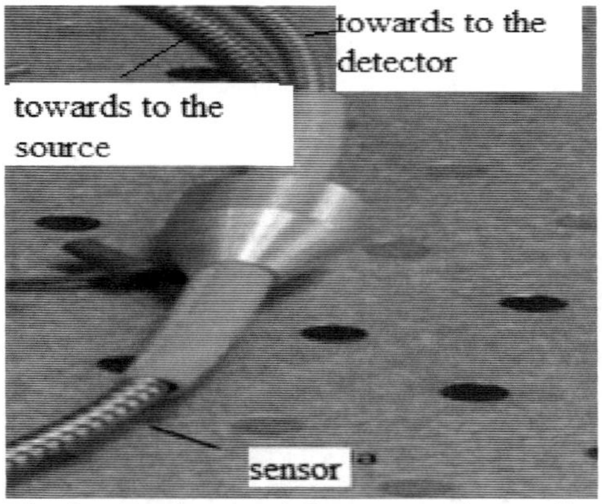

Fig.3 – Probe used for measuring reflexion of nails.

Monitoring the thermal response

For the actual monitoring of the thermal response we used the KrF (krypton - fluorine) excimer laser COMPEX F 205 Pro, Lambda Physik, radiating at a wavelength of 248 nm.
Heat distribution was scanned by a Fluke Ti 50 thermal imaging camera from Ahlborn. With the help of double sided tape we fixed the sample nail onto a piece of cardboard. The laser beam was narrowed down to an aperture diameter of 4 mm. Laser power was set to 188 mJ. The laser pulse length is fixed - 25 ns. Given that the thermal camera does not capture images in a synchronized way with the laser, the scanning interval was set at a frequency of 20 Hz. The laser was operated manually at the highest frequency. In at least one experiment we hoped to approach as closely as possible to the state of a nail after the laser beam reaches it. This solution is not ideal, however, for determining whether the nail was warmed up, this experiment sufficed.

Journal of Physics: Conference Series **497** (2014) 012022 doi:10.1088/1742-6596/497/1/012022

(Laser energy (E): 188 mJ. Effective surface beam (S): 0.126 cm^2. The energy density is given by: $\omega = E / S$, $\omega = 188 / 0.126$ which corresponds to $\omega = 1\ 492$ mJ / cm^2.)

Mold irradiation

T. Rubrum mold we irradiated using the excimer laser Lambda Physik, Compex Pro 248nm. To measure the energy applied in a pulse, we used a COHERENT FieldMax II device (Fig. 3). This value was particularly important to determine the energy density.
To document the irradiated samples, we used a Sony A200 DSLR camera with a Tamron 90/2.8. macro lens, and an Olympus microscope.
Irradiated samples were placed on the optical table using a stand (Fig. 4). A cover lid of the Petri dish with the mold Trichopfyton rubrum was uncovered just before the start of irradiation. The experimental setup was standard, used for similar experiments. We used a circular diaphragm of diameter 0.5 cm, i.e. 0.196 cm^2. The irradiated sample mold was placed behind the diaphragm as closely as possible (a contact position is not possible because of the sample parameters) so as to prevent irradiation from scattered radiation. For the full assembly please see Figure 4.
The energy in the pulse was varied in the range from 144 mJ to 304 mJ (higher values caused undesired heating of tissue). The corresponding energy density varies in an interval from 0.74 J/cm^2 to 1.55 J/cm^2. Repeating frequency was set to 10Hz and the number of pulses was changed for every value of energy - to 200, 500 and 1000. After irradiating a circular footprint, we observed it with our eyes and with a help of microscope for one month for bad or unwanted changes (growth of fungus).

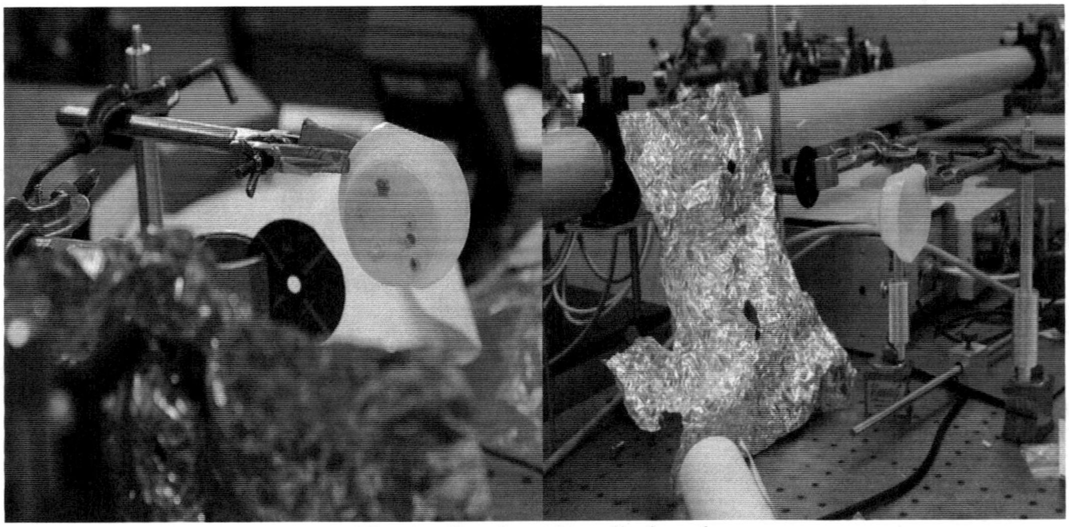

Fig.4 – Set for AF radiation therapy

Search for ablation thresholds on a sample of a human tissue

On some dissected and freshly frozen human skin we sought to determine ablation thresholds. Before and after irradiation, we took detailed photographs of the tissue. Parameters were the same throughout the whole experiment - a circular footprint of a 2.6 mm in diameter, i.e. 5.31 mm^2; and a repetition rate of 10 Hz.

Ablation threshold is the highest laser output, which does not interfere with the structure of a tissue, is for a toenail 128.37 mJ/cm^2. For therapeutic purposes, this extreme value must not be exceeded.

Results

Reflection, absorption and transmission of a nail sample are shown in the following graphs: (Figures 5 and 6)

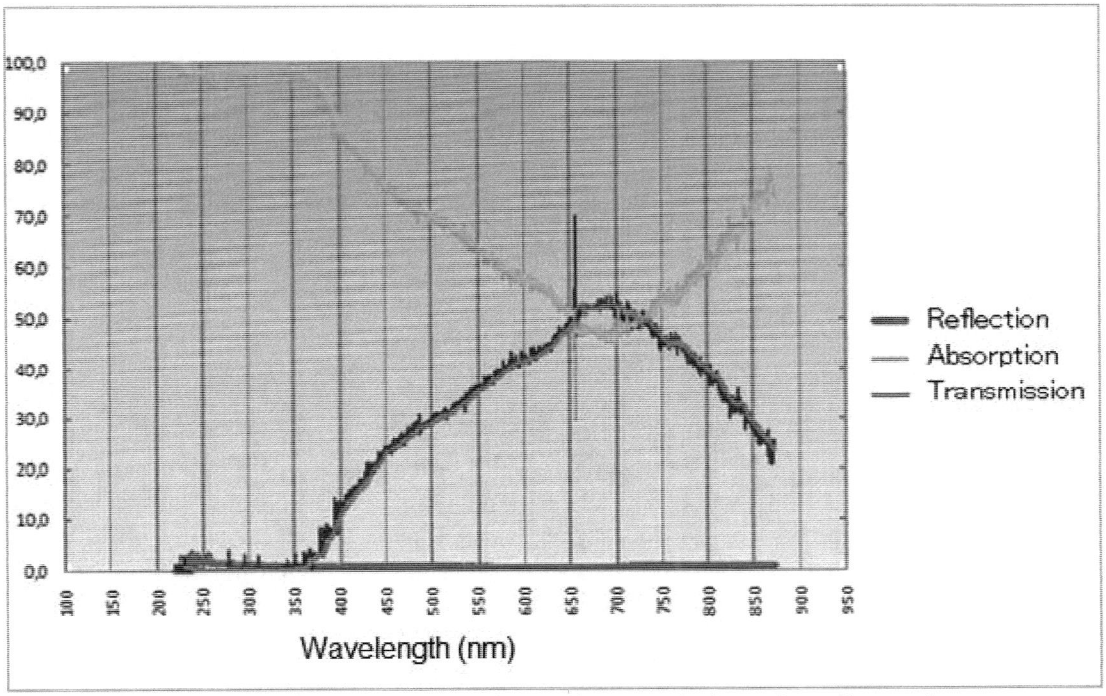

Fig.5 - Reflection, Absorption and Transmission of a nail sample
(Sample nail– thickness 0.43 mm)

Journal of Physics: Conference Series **497** (2014) 012022 doi:10.1088/1742-6596/497/1/012022

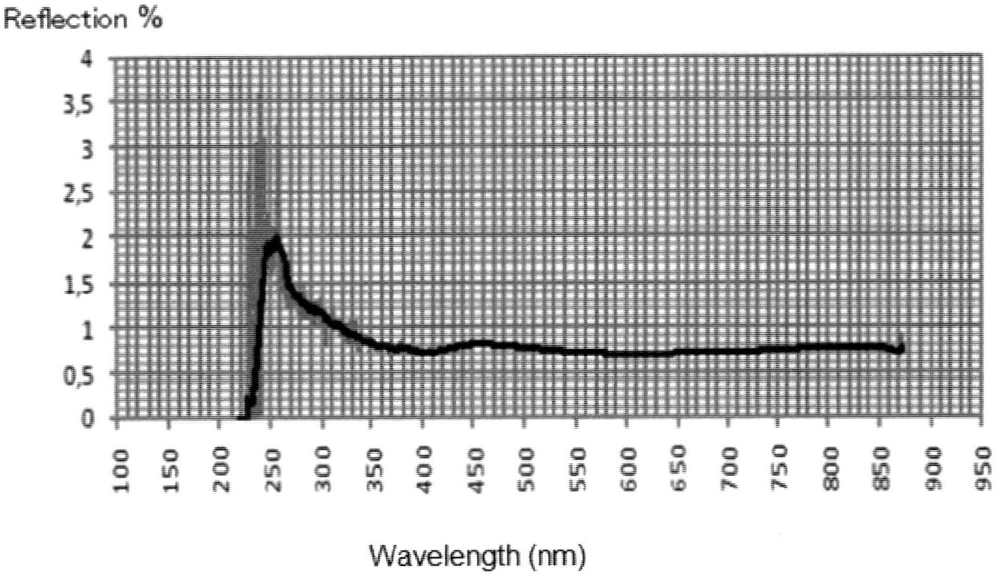

Fig.6 - Reflection of nails using different wavelengths
(Sample nail - thickness 0.43 mm)

As the graph above shows, reflection emitted by the source across the spectrum is low - around 1%. As we measured both - transmission and reflection, we can, using the following formula A = 100 - T - R [%] (where A is absorption, T is transmission and R is reflectivity), calculate the value of absorption. All figures are shown in percentages.

The measured results of transmission indicate that nails are impervious to radiation of wavelengths lower than approximately 400 nm. Results of this experiment clearly indicate an important fact - during nail laser irradiation in the UV spectrum (specifically UV-C) we do not need worry about the UV rays passing deeper into the skin, where it could cause undesirable thermal, biochemical and photo-physical effects. Reflection emitted by the source across the whole spectrum is low - around 1%.

Measurements of nail transmission for 248 nm using a high-performance KrF laser in pulse mode:

Sample number	1		2		3
Thickness (mm)	1,1		0,7		0,45
Energy before the impact (mJ)	37	50	37	50	37
Energy from the impact (mJ)	9	11	7	8	13
Transmission (%)	24	22	19	16	35

Table 1 - Measurements of nail transmission

Journal of Physics: Conference Series **497** (2014) 012022 doi:10.1088/1742-6596/497/1/012022

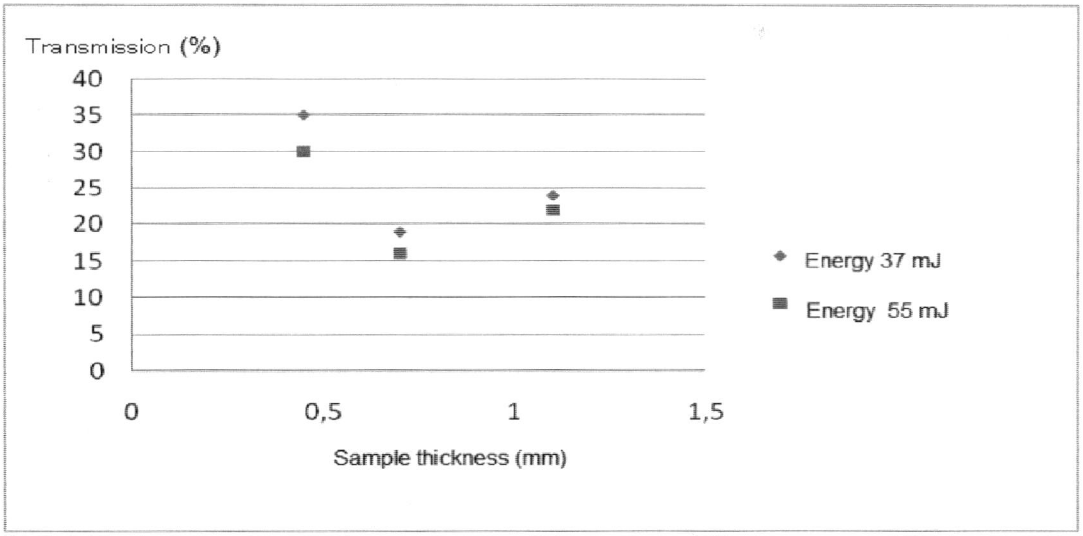

Fig.7 - Transmission graph

Verification of absorption of UVC radiation by thermal imaging camera.

The result of the heat area experiment after the use of the laser beam is that with values of energy needed for fungus elimination we do not reach temperatures in nail or underneath the nail tissue that would have unwanted effects. As inappropriate, we considered lasting heat exceeding 40oC, although we can exceed this temperature for a short time without the destruction of tissue. Here is a typical image of the heat impacted area, which we made using a thermo-camera.

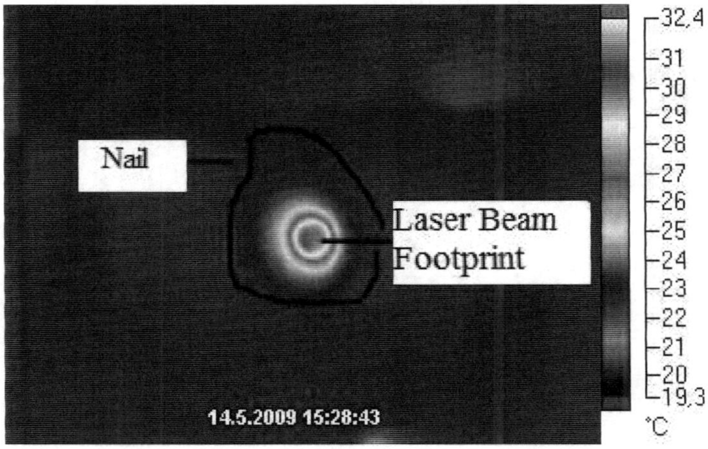

Fig.8 - Heat distribution after laser irradiation

Evaluation of the mold irradiation

Fungi are successfully destroyed even when using minimal values – energy in pulse 144 mJ and density of energy 0.74 J/cm^2, and they are not growing again. We have been observing the reaction of Trichophyton rubrum in relation to the number of pulses. For higher numbers of

pulses (500 and 1000) the destruction can be observed by the naked eye, for a lower number of pulses (200) we were able to observe the damage caused just through a microscope and we have only verified that there was no more growth of fungi. This value we consider as a limit. During our experiments all the fungi were destroyed even with these values. Because of the small size of the footprint compared to the size of a nail in medical use it will be necessary to make the footprint bigger, or (what is in our opinion better) move the footprint in two directions and in that case it would be better to use a lower number of pulses for each radiation and divide the footprint into layers.
Part of the experiment, we focused on the influence of the frequency of laser radiation to destroy fungi. We used a frequency of 10-40 Hz. From monitoring the samples, it is obvious that the change in frequency with the same number of pulses has no significant impact on the disposal of mold. The crucial factor is rather the actual number of pulses.

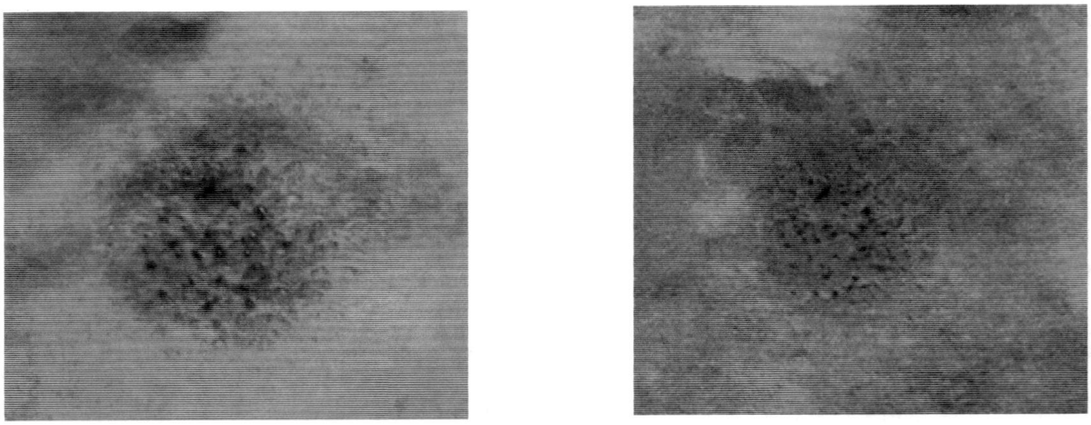

Fig.9 - Case photographs with samples of mold irradiated by higher value. The result is immediate mold destruction with no re-growth even after 4 weeks. (Parametres: f 20Hz, E 13mJ, T 5s, Ec = 1300mJ)

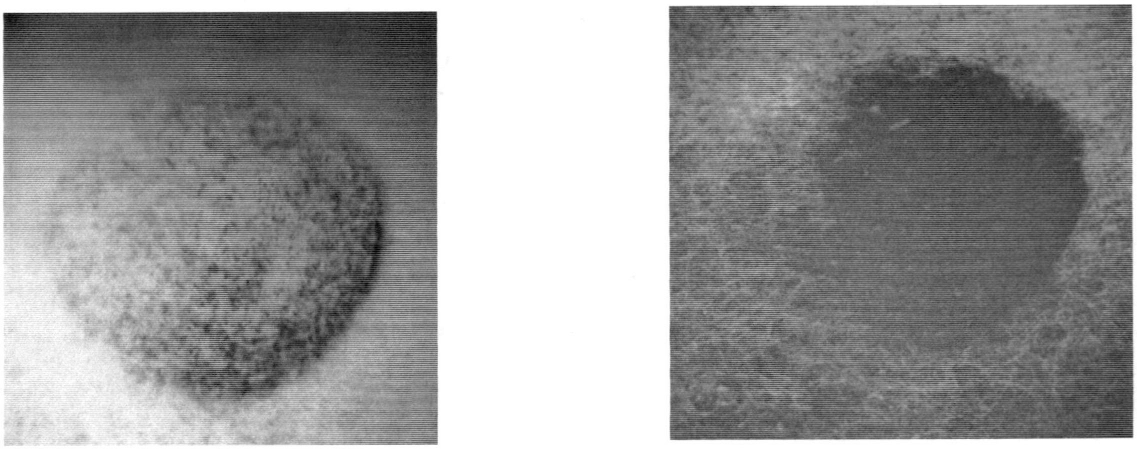

Fig.10 - Microscopic image of mold after laser irradiation

Evaluation of the thermal response of the irradiated tissue.

In other experiments performed in our department, frequency has proved to be an important factor. Even a small increase in frequency resulted in increased temperature and thus increased risk of tissue damage. For the treatment of onychomycosis, it is absolutely crucial to know the limit value adjustment of the laser, with which it is safe to use the laser for treating patients. Ablation threshold - the highest laser power, which does not interfere with the structure of tissue, we set out for the nail to 128.37 mJ/cm^2. This limit, therefore, cannot be exceeded during the therapy. The time of irradiation (irradiation product - power density in W cm^{-2} and radiation exposure time) is a fundamental parameter that determines whether the effects of radiation will be thermal (long exposure), acoustic, or other (very short pulses).In determining the ablation threshold of human skin the following parameters for excimer laser were used: a circular footprint of 2.6 mm in diameter, i.e. 5.31 mm^2; repetition rate of 10 Hz. Results are in Table 2 below.

Trial number	Energy in pulse [mJ]	Energy density [J/cm^2]	Number of pulses	Manifestation	Voltage/Tension (approximate value) [kV]
1	4,5	0,85	200	Without damage	18
2	7,5	1,42	200	Very mild smoke in later phase and very weak footprint	22
3	9,7	1,83	200	Smoke all the time and visible but weak footprint	26
4	7,5	1,42	1000	Occassional mild smoke and weak footprint	22
5	6,4	1,2	1000	Indistinct footprint (more „shiny“ at places)	20

Table 2 – Evaluation of the thermal response of irradiated tissue

The footprints were evaluated using the human eye and magnifying glass. "Smoke" mentioned in table 2 is probably caused by the ablation threshold of the surface layers, whether or not it is caused by the thermal effect, it needs to be verified in the future.

Discussion

Nails affected by mold vary considerably in their structure and thickness. In order to obtain a sufficient amount of information, it will be necessary to examine as many samples of different thicknesses as possible as well as nails in varying degrees of onychomycosis. In further tests it will be necessary to determine under what parameters damage to the skin during radiation might occur.

The results so far indicate that only a small dose of radiation is needed for eradication of nail fungus, which is favourable due to the fact that we must not damage the soft tissue under the nail. The objective of each single experiment was to determine the limits of the use of laser

Journal of Physics: Conference Series **497** (2014) 012022 doi:10.1088/1742-6596/497/1/012022

radiation in destroying fungi. As found and verified in previous experiments, fungi responds to wavelengths radiation in the UV spectrum positively and even relatively low power radiation is reliably destroying them.
However, for medical applications, it is necessary to know the minimal values required for total eradication of fungi, which is not needed to be known in laboratory conditions for experiments aimed at mere destruction of the mold.
In the search for the limiting values, it is necessary to take into account, firstly, that the transmitted radiation is substantially absorbed by the nail (relevant parameters of nails were examined in the previous experiments), and secondly, that the radiation is also absorbed by the tissue under the nail, not only by fungi, and therefore heat generation occurs, but temperature over 40 °C is undesirable.
In our experiment, we focused on determining such a value of the transmitted radiation that would ensure destruction of mold without an excessive increase in the tissue temperature.

Conclusion

The results of transmission measurements indicate that nails are impervious to radiation of wavelength less than about 400 nm. Reflection is low across the whole spectrum emitted by the source - at around 1%.
Measurements of nail transmission at wavelengths of 248 nm, using a high performance KrF laser in pulse mode, showed that the percentage of transmission depends on thickness of the nail, so for thicker nails the value is lower. Energy passing through the sample is very small and with usual nail thickness there will not be any undesired heat effects.
When verifying the absorption of UVC radiation by a thermal imaging camera, we found that the temperature of the irradiated nail sample does not exceed 40 ° C, and thus will not damage the soft tissue under the nail, and is not perceived as painful by the patient.
We have found out and subsequently verified that fungi are destroyed even when using energy values in pulse of 144 mJ and energy density 0.74 J/cm^2 and that they did not regrow. With those values we can use 200 pulses at a repeating frequency of 10 Hz.
Modulating the frequency of the laser radiation, according to our measurements, does not affect the destruction of mold.
The evaluation of monitoring the ablation threshold of human skin shows that this value does not exceed the ablation threshold of the nail.
Among factors that most influence the degree of tissue damage and thus are relevant to the determination of the parameters for the possible excimer laser treatment of onychomycosis include, according to our results, absorption, reflection and transmission of irradiated tissue, its size, laser wavelength, intensity of the laser beam and time of irradiation.
Determination of the precise parameters for excimer laser treatment requires additional tests but the results of our measurements show that this method could be one of the possible ways of treating onychomycosis, either on its own, or in combination with pharmacologic therapy.

References

1. Goldman, L., Rockwell, R.J.: Lasers in medicine. Ronald Waynant. [s.l.] CRCpress, 2002, 400 s.
2. Krutmann, J., Höniggsmann, H., Elmets, C.A.: Dermatological phototherapy and photodiagnostics methods. 2nd edition. Berlin: Springer, 2009, 447s.

3. Záruba, F.: Dermatovenerologie, 1. vyd. Praha : Science Press, 1994, 247 s, ISBN 80-85526-31-X.
4. KOLÁŘOVÁ, Hana a Dagmar DITRICHOVÁ. Laserové záření v medicíně. 1. vyd. Olomouc: Vydavatelství Univerzity Palackého, 1996, 37 s. ISBN 80-706-7608-6.
5. T. Dai, G.P. Tegos, G. Rolz-Cruz, W.E. Cumbie and M.R. Hamblin: Ultraviolet C inactivation of dermatophytes: implications for treatment of onychomycosis. British Journal of Dermatology 2008; 158, pp 1239–1246
6. D.T.Roberts, W.D.Taylor, J.Boyle: Introduction Guidelines for treatment of onychomycosis. British Journal of Dermatology 2003; 148, p 402
7. Adam S. Landsman and Alan H. Robbins:Treatment of Mild, Moderate, and Severe Onychomycosis Using 870- and 930-nm Light Exposure: Some Follow-up Observations at 270 Days, J. Am. Podiatr. Med. Assoc. 2012 102:169-171
8. Aditya K. Gupta and Fiona Simpson, Newly Approved Laser Systems for Onychomycosis J. Am. Podiatr. Med. Assoc. 2012 102:428-430

Main author's contact details:

MUDr. Jaroslava Kymplová, Ph.D.
Ústav biofyziky a informatiky 1.LF UK Praha
Salmovská 1
120 00 Praha 2
Tel:420 603 150 292
jkymplova@volny.cz

Journal of Physics: Conference Series **497** (2014) 012023 doi:10.1088/1742-6596/497/1/012023

Self-similar Shape Mode of Optical Pulse Propagation in de-focusing Medium with Two-Photon Absorption

Vycheslav A. Trofimov, Irina G. Zakharova, Mihail V. Fedotov

Lomonosov Moscow State University, Moscow 119992, Russia

e-mail vatro@cs,msu.ru

Abstract. We investigate the self-similar shape mode of laser pulse propagation in medium with two-photon and multi-photon absorption under action of either cubic nonlinearity or resonant nonlinearity. The annalitycal solution for laser pulse propagation under the detuning between carrier frequency of a wave packet and energy transition frequency for atom or molucule is developed. An existance of the self-similar shape mode for laser beam propagation in de-focusing medium is predicted if the nonlinear absoption takes place. This mode takes place along some distance of laser pulse propagation. We used as analytical approach for considered problem as well as computer simulation.

1. Introduction

Soliton or soliton-like profile of laser beam or shape of laser pulse plays a great role in modern nonlinear optics and attaches great attention of various scientist (see, for example [1-17]). In our opinion, an investigation of self-similar profile of laser beam, propagating in a medium with non-linear absorption is actual problem because powerful laser system produces high intensive laser pulse which induces a formation of plasma under its propagation in a gas.

Below we consider the possibility of laser pulse propagation with the self-similar profile in the medium with TPA or multi-photon absorption including a presence of either cubic nonlinear response of a medium (self-focusing or de-focusing) or resonant nonlinearity. To achieve this aim we find the corresponding beam profile from the eigenvalue problem. Early we have applied this method for various problems of nonlinear optics [18-27].

It should be stressed that each problem requires unique iterative method for finding the soliton from eigenvalue problem because the corresponding equation is nonlinear. Therefore, a particular method should be developed for each type of nonlinear response of the medium. This is also true for the medium with nonlinear absorption but in this case, the finding of a soliton (self-similar profile of intensity) is a more difficult task.

It should be stress, that in [28] in contract to current paper we discussed considering problem for both homogeneous and layered medium with TPA only. We does not consider an influence of a cubic nonlinear response on an existence of soliton-like mode at the laser pulse propagation in a medium with the nonlinear response.

2. Problem statement

Below we consider three cases of laser pulse interaction with nonlinear medium. First of them corresponds to laser pulse propagation in a medium with both two-photon absorption (TPA) and cubic nonlinear response. The aim of this consideration is an illustration of possibility to existence a mode of laser pulse propagation with self-similar shape even if laser beam de-focusing takes place.

In the second case we analyze a laser pulse interaction with medium, possessing a cubic nonlinear response and three-photon absorption (ThPA). The aim of this consideration is an investigation of absorption type influencing on laser pulse shaping. The last case of consideration deals with resonant nonlinear response with taking into account wave packet carrier frequency detuning from the energy transition frequency for medium molecules or atoms. We write an analytical solution for simplified problem for the case under consideration.

All laser pulse interactions with medium are considered in the frame-work of 1D dimensionless Schrödinger equation, written for layered medium, because our previous results showed that they are valid also for layered medium. Consequently, we will analyze laser pulse propagation in three types of medium:

$$\varepsilon(z)\frac{\partial A}{\partial t}+iD\frac{\partial^2 A}{\partial z^2}+\beta(i\varepsilon(z)+(\delta(z)+i\alpha(z))\,|\,A\,|^2)A=0,\quad 0<z<L_z,\quad 0<t\le L_t \tag{1}$$

- cubic nonlinear response with TPA,

$$\varepsilon(z)\frac{\partial A}{\partial t}+iD\frac{\partial^2 A}{\partial z^2}+\beta(i\varepsilon(z)+\delta(z)\,|\,A\,|^4+i\alpha(z)\,|\,A\,|^2)A=0,\quad 0<z<L_z,\quad 0<t\le L_t \tag{2}$$

- cubic nonlinear response with ThPA,

$$\varepsilon(z)\frac{\partial A}{\partial t}+iD\frac{\partial^2 A}{\partial z^2}+\beta(i\varepsilon(z)+\delta(z)(1+i\Delta)|A|^k)A=0,\quad 0<z<L_z,\quad 0<t\le L_t \tag{3}$$

- resonant nonlinear response of medium.

These equations are solved with zero-value boundary conditions

$$A(t,0)=A(t,L_z)=0,\quad 0\le t\le L_t \tag{4}$$

and initial distribution of complex amplitude

$$A(0,z)=A_0(z),\quad 0\le z\le L_z. \tag{5}$$

Above $A(t,z)$ is dimensionless complex amplitude that is slowly varying in time; z is coordinate, along which the laser pulse propagates, L_z is dimensionless length of the nonlinear medium; t is a time, L_t is dimensionless time interval during which the propagation of the laser pulse is investigated. Function $\varepsilon(z)$ describes a dielectric permittivity of medium. Parameter D characterizes a diffraction of laser beam and β relates with parameter D. Signs of these two parameters must be the same and they depend on choosing of wave propagation direction for a derivation of the Schrödinger equation from Maxwell's equations. Equations (1)-(3) take into account a wave reflected from any boundaries of medium layers.

Function $\delta(z)$ characterizes a nonlinear absorption (or amplification) of a medium. Sing of this function depends on medium type: a medium with absorption $\delta(z)>0$ or amplification $\delta(z)<0$ of optical radiation. Parameter k is equal to 2 for medium with TPA or 4 for medium with ThPA. Function $\alpha(z)$ describes the cubic nonlinear response of medium. In our notation its positive sign corresponds to compression of laser pulse at positive value of the diffraction coefficient. Opposite case corresponds to laser pulse decompression. It should be emphasized that the functions $\varepsilon(z)$, $\alpha(z)$ and $\delta(z)$ are equal to corresponding constants for the homogeneous medium ($\varepsilon(z)=\varepsilon=const$, $\alpha(z)=\alpha=const$, $\delta(z)=\delta=const$). Below we will consider a homogeneous medium.

It should be emphasized that the algorithm of finding the beam profile with the self-similar feature is valid for both homogeneous medium and layered one. However, for the layered medium the diffraction coefficient D and parameter β in our notation relate each with other in following way

$$D=-\frac{1}{4\pi\Omega},\quad \beta=-\pi\Omega,\quad \Omega=\frac{\omega}{\omega_{str}}. \tag{6}$$

Here Ω is ratio of laser light frequency to frequency of periodic structure. The sign "minus" can be changed to sign "plus" if we follow the wave propagating in the direction which is opposite to the direction chosen by us above for the layered medium. Below if we consider a homogeneous medium, we choose a value of the diffraction coefficient D and parameter β as

$$D=1, \beta=0,25. \tag{7}$$

If we will consider other values of these parameters then we will specify their new values.

3. Homogeneous medium with resonant nonlinear response

To find the requiring profile of the beam we consider the nonlinear eigenfunction problem (see equation (3)). With this aim let us represent the complex amplitude in a way

$$A(t,z)=B(z)e^{-i\lambda t}. \tag{8}$$

Substituting the function (6) into the Schrödinger equation (1), one can write the following eigenvalue and eigenfunction problem

$$\frac{D}{\varepsilon}\frac{d^2B}{dz^2}+\beta\left(1+\frac{\delta(\Delta-i)\,|B|^k}{\varepsilon}\left|e^{2\,\mathrm{Im}\,\lambda\cdot t}\right|\right)B=\lambda B,\quad B(0)=B(L_z)=0. \tag{9}$$

We can see that the equation (9) contains time-dependent factor $\left|e^{2\,\mathrm{Im}\,\lambda\cdot t}\right|$. Generally speaking, in this case we need to solve the problem for various moments of time because the factor $\mathrm{Im}\,\lambda\cdot t$ varies. It is obviously that we can consider the dependence $\left|e^{2\,\mathrm{Im}\,\lambda\cdot t}\right|$ from time as the variation of the absorption coefficients because at the propagation of laser beam its intensity varies with accordance of this factor. Thus, different eigenfunctions corresponding to various values of the absorption coefficients will correspond to eigenfunction at various moments of time. Taking into account these arguments we solve the following eigenfunction problem

$$\frac{D}{\varepsilon}\frac{d^2B}{dz^2}+\beta\left(1+\frac{\delta(\Delta-i)\,|\,B\,|^k}{\varepsilon}\right)B=\lambda B,\quad B(0)=B(L_z)=0 \tag{10}$$

with changing the intensity maximum.

For the homogeneous medium one can develop the analytical solution of this equation. With this aim let us consider the solution in the form

$$B(z)=\ a\ ch^{-m}(cz)e^{if(z)} \tag{11}$$

in which c a are a constant and $f(z)$ is a function, which describes the phase distribution along z – coordinate, m is a power of hyperbolic cosine. Substituting the expression (11) in the equation (10) we get two equations:

$$\begin{aligned}&\frac{D}{\varepsilon}\left[\left(m^2+m\right)c^2th^2(cz)-m\,c^2-\left(f'(z)\right)^2\right]+\beta[1+\frac{\Delta\delta}{\varepsilon}\frac{a^k}{ch^{mk}(cz)}]=\operatorname{Re}\lambda,\\&\frac{D}{\varepsilon}\left[f''(z)-2mc\cdot th(cz)\,f'(z)\right]-\frac{\beta\delta}{\varepsilon}\frac{a^k}{ch^{mk}(cz)}=\operatorname{Im}\lambda.\end{aligned} \tag{12}$$

Two important conclusions follow from set of the equations (12). First, we see that in both equations a factor $\delta\cdot a^k$. Therefore, we can choose the parameter a be equal to 1. and below (as well as above in (11)) we omit this parameter. Changing in amplitude a corresponds to changing in the absorption coefficient δ only. Second, for solution developing it is necessary to choose the parameter m depending from k in following way: $km=2$. In this case, first equation of the set of equations (12) can be rewritten as

$$th^2(cz)[\frac{D}{\varepsilon}\left(m^2+m\right)c^2-\frac{\delta\Delta\beta}{\varepsilon}]-\frac{D}{\varepsilon}\left(f'(z)\right)^2-m\,c^2\frac{D}{\varepsilon}+\beta[1+\frac{\delta\Delta}{\varepsilon}]=\operatorname{Re}\lambda. \tag{13}$$

Therefore, for the solution existence of this equation, the validity of the following set of equations is sufficient:

$$f'(z)=\sqrt{\frac{2c^2}{k}\left(\frac{2}{k}+1\right)-\frac{\delta\Delta\beta}{D}}\,th(cz),\quad \operatorname{Re}\lambda=\beta[1+\frac{\delta\Delta}{\varepsilon}]-\frac{2Dc^2}{\varepsilon k}. \tag{14}$$

Hence, the phase shift along the z - coordinate is described by

$$f(z)=\frac{1}{c}\sqrt{\frac{2c^2}{k}\left(\frac{2}{k}+1\right)-\frac{\delta\Delta\beta}{D}}\cdot\ln\left(ch(cz)\right)+c_1, \tag{15}$$

where c_1 is a constant, which defines the homogeneous phase shift and, obviously, does not influence on the intensity distribution of the laser beam.

Substitution of this relation in the second equation of (12) results in the following relation between coefficients of the problem

$$th^2(cz)\cdot\left[\frac{\beta\delta}{\varepsilon}-\frac{D}{\varepsilon}c\sqrt{\frac{2c^2}{k}\left(\frac{2}{k}+1\right)-\frac{\delta\Delta\beta}{D}}(1+\frac{4}{k})\right]+\frac{D}{\varepsilon}c\sqrt{\frac{2c^2}{k}\left(\frac{2}{k}+1\right)-\frac{\delta\Delta\beta}{D}}-\frac{\beta\delta}{\varepsilon}=\mathrm{Im}\lambda\ . \qquad (16)$$

For validity of this equation for any value of z coordinate it is sufficient that the following equalities take place

$$c\sqrt{\frac{2c^2}{k}\left(\frac{2}{k}+1\right)-\frac{\delta\Delta\beta}{D}}(1+\frac{4}{k})=\frac{\beta\delta}{D},\ \mathrm{Im}\lambda=-\frac{\beta\delta}{\varepsilon}\frac{4}{k+4}. \qquad (17)$$

From first relation of equations (17) one follows that a sign of parameter c depends from a sign of a right part of this equation. It takes place if a sign of δ changes to opposite sign, which corresponds to amplifying medium. The solution of this equation takes place for both signs of detuning Δ. As a result, we can write the following relations:

$$c=\pm k\sqrt{\frac{\beta|\delta|}{D}\frac{\Delta+\sqrt{\Delta^2+\frac{8(k+2)}{(k+4)^2}}}{4(k+2)}}\ . \qquad (18)$$

Choosing of a sign before square root is specified by a sign of the parameter δ.

Finally, a solution of the problem (12) is

$$m=\frac{2}{k},\ a=1,\ \lambda=\left(\beta[1+\frac{\delta\Delta}{\varepsilon}]-\frac{k\beta|\delta|}{\varepsilon}\frac{\Delta+\sqrt{\Delta^2+\frac{8(k+2)}{(k+4)^2}}}{2(k+2)};-\frac{\beta\delta}{\varepsilon}\frac{4}{(k+4)}\right), \qquad (19)$$

$$f(z)=\frac{1}{c}\sqrt{\frac{2c^2}{k}\left(\frac{2}{k}+1\right)-\frac{\delta\Delta\beta}{D}}\cdot\ln(ch(cz))+c_1,\ f'(z)=\sqrt{\frac{2c^2}{k}\left(\frac{2}{k}+1\right)-\frac{\delta\Delta\beta}{D}}\,th(cz)\ ,$$

$$c=\pm k\sqrt{\frac{\beta|\delta|}{D}\frac{\Delta+\sqrt{\Delta^2+\frac{8(k+2)}{(k+4)^2}}}{4(k+2)}}\ ,\qquad B(z)=(ch(cz))^{-2/k}e^{if(z)}\ .$$

From the expression of the intensity distribution (19) one follows a value of z coordinate, at which the intensity of laser beam decreases to two times ($|B(z)|^2=0{,}5$):

$$|c|\cdot z_h=arcch(2^{k/4})=\ln(2^{k/4}+1). \qquad (20)$$

Consequently, for the half-width of the intensity distribution we get the following ratio

$$a_z=2z_h=2\cdot\ln(2^{k/4}+1)/|c|. \qquad (21)$$

From(19) and (21) a dependence of the half-width of the intensity distribution follows. One can see

Journal of Physics: Conference Series **497** (2014) 012023 doi:10.1088/1742-6596/497/1/012023

that a positive value of detuning Δ leads to the half-width increasing. But its negative value results in the half-width decreasing.

4. Cubic nonlinearity and TPA

The influence of cubic nonlinear response on existence of the mode with self-similar beam profile propagation (see equation (1)) was investigated in [28] in detail. The main conclusion of this paper is an existence of such mode even for de-focusing medium. As example, a comparison of results of computer simulation with theoretical (analitycal) consideration is shown in Tables 1, 2. In these Tables an influence of cubic nonlinear response on characteristics of laser beam propagation is depicted also.

Table 1. Dependence of the first eigenvalue λ and of the half-width a_z from the cubic nonlinearity α for the medium with TPA and $\varepsilon(z) = 2{,}5, \delta = 10$.

α		2	5	10	-2	-2,5
Re λ	Numerical	0,16	0,36	0,66	-0,14	-0,18
	Theoretical	0,16	0,36	0,66	-0,14	-0,19
Im λ	Numerical	-0,67	-0,67	-0,67	-0,67	-0,67
	Theoretical	-0,67	-0,67	-0,67	-0,67	-0,67
a_z	Numerical	2,075	1,775	1,425	2,525	2,625
	Theoretical	2,07	1,78	1,44	2,55	2,57

Table 2. Dependence of the first eigenvalue λ and of the half-width a_z from the cubic nonlinearity α for the medium with TPA and $\varepsilon(z) = 2{,}5, \delta = 5$.

α	2	5	10	-2	-3
Re λ	0,27	0,45	0,72	-0,028	-0,11
Im λ	-0,33	-0,33	-0,33	-0,33	-0,33
a_z	2,625	2,025	1,525	3,975	4,375

We see good coincides between analitycal solution and computer simulation resultts. It should be noticed only that a convergence of our computer method disappeares at $\alpha < -2{,}5$ if $\delta = 10$ and at $\alpha < -3$ if $\delta = 5$. Nevertheless, we see clear dependendence of a beam width from a parameter α and the existence of a solution of the problem under consideration for de-focusing medium if the nonlinear absorption takes place.

5. Cubic nonlinear response and ThPA

The influence of cubic nonlinear response on existence of the mode with self-similar beam profile propagation in a medium with three-photon absorption (see equation (2) with k=4) is investigated numerically because an analitycal solution is absent nowadays. Computer simulation results are shown in Tables 3-5. We see an existence of beam profile corresponding to the self-similar profile mode for de-focusing medium if a nonlinear absorption takes place (Tables 3, 4). As well as for the case of a resonant nonlinearity an imaginary part of the eigenvalue depends only from the absorption coefficient and does depend from a value of the parameter α. Fig. 1 illustrates beam profile computed for for various relations between parameterscharacterizing a nonlinear response of medium. We see very good agreements between computed curve and approximmating function.

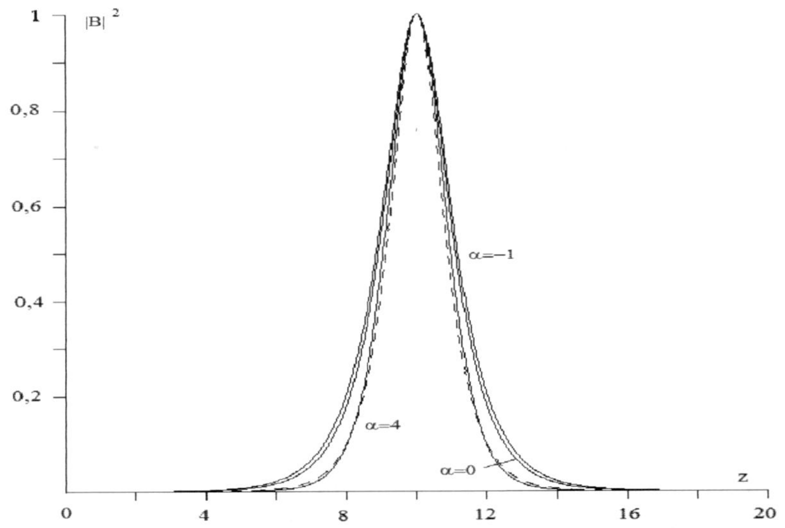

Fig. 1. Intensity profile computed for $\varepsilon = 2,5; \delta = 10$ and various values of α. Intensity profile computed for $\alpha = 4$ is approximated perfectly by the function $\text{ch}^{-1}(z/1,1)$ (dashed line).

Table 3. Dependence of the first eigenvalue λ and of the half-width a_z from the cubic nonlinear response α for the medium with ThPA $\varepsilon(z) = 2{,}5, \delta = 10$.

α	0	0,1	0,5	1	2	4	6	-0,1	-0,5	-1
Re λ	-0.039	-0,032	-0,006	0,026	0,087	0,2	0,3	-0,045	-0,073	-0,11
Im λ	-0,5	-0,5	-0,5	-0,5	-0,5	-0,5	-0,5	-0,5	-0,5	-0,5
a_z	2,175	2,155	2,125	2,075	1,975	1,775	1,575	2,225	2,275	2,325

Table 4. Dependence of the first eigenvalue λ and of the half-width a_z from the cubic nonlinear response α for the medium with ThPA $\varepsilon(z) = 2{,}5, \delta = 5$.

α	0; 0,1	1	2	4	6	-0,1	-0,5	-1	-2	-3
Re λ	0.11	0,17	0,22	0,32	0,4	0,1	0,07	0,03	-0,04	-0,12
Im λ	-0,25	-0,25	-0,25	-0,25	-0,25	-0,25	-0,25	-0,25	-0,25	-0,25
a_z	3,075	2,775	2,475	2,075	1,775	3,125	3,275	3,475	3,825	4,325

Table 5. Value of the first eigenvalue λ and of the half-width a_z for kerr solition propagating in transparent medium ($\delta = 0$) with $\varepsilon(z)=2{,}5$. Im λ is equal to 0 in this case.

α	0,1	1	2	4	6
Re λ	0,25	0,28	0,38	0,32	0,45
a_z	8,975	4,525	2,275	2,075	1,875

6. Conclusion

We showed that the self-similar mode of laser beam propagation in a medium with nonlinear absorption takes place even for a presence of de-focusing of laser beam. We developed an analitycal solution for the beam profile for laser pulse propagation in a medium with resonant nonlinear response for signs of detuning of a carrier frequency of a wave packet from energy transition frequence of atom or molecule.

7. Acknowledgmets

This work was financially supported in part by the Russian Foundation for Basic Research (grant number 13-01-92695-Ind-a).

References

[1] Akhmanov S A, Vysloukh V A and Chirkin A S 1992 *Optics of femtosecond laser pulses* (NY: American Institute of Physics)

[2] Shen Y R 1984 *The Principles of Nonlinear Optics* (NY: J. Wiley Press)

[3] Agrawal G P 1995 *Nonlinear fiber optics* (NY: Academic Press)

[4] Sukhorukov A P 1988 *Nonlinear Wave Interactions in Optics and Radiophysics* (Moscow: Nauka)

[5] Kivshar Y S and Agrawal G P 2003 *Optical Solitons: From Fibers to Photonic Crystals* (San Diego: Academic Press)

[6] Akhmediev N N and Ankiewicz A 1997 *Solitons: Nonlinear Pulses and beams* (London: Chapman and Hall Press)

[7] Brull L and Lange H 1986 Stationary, oscillatory and solitary wave type solution of singular nonlinear Schrödinger equations *Math. Meth. in the Appl. Sci.* **8** (4) 559–75

[8] Malomed B A 2006 *Soliton management in periodic systems* (NY: Springer)

[9] Berge L, Mezentsev V K, Rasmussen J J et al 2000 Self-guiding light in layered nonlinear media *Opt. Lett.* **25** (14) 1037–39

[10] Tower I and Malomed B A 2002 Stable (2+1)-dimensional solitons in a layered medium with sign-alternating Kerr nonlinearity *JOSA* B **19** (3) 537–43

[11] Yannopapas V, Modinos A and Stefonou N 2003 Anderson localization of light in inverted opals *Phys. Rev.* B **68** 193205

[12] Goodman R H, Holmes Ph J and Weinstein M I 2004 Strong NLS soliton-defect interactions *Physica* D **192** 215–42

[13] Driben R, Oz Y, Malomed B A et al. 2007 Mismatch management for optical and matter-wave quadratic solitons *Phys. Rev.* E **75** 026612

[14] Serkin V N and Hasegawa A 2002 Exactly integrable nonlinear Schrodinger equation models with varying dispersion, nonlinearity and gain: application for soliton dispersion *IEEE Journal of Selected Topics in Quantum Electronics* **8** (3) 418

[15] Serkin V N, Hasegawa A and Belyaeva T L 2007 Nonautonomous solitons in external potentials *Physics Review Letters* **98** 074102

[16] Baizakov B B, Malomed B A and Salerno M 2003 Multidimensional solitons in periodic potentials *Europhysics Letters* **63** (5) 642–48

[17] Al Khawaja U 2009 Integrability of a general Gross–Pitaevskii equation and exact solitonic solutions of a Bose–Einstein condensate in a periodic potential *Phys. Letters* A **373** 2710–16

[18] Trofimov V A and Varentsova S A 2005 Computational method for finding of soliton solutions of the nonlinear Schrödinger equation *Lectures Notes in Mathematics* ed L Vulkov (Berlin: Springer-Verlag) 3401 pp 550–557

[19] Varentsova S A and Trofimov V A 2005 On a difference method for finding the eigenmodes of the nonlinear Schrödinger equation *Moscow University Computational Mathematics and Cybernetics* **3** 1–8

[20] Matusevich O V and Trofimov V A 2008 Iterative method for finding the eigenfunctions of a

system of two Schrödinger equations with combined nonlinearity *Computational Mathematics and Mathematical Physics* **48** (4) 677–87

[21] Matusevich O V and Trofimov V A 2009 Numerical method for finding 3D solitons of the nonlinear Schrödinger equation in the axially symmetric case *Computational Mathematics and Mathematical Physics* **49** (11) 1902–12

[22] Matusevich O V and Trofimov V A 2009 A numerical method for calculating solitons of the nonlinear Schrödinger equation in the axially symmetric case *Moscow University Computational Mathematics and Cybernetics* **33** (3) 117–26

[23] Trofimov V A and Matusevich O V 2008 Numerical method for 2D soliton solution at SHG in media with time-dependent combined nonlinearity *Mathematical Modelling and Analysis* **13** (1) 123–34

[24] Trofimov V A, Lysak T M and Matysevich O V 2010 "Influence of transverse perturbation of soliton propagation direction on laser radiation evolution along the layered medium *Photonic Crystal Fibers IV* ed K Kalli and W Urbanczyk *Proc. of SPIE 7714* 77140K

[25] Trofimov V A, Lysak T M, Matysevich O V and Lan Sh 2010 Parameter control of optical soliton in one-dimensional photonic crystal *Mathematical modeling and analysis* **15** (4) 517–32

[26] Matusevich O V, Trofimov V A, Yudina E A and Malomed B A 2009 The evolution of two-frequency solitons in an optical fiber with a longitudinally nonuniform nonlinearity *Optics and Spectroscopy* **106** (1) 99–107

[27] Trofimov V A, Lysak T M and Matysevich O V 2010 Influence of transverse perturbation of soliton propagation direction on laser radiation evolution along the layered medium *Proc. of SPIE 7714* 77140K

[28] Trofimov V A, Matysevich O V and Smotrov D A 2011 Mode of propagation of optical radiation with self-similar pulse shape in layered medium with nonlinear absorption *Proc. of SPIE 8095* 80951K

22nd International Laser Physics Worksop (LPHYS 13) IOP Publishing
Journal of Physics: Conference Series **497** (2014) 012024 doi:10.1088/1742-6596/497/1/012024

Strong coupling study of spin 1 bosons in square and triangular optical lattices

Takashi Kimura
Department of Mathematics and Physics, Kanagawa University, 2946 Hiratsuka, Kanagawa 259-1293, apan

E-mail:
tkimura@kanagawa-u.ac.jp

Abstract. We examine the superfluid-Mott insulator (SF-MI) transition of antiferromagnetically interacting spin-1 bosons trapped in a square or triangular optical lattice. We perform a strong-coupling expansion up to the third order in the transfer integral t between the nearest-neighbor lattices. As expected from previous studies, an MI phase with an even number of bosons is considerably more stable against the SF phase than it is with an odd number of bosons. Results for the triangular lattice are similar to those for the square lattice, which suggests that the lattice geometry does not strongly affect the stability of the MI phase against the SF phase.

1. Introduction

The development of optical lattice systems based on laser technology has renewed interest in strongly correlated lattice systems. One of the most striking phenomena of the optical-lattice systems is the superfluid-Mott insulator (SF-MI) phase transition; the SF phase (i.e., the coherent-matter-wave phase) emerges when the kinetic energy is larger enough compared with the on-site repulsive interaction. Otherwise, the MI phase, i.e., the number-state phase without coherence emerges. The low-lying excitations of these optical-lattice systems can be described by using the Bose–Hubbard model. The temperature of trapped-atom systems can be extremely low, and hence, we hereafter assume the ground states of the system.

Journal of Physics: Conference Series **497** (2014) 012024 doi:10.1088/1742-6596/497/1/012024

Spin degrees of freedom also play an important role in optical-lattice systems. In theory, lots of analytical and numerical studies have been performed for the spin-1 Bose–Hubbard model [1], including rigorous results for a finite system [2]. In the case of antiferromagnetic spin-spin interactions, the perturbative mean-field approximation (PMFA) [3] indicates that when filling with an even number of bosons, the MI phase is considerably more stable against the SF phase than when filling with an odd number of bosons. This conjecture has been found by density matrix renormalization group (DMRG) [4] and quantum Monte Carlo (QMC) methods [5, 6] in one dimension (1D). Recently, QMC methods also confirmed that conjecture in a two-dimensional (2D) square lattice [8].

Another interesting property of the spin-1 Bose–Hubbard model with antiferromagnetic interactions is the first-order phase transition: the SF-MI phase transition is of the first order in a part of the SF-MI phase diagram. The first-order transition has also been studied by using the variational Monte Carlo [7] and QMC [8] methods in a 2D square lattice. The QMC results indicate that the phase transition can be of the first order, which is consistent with mean-field (MF) analysis [9, 10, 11]. However, the first-order transition disappears for strong antiferromagnetic interactions; a MF calculation similar to that of Ref. [9] and the QMC study [8] show that the first-order SF-MI transition from the Mott lobe with two bosons per site disappears when $U_2/U_0 \geq 0.32$ and $U_2/U_0 \geq 0.15$, respectively. Thus, we assume strong interactions where the SF-MI transition is of the second order.

For the second-order SF-MI transition, the strong-coupling expansion of kinetic energy [12] is excellent for obtaining the phase boundary. This method has been applied to the spinless [12, 13], extended [14], hardcore [15], and two-species models [16], and the results agree well with QMC results [13, 15]. Thus, in this study, we perform the strong-coupling expansion with the spin-1 Bose–Hubbard model. In another publication [17], we examined the case of hypercubic lattices. In this study, we examine the triangular lattice and compare the results with those of a square lattice to clarify whether the lattice structure plays a key role for the SF-MI transition. The triangular lattice is intriguing because it frustrates the spin systems or spinful Fermi systems.

The rest of this paper is organized as follows: Section II briefly introduces the spin-1 Bose–Hubbard model and the strong-coupling expansion. Section III provides our results. A summary of the results is given in Section IV. Some long equations that result from the strong-coupling expansion are summarized in Appendix A.

2. Spin 1 Bose–Hubbard model and strong coupling expansion

The spin-1 Bose–Hubbard model is given by $H = H_0 + H_1$, where

$$H_0 = -t \sum_{\langle i,j\rangle,\alpha} (a^{\dagger}_{i\alpha}a_{j\alpha} + a^{\dagger}_{j\alpha}a_{i\alpha}),$$

Journal of Physics: Conference Series **497** (2014) 012024 doi:10.1088/1742-6596/497/1/012024

$$
\begin{aligned}
H_1 = & -\mu \sum_{i,\alpha} a_{i\alpha}^\dagger a_{i\alpha} + \frac{1}{2} U_0 \sum_{i,\alpha,\beta} a_{i\alpha}^\dagger a_{i\beta}^\dagger a_{i\beta} a_{i\alpha} \\
& + \frac{1}{2} U_2 \sum_{i,\alpha,\beta,\gamma,\delta} a_{i\alpha}^\dagger a_{i\gamma}^\dagger \mathbf{F}_{\alpha\beta} \cdot \mathbf{F}_{\gamma\delta} a_{i\delta} a_{i\beta}. \\
= & \sum_i \Big[-\mu \hat{n}_i + \frac{1}{2} U_0 \hat{n}_i (\hat{n}_i - 1) + \frac{1}{2} U_2 (\hat{\mathbf{S}}_i^2 - 2\hat{n}_i) \Big]. \qquad (1)
\end{aligned}
$$

Here, μ and $t(> \ 0)$ are the chemical potential and the hopping matrix element, respectively. U_0 (U_2) is the spin-independent (spin-dependent) interaction between bosons. We assume repulsive ($U_0 > 0$) and antiferromagnetic ($U_2 > 0$) interaction. $a_{i\alpha}$ ($a_{i\alpha}^\dagger$) annihilates (creates) a boson at site i with spin-magnetic quantum number $\alpha = 1, 0, -1$. $\hat{n}_i \equiv \sum_\alpha n_{i\alpha}$ ($n_{i\alpha} \equiv a_{i\alpha}^\dagger a_{i\alpha}$) is the number operator at site i. $\hat{\mathbf{S}}_i \equiv \sum_{\alpha,\beta} a_{i\alpha}^\dagger \mathbf{F}_{\alpha\beta} a_{i\beta}$ is the spin operator at site i, where $\mathbf{F}_{\alpha\beta}$ represents the spin-1 matrices. In this study, we assume a tight-binding model with only nearest-neighbor hopping and $\langle i, j \rangle$ expresses sets of adjacent sites i and j.

When $t \to 0$, the ground state is the MI phase with the lowest interaction energy. The number n_0 of bosons per site is odd when $U_0(n_0 - 1) < \mu < U_0 n_0 - 2U_2$, whereas it is even when $U_0(n_0 - 1) - 2U_2 < \mu < U_0 n_0$. The MI phase with an even number of bosons is

$$\Psi_{\text{even}} = \prod_k |n_0, 0, 0\rangle_k. \qquad (2)$$

Here, $|n_0, 0, 0\rangle_k$ implies the boson number n_0, the spin magnitude $S = 0$, and the spin magnetic quantum number $S_z = 0$ at site k. However, for the MI phase with an odd number of bosons per site, we define a nematic state with $S_z = 0$:

$$\Psi_{\text{odd}} = \prod_k |n_0, 1, 0\rangle_k \qquad (3)$$

because we assume antiferromagnetic interactions. The dimerized state is degenerate with Ψ_{odd} for $t = 0$ and is considered to be the ground state for finite t in 1D. Therefore, the results based on Ψ_{odd} are basically limited to 2D or larger dimensions.

Next, we define the defect states by doping an extra particle or hole into Ψ_{even} and Ψ_{odd} as follows:

$$\Psi_{\text{even}}^{\text{part}} = \frac{1}{\sqrt{N}} \sum_i \Big[|n_0 + 1, 1, 0\rangle_i \otimes \prod_{k \neq i} |n_0, 0, 0\rangle_k \Big], \qquad (4)$$

$$\Psi_{\text{even}}^{\text{hole}} = \frac{1}{\sqrt{N}} \sum_i \Big[|n_0 - 1, 1, 0\rangle_i \otimes \prod_{k \neq i} |n_0, 0, 0\rangle_k \Big], \qquad (5)$$

$$\Psi_{\text{odd}}^{\text{part}} = \frac{1}{\sqrt{N}} \sum_i \Big[|n_0 + 1, 0, 0\rangle_i \otimes \prod_{k \neq i} |n_0, 1, 0\rangle_k \Big], \qquad (6)$$

Journal of Physics: Conference Series **497** (2014) 012024 doi:10.1088/1742-6596/497/1/012024

$$\Psi_{\rm odd}^{\rm hole} = \frac{1}{\sqrt{N}} \sum_i \Big[|n_0 - 1, 0, 0\rangle_i \otimes \prod_{k \neq i} |n_0, 1, 0\rangle_k \Big]. \tag{7}$$

Here, N is the number of lattice sites. We assume that these defect states can be regarded as the SF states doped with infinitesimal numbers of particles or holes. By applying the Rayleigh–Schrödinger perturbation theory to these MI and defect states, we obtain the energy of the MI state and that of the defect states up to the third order in t.

3. Results

The results for the energy per site of the MI state and of defect states in the square lattice are summarized in Appendix A. The phase can be determined by these energies. Specifically, if $E_{\rm MI}(n_0, \mu, t) > (<) \min\left(E_{\rm def}^{\rm part}(n_0, \mu, t), E_{\rm def}^{\rm hole}(n_0, \mu, t)\right)$, then the phase is SF (MI), where $E_{\rm MI}(n_0, \mu, t)$ is the energy of the MI state and $E_{\rm def}^{\rm part}(n_0, \mu, t)$ [$E_{\rm def}^{\rm hole}(n_0, \mu, t)$] is the energy of the defect state with one extra particle (hole). The SF–MI phase boundary is thus determined by

$$E_{\rm MI}(n_0, \mu, t) = E_{\rm def}^{\rm part}(n_0, \mu, t) \tag{8}$$

or

$$E_{\rm MI}(n_0, \mu, t) = E_{\rm def}^{\rm hole}(n_0, \mu, t). \tag{9}$$

Figures 1 and 2 show the phase diagram obtained from the strong-coupling expansion for the square lattice and for the triangular lattice, respectively. In both lattices, the MI phase for even-boson filling is considerably more stable against the SF phase than for odd-boson filling, which is consistent with QMC studies for the square lattice and MF studies. The area of the MI phase for even-boson (odd-boson) filling increases more (decreases more) for $U_2/U_0 = 0.4$ than for $U_2/U_0 = 0.2$. The convergence of the strong-coupling expansion from the first to the third order in t is fairly good. We also find that higher-order terms mostly render the SF phase more stable against the MI phase because the area of the MI phase mostly becomes smaller for higher-order expansions, which is similar to the case of spinless Bose–Hubbard model.

The phase-boundary curve obtained by the strong-coupling expansion up to the third order in t has a cusp at the peak of the Mott lobe. However, the chemical potential in infinite order in t should follow a power-law scaling near $t_{\rm C}$, which is the value of t at the tip of the Mott lobe. Following Ref. [12] to the spinless Bose–Hubbard model in 2D, we perform a chemical-potential fitting method which assumes

$$\mu = A(t) \pm B(t)(t_{\rm C} - t)^{z\nu}. \tag{10}$$

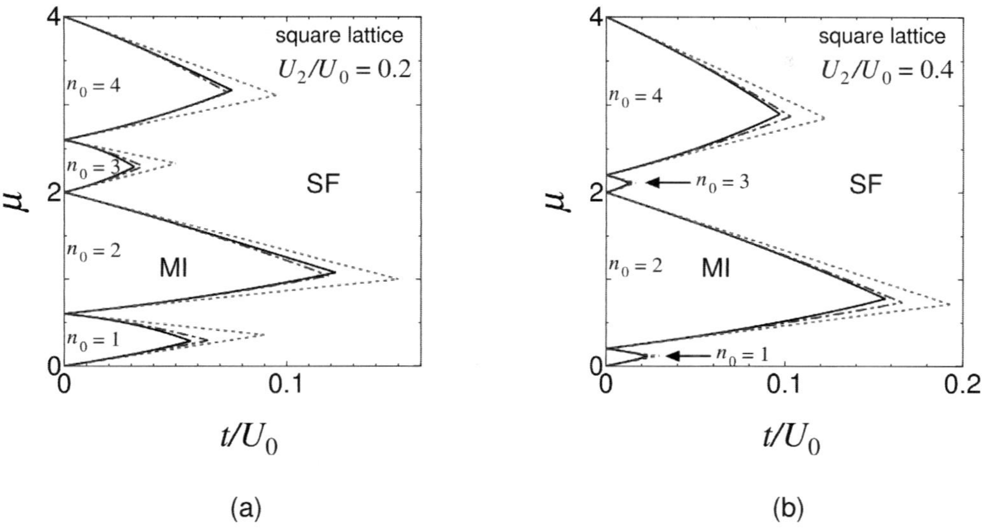

Figure 1. (Color online) Phase diagrams obtained by strong-coupling expansion for (a) $U_2/U_0 = 0.2$ and for (b) $U_2/U_0 = 0.4$ in the square lattice. The solid curves show the results up to third order in t. Results up to first order (second order) in t are also shown by the green dashed (blue dot-dashed) curve.

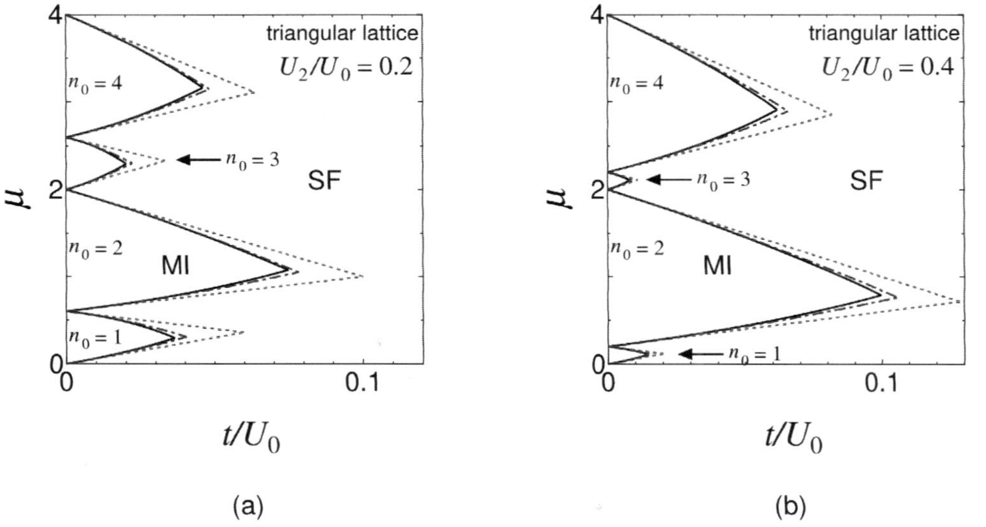

Figure 2. (Color online) Same plot as in Fig. 1 but for (a) $U_2/U_0 = 0.2$ and (b) $U_2/U_0 = 0.4$ in the triangular lattice.

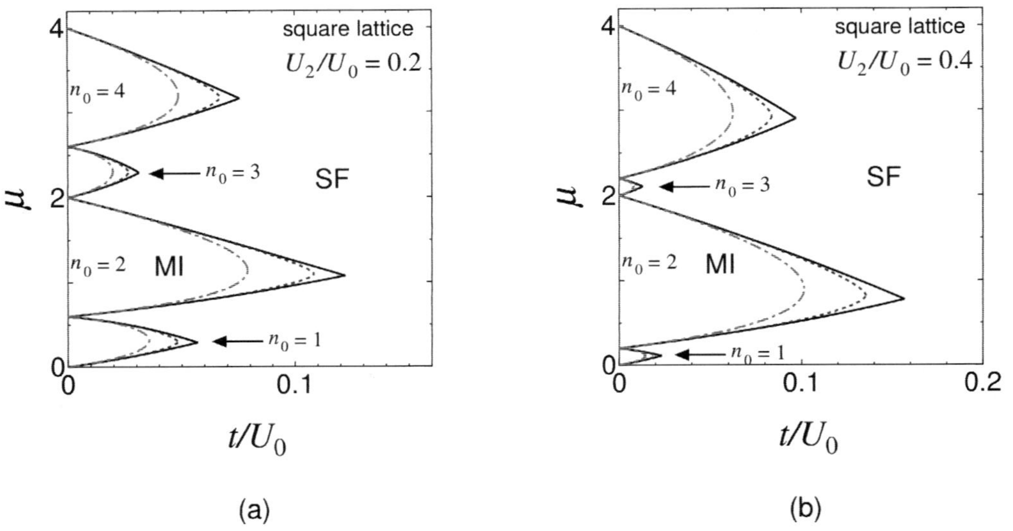

Figure 3. (Color Online) Phase diagrams for (a) $U_2/U_0 = 0.2$ and for (b) $U_2/U_0 = 0.4$ obtained by the strong-coupling expansion (solid curve) and its chemical-potential fitting (brown dashed curve) in the square lattice. The red dot-dashed curve shows the PMFA results.

with a critical exponent $z\nu \simeq 2/3$. Here $A(t) \approx a + bt + ct^2 + dt^3$ and $B(t) \approx \alpha + \beta t + \gamma t^2$ are the regular polynomial functions of t. By using our strong-coupling expansion up to the third order in t, we immediately determine a, b, c, and d by setting $A(t) = [\mu^{\text{part}}(t) + \mu^{\text{hole}}(t)]/2$. In order to determine α, β, γ, and t_C we compare the Taylor expansion of t in $B(t)(t_C - t)^{z\nu}$ with $[\mu^{\text{part}}(t) - \mu^{\text{hole}}(t)]/2$. Figures 3 and 4 compare the phase-boundary curves obtained by the third-order strong-coupling expansion with those obtained by the chemical-potential fitting in the square lattice and in the triangular lattice, respectively. The chemical-potential fitting makes the phase-boundary curves smooth and natural, as expected. The results obtained by the PMFA are also presented in both figures.

For the square lattice, the QMC data show, for the $n_0 = 2$ Mott lobe, $t_C/U_0 \simeq 0.09(0.13)$ when $U_2/U_0 = 0.25(0.5)$ (as per our interpretation of Fig. 15 of Ref. [8]). Our third-order strong-coupling results and the chemical-potential fitting show $t_C/U_0 = 0.132(0.187)$ when $U_2/U_0 = 0.25(0.5)$ and $t_C/U_0 = 0.114(0.151)$ when $U_2/U_0 = 0.25(0.5)$, respectively. This shows that the chemical-potential fitting clearly improves the results obtained by the strong-coupling expansion up to third order in t. On the other hand, $t_C/U_0 = 0.084(0.113)$ is obtained by the PMFA when $U_2/U_0 = 0.25(0.5)$. Hence the results obtained by QMC methods are in between those obtained by the chemical-potential fitting and PMFA, at least in the case of the square lattice where QMC has been performed. This seems to reflect the fact that the PMFA includes a part of the infinite-order terms in t but neglects the quantum fluctuations that possibly stabilize the MI phase, whereas the chemical-potential fitting based on the strong-coupling expansion up to third order in t includes the quantum fluctuations but does

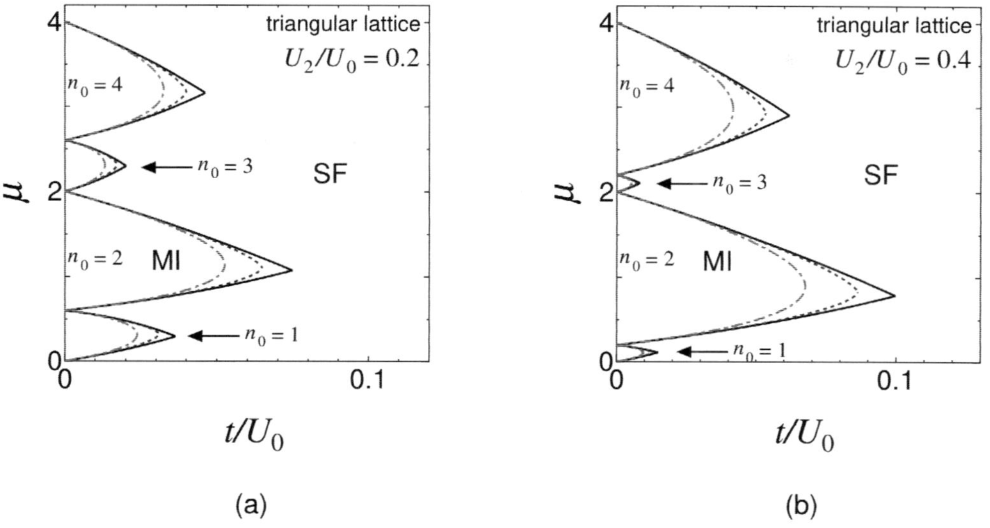

Figure 4. (Color online) Same plot as in Fig. 3 but for (a) $U_2/U_0 = 0.2$ and (b) $U_2/U_0 = 0.2$ in the triangular lattice.

not directly include the higher-order terms in t that possibly stabilize the SF phase.

Comparing the results obtained in the square lattice and those obtained in the triangular lattice, we find that t_C is roughly $6/4 = 3/2$ times greater in the square lattice than that in the triangular lattice, because of the difference between the number of nearest-neighbor sites z [$z = 4(6)$ in the square (triangular) lattice]. However, there seems to be no significant qualitative difference between the results obtained in the square lattice and those obtained in the triangular lattice. These results suggest that there the SF-MI transition has no significant dependence on the lattice structure.

4. Summary

In this paper, we present the strong-coupling expansion for the spin-1 Bose–Hubbard model for the square and triangular lattices. In both lattices, we confirm that for filling with an even number of bosons, the MI phase is considerably more stable against the SF phase than for filling with an odd number of bosons. We also fit the phase-boundary curves obtained by the strong-coupling expansion to the scaling form of the curves by using the chemical-potential fitting method. The fitting curves smooth and natural as expected, and the values of t_C are clearly improved.

Overall, there has been no essential difference in the phase boundary curves for the square and triangular lattices. This might be due to our assumption that the zeroth order in t MI state is a nematic state. However, QMC study [8] of the square lattice indicates that the magnetic structure factor shows no trace of magnetic order anywhere in the phase diagram. Thus, in contrast to the usual antiferromagnetic spin systems, there might be no significant dependence on the lattice structure for spin-1 bosons in

Journal of Physics: Conference Series **497** (2014) 012024 doi:10.1088/1742-6596/497/1/012024

S two or more dimensions.

Acknowledgments

The author acknowledges Makoto Yamashita and Yuta Toga for fruitful discussions about the spin-1 Bose–Hubbard model.

4.1. Appendices

Appendix A. Energies of Mott insulator and defect states in square and triangular lattices

By using $\Psi_{\rm even}$ and $\Psi_{\rm odd}$ [Eqs. (2) and (3)], the energy per site of the MI state in the square lattice is

$$\frac{E_{\rm square,MI,even}(n_0)}{N} = \frac{U_0}{2}n_0(n_0-1) - U_2 n_0 - n_0\mu - \frac{z_s t^2}{3}\frac{n_0(n_0+3)}{U_0+2U_2}, \quad (A.1)$$

$$\frac{E_{\rm square,MI,odd}(n_0)}{N} = \frac{U_0}{2}n_0(n_0-1) - U_2(n_0-1) - n_0\mu - \frac{1}{9}z_s t^2\Big[\frac{34}{25}\frac{(n_0+4)(n_0-1)}{U_0+4U_2} + \frac{4}{5}\frac{2n_0^2+6n_0+7}{U_0+U_2} + \frac{(n_0+1)(n_0+2)}{U_0-2U_2}\Big] \quad (A.2)$$

up to the third order in t, and where $z_s = 4$ is the number of nearest-neighbor sites in the square lattice. The corresponding MI-state energy in the triangular lattice is

$$\frac{E_{\rm triangular,MI,even}(n_0)}{N} = \frac{U_0}{2}n_0(n_0-1) - U_2 n_0 - n_0\mu - \frac{z_t t^2}{3}\frac{n_0(n_0+3)}{U_0+2U_2} - \frac{2}{9}z_t t^3\frac{n_0(n_0+3)(2n_0+3)}{(U_0+2U_2)^2}, \quad (A.3)$$

$$\begin{aligned}\frac{E_{\rm triangular,MI,odd}(n_0)}{N} &= \frac{U_0}{2}n_0(n_0-1) - U_2(n_0-1) - n_0\mu \\ &- \frac{1}{9}z_t t^2\Big[\frac{34}{25}\frac{(n_0-1)(n_0+4)}{U_0+4U_2} + \frac{4}{5}\frac{2n_0^2+6n_0+7}{U_0+U_2} + \frac{(n_0+1)(n_0+2)}{U_0-2U_2}\Big] \\ &- \frac{2}{27}z_t t^3\Big[\frac{118}{125}\frac{(n_0-1)(n_0+4)(2n_0+3)}{(U_0+4U_2)^2} + \frac{32}{25}\frac{(n_0-1)(n_0+4)(2n_0+3)}{(U_0+4U_2)(U_0+U_2)} \\ &+ \frac{4}{25}\frac{(n_0^2-1)(9n_0+1)+(n_0+2)(n_0+4)(9n_0+26)}{(U_0+U_2)^2} \\ &+ \frac{8}{5}\frac{(n_0+1)(n_0+2)(2n_0+3)}{(U_0+U_2)(U_0-2U_2)} + \frac{(n_0+1)(n_0+2)(2n_0+3)}{(U_0-2U_2)^2}\Big]\end{aligned} \quad (A.4)$$

up to the third order in t, and where $z_t = 6$ is the number of nearest-neighbor sites in the triangular lattice. Up to the second order in t, the expressions for the MI energy is the same for both lattices except for the difference between z_s and z_t. However, in contrast to the square lattice, the third-order terms in t exist for the triangular lattice.

The energies of the defect states can be obtained by using $\Psi^{\rm part}_{\rm even}$, $\Psi^{\rm hole}_{\rm even}$, $\Psi^{\rm part}_{\rm odd}$, and $\Psi^{\rm hole}_{\rm odd}$ [Eqs. (4)–(7)] up to the third order in t. For the square lattice, we obtain

$$\begin{aligned}
&E^{\rm part}_{\rm square,def,even}(n_0) - E_{\rm square,MI,even}(n_0)\\
&= U_0 n - \mu - z_s t\frac{n_0+3}{3}\\
&\quad - \frac{z_s(z_s-7)t^2}{9}\frac{n_0(n_0+3)}{U_0+2U_2} - \frac{z_s t^2 n_0}{9}\Big[2\Big(\frac{n_0+5}{2U_0+3U_2}+\frac{n_0+3}{3U_2}\Big)+\frac{n_0+2}{2U_0}\Big]\\
&\quad - \frac{z_s t^3}{27}n_0(n_0+3)\Big\{(z_s-1)\Big[\frac{(2n_0+3)z_s-3(3n_0+8)}{(U_0+2U_2)^2}\\
&\quad + \frac{2}{U_0+2U_2}\Big(2\frac{n_0+5}{2U_0+3U_2}+\frac{n_0+2}{2U_0}\Big)+\frac{4(n_0+3)}{3U_2(U_0+2U_2)}\Big]\\
&\quad - z_s\Big[2\Big(\frac{n_0+5}{(2U_0+3U_2)^2}+\frac{n_0+3}{(3U_2)^2}\Big)+\frac{n_0+2}{(2U_0)^2}\Big]+\frac{4}{3U_2}\Big(\frac{1}{5}\frac{n_0+5}{2U_0+3U_2}+\frac{n_0+2}{2U_0}\Big)\Big\}, \qquad \text{(A.5)}
\end{aligned}$$

$$\begin{aligned}
&E^{\rm hole}_{\rm square,def,even}(n_0) - E_{\rm square,MI,even}(n_0)\\
&= -U_0(n_0-1)+2U_2+\mu - z_s t\frac{n_0}{3}\\
&\quad - \frac{z_s(z_s-7)t^2}{9}\frac{n_0(n_0+3)}{U_0+2U_2} - \frac{z_s t^2(n_0+3)}{9}\Big[2\Big(\frac{n_0-2}{2U_0+3U_2}+\frac{n_0}{3U_2}\Big)+\frac{n_0+1}{2U_0}\Big]\\
&\quad - \frac{z_s t^3}{27}n_0(n_0+3)\Big\{(z_s-1)\Big[\frac{(2n_0+3)z_s-3(3n_0+1)}{(U_0+2U_2)^2}\\
&\quad + \frac{2}{U_0+2U_2}\Big(2\frac{n_0-2}{2U_0+3U_2}+\frac{n_0+1}{2U_0}\Big)+\frac{4n_0}{3U_2(U_0+2U_2)}\Big]\\
&\quad - z_s\Big[2\Big(\frac{n_0-2}{(2U_0+3U_2)^2}+\frac{n_0}{(3U_2)^2}\Big)+\frac{n_0+1}{(2U_0)^2}\Big]+\frac{4}{3U_2}\Big(\frac{1}{5}\frac{n_0-2}{2U_0+3U_2}+\frac{n_0+1}{2U_0}\Big)\Big\}, \qquad \text{(A.6)}
\end{aligned}$$

$$\begin{aligned}
&E^{\rm part}_{\rm square,def,odd}(n_0) - E_{\rm square,MI,odd}(n_0)\\
&= U_0 n_0 - 2U_2 - z_s t\frac{n_0+1}{3} - \mu\\
&\quad - \frac{z_s(z_s-3)t^2}{9}(n_0+1)\Big[\frac{n_0+2}{U_0-2U_2}+\frac{4}{5}\frac{n_0-1}{U_0+U_2}\Big]\\
&\quad - \frac{z_s t^2}{9}(n_0+4)\Big[2\frac{n_0-1}{2U_0+3U_2}+\frac{n_0+2}{2U_0}-\frac{68}{25}\frac{n_0-1}{U_0+4U_2}-\frac{8}{5}\frac{n_0+2}{U_0+U_2}\Big]\\
&\quad - \frac{2}{45}z_s(2z_s+3)t^2\frac{(n_0+1)(n_0+4)}{3U_2}\\
&\quad - \frac{z_s(z_s-1)t^3}{27}\frac{(n_0+1)(n_0+2)}{(U_0-2U_2)^2}[(2n_0+3)z_s-(5n_0+6)]\\
&\quad - \frac{4}{675}z_s(z_s-1)t^3\frac{n_0+1}{(U_0+U_2)^2}[(n_0-1)(9n_0+1)z_s-2(17n_0^2+26n_0+32)]\\
&\quad - \frac{z_s(z_s-1)^2t^3}{27}\frac{n_0+1}{U_0+U_2}\Big[\frac{32}{25}\frac{(n_0-1)(n_0+4)}{U_0+4U_2}+\frac{8}{5}\frac{(n_0+2)(2n_0+3)}{U_0-2U_2}\Big]
\end{aligned}$$

$$
\begin{aligned}
&- \frac{2}{27} z_s(z_s-1)t^3(n_0+1)(n_0+4)\Big\{\frac{n_0-1}{2U_0+3U_2}\Big[\frac{34}{25}\frac{1}{U_0+4U_2}+\frac{4}{5}\frac{1}{U_0+U_2}\Big] \\
&+ \frac{n_0+2}{2U_0}\Big[\frac{1}{U_0-2U_2}+\frac{4}{5}\frac{1}{U_0+U_2}\Big]+\frac{1}{3U_2}\Big[\frac{2}{25}(8z_s+9)\frac{n_0-1}{U_0+4U_2}+\frac{4}{5}z_s\frac{n_0+2}{U_0+U_2}\Big]\Big\} \\
&- \frac{4}{135} z_s(2z_s+3)t^3\frac{(n_0+1)(n_0+4)}{3U_2}\Big[\frac{1}{5}\frac{n_0-1}{2U_0+3U_2}+\frac{n_0+2}{2U_0}\Big] \\
&+ \frac{2}{675} z_s t^3\frac{(n_0+1)(n_0+4)}{(3U_2)^2}[2(n_0-11)z_s^2+9(3n_0+7)z_s-9(n_0+4)] \\
&+ \frac{z_s t^3}{27}(n_0+1)(n_0+4)\Big[\frac{68}{25}(z_s-1)\frac{n_0-1}{(U_0+4U_2)^2} \\
&+ 2z_s\frac{n_0-1}{(2U_0+3U_2)^2}+z_s\frac{n_0+2}{(2U_0)^2}\Big], \qquad \text{(A.7)}
\end{aligned}
$$

$$
\begin{aligned}
&E^{\text{hole}}_{\text{square,def,odd}}(n_0) - E_{\text{square,MI,odd}}(n_0) \\
=& - U_0(n_0-1) - z_s t\frac{n_0+2}{3} + \mu \\
&- \frac{z_s(z_s-3)t^2}{9}(n_0+2)\Big[\frac{n_0+1}{U_0-2U_2}+\frac{4}{5}\frac{n_0+4}{U_0+U_2}\Big] \\
&- \frac{z_s t^2}{9}(n_0-1)\Big[2\frac{n_0+4}{2U_0+3U_2}+\frac{n_0+1}{2U_0}-\frac{68}{25}\frac{n_0+4}{U_0+4U_2}-\frac{8}{5}\frac{n_0+1}{U_0+U_2}\Big] \\
&- \frac{2}{45} z_s(2z_s+3)t^2\frac{(n_0-1)(n_0+2)}{3U_2} \\
&- \frac{z_s(z_s-1)t^3}{27}\frac{(n_0+1)(n_0+2)}{(U_0-2U_2)^2}[(2n_0+3)z_s-(5n_0+9)] \\
&- \frac{4}{675} z_s(z_s-1)t^3\frac{n_0+2}{(U_0+U_2)^2}[(n_0+4)(9n_0+26)z_s-2(17n_0^2+76n_0+107)] \\
&- \frac{z_s(z_s-1)^2t^3}{27}\frac{n_0+2}{U_0+U_2}\Big[\frac{32}{25}\frac{(n_0-1)(n_0+4)}{U_0+4U_2}+\frac{8}{5}\frac{(n_0+1)(2n_0+3)}{U_0-2U_2}\Big] \\
&- \frac{2}{27} z_s(z_s-1)t^3(n_0-1)(n_0+2)\Big\{\frac{n_0+4}{2U_0+3U_2}\Big[\frac{34}{25}\frac{1}{U_0+4U_2}+\frac{4}{5}\frac{1}{U_0+U_2}\Big] \\
&+ \frac{n_0+1}{2U_0}\Big[\frac{1}{U_0-2U_2}+\frac{4}{5}\frac{1}{U_0+U_2}\Big]+\frac{1}{3U_2}\Big[\frac{2}{25}(8z_s+9)\frac{n_0+4}{U_0+4U_2}+\frac{4}{5}z_s\frac{n_0+1}{U_0+U_2}\Big]\Big\} \\
&- \frac{4}{135} z_s(2z_s+3)t^3\frac{(n_0-1)(n_0+2)}{3U_2}\Big[\frac{1}{5}\frac{n_0+4}{2U_0+3U_2}+\frac{n_0+1}{2U_0}\Big] \\
&+ \frac{2}{675} z_s t^3\frac{(n_0-1)(n_0+2)}{(3U_2)^2}[2(n_0+14)z_s^2+9(3n_0+2)z_s-9(n_0-1)] \\
&+ \frac{z_s t^3}{27}(n_0-1)(n_0+2)\Big[\frac{68}{25}(z_s-1)\frac{n_0+4}{(U_0+4U_2)^2} \\
&+ 2z_s\frac{n_0+4}{(2U_0+3U_2)^2}+z_s\frac{n_0+1}{(2U_0)^2}\Big]. \qquad \text{(A.8)}
\end{aligned}
$$

For the triangular lattice, we obtain

$$
\begin{aligned}
&E^{\text{part}}_{\text{triangular,def,even}}(n_0) - E_{\text{triangular,MI,even}}(n_0) \\
&= U_0 n - \mu - z_t t\frac{n_0+3}{3}
\end{aligned}
$$

$$
\begin{aligned}
&- \frac{z_t(z_t-7)t^2}{9}\frac{n_0(n_0+3)}{U_0+2U_2} - \frac{z_t t^2 n_0}{9}\Big[2\Big(\frac{n_0+5}{2U_0+3U_2}+\frac{n_0+3}{3U_2}\Big)+\frac{n_0+2}{2U_0}\Big] \\
&- \frac{z_t t^3}{27} n_0\Big\{\frac{n_0+3}{(U_0+2U_2)^2}[(2n_0+3)z_t^2 - 2(4n_0+9)z_t - 3(17n_0+32)] \\
&- 2\frac{n_0+5}{(2U_0+3U_2)^2}[(n_0+3)z_t - 2n_0] - \frac{n_0+2}{(2U_0)^2}[(n_0+3)z_t - 2n_0] - 2(z_t-2)\frac{(n_0+3)^2}{(3U_2)^2} \\
&+ 2(z_t+1)\frac{n_0+3}{U_0+2U_2}\Big[2\frac{n_0+5}{2U_0+3U_2}+\frac{n_0+2}{2U_0}\Big] \\
&+ 4\frac{n_0+3}{3U_2}\Big[\frac{(n_0+3)z+n_0-3}{U_0+2U_2}+\frac{1}{5}\frac{n_0+5}{2U_0+3U_2}+\frac{n_0+2}{2U_0}\Big]\Big\} \qquad \text{(A.9)}
\end{aligned}
$$

$$
\begin{aligned}
&E^{\text{hole}}_{\text{triangular,def,even}}(n_0) - E_{\text{triangular,MI,even}}(n_0) \\
={}& - U_0(n_0-1) + 2U_2 + \mu - z_t t\frac{n_0}{3} \\
&- \frac{z_t(z_t-7)t^2}{9}\frac{n_0(n_0+3)}{U_0+2U_2} - \frac{z_t t^2(n_0+3)}{9}\Big[2\Big(\frac{n_0-2}{2U_0+3U_2}+\frac{n_0}{3U_2}\Big)+\frac{n_0+1}{2U_0}\Big] \\
&- \frac{z_t t^3}{27}(n_0+3)\Big\{\frac{n_0}{(U_0+2U_2)^2}[(2n_0+3)z_t^2 - 2(4n_0+3)z_t - 3(17n_0+19)] \\
&- 2\frac{n_0-2}{(2U_0+3U_2)^2}[n_0 z_t - 2(n_0+3)] - \frac{n_0+1}{(2U_0)^2}[n_0 z_t - 2(n_0+3)] - 2(z_t-2)\frac{n_0^2}{(3U_2)^2} \\
&+ 2(z_t+1)\frac{n_0}{U_0+2U_2}\Big[2\frac{n_0-2}{2U_0+3U_2}+\frac{n_0+1}{2U_0}\Big] \\
&+ 4\frac{n_0}{3U_2}\Big[\frac{n_0 z+n_0+6}{U_0+2U_2}+\frac{1}{5}\frac{n_0-2}{2U_0+3U_2}+\frac{n_0+1}{2U_0}\Big]\Big\} \qquad \text{(A.10)}
\end{aligned}
$$

$$
\begin{aligned}
&E^{\text{part}}_{\text{triangular,def,odd}}(n_0) - E_{\text{triangular,MI,odd}}(n_0) \\
={}& U_0 n_0 - 2U_2 - z_t t\frac{n_0+1}{3} - \mu \\
&- \frac{z_t(z_t-3)t^2}{9}(n_0+1)\Big[\frac{n_0+2}{U_0-2U_2}+\frac{4}{5}\frac{n_0-1}{U_0+U_2}\Big] \\
&- \frac{z_t t^2}{9}(n_0+4)\Big[2\frac{n_0-1}{2U_0+3U_2}+\frac{n_0+2}{2U_0}-\frac{68}{25}\frac{n_0-1}{U_0+4U_2}-\frac{8}{5}\frac{n_0+2}{U_0+U_2}\Big] \\
&- \frac{2}{45}z_t(2z_t+3)t^2\frac{(n_0+1)(n_0+4)}{3U_2} \\
&+ \frac{2z_t t^3}{3375}\frac{(n_0-1)(n_0+4)}{(U_0+4U_2)^2}[85(n_0+1)z_t + 2(524n_0+701)] \\
&+ \frac{4}{675}\frac{z_t t^3}{(U_0+U_2)^2}\Big\{(n_0^2-1)[-(9n_0+1)z_t^2 + 4(7n_0+3)z_t + 8(11n_0+4)] \\
&+ (n_0+2)(n_0+4)[5(n_0+1)z_t + 2(37n_0+88)]\Big\} \\
&+ \frac{z_t t^3}{27}\frac{(n_0+1)(n_0+2)}{(U_0-2U_2)^2}[-(2n_0+3)z_t^2 + 2(3n_0+4)z_t + 19n_0 + 26] \\
&+ \frac{2z_t t^3}{675}\frac{(n_0-1)(n_0+4)}{(2U_0+3U_2)^2}[25(n_0+1)z_t - 34(n_0-1)]
\end{aligned}
$$

$$
\begin{aligned}
&+\frac{z_t t^3}{27}\frac{(n_0+2)(n_0+4)}{(2U_0)^2}[(n_0+1)z_t-2(n_0+2)]\\
&+\frac{2z_t t^3}{675}\frac{(n_0+1)(n_0+4)}{(3U_2)^2}[2(n_0-11)z_t^2+9(3n_0+7)z_t-27(n_0+4)]\\
&-\frac{32z_t t^3}{675}\frac{(n_0-1)(n_0+4)}{(U_0+4U_2)(U_0+U_2)}[(n_0+1)z_t^2-2(n_0+1)z_t-(13n_0+19)]\\
&-\frac{8}{135}z_t(z_t^2-2z_t-7)t^3\frac{(n_0+1)(n_0+2)(2n_0+3)}{(U_0+U_2)(U_0-2U_2)}-\frac{16z_t t^3}{135}\frac{(n_0-1)(n_0+2)(n_0+4)}{2U_0(2U_0+3U_2)}\\
&-\frac{68z_t t^3}{675}\frac{(n_0-1)(n_0+4)}{(2U_0+3U_2)(U_0+4U_2)}[(n_0+1)z_t+n_0+7]\\
&-\frac{8}{135}z_t(z_t+1)t^3\frac{(n_0^2-1)(n_0+4)}{(2U_0+3U_2)(U_0+U_2)}-\frac{8z_t t^3}{135}\frac{(n_0+2)(n_0+4)}{2U_0(U_0+U_2)}[(n_0+1)z_t+n_0+7]\\
&-\frac{2}{27}z_t(z_t+1)t^3\frac{(n_0+1)(n_0+2)(n_0+4)}{2U_0(U_0-2U_2)}-\frac{4}{675}z_t(8z_t^2+z_t+9)t^3\frac{(n_0^2-1)(n_0+4)}{3U_2(U_0+4U_2)}\\
&-\frac{8}{135}z_t^2(z_t-1)t^3\frac{(n_0+1)(n_0+2)(n_0+4)}{3U_2(U_0+U_2)}-\frac{4}{675}z_t(2z_t+3)t^3\frac{(n_0^2-1)(n_0+4)}{3U_2(2U_0+3U_2)}\\
&-\frac{4}{135}z_t(2z_t+3)t^3\frac{(n_0+1)(n_0+2)(n_0+4)}{3U_2\cdot 2U_0}
\end{aligned}
\qquad (A.11)
$$

$$
\begin{aligned}
&E^{\text{hole}}_{\text{triangular,def,odd}}(n_0)-E_{\text{triangular,MI,odd}}(n_0)\\
=&-U_0(n_0-1)-z_t t\frac{n_0+2}{3}+\mu\\
&-\frac{z_t(z_t-3)t^2}{9}(n_0+2)\Big[\frac{n_0+1}{U_0-2U_2}+\frac{4}{5}\frac{n_0+4}{U_0+U_2}\Big]\\
&-\frac{z_t t^2}{9}(n_0-1)\Big[2\frac{n_0+4}{2U_0+3U_2}+\frac{n_0+1}{2U_0}-\frac{68}{25}\frac{n_0+4}{U_0+4U_2}-\frac{8}{5}\frac{n_0+1}{U_0+U_2}\Big]\\
&-\frac{2}{45}z_t(2z_t+3)t^2\frac{(n_0-1)(n_0+2)}{3U_2}\\
&+\frac{2z_t t^3}{3375}\frac{(n_0-1)(n_0+4)}{(U_0+4U_2)^2}[85(n_0+2)z_t+2(524n_0+871)]\\
&+\frac{4}{675}\frac{z_t t^3}{(U_0+U_2)^2}\Big\{(n_0+2)(n_0+4)[-(9n_0+26)z_t^2+4(7n_0+18)z_t+8(11n_0+29)]\\
&+(n_0^2-1)[5(n_0+2)z_t+2(37n_0+23)]\Big\}\\
&+\frac{z_t t^3}{27}\frac{(n_0+1)(n_0+2)}{(U_0-2U_2)^2}[-(2n_0+3)z_t^2+2(3n_0+5)z_t+19n_0+31]\\
&+\frac{2z_t t^3}{675}\frac{(n_0-1)(n_0+4)}{(2U_0+3U_2)^2}[25(n_0+2)z_t-34(n_0+4)]\\
&+\frac{z_t t^3}{27}\frac{n_0^2-1}{(2U_0)^2}[(n_0+2)z_t-2(n_0+1)]\\
&+\frac{2z_t t^3}{675}\frac{(n_0-1)(n_0+2)}{(3U_2)^2}[2(n_0+14)z_t^2+9(3n_0+2)z_t-27(n_0-1)]
\end{aligned}
$$

$$
\begin{aligned}
&-\frac{32z_t t^3}{675}\frac{(n_0-1)(n_0+4)}{(U_0+4U_2)(U_0+U_2)}[(n_0+2)z_t^2-2(n_0+2)z_t-(13n_0+20)]\\
&-\frac{8}{135}z_t(z_t^2-2z_t-7)t^3\frac{(n_0+1)(n_0+2)(2n_0+3)}{(U_0+U_2)(U_0-2U_2)}-\frac{16z_t t^3}{135}\frac{(n_0^2-1)(n_0+4)}{2U_0(2U_0+3U_2)}\\
&-\frac{68z_t t^3}{675}\frac{(n_0-1)(n_0+4)}{(2U_0+3U_2)(U_0+4U_2)}[(n_0+2)z_t+n_0-4]\\
&-\frac{8}{135}z_t(z_t+1)t^3\frac{(n_0-1)(n_0+2)(n_0+4)}{(2U_0+3U_2)(U_0+U_2)}-\frac{8z_t t^3}{135}\frac{n_0^2-1}{2U_0(U_0+U_2)}[(n_0+2)z_t+n_0-4]\\
&-\frac{2}{27}z_t(z_t+1)t^3\frac{(n_0^2-1)(n_0+2)}{2U_0(U_0-2U_2)}-\frac{4}{675}z_t(8z_t^2+z_t+9)t^3\frac{(n_0-1)(n_0+2)(n_0+4)}{3U_2(U_0+4U_2)}\\
&-\frac{8}{135}z_t^2(z_t-1)t^3\frac{(n_0^2-1)(n_0+2)}{3U_2(U_0+U_2)}-\frac{4}{675}z_t(2z_t+3)t^3\frac{(n_0-1)(n_0+2)(n_0+4)}{3U_2(2U_0+3U_2)}\\
&-\frac{4}{135}z_t(2z_t+3)t^3\frac{(n_0^2-1)(n_0+2)}{3U_2\cdot 2U_0}
\end{aligned}\qquad \text{(A.12)}
$$

As for the MI energy, the expressions for the MI energy up to the second order in t are the same for both lattices except for the difference between z_s and z_t but the third-order terms in t are different from each lattice. This difference between the third-order terms originates from particles or holes that can (cannot) return to their original site through three hopping processes in the triangular (square) lattice.

References

[1] Stamper-Kurn D M and Ueda M 2013 *Rev. Mod. Phys.* **85** 1191; ewenstein M, Sanpera A and Ahufinger V 2012 *Ultracold Atoms in Optical Lattices: Simulating Quantum Many-Body Systems* (Oxford Univ. Pr.)
[2] Katsura H and Tasaki H 2013 *Phys. Rev. Lett.* **110** 130405
[3] Tsuchiya S, Kurihara S and Kimura T 2004 *Phys. Rev. A* **70** 043628
[4] Rizzi M, Rossini D, De Chiara G, Montangero S and Fazio R 2005 *Phys. Rev. Lett.* **95** 240404
[5] Batrouni G G, Rousseau V G and Scalettar R T 2009 *Phys. Rev. Lett.* **102** 140402
[6] Apaja V and Syljuåsen O F 2006 *Phys. Rev. A* **74** 035601
[7] Toga Y, Tsuchiura H, Yamashita M, Inaba K and Yokoyama H 2012 *J. Phys. Soc. Jpn.* **81** 063001
[8] de Forges de Parny , Hébert F, Rousseau V G and Batrouni G G arXiv:1306.5515
[9] Kimura T, Tsuchiya S and Kurihara S 2005 *Phys. Rev. Lett.* **94** 110403
[10] Krutitsky K V and Graham R 2004 *Phys. Rev. A* **70** 063610
[11] Krutitsky K V, Timmer M and Graham R 2005 *Phys. Rev. A* **71** 033623
[12] Freericks K and Monien H 1996 *Phys. Rev. B* **53** 2691; Freericks K and Monien H 1994 *Europhys. Lett.* **26**, 545
[13] Freericks K, Krishnamurthy H R, Kato Y, Kawashima N and Trivedi N 2009 *Phys. Rev. A* **79** 053631
[14] Iskin M and Freericks K 209 *Phys. Rev. A* **79** 053634
[15] Hen I, Iskin M and Rigol M 2010 *Phys. Rev. B* **81** 064503
[16] Iskin M 2010 *Phys. Rev. A* **82** 033630
[17] Kimura T 2013 *Phys. Rev. A* **87** 043624

Journal of Physics: Conference Series **497** (2014) 012025 doi:10.1088/1742-6596/497/1/012025

Vortices in the two-dimensional dipolar Bose gas

B. C. Mulkerin[1], D. H. J. O'Dell[2], A. M. Martin[1] and N. G. Parker[3]

[1] School of Physics, University of Melbourne, Victoria 3010, Australia

[2] Department of Physics and Astronomy, McMaster University, Hamilton, Ontario, L8S 4M1, Canada

[3] Joint Quantum Centre Durham–Newcastle, School of Mathematics and Statistics, Newcastle University, Newcastle upon Tyne, NE1 7RU, United Kingdom

E-mail: nick.parker@newcastle.ac.uk

Abstract. We present vortex solutions for the homogeneous two-dimensional Bose-Einstein condensate featuring dipolar atomic interactions, mapped out as a function of the dipolar interaction strength (relative to the contact interactions) and polarization direction. Stable vortex solutions arise in the regimes where the fully homogeneous system is stable to the phonon or roton instabilities. Close to these instabilities, the vortex profile differs significantly from that of a vortex in a nondipolar quantum gas, developing, for example, density ripples and an anisotropic core. Meanwhile, the vortex itself generates a mesoscopic dipolar potential which, at distance, scales as $1/\rho^2$ and has an angular dependence which mimics the microscopic dipolar interaction.

1. Introduction

A landmark experiment in quantum gas research was the achievement in 2005 of a dipolar Bose-Einstein condensate (BEC) made of ^{52}Cr atoms [1]. Other groups have subsequently used novel methods to make ^{52}Cr condensates [2], as well as dipolar condensates composed of ^{164}Dy [3] and ^{168}Er [4]. The dipolar nature of these atoms arises from their sizeable magnetic dipole moment which for ^{52}Cr is $d = 6\mu_B$, for ^{168}Er is $d = 7\mu_B$, and for ^{164}Dy is $d = 10\mu_B$, where μ_B denotes the Bohr magneton. These numbers should be compared to the value $d = 1\mu_B$ found in the alkali atoms which are the standard workhorses used in most ultracold atomic experiments. While the dominant interatomic interactions in alkali atom condensates are the isotropic van der Waals interactions which fall off as $1/r^6$, the effects of magnetic dipole-dipole interactions, which are anisotropic and fall off as $1/r^3$, have been seen in the BECs composed of the more strongly magnetic atoms listed above (note that the strength of the magnetic dipole-dipole interaction goes as d^2). In particular, magnetostriction of a condensate has been observed [5] as well as d-wave collapse and explosion [6]. Of particular relevance to this paper are theoretical studies of dipolar condensates which predict roton excitations [7, 8, 9], akin to those observed in Helium II. Although the experimental observation of rotons in dipolar BECs with static dipole-dipole interactions remains illusive, their effects have been seen in BECs with laser-induced dipole-dipole interactions [10]. For a comprehensive review of the field of dipolar BECs up to 2009 we refer the reader to the article [11].

Despite the enhanced dipole-dipole interactions in ^{52}Cr relative to alkali atoms, the dipolar interactions in ^{52}Cr still play a secondary role compared to the van der Waals interactions unless something is done to suppress the latter. Fortuitously, this has proved possible using

Journal of Physics: Conference Series **497** (2014) 012025 doi:10.1088/1742-6596/497/1/012025

the technique of Feshbach resonance [12] which has allowed the creation of a purely dipolar condensate [13]. This method has also been applied to ^{168}Er [4]. Polar molecules are another promising system for observing (electric) dipolar interactions due to their potentially very large electric dipole moments. These systems are now close to quantum degeneracy [14].

The quantum-coherent nature of BECs means that they support vortices characterised by a core of vanishing density and a 2π-quantized singularity of the quantum mechanical phase. Diverse vortex structures have been generated in BECs, from single vortices to vortex rings and vortex pairs, to vortex lattices and turbulent states [15, 16]. While vortices have not yet been observed in dipolar BECs, simulations indicate that vortex rings were formed in the *d*-wave collapse experiment [6]. Theoretical work has also revealed that dipolar interactions modify the vortex properties, introducing density ripples about the core and an elliptical core [17, 18, 19, 20], as well as affecting the critical rotation frequency for vortex generation [21] and the associated dynamical instability [22, 23, 24]. The critical obstacle velocity for vortex nucleation is likewise modified [25] as well as the structure of vortex lattices [26].

Here we examine the single vortex solutions in an infinite two-dimensional dipolar Bose gas where the dipoles are aligned by an external field, expanding on some aspects of our recent work [20]. We begin by establishing the stability regimes of the homogeneous 2D system. We then explore the vortex solutions as a function of dipolar strength and polarization angle, and end with an analysis of the dipolar potential generated by the vortex.

2. Mean-Field Model

We consider a dilute BEC composed of atoms of mass m with dipole moment d. These atomic dipoles are polarized in a common direction by an external magnetic field, with the polarization axis taken to lie at angle α to the z-axis and in the xz plane. For the purposes of Gross-Pitaevskii mean-field theory the total interaction between two atoms at $\mathbf{r}$ and $\mathbf{r}'$ can be represented by the potential [11, 27],

$$U(\mathbf{r}-\mathbf{r}') = g_{3\mathrm{D}}\delta(\mathbf{r}-\mathbf{r}') + \frac{C_{\mathrm{dd}}}{4\pi}\frac{1-3\cos^2\theta}{|\mathbf{r}-\mathbf{r}'|^3}. \tag{1}$$

The first term is a pseudo-potential which models the van der Waals interactions through a contact potential with coupling strength $g_{3\mathrm{D}} = 4\pi\hbar^2 a_\mathrm{s}/m$, where a_s is the s-wave scattering length. The second term is the bare dipole-dipole interaction, where θ is the angle between the polarization direction and the inter-particle vector $\mathbf{r}-\mathbf{r}'$. For magnetic dipoles the dipolar coupling strength is $C_{\mathrm{dd}} = \mu_0 d^2$, where μ_0 is the permeability of free space.

Working in the limit of zero temperature, we describe the condensate by the mean-field wavefunction $\Psi(\mathbf{r},t)$, which specifies the atomic density $n(\mathbf{r},t) = |\Psi(\mathbf{r},t)|^2$. The wavefunction obeys the dipolar Gross-Pitaevskii equation,

$$i\hbar\frac{\partial\Psi}{\partial t} = \left[-\frac{\hbar^2}{2m}\nabla^2 + V(\mathbf{r}) + g_{3\mathrm{D}}|\Psi|^2 + \Phi_{3\mathrm{D}}(\boldsymbol{r},t) - \mu_{3\mathrm{D}}\right]\Psi, \tag{2}$$

where $V(\mathbf{r})$ is the external potential acting on the condensate and $\mu_{3\mathrm{D}}$ is the chemical potential. The van der Waals interactions lead to the purely local term $g_{3\mathrm{D}}|\Psi|^2$ in the above Hamiltonian, while the dipolar interactions are incorporated via the non-local potential $\Phi_{3\mathrm{D}}$ given by

$$\Phi_{3\mathrm{D}}(\boldsymbol{r},t) = \int U_{\mathrm{dd}}(\boldsymbol{r}-\boldsymbol{r}')n(\boldsymbol{r}',t)\,d\boldsymbol{r}', \tag{3}$$

where U_{dd} is the dipolar interaction potential, the second term in Eq. (1).

The reduction to an effectively 2D BEC, homogeneous in one plane, is performed by considering the external potential to be harmonic trapping in the axial direction only, $V(\mathbf{r}) =$

$(1/2)m\omega_z^2 z^2$, where ω_z is the corresponding trap frequency. The axial trapping is assumed to be so strong that Ψ becomes frozen into the axial ground harmonic state, a Gaussian of width $l_z = \sqrt{\hbar/m\omega_z}$. Formally, this "2D mean-field regime" is reached if $(1 + 2\varepsilon_{\rm dd})a_s l_z n \ll 1$, where n is the average 2D density [28]. This scenario of an effectively 2D homogeneous dipolar BEC is illustrated in Fig. 1.

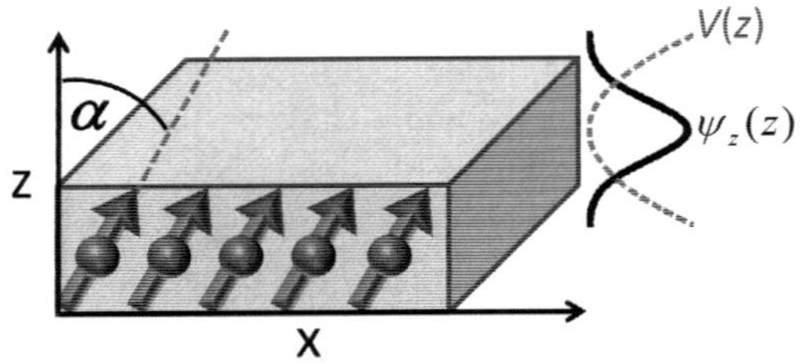

Figure 1. Schematic of our homogeneous 2D dipolar gas. The dipoles are taken to be polarized at angle α to the z-axis in the xz-plane.

Within the 2D mean-field regime one can decompose the wavefunction according to $\Psi(\mathbf{r},t) = \psi(\boldsymbol{\rho},t)\psi_z(z)$, where ψ is the 2D wavefunction in the xy-plane. Integrating over z then leads to the effective 2D dipolar GPE [25], equivalent to Eq. (2) with the following 3D quantities replaced by their 2D analogs: $\mu_{\rm 3D} \mapsto \mu$, $g_{\rm 3D} \mapsto g = g_{\rm 3D}/\sqrt{2\pi}l_z$, $\Phi_{\rm 3D} \mapsto \Phi$ and $\Psi(\mathbf{r},t) \mapsto \psi(\boldsymbol{\rho},t)$. The reduced dipolar potential can be expressed as,

$$\Phi(\boldsymbol{\rho},t) = \int U_{\rm dd}^{\rm 2D}(\boldsymbol{\rho}-\boldsymbol{\rho}')n(\boldsymbol{\rho}',t)\, d\boldsymbol{\rho}', \tag{4}$$

where $U_{\rm dd}^{\rm 2D}$ is the effective 2D dipolar interaction potential. Rather than evaluating Φ directly through Eq. (4), it is more convenient to evaluate it using the convolution theorem as $\Phi(\boldsymbol{\rho},t) = \mathcal{F}^{-1}\left[\tilde{U}_{\rm dd}^{\rm 2D}(\mathbf{k})\tilde{n}(\mathbf{k},t)\right]$. Here the Fourier transform of $U_{\rm dd}^{\rm 2D}$ has been shown elsewhere to be [25, 30],

$$\tilde{U}_{\rm dd}^{\rm 2D}(\mathbf{q}) = \frac{4\pi g_{\rm dd}}{3}\left[F_{\parallel}(\mathbf{q})\sin^2\alpha + F_{\perp}(\mathbf{q})\cos^2\alpha\right], \tag{5}$$

with $g_{\rm dd} = C_{\rm dd}/3\sqrt{2\pi}l_z$, $\mathbf{q} = \mathbf{k}l_z/\sqrt{2}$, $F_{\parallel}(\mathbf{q}) = -1 + 3\sqrt{\pi}\frac{q_x^2}{q}e^{q^2}\mathrm{erfc}(q)$ and $F_{\perp}(\mathbf{q}) = 2 - 3\sqrt{\pi}qe^{q^2}\mathrm{erfc}(q)$. The axial width of the condensate l_z plays a more important role for 2D dipolar condensates than their non-dipolar counterparts since it specifies how much of the out-of-plane dipolar interaction is experienced by the condensate.

The strength of the dipolar interactions can be parameterized via the ratio [11]

$$\varepsilon_{\rm dd} = g_{\rm dd}/g\ . \tag{6}$$

We note that g can be tuned between $-\infty$ and $+\infty$ via a Feshbach resonance while $g_{\rm dd}$ can be varied below its natural value, including to negative values, via magnetic field rotation [11]. Hence it is feasible to consider $-\infty < \varepsilon_{\rm dd} < \infty$ and both negative and positive $g_{\rm dd}$.

Numerical solutions are obtained by propagating the 2D dipolar GPE in imaginary time using a Runga-Kutta algorithm on a 256^2 grid extending over a region $[-50,50]\xi_0 \times [-50,50]\xi_0$, where ξ_0 is the healing length (defined in the following section). At the boundary of the numerical box, the value of the wavefunction is fixed to the uniform value to mimic an infinite system (Dirichlet boundary conditions).

3. The Homogeneous 2D Dipolar Gas

In the absence of any potential in the plane the stable ground state solutions have uniform density, which we denote n_0. This generates a uniform dipolar potential given by $\Phi_0 = n_0 g_{\rm dd}(3\cos^2\alpha - 1)$. The dipolar GPE then reduces to [25],

$$\mu_0 = n_0\left(g + g_{\rm dd}[3\cos^2\alpha - 1]\right), \tag{7}$$

where μ_0 is the chemical potential of the homogeneous system. This relation is similar to the result for a homogeneous 2D non-dipolar BEC $\mu = n_0 g$, but where the dipolar interactions effectively introduce an additional contribution to the contact interaction $g_{\rm dd}[3\cos^2\alpha - 1]$. The angular dependence accounts for the "tilt" of the dipoles. Taking the dipoles to be conventional in the sense that $g_{\rm dd} > 0$, then for $\alpha = 0$ the dipoles lie side-by-side and experience the repulsive part of the dipole-dipole interaction, while for $\alpha = \pi/2$, the dipoles lie head-to-tail in the plane and experience the attractive part of the interaction. For $g_{\rm dd} < 0$, the sign of the dipole-dipole interaction becomes reversed, such that the $\alpha = 0$ arrangement of dipoles becomes *attractive* and the $\alpha = \pi/2$ arrangement becomes *repulsive*. Note that, at the "magic angle" $\alpha_0 = \arccos(1/\sqrt{3}) \approx 54.7°$, the in-plane dipolar interaction is zero (for either sign of $g_{\rm dd}$).

A natural length scale for the homogeneous system is the dipolar healing length, $\xi_0 = \hbar/\sqrt{m\mu_0}$. In our analysis, we set the axial length scale of the condensate to be $l_z = 0.5\xi_0$, which satisfies the criteria for the system to be two-dimensional $l_z/\xi_0 < 1$.

3.1. The phonon instability

Perturbations in the density and velocity of the condensate, with frequency ω and wavevector k, satisfy the Bogoluibov excitation spectrum [11],

$$E(k) = \hbar k\sqrt{\frac{\mu_0}{m} + \frac{\hbar^2 k^2}{4m^2}}.$$

When μ_0 is less than zero (i.e. the net local contract interactions are attractive), then in the phonon limit $k \to 0$, the perturbations possess imaginary energies, signifying the unstable exponential growth of the perturbations over time. This is the *phonon instability*, familiar from the conventional condensate with attractive contact interactions [29].

From Eq. (7), the instability condition $\mu_0 < 0$ means that [8, 30, 11, 33],

$$g + g_{\rm dd}[3\cos^2\alpha - 1] < 0. \tag{8}$$

Let us consider the implications for the cases of $g > 0$ and $g < 0$ separately, with the results summarized in Fig. 2(a) and (b) respectively. For $g > 0$ the system undergoes the phonon instability (PI) when the in-plane dipolar contact interactions $g_{\rm dd}[3\cos^2\alpha - 1]$ are attractive (when either $g_{\rm dd} > 0$ and $\alpha > \alpha_0$, or $g_{\rm dd} < 0$ and $\alpha < \alpha_0$) and of sufficient magnitude to overcome the repulsive vdW contact interactions (characterized by g). Dividing Eq. (8) by g (> 0) and introducing $\varepsilon_{\rm dd} = g_{\rm dd}/g$ we obtain the condition for the phonon instability to be $1 > -\varepsilon_{\rm dd}[3\cos^2\alpha - 1]$. We define the threshold value of $\varepsilon_{\rm dd}$ as $\varepsilon_{\rm dd}^{\rm PI} = [1 - 3\cos^2\alpha]^{-1}$ (red lines in Fig. 2). Note the importance of the magic angle α_0, at which the instability threshold diverges. It then follows that (for $g > 0$) the phonon instability arises for $\varepsilon_{\rm dd} < \varepsilon_{\rm dd}^{\rm PI}$ when $\alpha < \alpha_0$, and for $\varepsilon_{\rm dd} > \varepsilon_{\rm dd}^{\rm PI}$ when $\alpha > \alpha_0$, as depicted in Fig. 2(a) (pink shaded regions).

Meanwhile, for $g < 0$ the system will be stable to the phonon instability when in-plane dipolar interactions are repulsive and of greater magnitude than the attractive vdW interactions. The instability criterion is then $\varepsilon_{\rm dd} > \varepsilon_{\rm dd}^{\rm PI}$ for $\alpha < \alpha_0$ and $\varepsilon_{\rm dd} < \varepsilon_{\rm dd}^{\rm PI}$ for $\alpha > \alpha_0$, as depicted by in Fig. 2(b) (pink shaded regions).

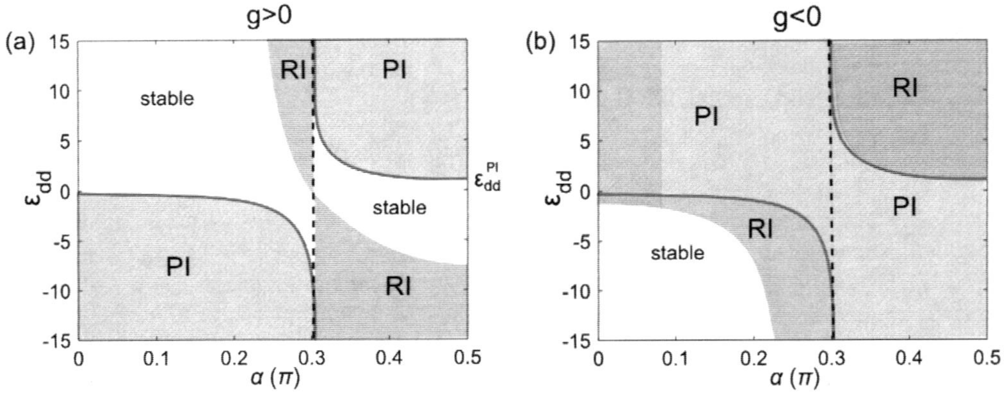

Figure 2. Stability diagram in $\varepsilon_{\rm dd} - \alpha$ space for the homogeneous 2D dipolar BEC ($\sigma = 0.5$) with (a) $g > 0$ and (b) $g < 0$. Shown are the regions of stability (white), phonon instability (pink) and roton instability (blue). The threshold for the phonon instability $\varepsilon_{\rm dd}^{\rm PI} = [1-3\cos^2\alpha]^{-1}$ is shown by the red lines. The vertical dashed line indicates the magic angle α_0. Note that the plots only extend up to $\alpha = \pi/2$; beyond this the results are the mirror image of the presented region.

3.2. The roton instability

In addition to the phonon instability, a dipolar BEC also experiences the roton instability. Dipolar interactions in a BEC can induce a roton dip in the excitation spectrum at finite momenta [7, 8, 18], reminiscent of rotons in superfluid Helium [35]. For certain parameters, the roton "softens" to zero energy, inducing a collapse instability at finite momentum, associated with the runaway alignment of dipoles in the head-to-tail configuration (for $g_{\rm dd} > 0$) or side-by-side (for $g_{\rm dd} < 0$). Close to the RI, ripples in the BEC density arise when the roton mode mixes into the ground state [17, 18].

We assess the regimes of roton stability/instability by numerically seeking solutions of the 2D dipolar GPE where the initial condition is a homogeneous wavefunction modulated with sinusoidal perturbations in x and y, $\psi(\boldsymbol{\rho}, t) = \sqrt{n_0}[1 + 0.2\sin(x/\xi_0) + 0.2\sin(y/\xi_0)]$. This perturbation facilitates the development of the roton instability, where it exists. Where stationary solutions are not obtained (and the system is in a regime which does not suffer the phonon instability) it is deemed to possess the roton instability. The roton instability regimes are depicted in the space of $\varepsilon_{\rm dd} - \alpha$ in Fig. 2(a) and (b) (blue shaded regions).

Again, we interpret the $g > 0$ and $g < 0$ cases separately. For $g > 0$ the roton instability is induced by the attractive part of the dipolar interaction. For $\alpha = 0$ (dipoles polarized parallel to the z-axis) the condensate, confined to a narrow region about $z = 0$, cannot probe this attractive part of $U_{\rm dd}$ and no roton instability exists [30]. For increased polarization angle α (while still in the regime $\alpha < \alpha_0$), the attractive part of $U_{\rm dd}$ can be experienced by the condensate and so a regime of roton instability does occur. This falls in the parameter regime $\alpha \gtrsim 0.23$ and $\varepsilon_{\rm dd} > 0$ (i.e. $g_{\rm dd} > 0$). Meanwhile, for $\alpha > \alpha_0$, a region of roton instability occurs for negative $\varepsilon_{\rm dd}$, i.e. $g_{\rm dd} = g\varepsilon_{\rm dd} < 0$. Although the dipoles are approaching an head-to-tail configuration, since $g_{\rm dd} < 0$ the dipoles *repel* in the plane and *attract* axially, with the latter attraction inducing the roton instability.

Now we consider the roton instability for $g < 0$. For $\alpha = 0$ the presence of attractive contact interactions can support a roton and an associated roton instability [34]. In contrast to the $g > 0$ case, for $g < 0$ the roton instability exists for all α (barring the magic angle). For $\alpha < \alpha_0$ the roton instability occurs in a band of ε_{dd} in the negative ε_{dd} half-plane, i.e. $g_{dd} = g\varepsilon_{dd} > 0$. As for $\alpha = 0$, the roton instability here is induced by the attractive contact interactions. As ε_{dd} is made increasingly negative (corresponding to increasingly positive g_{dd}) the roton instability is removed due to the stabilizing effect of the large, repulsive in-plane dipolar interactions. For $\alpha > \alpha_0$, a region of roton instability occurs for positive ε_{dd}, i.e. negative g_{dd}. As for $g > 0$, $\varepsilon_{dd} < 0$ case discussed above, the roton instability here is driven by the attractive axial component of the dipolar interactions.

Looking at the overall stability/instability of the system depicted in Fig. 2, it is noteworthy that, for $g > 0$, regimes of stability exist for all α, while for $g < 0$ the only stable regime lies for small α and negative ε_{dd}.

3.3. Role of σ

It follows from the analysis in Section 3.1 that the regimes of phonon stability/instability, and the threshold between them, is independent of the axial size of the condensate, σ. However, the roton instability is affected by σ. For a narrower (wider) condensate, the regimes of roton instability in the phase diagram shrink (expand). This is because the atoms experience less (more) of the out-of-plane component of U_{dd}. To illustrate the quantitative dependence, take for example $\alpha = 0$. Denoting the threshold for the roton instability as ε_{dd}^{RI} the solutions are unstable for $\varepsilon_{dd}^{RI}(\sigma) < \varepsilon_{dd} < \varepsilon_{dd}^{PI}$, where $\varepsilon_{dd}^{PI} = -0.5$. For $\sigma = 0.2$, 0.5 and 1 then $\varepsilon_{dd}^{RI} \approx -0.7$, -1.2 and -2.7, respectively.

4. Vortex Solutions

We generate singly-charged vortices at the origin of our otherwise homogeneous system by setting the phase $S(\boldsymbol{\rho}) = \arctan[\mathrm{Im}(\psi)/\mathrm{Re}(\psi)]$ of the wavefunction to be $S = \arctan[y/x]$ (corresponding to a singly-charged vortex) during imaginary time propagation of the dipolar GPE.

In the non-dipolar system, a vortex has a size characterized by the conventional healing length $\xi_0 = \hbar/\sqrt{m\mu_0}$, where $mu_0 = n_0 g$. In the dipolar system, the vortex core size is of the order of the dipolar healing length ξ_0. This is the motivation for expressing length in these units. However, it should be noted that this length scale is a function of the system parameters, and even diverges as $\mu_0 = n_0(g + g_{dd}[3\cos^2\alpha - 1]) \to 0$.

We examine the structure of the vortex solutions for three representative polarization angles: (a) $\alpha = 0$, (b) $\pi/4$ and (c) $\pi/2$ (each of which fall into distinct regimes of the stability diagram). Figures 3(a), (b) and (c) plot the vortex solutions for the respective polarization angle, as a function of ε_{dd}. The left-hand side of the main plots shows the density profile along x (with $y = 0$) and the right-hand side shows the density profile along y (with $x = 0$). Note that we find that the introduction of a vortex has no observable effect on the stability of the system, which maintains the stability diagram plotted in Fig. 2, and hence regions with no vortex solutions occur in the plot.

(a) $\alpha = 0$: Since the dipoles are perpendicular to the plane, the dipolar potential, and hence density profile, are circularly-symmetric [2(a)]. For $\varepsilon_{dd} = 0$ the vortex has the standard appearance familiar from non-dipolar condensates [see left inset of Fig. 2(a)], consisting of a circularly-symmetric core of vanishing density of width ξ_0 [29].
For $\varepsilon_{dd} \neq 0$ and for most of the parameter space the vortex solution is essentially identical to that for $\varepsilon_{dd} = 0$ (due to our choice of ξ_0 as the length scale). Here the system behaves like a non-dipolar system but with modified contact interactions, and the non-local effects of the dipolar interaction are insignificant. However, there exist two limiting regimes where the vortex solutions deviate significantly. Firstly, as one approaches $\varepsilon_{dd} = -0.5$ from above, the vortex core becomes increasingly narrow (in units of ξ_0) and takes on a modified profile

[see middle inset of Fig. 2(a)]. This is associated with the cancellation of explicit contact interactions in the system—the vdW interactions cancel the contact contribution from the dipolar interactions. Indeed, the total contact interaction disappears altogether at the PI threshold, $\varepsilon_{\rm dd} = -0.5$). Secondly, as the RI is approached from below, circularly-symmetric density ripples appear around the vortex [see middle inset of Fig. 2(c)]. The ripples decay with distance from the core and have an amplitude of up to $\sim 20\% n_0$. These ripples are associated with the roton mode mixing into the ground state [17, 18], and are physically associated with the energetic favourability of dipoles aligning head-to-tail, but where the van der Waals interactions and axial trapping prevent collapse.

Such ripples have been predicted to appear around vortices in trapped 3D dipolar condensates [17, 18, 19], and around more general localized density perturbations [25]. The ripple wavelength, which reflects the roton wavelength, is approximately $4\xi_0$.

(b) $\alpha = \pi/4$: The dipoles are now polarized off-axis, i.e. tipped in the direction of the x-axis. As a result, the circular-symmetry of the dipolar potential and density is lost, as evident from the asymmetry of the family of solutions in Fig. 3(b). For $g_{\rm dd} > 0$ (< 0) the dipoles attract (repel) along x and repel (attract) along y. The local dipolar potential is still net positive throughout (since $\alpha < \alpha_0$). The vortex core is elongated along x for $\varepsilon_{\rm dd} > 0$, and reverses for $\varepsilon_{\rm dd} < 0$. Close to the RI, density ripples form about the vortex (of large amplitude up to $\sim 40\% n_0$) with wavefronts parallel to the x-axis. This alignment is due to the preferred head-to-tail alignment of dipoles which are tipped along x, and related to the *anisotropic* mixing of the roton into the ground state [25].

(c) $\alpha = \pi/2$: For $g_{\rm dd} > 0$ (< 0) the dipoles are net attractive (repulsive) along x and repulsive (attractive) along y. The anisotropy of the vortex core and ripples is reversed compared to $\alpha = \pi/4$.

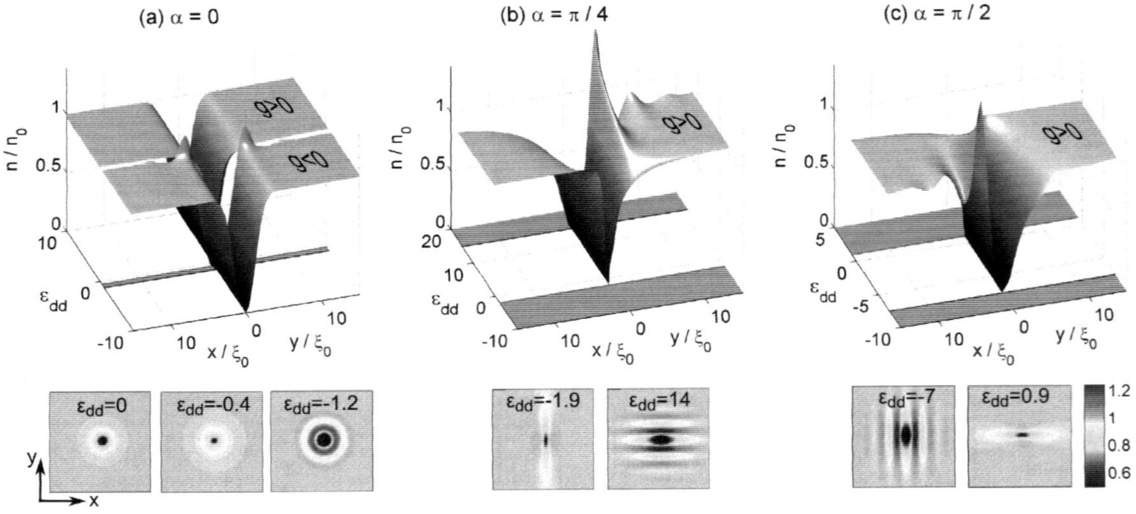

Figure 3. Vortex density profile in the presence of dipolar interactions $\varepsilon_{\rm dd}$ (with $\sigma = 0.5$). The left (right) side of the abcissa shows the profile along x (y). (a) Dipoles polarized along z ($\alpha = 0$). (b) Dipoles polarized off-axis at $\alpha = \pi/4$. (c) Dipoles polarized off-axis at $\alpha = \pi/2$. Grey bands indicate the unstable regimes of $\varepsilon_{\rm dd}$. Insets: vortex density $n(x, y)$ for example values of $\varepsilon_{\rm dd}$ over an area $(40\xi_0)^2$.

5. Dipolar Potential of a Vortex

5.1. Dipoles Perpendicular to the Plane (α=0)

A region of depleted density due to the presence of a vortex in an otherwise uniform system may be alternatively viewed as a lump of "anti-dipoles" (whose directions have been reversed) sitting in empty space [32, 21]. From this point of view a vortex treated as an object in its own right will generate its own mesoscopic dipolar potential. We now seek to estimate this dipolar potential analytically and compare our estimate against exact numerical calculations.

Consider first the simplified case of dipoles polarized perpendicular to the plane, $\alpha = 0$. In Fig. 4(a) we show the dipolar potential $\Phi(\rho)$ in the vicinity of the vortex core for various values of $\varepsilon_{\rm dd}$. These values range from large negative $\varepsilon_{\rm dd}$ ($\varepsilon_{\rm dd} = -5$, black line) to large positive $\varepsilon_{\rm dd}$ ($\varepsilon_{\rm dd} = 5$, red line). In between we include a case close to the roton instability where the vortex possesses large ripples ($\varepsilon_{\rm dd} = -1.2$, blue line) and close to the phonon instability, where the vortex core becomes narrow ($\varepsilon_{\rm dd} = -0.4$, green line).

Far away from the core Φ approaches $\Phi_0 = n_0 g_{\rm dd}(3\cos^2\alpha - 1)$, the homogeneous result. For $\rho \lesssim 5\xi_0$, Φ is dominated by the core structure and ripples (where present). Figure 4(b) plots, on a log-log scale, the decay of Φ towards the homogeneous value. Far away from the core, Φ decays as $1/\rho^2$ (grey line).

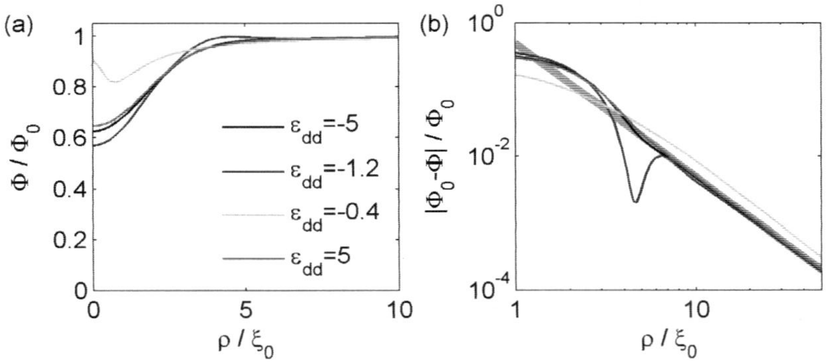

Figure 4. Vortex dipolar potential $\Phi(\rho)$, rescaled by the homogeneous background value Φ_0, for $\alpha = 0$ and various values of $\varepsilon_{\rm dd}$ (and $\sigma = 0.5$). The derived $1/\rho^2$ result of Eq. (10) is shown (thick grey line). The results for $\varepsilon_{\rm dd} = -5$ and $\varepsilon_{\rm dd} = 5$ closely match since the vortex profiles are practically identical to each other (approximating the non-dipolar vortex). The behaviour differs for $\varepsilon_{\rm dd} = -0.4$ and $\varepsilon_{\rm dd} = -1.2$, close to the phonon and roton instabilities respectively, where the vortex develops a modified structure. In (b) we see that, for all cases, the dipolar potential decays as $1/\rho^2$ at long-range.

The long-range behaviour of the dipolar potential can be obtained analytically from the vortex ansatz [38, 21],

$$n(\rho) = n_0\left(1 - \frac{1}{1+\rho'^2}\right), \tag{9}$$

where $\rho' = \rho/\xi_0$. This ansatz gives both the correct ρ'^2 scaling of the density as $\rho' \to 0$ and the correct $1/\rho'^2$ decay to the homogeneous density as $\rho' \to \infty$ [29]. We evaluate the dipolar potential via the convolution result $\Phi(\boldsymbol{\rho}, t) = \mathcal{F}^{-1}\left[\tilde{U}_{\rm dd}^{\rm 2D}(\mathbf{k})\tilde{n}(\mathbf{k}, t)\right]$ and by taking advantage of the cylindrical symmetry afforded by the $\alpha = 0$ scenario to employ Hankel transformations. The

Journal of Physics: Conference Series **497** (2014) 012025 doi:10.1088/1742-6596/497/1/012025

Hankel transform of Eq. (9) is,

$$\tilde{n}(\mathbf{k}) = n_0 \left(\frac{\delta(k)}{k} + \frac{K_0(k/\sqrt{2})}{2} \right),$$

where $K_0(k)$ is a modified Bessel function of the second kind. We expand the dipolar interaction $\tilde{U}_{\rm dd}^{\rm 2D}(\mathbf{k})$ in terms of the condensate width parameter σ up to first-order giving,

$$\tilde{U}_{\rm dd}^{\rm 2D}(\mathbf{k}) = g_{\rm dd} \left(2 - \sqrt{\frac{9\pi}{2}} k\sigma \right) + \mathcal{O}\left(\sigma^2\right).$$

Then, up to first order in σ and third order in $1/\rho'$, the dipolar potential generated by the vortex ansatz is,

$$\frac{\Phi(\rho)}{\Phi_0} = \left(1 - \frac{1}{\rho'^2} \right) + \left(\frac{A \ln \rho' + B}{\rho'^3} \right) \sigma, \tag{10}$$

with constants $A = -\sqrt{9\pi/8} \approx -1.88$ and $B = (\ln 2 - 1)A \approx 0.577$. Recall Φ_0 is the dipolar potential at infinity.

To analyse this result it is useful to consider Φ as the sum of a local term $\Phi_{\rm L}$, proportional to density and given by,

$$\Phi_{\rm L}(\boldsymbol{\rho}) = n(\boldsymbol{\rho}) g_{\rm dd} (3\cos^2\alpha - 1), \tag{11}$$

and non-local term $\Phi_{\rm NL}$. The non-local term can be viewed as arising from a fictitious electrostatic potential associated with the dipoles [37, 28] and has a complicated form given elsewhere [31]. It depends on *variations* of density, becoming zero for a homogeneous system.

The first bracketed term in Eq. (10) corresponds to the *local* contribution to the vortex dipolar potential. This is evident since, at long-range ($\rho' \gg 1$), the vortex ansatz becomes $n(\rho) = n_0(1 - 1/\rho'^2)$, i.e. it decays as $1/\rho'^2$ to the background density. Note that the s-wave interactions generate a local potential with identical form (proportional to density) [36]. The second bracketed term in Eq. (10) describes the *non-local* contribution, i.e. the contribution to the potential arising from the long-range contributions from all anti-dipoles in the vortex. This vanishes in the true 2D limit $\sigma = 0$ since the volume of anti-dipoles in the vortex core vanishes. This illustrates the point that the dipolar potential generated by a vortex is not a topological quantity like the potential associated with the hydrodynamic flow around the vortex (which leads to the logarithmic interaction between two vortices), but instead depends on the number of dipoles excluded from the region by the presence of the vortex. Also, the fact that the non-local vortex potential scales dominantly as $\ln \rho'/\rho'^3$, and not $1/\rho'^3$, informs us that the vortex does not strictly behave as a point-like collection of dipoles at long-range. This is due to the slow power-law recovery of the vortex density to n_0. This point is reinforced by noting that for sharply-truncated density profiles, e.g. a circular "hole", and exponentially-decaying density profiles, e.g. $1 - \exp(-\rho'^2)$ or $\tanh^2(\rho')$, the dipolar potential *does* decay as $1/\rho'^3$ at long-range. Due to the fast decay of the density, the inhomogeneity then really does resemble a giant dipole at long-range. This further confirms the subtle but critical role of the slow power-law decay of the vortex density. Nevertheless, whatever the subtleties of the non-local part of Φ, its long range behaviour is dominated by the local $1/\rho'^2$ scaling. This is confirmed by our numerical solutions, whose Φ decays to the background value as $1/\rho'^2$ at long-range, as shown in Fig. 4(b).

5.2. Dipoles with a Projection in the Plane ($\alpha \neq 0$)

When the polarization direction is tilted away from the vertical, the dipolar interaction becomes anisotropic in the plane, resulting in the anisotropic vortex cores discussed in Section 4 above. The dipolar potential Φ will then be anisotropic and have three contributions: an isotropic contribution from the homogeneous background density, a local contribution from the anisotropic density profile of the vortex, and a non-local anisotropic contribution arising from the anti-dipoles associated with the rest of the vortex.

We analyse the ensuing dipolar potential through the exemplar of $\alpha = \pi/4$ and $\varepsilon_{\rm dd} = 5$. The density profile of this vortex [Fig. 5(a)] shows a core elongated in x with adjacent density ripples aligned in x. The dipolar potential of the vortex [Fig. 5(b)] is indeed anisotropic, with a raised potential along x and a reduced potential along y (relative to the background dipolar potential Φ_0). Indeed, the angular dependence of Φ, relative to Φ_0, resembles that of the dipole-dipole interaction itself $\sim 1 - 3\cos^2\theta$. Thus, at least in its angular dependence, the vortex shares qualitatively the characteristics of a mesoscopic dipole.

Away from the core, Φ decays towards the homogeneous value $\Phi_0 = \mu_0/2$. At long-range $\rho' \gg 1$, we again see the decay of Φ developing a $1/\rho'^2$ behaviour [Fig. 5(c)], consistent with the local contribution to Φ from the vortex core. The same behaviour is achieved along the x and y directions; up to mid-scale distances $\rho' \lesssim 10$, Φ is dominated by the core structure, while at longer ranges, a common $1/\rho'^2$ behaviour is obtained.

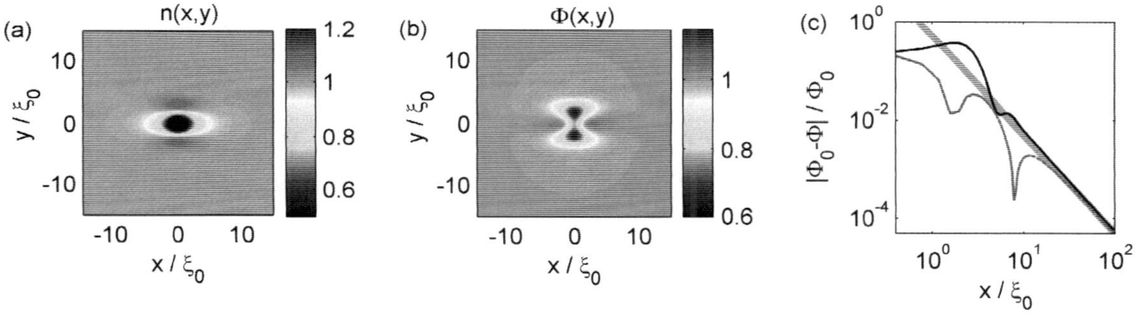

Figure 5. Vortex with dipoles polarized off-axis ($\alpha = \pi/4$, $\varepsilon_{\rm dd} = 5$ and $\sigma = 0.5$). (a) The density profile $n(x, y)$ is anisotropic, with density ripples polarized along x. (b) The dipolar potential $\Phi(x, y)$, rescaled by the homogeneous value Φ_0, is modified from the background value due to the presence of the vortex. In particular, the collection of anti-dipoles which can be associated with the vortex generates an anisotropic modulation of the dipolar potential which mimics the dipolar potential itself $\sim 1 - 3\cos^2\theta$. (c) The decay of Φ with distance along $x = 0$ (black line) and $y = 0$ (red line), on a logarithmic scale. At short and intermediate scales, the decay varies due to the density variations associated with the vortex, while at long-range the decay approaches the $1/\rho'^2$ form. The line $1/\rho'^2$ (grey line) is shown for reference.

6. Conclusions

In this paper we have explored the stability and vortex solutions of the infinite two-dimensional dipolar Bose-Einstein condensate as a function of the strength and polarization angle of the dipoles. We mapped out the parameter regimes where a homogeneous density distribution in the plane is unstable towards either long wavelength phonon fluctuations or finite wavelength roton fluctuations. The instability is always towards configurations where the dipoles are arranged head-to-tail in order to benefit from the attractive portion of the dipolar interaction. The long-range and anisotropic nature of the dipolar interaction can lead to significant modifications to the structure of a single vortex, particularly in and around the core, including the formation of ripples in the core periphery and, for dipoles polarized off-axis, an anisotropic density profile. Furthermore, the inhomogeneous density of dipoles associated with the vortex generates a potential which decays slowly from the core and, for dipoles polarized off-axis, varies anisotropically in space. At long-range, this potential has $1/\rho^2$ and $\ln(\rho)/\rho^3$ contributions, and an angular dependence $\sim 1 - 3\cos^2\theta$. The long-range dipolar potential emanating from a single vortex in turn modifies the vortex-vortex interaction [20] and is expected to support new regimes of vortex dynamics and collective structures in many-vortex systems.

DHJO acknowledges support from NSERC (Canada).

[1] Griesmaier A, Werner J, Hensler S, Stuhler J and Pfau T 2005 *Phys. Rev. Lett.* **94** 160401
[2] Beaufils Q, Chicireanu R, Zanon T, Laburthe-Tolra B, Maréchal E, Vernac L, Keller J C and Gorceix O 2008 *Phys. Rev. A* **77** 061601(R)
[3] Lu M, Burdick N Q, Youn S H and Lev B L 2011 *Phys. Rev. Lett.* **107** 190401
[4] Aikawa K *et al.* 2012 *Phys. Rev. Lett.* **108** 210401
[5] Lahaye T *et al.* 2007 *Nature* **448** 672
[6] Lahaye T *et al.* 2008 *Phys. Rev. Lett.* **101** 080401
[7] O'Dell D H J, Giovanazzi S and Kurizki G 2003 *Phys. Rev. Lett.* **90** 110402
[8] Santos L, Shlyapnikov G V and Lewenstein M 2003 *Phys. Rev. Lett.* **90** 250403
[9] Ronen S, Bortolotti D C E and Bohn J L 2007 *Phys. Rev. Lett.* **98** 030406
[10] Mottl R, Brennecke F, Baumann K, Landig R, Donner T and Esslinger T 2012 *Science* **336** 1570
[11] Lahaye T, Menotti C, Santos L, Lewenstein M and Pfau T (2009) *Rep. Prog. Phys.* **72** 126401
[12] Werner J, Griesmaier A, Hensler S, Stuhler J, Pfau T, Simoni A and Tiesinga E 2005 *Phys. Rev. Lett.* **94** 183201
[13] Koch T, Lahaye T, Metz J, Frölich B, Griesmaier A and Pfau T (2008) *Nature Physics* **4** 218
[14] Ni K K *et al.* 2010 *Nature* **464** 1324
[15] Kevrekidis P G, Frantzeskakis D J and Carretero-González R (Eds.) 2008 *Emergent Nonlinear Phenomena in Bose-Einstein condensates* (Springer, Berlin)
[16] Fetter A L 2010 *J. Low Temp. Phys.* **161** 445
[17] Yi S and Pu H 2006 *Phys. Rev. A* **73** 061602(R)
[18] Wilson R M, Ronen S, Bohn J L and Pu H 2008 *Phys. Rev. Lett.* **100** 245302; Wilson R M, Ronen S and Bohn J L 2009 *Phys. Rev. A* **79** 013621
[19] Abad M, Guilleumas M, Mayol R, Pi M and Jezek D M 2009 *Phys. Rev. A* **79** 063622
[20] Mulkerin B C, van Bijnen R M W, Martin A M and Parker N G 2013 *Phys. Rev. Lett.* **111** 170402
[21] O'Dell D H J and Eberlein C 2007 *Phys. Rev. A* **75** 013604
[22] van Bijnen R M W, O'Dell D H J, Parker N G and Martin A M 2007 *Phys. Rev. Lett.* **98** 150401
[23] van Bijnen R M W, Dow A J, O'Dell D H J, Parker N G and Martin A M 2009 *Phys. Rev. A* **80** 033617
[24] Kishor Kumar R and Muruganandam P 2012 *J. Phys. B* **45** 215301
[25] Ticknor C, Wilson R M and Bohn J L 2011 *Phys. Rev. Lett.* **106** 065301
[26] Cooper N R, Rezayi E H and Simon S H 2005 *Phys. Rev. Lett.* **95** 200402; Zhang J and Zhai H 2005 *Phys. Rev. Lett.* **95** 200403
[27] Yi S and You L 2001 *Phys. Rev. A* **63** 053607
[28] Parker N G and O'Dell D H J 2008 *Phys. Rev. A* **78** 041601(R)
[29] Pethick C J and Smith H 2002 *Bose-Einstein Condensation in Dilute Gases* (Cambrige University Press, Cambridge)
[30] Fischer U R 2006 *Phys. Rev. A* **73** 031602
[31] Cai Y, Rosenkranz M, Lei Z and Bao W 2010 *Phys. Rev. A* **82** 043623
[32] Klawunn M, Nath R, Pedri P and Santos L 2008 *Phys. Rev. Lett.* **100** 240403
[33] van Bijnen R M W, Parker N G, Kokkelmans S J J M F, Martin A M and O'Dell D H J 2010 *Phys. Rev. A* **82** 033612
[34] Klawunn M and Santos L 2009 *Phys. Rev. A* **80** 013611
[35] Donnelly R J 1991 *Quantized Vortices in Helium II* (Cambridge University Press, Cambridge)
[36] Wu H and Sprung D W L 1994 *Phys. Rev. A* **49** 4305
[37] O'Dell D H J, Giovanazzi S and Eberlein C 2004 *Phys. Rev. Lett.* **92** 250401
[38] Fetter A L 1969 *Lectures in Theoretical Physics* ed Mahanthappa K T and Brittin W E (Gordon and Breach, New York).

Journal of Physics: Conference Series **497** (2014) 012026 doi:10.1088/1742-6596/497/1/012026

Fermionic condensation in ultracold atoms, nuclear matter and neutron stars

Luca Salasnich
Dipartimento di Fisica e Astronomia "Galileo Galilei" and CNISM, Università di Padova, Via Marzolo 8, 35131 Padova, Italy

E-mail: luca.salasnich@unipd.it

Abstract. We investigate the Bose-Einstein condensation of fermionic pairs in three different superfluid systems: ultracold and dilute atomic gases, bulk neutron matter, and neutron stars. In the case of dilute gases made of fermionic atoms the average distance between atoms is much larger than the effective radius of the inter-atomic potential. Here the condensation of fermionic pairs is analyzed as a function of the s-wave scattering length, which can be tuned in experiments by using the technique of Feshbach resonances from a small and negative value (corresponding to the Bardeen-Cooper-Schrieffer (BCS) regime of Cooper Fermi pairs) to a small and positive value (corresponding to the regime of the Bose-Einstein condensate (BEC) of molecular dimers), crossing the unitarity regime where the scattering length diverges. In the case of bulk neutron matter the s-wave scattering length of neutron-neutron potential is negative but fixed, and the condensate fraction of neutron-neutron pairs is studied as a function of the total neutron density. Our results clearly show a BCS-quasiunitary-BCS crossover by increasing the neutron density. Finally, in the case of neutron stars, where again the neutron-neutron scattering length is negative and fixed, we determine the condensate fraction as a function of the distance from the center of the neutron star, finding that the maximum condensate fraction appears in the crust of the neutron star.

1. Introduction

In 1951 Penrose [1] and Onsager [2] introduced the idea of off-diagonal long-range order (ODLRO) of the one-body density matrix to determine the Bose-Einstein condensate fraction in a system of interacting bosons. In 1962 Yang [3] proved that for attractive fermions the Bose-Einstein condensation of fermionic pairs is instead related to the ODLRO of the two-body density matrix.

It is now established that at zero temperature the Bose-Einstein condensate fraction of bosonic liquid ^{4}He is below 10% [5], while for dilute and ultracold bosonic alkali-metal atoms it can reach 100% [4]. Some years ago, by using the ODLRO of the two-body density matrix, Salasnich, Manini and Parola [6] calculated the condensate fraction of fermionic pairs in the crossover from the Bardeen-Cooper-Schrieffer (BCS) state of Cooper Fermi pairs to the Bose-Einstein condensate (BEC) of molecular dimers at zero temperature (later in the same year there were other two papers [7, 8] on the same topic). In particular, it was found that the condensate fraction grows from zero to one in the BCS-BEC crossover [6, 7, 8]. These theoretical predictions are in quite good agreement with the data obtained in two experiments [9, 10] with Fermi vapors of ^{6}Li atoms. Recently, the condensate fraction of bulk neutron matter has been calculated by Wlazlowski and Magierski [11, 12] and also by Salasnich [13].

In this paper we review the zero-temperature mean-field theory we have used to extract the fermionic condensate fraction in ultracold atoms [6] and bulk neutron matter [13]. In addition we determine the condensate fraction of neutron-neutron pairs as a function of the distance from the center of a neutron star [14, 15]. In particular, we find the maximum condensate fraction at the distance $r/R \simeq 0.96$, where R is the star radius.

2. Bosonic and fermionic condensation

A quantum system of interacting identical bosons can be described by the bosonic field operator $\hat{\phi}(\mathbf{r})$, which satisfies the familiar commutation rules [16]

$$\left[\hat{\phi}(\mathbf{r}), \hat{\phi}^+(\mathbf{r}')\right] = \delta(\mathbf{r}-\mathbf{r}') \; , \qquad \left[\hat{\phi}(\mathbf{r}), \hat{\phi}(\mathbf{r}')\right] = \left[\hat{\phi}^+(\mathbf{r}), \hat{\phi}^+(\mathbf{r}')\right] = 0 \; , \tag{1}$$

where $\left[\hat{A}, \hat{B}\right] = \hat{A}\hat{B} - \hat{B}\hat{A}$ and $\delta(\mathbf{r})$ is the Dirac delta function. The bosonic one-body density matrix is given by

$$n(\mathbf{r},\mathbf{r}') = \langle \hat{\phi}^+(\mathbf{r})\, \hat{\phi}(\mathbf{r}') \rangle \; , \tag{2}$$

where the average $\langle \cdots \rangle$ can be a thermal average or a zero-temperature average. Its diagonal part is the average local density of bosons, i.e. $n(\mathbf{r}) = n(\mathbf{r},\mathbf{r}) = \langle \hat{\phi}^+(\mathbf{r})\, \hat{\phi}(\mathbf{r}) \rangle$, while the average total number of bosons reads

$$N = \int \langle \hat{\phi}^+(\mathbf{r})\, \hat{\phi}(\mathbf{r}) \rangle \, d^3\mathbf{r} \; . \tag{3}$$

As previously discussed, Penrose and Onsager [1, 2] used the ODLRO of the one-body density matrix of a uniform bosonic system to determine the condensate number N_0 of bosons. For a non-uniform bosonic system this condensate number N_0 is nothing else than the largest eigenvalue of the one-body density matrix [16]. As shown by Yukalov [17], in the thermodynamic limit this is equivalent to the spontaneous symmetry breaking of $U(1)$ gauge symmetry, which gives

$$N_0 = \int |\langle \hat{\phi}(\mathbf{r}) \rangle|^2 \, d^3\mathbf{r} \tag{4}$$

as the average number of condensed bosons.

A quantum system of interacting identical fermions with two spin components ($\sigma = \uparrow, \downarrow$) can be described by the fermionic field operator $\hat{\psi}_\sigma(\mathbf{r})$, which satisfies the familiar anti-commutation rules [16]

$$\left\{\hat{\psi}_\sigma(\mathbf{r}), \hat{\psi}^+_{\sigma'}(\mathbf{r}')\right\} = \delta(\mathbf{r}-\mathbf{r}')\, \delta_{\sigma,\sigma'} \; , \qquad \left\{\hat{\psi}_\sigma(\mathbf{r}), \hat{\psi}_{\sigma'}(\mathbf{r}')\right\} = \left\{\hat{\psi}^+_\sigma(\mathbf{r}), \hat{\psi}^+_{\sigma'}(\mathbf{r}')\right\} = 0 \; , \tag{5}$$

where $\left\{\hat{A}, \hat{B}\right\} = \hat{A}\hat{B} + \hat{B}\hat{A}$ and $\delta_{\sigma,\sigma'}$ is the Kronecher symbol. The fermionic one-body density matrix is given by

$$n_{\sigma,\sigma'}(\mathbf{r},\mathbf{r}') = \langle \hat{\psi}^+_\sigma(\mathbf{r})\, \hat{\psi}_{\sigma'}(\mathbf{r}') \rangle \; . \tag{6}$$

Its diagonal part is the average local density of fermions with spin σ, i.e. $n_\sigma(\mathbf{r}) = n_{\sigma,\sigma}(\mathbf{r},\mathbf{r}) = \langle \hat{\psi}^+_\sigma(\mathbf{r})\, \hat{\psi}_\sigma(\mathbf{r}) \rangle$, while the average total number of fermions reads

$$N = \sum_{\sigma=\uparrow,\downarrow} \int \langle \hat{\psi}^+_\sigma(\mathbf{r})\, \hat{\psi}_\sigma(\mathbf{r}) \rangle \, d^3\mathbf{r} \; . \tag{7}$$

Yang [3] suggested that for a uniform strongly-interacting fermionic system the number N_0 of Bose-condensed fermions, that is twice the number of condensed fermionic pairs, is related by the ODLRO of the fermionic two-body density matrix, given by

$$n_{\sigma_1,\sigma_2,\sigma_1',\sigma_2'}(\mathbf{r}_1,\mathbf{r}_2,\mathbf{r}_1',\mathbf{r}_2') = \langle \hat{\psi}^+_{\sigma_1}(\mathbf{r}_1)\, \hat{\psi}^+_{\sigma_2}(\mathbf{r}_2)\, \hat{\psi}_{\sigma_2'}(\mathbf{r}_2')\, \hat{\psi}_{\sigma_1'}(\mathbf{r}_1') \rangle \; . \tag{8}$$

For a generic (non-uniform) strongly-interacting system of identical fermions this condensate number N_0 is nothing else than twice the largest eigenvalue of the fermionic two-body density matrix [16]. In the thermodynamic limit this is equivalent [18] to the spontaneous symmetry breaking of $SU(2)$ gauge symmetry, which gives

$$N_0 = 2 \sum_{\sigma,\sigma'=\uparrow,\downarrow} \int |\langle \hat{\psi}_\sigma(\mathbf{r})\, \hat{\psi}_{\sigma'}(\mathbf{r}')\rangle|^2 \, d^3\mathbf{r}\, d^3\mathbf{r}' \tag{9}$$

as the average number of condensed fermions. It is important to stress that within the BCS theory of superconductors and fermionic superfluids the condensate number N_0 fermions can be much smaller than the average total number of fermions N. Moreover, in a uniform system N_0 is twice the average number of fermionic pairs in the (pseudo) Bose-Einstein condensate, i.e. the number of fermionic pairs which have their center of mass with zero linear momentum.

3. Ultracold and dilute atomic gases

The shifted Hamiltonian of the uniform two-spin-component Fermi superfluid made of ultracold atoms is given by

$$\hat{H}' = \int d^3\mathbf{r} \sum_{\sigma=\uparrow,\downarrow} \hat{\psi}_\sigma^+(\mathbf{r}) \left(-\frac{\hbar^2}{2m}\nabla^2 - \mu \right) \hat{\psi}_\sigma(\mathbf{r}) + g\, \hat{\psi}_\uparrow^+(\mathbf{r})\, \hat{\psi}_\downarrow^+(\mathbf{r})\, \hat{\psi}_\downarrow(\mathbf{r})\, \hat{\psi}_\uparrow(\mathbf{r}) , \tag{10}$$

where $\hat{\psi}_\sigma(\mathbf{r})$ is the field operator that annihilates a fermion of spin σ in the position $\mathbf{r}$, while $\hat{\psi}_\sigma^+(\mathbf{r})$ creates a fermion of spin σ in $\mathbf{r}$. Here $g < 0$ is the strength of the attractive fermion-fermion interaction, which is approximated by a contract Fermi pseudo-potential [4] because for ultracold and dilute gases the average distance between atoms is much larger than the effective radius of the inter-atomic potential [4, 16]. The ground-state average of the number of fermions is given by Eq. (7). This total number N is fixed by the chemical potential μ which appears in Eq. (10).

Within the Bogoliubov approach the mean-field Hamiltonian derived from Eq. (10) can be diagonalized by using the Bogoliubov-Valatin representation of the field operator $\hat{\psi}_\sigma(\mathbf{r})$ in terms of the anti-commuting quasi-particle Bogoliubov operators $\hat{b}_{\mathbf{k}\sigma}$ with amplitudes $u_\mathbf{k}$ and $v_\mathbf{k}$ and the quasi-particle energy $E_\mathbf{k}$. In this way one finds familiar expressions for these quantities:

$$E_\mathbf{k} = \left[(\epsilon_\mathbf{k} - \mu)^2 + \Delta^2 \right]^{1/2} \tag{11}$$

and

$$u_\mathbf{k}^2 = \left(1 + (\epsilon_\mathbf{k} - \mu)/E_\mathbf{k}\right)/2 \tag{12}$$

$$v_\mathbf{k}^2 = \left(1 - (\epsilon_\mathbf{k} - \mu)/E_\mathbf{k}\right)/2 , \tag{13}$$

where $\epsilon_\mathbf{k} = \hbar^2 k^2/(2m)$ is the single-particle energy. The parameter Δ is the pairing gap, which satisfies the gap equation

$$-\frac{1}{g} = \frac{1}{\Omega} \sum_\mathbf{k} \frac{1}{2E_\mathbf{k}} , \tag{14}$$

where Ω is the volume of the uniform system. Notice that this equation is ultraviolet divergent and it must be regularized. The equation for the total density $n = N/\Omega$ of fermions is obtained from Eq. (7) as

$$n = \frac{2}{\Omega} \sum_\mathbf{k} v_\mathbf{k}^2 . \tag{15}$$

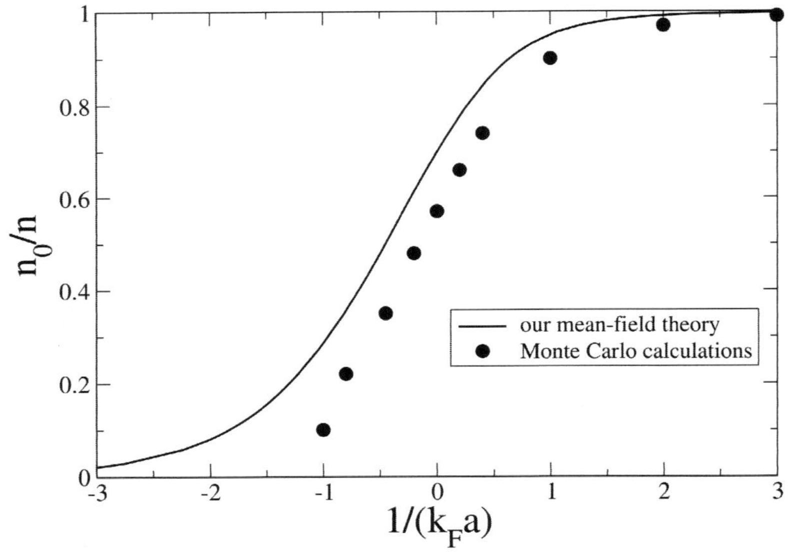

Figure 1. Condensate fraction of fermionic atoms as a function of the inverse interaction strength $1/(k_F a)$: our mean-field theory [6] (solid line); fixed-node diffusion Monte Carlo results [8] (filled circles). Here $k_F = (3\pi^2 n)^{1/3}$ is the Fermi wavenumber, with n the total number density of atoms, and a is the s-wave scattering length of the inter-atomic potential.

Finally, from Eq. (9) one finds that the condensate density $n_0 = N_0/\Omega$ of paired fermions is given by [6, 18]

$$n_0 = \frac{2}{\Omega} \sum_{\mathbf{k}} u_{\mathbf{k}}^2 v_{\mathbf{k}}^2 \, . \tag{16}$$

In three dimensions, a suitable regularization [19] of the gap equation is obtained by introducing the s-wave scattering length a via the equation

$$-\frac{1}{g} = -\frac{m}{4\pi\hbar^2 a} + \frac{1}{\Omega} \sum_{\mathbf{k}} \frac{m}{\hbar^2 k^2} \, , \tag{17}$$

and then subtracting this equation from the gap equation (14). In this way one obtains the three-dimensional regularized gap equation

$$-\frac{m}{4\pi\hbar^2 a} = \frac{1}{\Omega} \sum_{\mathbf{k}} \left(\frac{1}{2E_k} - \frac{m}{\hbar^2 k^2} \right) , \tag{18}$$

which can be used to study the full BCS-BEC crossover [6] by changing the amplitude and sign of the s-wave scattering length a.

Taking into account the functional dependence of the amplitudes u_k and v_k on μ and Δ, one finds [6] the very nice formula

$$n_0 = \frac{m^{3/2}}{8\pi\hbar^3} \Delta^{3/2} \sqrt{\frac{\mu}{\Delta} + \sqrt{1 + \frac{\mu^2}{\Delta^2}}} \, , \tag{19}$$

which shows the not trivial relationship between the energy gap Δ and the condensate density n_0. By the same techniques, also the two BCS-BEC equations can be written in a more compact form as

$$-\frac{1}{a} = \frac{2(2m)^{1/2}}{\pi\hbar^3}\,\Delta^{1/2}\,I_1\Big(\frac{\mu}{\Delta}\Big)\,, \tag{20}$$

$$n = \frac{(2m)^{3/2}}{2\pi^2\hbar^3}\,\Delta^{3/2}\,I_2\Big(\frac{\mu}{\Delta}\Big)\,, \tag{21}$$

where $I_1(x)$ and $I_2(x)$ are two monotonic functions [19] given by

$$I_1(x) \;=\; \int_0^{+\infty} y^2\left(\frac{1}{\sqrt{(y^2-x)^2+1}} - \frac{1}{y^2}\right) dy\,, \tag{22}$$

$$I_2(x) \;=\; \int_0^{+\infty} y^2\left(1 - \frac{y^2-x}{\sqrt{(y^2-x)^2+1}}\right) dy\,. \tag{23}$$

In Fig. 1 we report the condensate fraction n_0/n of fermionic atoms in the BCS-BEC crossover as a function of the inverse interaction strength $1/(k_F a)$ obtained with this mean-field theory [6, 7] (solid line). In the figure we compare our calculations [6] with the fixed-node diffusion Monte Carlo results (filled circles) obtained by Astrakharchik, Boronat, Casulleras, and S. Giorgini [8] with $N = 66$ fermions and a tunable square-well potential. Remarkably, the agreement between the two theoretical approaches is better in the BEC side of the crossover. On the other hand, as discussed in Refs. [6, 20], our Eq. (19) is in full agreement with the experimental data of the MIT group [9] in the BCS side of the crossover.

4. Nuclear matter

Let us now consider the nuclear matter, and in particular the bulk neutron matter, which is a dense Fermi liquid made of two-component (spin up and down) neutrons. The shifted Hamiltonian of the uniform neutron matter can be written as

$$\begin{aligned} \hat{H}' \;&=\; \int d^3\mathbf{r} \sum_{\sigma=\uparrow,\downarrow} \hat{\psi}_\sigma^+(\mathbf{r})\left(-\frac{\hbar^2}{2m}\nabla^2 - \mu\right)\hat{\psi}_\sigma(\mathbf{r}) \\ &+\; \int d^3\mathbf{r}\, d^3\mathbf{r}'\, \hat{\psi}_\uparrow^+(\mathbf{r})\, \hat{\psi}_\downarrow^+(\mathbf{r}')\, V(\mathbf{r}-\mathbf{r}')\, \hat{\psi}_\downarrow(\mathbf{r}')\, \hat{\psi}_\uparrow(\mathbf{r})\,, \end{aligned} \tag{24}$$

where $\hat{\psi}_\sigma(\mathbf{r})$ is the field operator that annihilates a neutron of spin σ in the position $\mathbf{r}$, while $\hat{\psi}_\sigma^+(\mathbf{r})$ creates a neutron of spin σ in $\mathbf{r}$. Here $V(\mathbf{r}-\mathbf{r}')$ is the neutron-neutron potential characterized by s-wave scattering length $a = -18.5$ fm and effective range $r_e = 2.7$ fm [21].

One can apply the familiar Bogoliubov approach to diagonalize the mean-field quadratic Hamiltonian derived from Eq. (24), but now the paring gap $\Delta_\mathbf{k}$ depends explicitly on the wave number $\mathbf{k}$ and satisfies the integral equation

$$\Delta_\mathbf{q} = \sum_\mathbf{k} V_{\mathbf{qk}}\,\frac{\Delta_\mathbf{k}}{2E_\mathbf{k}}\,, \tag{25}$$

where $V_{\mathbf{qk}} = \langle \mathbf{q}, -\mathbf{q}|V|\mathbf{k}, -\mathbf{k}\rangle$ is the wave-number representation of the neutron-neutron potential, and

$$E_\mathbf{k} = \sqrt{\left(\frac{\hbar^2k^2}{2m} - \mu\right)^2 + |\Delta_\mathbf{k}|^2}\,. \tag{26}$$

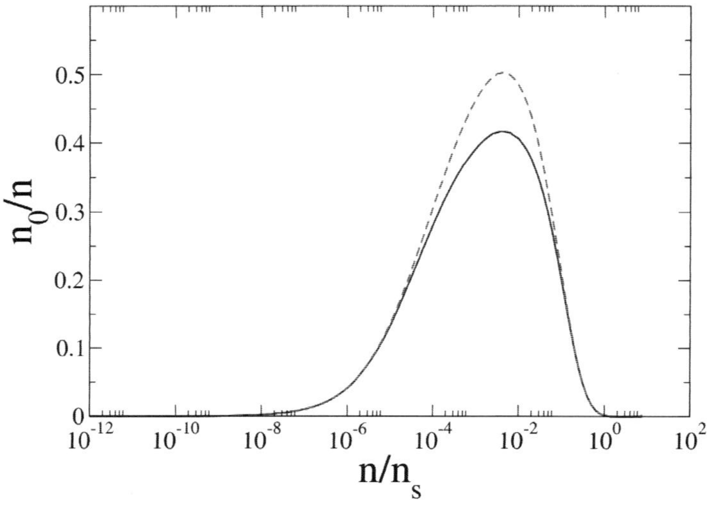

Figure 2. Condensate fraction n_0/n of neutron pairs in neutron matter as a function of the scaled neutron number density n/n_s, where $n_s = 0.16$ fm^{-3} is the nuclear saturation density (see also [13]). The solid line is obtained by using Eqs. (27) and (29). The dashed line is obtained by using Eqs. (28) and (29).

Under the simplifying assumptions $\mu \simeq \epsilon_F = \frac{\hbar^2}{2m}(3\pi^2 n)^{2/3}$ and $\Delta_{\mathbf{k}} \simeq \Delta$, in the continuum limit we determine the condensate fraction as [13]

$$\frac{n_0}{n} = \frac{\pi}{2^{5/2}} \frac{\sqrt{\frac{\epsilon_F}{\Delta} + \sqrt{1 + \frac{\epsilon_F^2}{\Delta^2}}}}{I_2(\frac{\epsilon_F}{\Delta})} \tag{27}$$

Notice that in the deep BCS regime where $\Delta/\epsilon_F \ll 1$ one finds

$$\frac{n_0}{n} = \frac{3\pi}{8} \frac{\Delta}{\epsilon_F} . \tag{28}$$

Fitting the numerical data of Δ/ϵ_F vs k_F obtained by Matsuo [21] from realistic neutron-neutron potentials we get the formula

$$\frac{\Delta}{\epsilon_F} = \frac{\beta_0 k_F^{\beta_1}}{\exp(k_F^{\beta_2}/\beta_3) - \beta_3} \tag{29}$$

with the following fitting parameters: $\beta_0 = 2.851$, $\beta_1 = 1.942$, $\beta_2 = 1.672$, $\beta_3 = 0.276$, $\beta_4 = 0.975$. By using this fitting formula and Eq. (27) we finally get the condensate fraction of neutron matter as a function of the neutron density n [13].

The condensate fraction n_0/n of neutron pairs is shown in Fig. 2 as a function of the scaled density n/n_s, where $n_s = 0.16$ fm^{-3} is the nuclear saturation density. Notice that the horizontal axis is in logarithmic scale. At very low neutron density n the neutron matter behaves like a quasi-ideal Fermi gas with weakly correlated Cooper pairs and the condensate fraction n_0/n is exponentially small. By increasing the neutron density n the attractive tail of the neutron-neutron potential becomes relevant and the condensate fraction n_0/n grows significantly. The maximum of the condensate fraction is $(n_0/n)_{max} = 0.42$ at the neutron density $n = 5.3 \cdot 10^{-4}$

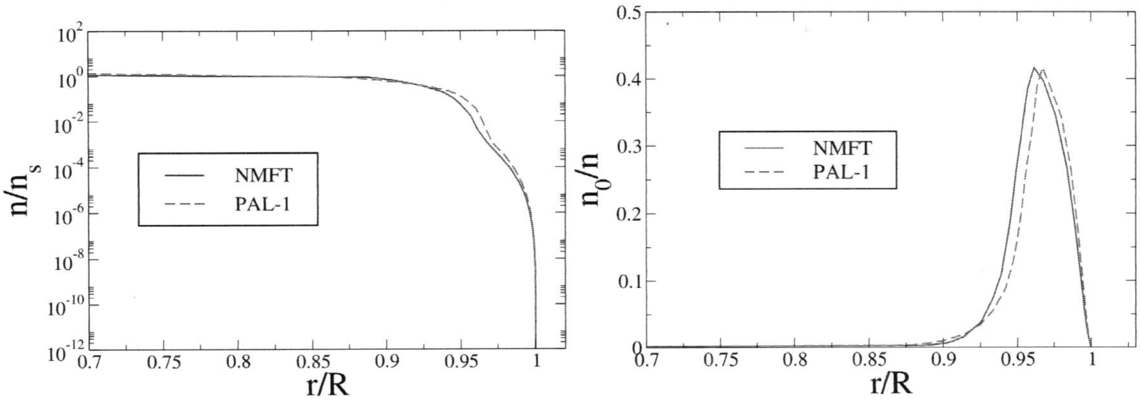

Figure 3. Properties of the 1.4 solar mass neutron star. Left panel: Scaled density profile n/n_s vs scaled distance r/R. Here $n_s = 0.16$ fm^{-3} is the nuclear saturation density and R is the radius of the star. Right panel: condensate fraction n_0/n of neutron pairs vs scaled distance r/R. Solid lines are obtained with the Walecka model of bulk neutron matter [23]. Dashed lines are obtained with the more sophisticated model 1 of Prakash, Ainsworth, and Lattimer [24].

fm^{-3} which corresponds to the Fermi wave number $k_F = (3\pi^2 n)^{1/3} = 0.25$ fm^{-1}. By further increasing the density n the repulsive core of the neutron-neutron potential plays an important role in destroying the correlation of Cooper pairs and the condensate fraction n_0/n slowly goes to zero. Remarkably, the results of Fig. 2 are fully consistent with the Monte Carlo value $n_0/n \simeq 0.35$ at $n = 0.003$ fm^{-3} one extracts from the finite-temperature Path Integral Monte Carlo data of Wlazlowski and Magierski [11, 12].

5. Neutron stars

Neutron stars are astronomical compact objects which can result from the gravitational collapse of a massive star during a supernova event. Such stars are mainly composed of neutrons. Neutron stars are very hot and are supported against further collapse by Fermi pressure. A typical neutron star has a mass M between 1.35 and about 2.0 solar masses with a corresponding radius R of about 12 km. Notice that in the crust of neutron stars one estimates [15] a temperature $T \simeq 10^8$ K, while the critical temperature of the normal-superfluid transition is $T_c \simeq 10^{10}$ K. Thus the crust of neutron stars is superfluid.

Some years ago, Datta, Thampan, and Bhattacharya [22] have calculated several mass density profiles $\rho(r)$ of spherical and non-rotating neutron stars by solving the Tolman- Oppenheimer-Volkoff (TOV) equation, which describes the interplay between the expulsive kinetic pressure of the star and its gravitational self-attraction [14, 15]. Datta, Thampan, and Bhattacharya [22] have solved the TOV equation by using various equations of state (EOS) of the nuclear matter. In the left panel of Fig. 3 we plot their results in the case of 1.4 solar mass neutron star. In particular, we report the scaled density profile $n(r)/n_s$ of the neutron star as a function of the scaled distance r/R from the center of the star, where n_s is the nuclear saturation density and R is the radius of the star. Notice that $n(r) = \rho(r)/m_N$ with m_N the neutron mass. The solid line is obtained [22] solving the TOV equation with the simple nuclear EOS of Walecka [23], while the dashed line is obtained [22] solving the TOV equation with a more sophisticated EOS, called model 1, of Prakash, Ainsworth and Lattimer [22, 24].

In the previous section we have found a fitting formula for the condensate fraction n_0/n of neutron matter as a function of the bulk neutron density n. Knowing the density profile $n(r)$ of a neutron star, i.e. the neutron density n as a function of the distance r from the center of a neutron star, we can determine (local density approximation) the condensate fraction n_0/n of the neutron star as a function of the distance r.

The results are shown in the right panel of Fig. 3. The figure shows that the two EOS give very similar results and that a relevant condensate fraction appears only in the crust of the neutron star in the region between $r/R = 0.85$ and $r/R = 1$ with a maximum value $\simeq 0.4$ at $r/R \simeq 0.96$. We stress that this suggestive plot can be certainly improved because in neutron stars the hadronic matter is not made only of neutrons [14, 15]. On the other hand, it is exactly in the crust of neutron stars that it is expected to find the dilute neutron matter we have considered to derive Eq. (27).

6. Conclusions

We have seen that the condensate fraction of Cooper pairs can be calculated in various superfluid fermionic systems: dilute atomic gases, dense neutron matter and neutron stars. We observe that, while the condensate fraction in ultracold gases of fermionic atoms has been measured in two sophisticated experiments by measuring the momentum distribution of pairs [9, 10], it remains open the exciting problem of finding reliable observational signatures of the condensate fraction of neutron-neutron pairs in atomic nuclei and in neutron stars. In conclusion, we point out that recently we have studied the behavior of the condensate fraction in ultracold and dilute gases of fermionic atoms not only in the 3D uniform system but also in other configurations: 2D uniform system [25], 2D system on a square lattice [26], 3D and 2D uniform system with spin-orbit coupling [27, 28], 3D and 2D uniform system with three-spin components [29], and 3D uniform system with a narrow Feschbach resonance [30].

Acknowledgments

The author thanks Masayuki Matsuo for making available his numerical data and Roberto Turolla for useful discussions. The author acknowledges research grants from Università di Padova (Progetto di Ricerca di Ateneo 2012-201), Fondazione CARIPARO (Progetto di Eccellenza 2012-2013), and Ministero dell'Istruzione Università e Ricerca (Progetto PRIN call 2011-2012).

References

[1] Penrose O 1951 Phil. Mag. **42** 1373
[2] Penrose O and Onsager L 1956 Phys. Rev. **104** 576
[3] Yang CN 1962 Rev. Mod. Phys. **34** 694
[4] Pethick C and Smith H 2002, Bose-Einstein Condensation in Dilute Gases (Cambridge: Cambridge University Press)
[5] Ceperly DM 1995 Rev. Mod. Phys. **67** 2
[6] Salasnich L, Manini N, and Parola A 2005 Phys. Rev. A **72**, 023621
[7] Ortiz G and Dukelsky J 2005 Phys. Rev. A **72** 043611
[8] Astrakharchik GE, Boronat J, Casulleras J, and Giorgini S 2005 Phys. Rev. Lett. **95** 230405
[9] Zwierlein MW, Schunck CH, Stan CA, Raupach SMF, and Ketterle W 2005 Phys. Rev. Lett. **94** 180401
[10] Inada Y, Horikoshi M, Nakajima S, Kuwata-Gonokami M, Ueda M, and Mukaiyama T 2008 Phys. Rev. Lett. **101** 180406
[11] Wlazlowski G and Magierski P 2011 Phys. Rev. C **83** 012801(R)
[12] Wlazlowski G and Magierski P 2011 Int. J. Mod. Phys. E **20** 569
[13] Salasnich L 2011 Phys. Rev. C **84**, 067301
[14] Pethick CJ and Ravenhall DG 1995 Ann. Rev. Nucl. Part. Sci. **45** 429
[15] Zane S, Turolla R, and Page D 2007 Isolated Neutron Stars: From the Surface to the Interior (Dordrecht: Springer)
[16] Leggett AJ 2006 Quantum liquids. Bose condensation and Cooper pairing in condensed-matter systems (Oxford: Oxford University Press)
[17] Yukalov VI 2007 Laser Phys. Lett. **4** 632
[18] Campbell CE 1997 in Condensed Matter Theories, vol. **12**, 131 (New York: Nova Science)

[19] Marini M, Pistolesi P, and Strinati GC 1998 Eur. Phys. J. B **1** 151
[20] Ketterle W and Zwierlein MW 2008 in Ultracold Fermi Gases, Proceedings of the International School of Physics "Enrico Fermi", Course CLXIV, Varenna, 20 - 30 June 2006, edited by M. Inguscio, W. Ketterle, and C. Salomon (Amsterdam: IOS Press); in particular, see Fig. 46 and its analysis
[21] Matsuo M 2006 Phys. Rev. C **73** 044309
[22] Datta B, Thampan AV, and Bhattacharya D 1995 J. Astrophys. Astr. **16** 375
[23] Walecka JD 1974 Ann. Phys. (N.Y.) **83** 491
[24] Prakash M, Ainsworth TL, and Lattimer JM 1988 Phys. Rev. Lett. **61** 2518
[25] Salasnich L 2007 Phys. Rev. A **76** 015601
[26] Salasnich L and Toigo F 2012 PRA **86** 023619
[27] Dell'Anna L, Mazzarella G, and Salasnich L 2011 Phys. Rev. A **84** 033633
[28] Dell'Anna L, Mazzarella G, and Salasnich L 2012 Phys. Rev. A **86** 053632
[29] Salasnich L 2011 Phys. Rev. A **83** 033630
[30] Salasnich L 2012 Phys. Rev. A **86** 055602

Journal of Physics: Conference Series **497** (2014) 012027 doi:10.1088/1742-6596/497/1/012027

Resonances in ultracold collisions con ned by atomic traps

Vladimir S. Melezhik

Bogoliubov Laboratory of Theoretical Physics, Joint Institute for Nuclear Research, Dubna, Moscow Region 141980, Russian Federation

E-mail: `melezhik@theor.jinr.ru`

Abstract. We have developed a method for treating ultracold pair collisions in con ned geometry of tight atomic waveguides. With this approach we have predicted and analyzed a few novel resonant and multichannel e ects induced by con ning traps in ultracold atomic collisions. In this report we discuss our model yielding the shifts and widths of magnetic Feshbach resonances in tight atomic waveguides. We also briefly discuss a resonant mechanism of molecule formation in atomic traps with transferring the energy release to the center-of-mass excitation while forming molecule and dipolar con nement-induced resonances in waveguides. These results could potentially be used experimentally.

1. Introduction

Impressive progress of the physics of ultracold quantum gases have opened new pathways for the study of low-dimensional few-body systems (see, for example [1]) as well as strongly correlated many-body systems [2]. It has stimulated the necessity of detailed and comprehensive investigations of collisional processes in the confined geometry of atomic traps. It soon became clear that here the free-space scattering theory is no longer valid and the development of the low-dimensional theory including the influence of the confinement is needed [3]. In our works [4, 5, 6, 7] we have developed a computational method for pair collisions in tight atomic waveguides and have found several novel effects in its application: the confinement-induced resonances (CIRs) in multimode regimes including effects of transverse excitations and deexcitations [5], the so-called dual CIR yielding a complete suppression of quantum scattering [4, 8], and resonant molecule formation with a transferred energy to center-of-mass (CM) excitation while forming molecules [9]. Our calculations have also been used for planning and interpretation of the Innsbruck experiment on investigation of CIRs in ultracold Cs gas [10]. Our computational schemes are based on the works [11, 12], where a method for solving stationary few-dimensional scattering problem as a boundary value problem was suggested, and [13] devoted to a component-by-component splitting up method for three-dimensional time-dependent Schrödinger equation. A key element of the developed economic computational schemes is the non-direct product 2D discrete-variable representation we suggested for approximation angular variables [14].

In this report, we discuss the most important results obtained recently in our study of resonances in ultracold collisions confined by atomic waveguides. It is the development of a theoretical model for describing Feshbach resonance shift and width induced by an atomic waveguide [15]. It is a resonant mechanism of molecule formation in an atomic waveguide with transferring the energy release to the CM excitation while forming molecule we suggested in [9]. We briefly discuss this model, which becomes actual in connection with recent experiments in Heidelberg on the resonant formation of i molecules in atomic traps [16]. We also briefly discuss our nonperturbative model to treat pair collisions with long-range anisotropic interactions in quasi-one-dimensional geometries [17]. This model avoids the limitations of pseudopotential theory [18, 19] and gives exact prediction of the positions of dipolar CIRs in waveguides, which may pave the way for the experimental realization of special phases in one-dimensional dipolar gases (see, for example [20]).

2. Shifts and widths of Feshbach resonances in atomic waveguides

Already at the initial stage of the investigation of ultracold pair collisions in atomic traps, it was shown that the confining geometry can drastically change the scattering properties of ultracold atoms and induce resonances in the collisions (confinement-induced resonances (CIRs)) [3]. Further research in this field has shown that a necessary ingredient for the experimental realization of the CIR in a confining trap is the existence of magnetic Feshbach resonances in free-space [1, 2, 10, 21]. However, until recently, single-channel potential models with zero-energy bound states were used for simulating the magnetic Feshbach resonances in the 3D free-space (see, for example, the papers [3, 4, 5, 7, 9, 16, 17, 18, 19, 22, 23]). The simplified models were used despite that the single-channel interatomic interaction approach permits to explore only the main attribute of the Feshbach resonances in the 3D free-space, namely the appearance of a singularity in the s-wave scattering length $a \to \pm\infty$ when the molecular bound state with energy E_c crosses the atom-atom scattering threshold at energy $E = 0$ in the entrance channel. Other important parameters of the Feshbach resonance, such as the rotational and spin structure of the molecular bound state in the closed channel, background scattering length a_{bg} as well as the width   of the resonance characterizing the coupling Γ of the molecular state with the entrance channel [1], were ignored.

The main goal of our work [15] was to exit from the frame of the single-channel theoretical approaches developed earlier for the CIRs and take into account the effects of different rotational structure of the Feschach resonance, the resonance width and the background scattering. We have developed and a theoretical model [15] which yields the shifts and widths of Feshbach resonances (having different tensorial structure) in an atomic waveguide. It is based on our multichannel approach for CIRs and atomic transitions in the waveguides in the multimode regime [5]. We replace in this scheme the single-channel scalar interatomic interaction by the four-channel tensorial potential $\hat{V}(r)$ modeling resonances of broad, narrow and overlapping character according to the two-channel parametrization of A.D. ange et. al. [24].

In this approach the scattering amplitude $f(k,B)$ describing a pair collision in free-space ($\omega_\perp = 0$) is calculated for different magnetic field strengths B and varying parameters of the potential $\hat{V}$ by solving the Schrödinger equation

$$\left([-\frac{\hbar^2}{2\mu}\nabla^2 + \frac{1}{2}\mu\omega_\perp^2\rho^2]\hat{I} + \hat{B} + \hat{V}(r)\right)|\psi\rangle = E|\psi\rangle \tag{1}$$

with the scattering boundary conditions

$$\psi_e(\mathbf{r}) \to \exp\{ikz\} + f(k,B)/r\exp\{ikr\}\,, \quad \psi_{c,i}(\mathbf{r}) \to 0$$

at $kr \to \infty$ for the fixed $E \to 0$ ($k = \sqrt{2\mu E}/\hbar \to 0$) [12]. The diagonal matrix $\hat{B}$ in (1) is defined as $B_{ii} = \delta\mu_i(B - B_i)$ (here $\delta\mu_i$ is the relative magnetic moment of the channel i, $i = 1,2,3$)

and $B_{ee} = 0$. After separation of the angular part in (1) we come to the system of four coupled radial equations

$$[-\frac{\hbar^2}{2\mu}\frac{d^2}{dr^2} + \frac{\hbar^2 l_\alpha(l_\alpha+1)}{2\mu r^2} + B_{\alpha\alpha}]\phi_\alpha(r) + \sum_\beta V_{\alpha\beta}(r)\phi_\beta(r) = E\phi_\alpha(r) \qquad (2)$$

for the radial part $\phi_\alpha(r)$ of the desired wave function $|\psi\rangle = \sum_\alpha \psi_\alpha(\mathbf{r})|\alpha\rangle = \sum_\alpha \phi_\alpha(r)Y_{l_\alpha 0}(\hat{r})|\alpha\rangle$, where $\alpha = \{e, i = 1,2,3\}$, $(l_e = 0, l_1 = 0, l_2 = 2, l_3 = 4)$ and the matrix elements $V_{\alpha\beta}(r)$ are defined by four diagonal parameters $V_{c,i}$, V_e and three non-diagonal ones Ω_i [15]. The centrifugal barrier $\frac{\hbar^2 l_\alpha(l_\alpha+1)}{2\mu r^2}$ in (2) models at $r \to 0$ the correct asymptotic behavior $\phi_i(r) \sim r^{l_i+1}$ of the molecular bound states $|c,i\rangle$ in the closed channels which couple to the entrance s-wave $(l_e = 0)$ channel $|e\rangle$ by the nondiagonal terms $V_{\alpha\beta}(r)(\alpha \neq \beta)$.

By varying the $V_{c,i}$, V_e and Ω_i we have obtained an excellent agreement of the calculated s-wave scattering length $a(B) = -f(k \to 0, B)$ with the analytical results from [24] for Cs atoms in the hyperfine state $|F = 3, m_F = 3\rangle$ for the considered magnetic field regime $-40G < B < 60G$, corresponding to the conditions of Innsbruck experiments [10]. In this regime, we observe three resonances B_i in $a(B)$, which correspond to the coupling to the s-, d-, and g-wave molecular states in the closed channels $|c,i\rangle$.

Then, after finding the parameters of interparticle interaction $\hat{V}(r)$, we have analyzed the scattering properties of the s-, d- and g-wave magnetic Feshbach resonances in harmonic waveguides by integrating the Schrödinger equation (1) for $\omega_\perp \neq 0$ with the scattering boundary conditions

$$\psi_e(\mathbf{r}) = \left(\cos(k_{||}z) + f_e(B)\exp\{ik_{||} \mid z \mid\}\right)\Phi_0(\rho)\ , \quad \psi_{c,i}(\mathbf{r}) \to 0 \qquad (3)$$

at $\mid z \mid = \mid r\cos\theta \mid \to \infty$ adopted for a confining trap [5]. $f_e(B,E)$ is the scattering amplitude, corresponding to symmetric states with respect to the exchange $z \to -z$ (we consider collisions of identical bosonic Cs atoms), $\Phi_0(\rho)$ is the wave function of the ground-state of the two-dimensional harmonic oscillator and $k_{||} = \sqrt{2\mu(E - \hbar\omega_\perp)}/\hbar = \sqrt{2\mu E_{||}}/\hbar$. In the presence of of the harmonic trap $(\omega_\perp \neq 0)$ the problem (1),(3) becomes non-separable in the plane $\{\rho, z\}$, i.e. the azimuthal angular part is separated in the wave function $|\psi\rangle$ and Eq.(1) is reduced to the coupled system of four 2D Schrödinger-type equations. To integrate this coupled channel 2D scattering problem in the plane $\{r, \theta\}$ we have applied the extended version of the computational scheme [5]. The computations have been performed in a range of variation of $\omega_\perp$ close to the experimental values of the trap frequencies $\omega_\perp \sim 2\pi \times 14.5$kHz [10]. We have integrated Eq.(1) for varying B and fixed longitudinal colliding energy $E_{||} = E - \hbar\omega_\perp \to 0$.

As an input the experimentally known parameters of Feshbach resonances in the absence of the waveguide [24], namely the resonant energies $E_{c,i}$ (or the corresponding values of the field strengths B_i of the external magnetic field), the widths of the resonances $_i(\Gamma_i)$, spin characteristics and the background scattering length a_{bg}, were used. The output of the model is the calculated scattering amplitude $f_e(B)$ in the confining trap $\omega_\perp$ and the transmission $\mathrm{T}(B) = \mid 1 + f_e(B) \mid^2$.

We have calculated the shifts and widths of s-, d- and g-wave magnetic Feshbach resonances of Cs atoms emerging in harmonic waveguides as CIRs (the minimums in the transmission $\mathrm{T}(B)$) and resonant enhancement of transmission T(B) at zeros $a(B) = 0$ of the free-space scattering length. The results are illustrated by the curves T(B) calculated near the d-wave resonance for different $\omega_\perp$ (Fig.1) corresponding to the transverse frequencies of the optical trap being used in the experiment [10]. It is shown that the position of B_{min} of the transmission coefficient T(B) minimum is dependent on the transverse trap width $a_\perp = \sqrt{\hbar/(\mu\omega_\perp)}$ according to simplified Olshanii model [3]. Actually, the corresponding scattering length $a(B_{min})$ at the point B_{min} is accurately described by the formulas $a(B_{min}) = a_\perp/C$ obtained in [3] for the position of the CIR in s-wave case. Our analysis has shown that this law is fulfilled with high accuracy for Feshbach resonances of different tensorial structure which holds in spite of the fact that the formulas defining the position of CIR was originally obtained in the framework of the s-wave single-channel pseudopotential approach (see Fig.2). This property was experimentally confirmed for d-wave Feshbach resonances in a gas of Cs atoms [10].

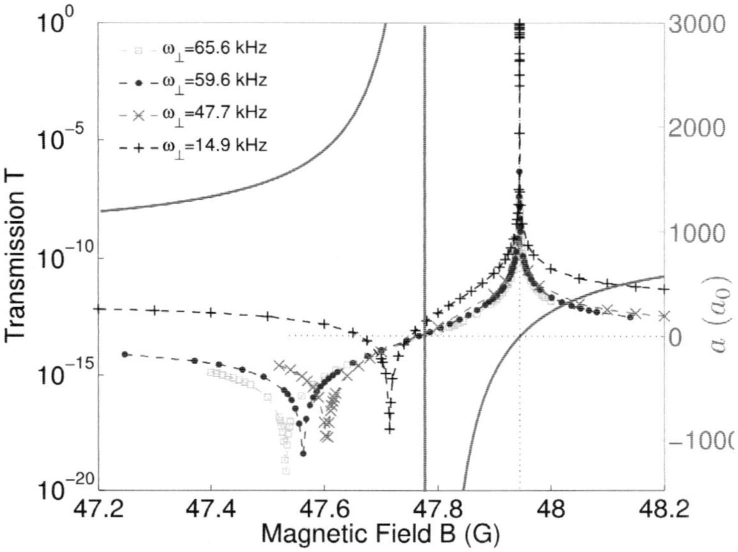

Figure 1. The transmission T(B) as a function of the strength of the magnetic field for several trap frequencies $\omega_\perp$. Here the s-wave free-space scattering length $a(B)$ as a function of B is also given in atomic units a_0 (solid curve).

We have also found the linear dependence of the width of the resonance in the transmission $\mathrm{T}(B^*)$ at the point $a(B^*) = 0$ on the longitudinal atomic momentum $k_{||} = \sqrt{2\mu(E - \hbar\omega_\perp)}/\hbar = \sqrt{2\mu E_{||}}/\hbar$ and quadratic dependence on the trap width $a_\perp$: $\Gamma_i^* = \ _i k_{||} a_\perp^2/a_{bg}$ (here $\ _i = B^* - B_i$ is the width of the Feshbach resonance at B_i and a_{bg} is the background scattering length in free-space). This formula was confirmed by numerical results at $k_{||} \to 0$. The found effect could potentially be used experimentally. Actually, one can control the width $\Gamma_i^* = \ _i k_{||} a_\perp^2/a_{bg}$ of the resonance by varying the trap width $a_\perp$, an increase of $\omega_\perp$ leads to a narrowing of the resonance. From the other side, by measuring the width Γ_i^* one can extract from the obtained formulae [15] important information about the longitudinal colliding energy $E_{||} = E - \hbar\omega_\perp$ and estimate the temperature of the atomic cloud in the trap.

Finally, we have calculated the populations of the molecular states in closed channels at pair atomic collisions in a trap. It was found that the molecule formation rates in a waveguide show an enhancement for the case of a corresponding zero of the s-wave scattering length $a(B^*) = 0$. The effect is illustrated by Fig.3. We have also shown that the positions of these resonances are stable w.r.t. the variation of the confining frequency $\omega_\perp$ of the waveguide.

3. Resonance mechanism of molecule formation in 1D trap with CM excitation

In our paper [9] a resonant mechanism for molecule formation in heteronuclear atomic collisions in waveguides with transferring the energy release to the CM excitation was suggested. The key point of this mechanism is the confinement-induced mixing of the relative and CM motions in the atomic collision process due to the coupling term

$$W(\rho_{CM}, \mathbf{r}) = \mu(\omega_1^2 - \omega_2^2) r \rho_{CM} \sin\theta \cos\phi \tag{4}$$

in the Hamiltonian of the atomic pair in the confining potential of the trap. Here ρ_{CM} is the radial variable of the CM, $\mathbf{r}$ are the relative variables and ω_i is the frequency of the confining trap

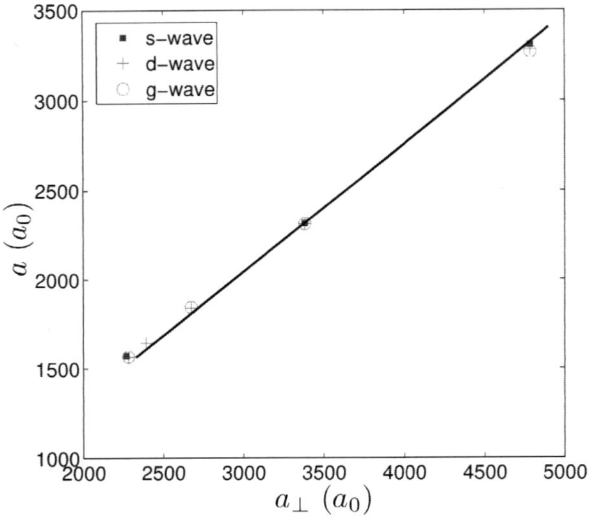

Figure 2. The dependence of the s-wave scattering length $a(B)$ on the trap width $a_\perp$ at the points B_{min} of the minimum of the transmission $\mathrm{T}(a(B), a_\perp)$ (see Fig.1). The dots, pluses and cycles correspond to the calculated points near the s-,d- and g-wave magnetic Feshbach resonances, respectively. The solid line corresponds to the formula $a = a_\perp/C$ (where C=1.4603) from [3]. The values $a(B)$ and $a_\perp$ are given in atomic units a_0.

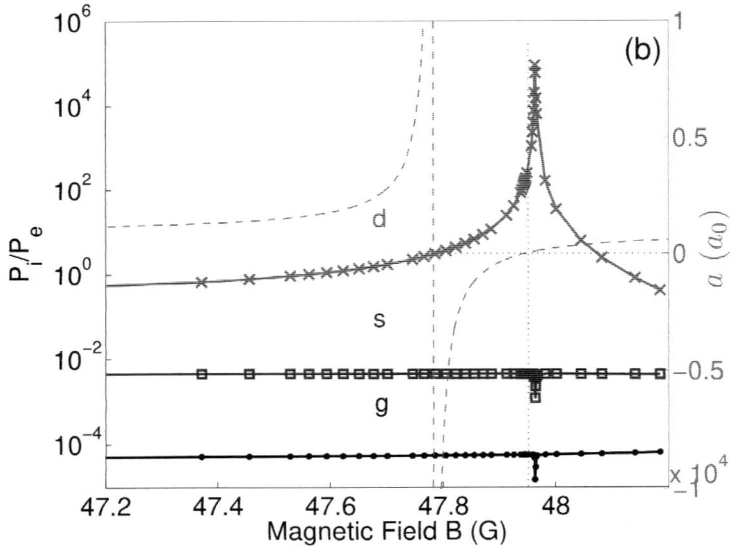

Figure 3. The relative population P_i/P_e of the molecular states i = s, d and g calculated as a function of magnetic field B near the d-wave Feshbach resonance in Cs for pair atomic collisions in the harmonic waveguide $\omega_\perp$ =59.6 kHz. The free-space scattering length $a(B)$ is given in atomic units.

corresponding to the atom i. This term leads to a coupling of the diatomic continuum to the CM excited molecular states in closed transverse channels. By using our wave-packet propagation method we have modeled the collisional dynamics of two atoms in the ground transverse state $n_1 = n_2 = 0$ ($n = N = 0$) of the harmonic waveguide near the CIRs (n_i is the quantum number of the harmonic spectrum of the atom i, the quantum numbers n and N numerates the harmonic spectrums of the relative and the CM transverse eigenstates). The possibilities of the resonance molecule formation in a harmonic wavegaide under the scheme

$$A_{n_1=0} + B_{n_2=0} \rightarrow (AB)_{n=0,N=1}$$

with the CM excitation $n = N = 0 \rightarrow n = 0, N = 1$ was investigated in details in [9] and the crucial role of the coupling term was demonstrated. It was shown that the molecular formation probabilities can be tuned by changing the trap frequencies ω_i, that characterize the transverse modes of the atomic species, and by varying the longitudinal colliding energy $E_{\|}$.

It is known that for identical atoms (when $\omega_1 = \omega_2$) the coupling term (4) is absent but the mixing of the relative and CM variables is, nevertheless, present in the two-body Hamiltonian in anharmonic traps (see, for example, [25, 26]). In the work [26] the mechanism of molecule formation with transferring the energy release to CM excitation of forming molecule was considered for an anharmonic lattice.

Experimental confirmation of the resonate molecule formation in an atomic waveguide with transferring the energy release to CM excitation was recently obtained [16], what opens iteresting possibilities for investigations in this field.

4. Dipolar CIRs of ultracold gases in waveguides

By using our method [5, 6, 12] we develop a nonperturbative theoretical framework to treat collisions with generic anisotropic interactions in quasi-one-dimensional geometries [17]. The developed scheme avoids the limitations of pseudopotential theory [8, 19] and allows us to include accurately long-range anisotropic interactions. For ultracold dipolar collisions in a harmonic waveguide we predict dipolar confinement-induced resonances (DCIRs) which are attributed to different angular momentum states. The analytically derived resonance condition reveals in detail the interplay of the confinement with the anisotropic nature of the dipole-dipole interactions. The results are in excellent agreement with ab initio numerical calculations confirming the robustness of the presented approach. The exact knowledge of the positions of DCIRs may pave the way for the experimental realization of different exotic phases in effective one-dimensional dipolar gases [20].

5. Conclusion

In this report we have presented three novel resonant effects induced by confining waveguides in ultracold atomic collisions which we have analyzed recently with the developed computational method. These results could potentially be used experimentally. The high efficiency and flexibility of the method we observed in theses studies make it very promising in application to other resonant processes in atomic traps of different geometry.

Possible actual problems can be the fermionic atom collisions in waveguides, ultacold pair processes in pancake-like traps, triple ultracold collisions.

Acknowledgments

Presented results were obtained in collaborative work with P. Schmelcher, S. Saeidian, and P. Giannakeas, to whom I express my deep gratitude. I also acknowledge financial support by the Deutsche Forschungsgemeinschaft and the Heisenberg- andau Program.

References

[1] Chin C, Grimm R, Julienne P S and Tiesinga E 2010 *Rev. Mod. Phys.* **82** 1225
[2] Bloch I, Dalibard J and werger W 2008 *Rev. Mod. Phys.* **80** 885
[3] Olshanii M 1998 *Phys. Rev. Lett.* **81** 938
[4] Melezhik V S, Kim J I and Schmelcher P 2007 *Phys. Rev.* A **76** 053611
[5] Saeidian S, Melezhik V S and Schmelcher P 2008 *Phys. Rev.* A **77** 042721
[6] Melezhik V S 2012 Multi-channel computations in low-dimensional few-body physics *Lecture Notes in Computer Science* vol 7125 (Berlin Heidelberg: Springer-Verlag) p 94 (ar iv:1110.3919)
[7] Melezhik V S 2013 *Phys. Atom. Nucl.* **76** 139
[8] Kim J I, Melezhik V S and Schmelcher P 2006 *Phys. Rev. Lett.* **97** 193203
[9] Melezhik V and Schmelcher P 2009 Quantum dynamics of resonant molecule formation in waveguides *New J. Phys.* **11** 073031
[10] Haller E, Mark M J, Hart R, Danzl J G, Reichsollner L, Melezhik V, Schmelcher P and Nagerl H C 2010 *Phys. Rev. Lett.* **104** 153203

Journal of Physics: Conference Series **497** (2014) 012027 doi:10.1088/1742-6596/497/1/012027

[11] Melezhik V S 1991 *J. Comput. Phys.* **92** 67
[12] Melezhik V S and Hu Chi-Yu 2003 *Phys. Rev. Lett.* **90** 083202
[13] Melezhik V S 1997 *Phys. Lett.* A **230** 203
[14] Melezhik V S 2012 *Int. Conf. Numerical Analysis and Applied Mathematics ICNAAM2012 (Kos, Greece)* AIP Conf. Proc. vol 1479 (American Institute of Physics) p 1200
[15] Saeidian S, Melezhik V S and Schmelcher P 2012 *Phys. Rev.* A **86** 062713
[16] S Sala, urn G, Lompre T, Wenz A N, Murmann S, Serwane F, Jochim S and Saenz A 2013 *Phys. Rev. Lett.* **110** 203202
[17] Giannakeas P, Melezhik V S and Schmelcher P 2013 *Phys. Rev. Lett.* **111** 183201
[18] Sinha S and Santos L 2007 *Phys. Rev. Lett.* **99** 140406
[19] Deuretzbacher F, Cremon J C and Reimann S M 2010 *Phys. Rev.* A **81** 063616
[20] Lahaye T, Menotti C, Santos L, Lewenstein M and Pfau T 2009 *Rep. Prog. Phys.* **72** 126401
[21] Lamporesi G, Catani J, Barontini G, Nishida Y, Inguscio M and Minardi F 2010 *Phys. Rev. Lett.* **104** 153202
[22] Melezhik V S and Schmelcher P 2011 *Phys. Rev.* A **84** 042712
[23] Bergeman T, Moore M G and Olshanii M 2003 *Phys. Rev. Lett.* **91** 163201
[24] Lange A D, Pilch K, Prantner A, Ferlaino F, Engesser B, Nagerl H C, Grimm R and Chin C *Phys. Rev.* A **79** 013622
[25] Peano V, Thorwart M, Mora C and Egger R 2005 Con nement-induced resonances for a two-component ultracold atom gas in arbitrary quasi-one-dimensional traps *New J. Phys* **7** 192
[26] Bolda E L, Tiesinga E and Julienne P S 2005 *Phys. Rev. A* **71** 033404

Journal of Physics: Conference Series **497** (2014) 012028 doi:10.1088/1742-6596/497/1/012028

Collective modes of three- and two-dimensional trapped Fermi gases in the normal phase

M. Urban[1], S. Chiacchiera[2], D. Davesne[3], T. Enss[4] and P.-A. Pantel[3]

[1] Institut de Physique Nucléaire, CNRS/IN2P3 and Univ. Paris Sud, 91406 Orsay cedex, France

[2] Centro de Física Computacional, Department of Physics, University of Coimbra, 3004-516 Coimbra, Portugal

[3] IPN Lyon, Université de Lyon, Université Lyon 1, CNRS/IN2P3, 69622 Villeurbanne cedex, France

[4] Institut für Theoretische Physik, Universität Heidelberg, D-69120 Heidelberg, Germany

E-mail: urban@ipno.in2p3.fr

Abstract. We study the transition from the hydrodynamic to the collisionless regime in collective modes of three- and two-dimensional Fermi gases by using the semiclassical Boltzmann equation. We use direct numerical simulations as well as the method of phase-space moments to solve the Boltzmann equation and show that the restriction to second-order moments is not accurate enough. By including higher-order moments, we can successfully describe the hydrodynamic to collisionless transition observed in the quadrupole mode in three-dimensional Fermi gases and the frequency shift and damping of the sloshing mode due to the anharmonic shape of the experimental trap potential. In the case of two-dimensional Fermi gases, however, the strong damping of the quadrupole mode observed in a recent experiment remains unexplained.

1. Introduction

By studying collective motion in ultracold atoms, one can address some questions that cannot be answered from the measurement of static (equilibrium) quantities alone. Concentrating here on the case of unpolarized Fermi gases, let us mention the superfluid-normal phase transition, which can be seen most convincingly by looking at vortices in rotating traps [1]. It also shows up, somewhat less impressively, in the radial quadrupole mode [2, 3]. Also in the normal phase, above the critical temperature, Fermi gases in the strongly interacting regime reveal interesting dynamical properties. For instance, in the anisotropic expansion of normal-fluid Fermi gases in the unitary limit (i.e., exactly on a Feshbach resonance, where the scattering length a diverges) it was observed that the viscosity of these systems is surprisingly low [4]. With increasing temperature, however, the system expands, gets more and more dilute, and finally it becomes essentially collisionless [5]. As one approaches the collisionless regime, the concept of viscous hydrodynamics breaks down and a more microscopic description in the framework of the Boltzmann equation is needed.

In the present work we concentrate on collective oscillations with small amplitude. Some of these collective modes are depicted schematically in Fig. 1. The first one (a) is the sloshing mode, an oscillation of the atom cloud as a whole with the trap frequency. This mode is very

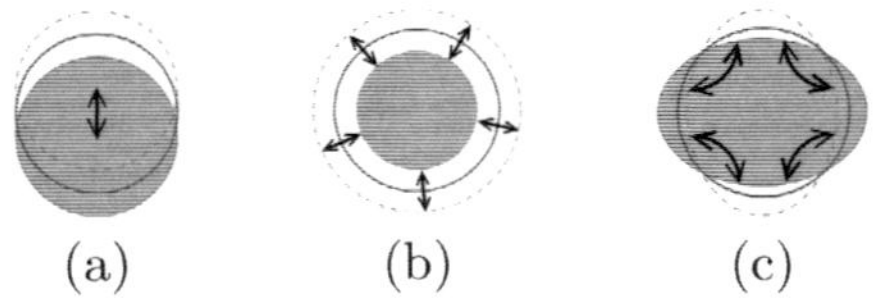

Figure 1. Schematic representation of different collective modes, as seen in the xy plane: (a) sloshing mode, (b) breathing mode, (c) radial quadrupole mode.

useful since it allows the experimentalists to determine the trap frequency. However, as discussed later, in a realistic (Gaussian) trap one has to be careful to account for anharmonicity effects. The second example (b) shows the radial breathing mode. Breathing modes are of interest because they give information about the equation of state. While in the usual three-dimensional traps the axial breathing mode behaves always hydrodynamically because of its low frequency, the frequency and damping rate of the radial breathing mode are sensitive to the transition from the hydrodynamic to the collisionless regime. This is even more pronounced in the case of the radial quadrupole mode (c), whose frequency changes from $\sqrt{2}\omega_r$ (ω_r being the radial trap frequency) in the hydrodynamic limit to approximately $2\omega_r$ in the collisionless limit. The damping is maximal in the intermediate regime between the two limits.

2. Formalism

We consider a two-component ($\uparrow,\downarrow$) Fermi gas with equal distribution functions $f_\uparrow(\mathbf{r},\mathbf{p},t) = f_\downarrow(\mathbf{r},\mathbf{p},t) = f(\mathbf{r},\mathbf{p},t)$. The Boltzmann equation describing the time dependence of the common distribution function $f(\mathbf{r},\mathbf{p},t)$ reads

$$\dot{f} + \frac{\mathbf{p}}{m}\cdot\boldsymbol{\nabla}_r f - \boldsymbol{\nabla}_r V\cdot\boldsymbol{\nabla}_p f = -I[f] \tag{1}$$

where m is the atom mass, $V(\mathbf{r}) = V_{trap}(\mathbf{r}) + U(\mathbf{r})$ is the sum of trap potential and mean field. The mean field $U(\mathbf{r})$ is obtained in local-density from the real part of the self-energy, $\mathrm{Re}\,\Sigma$, calculated in ladder approximation (see Fig. 2) [6]. The collision term I on the rhs. of Eq. (1) reads

$$I[f] = \int\frac{d^3p_1}{(2\pi\hbar)^3}\int d\Omega\frac{d\sigma}{d\Omega}|\mathbf{v}-\mathbf{v}_1|\,(ff_1\bar{f}'\bar{f}_1' - f'f_1'\bar{f}\bar{f}_1) \tag{2}$$

where the factors $\bar{f} = (1-f)$ etc. are included to take Pauli blocking into account, $\mathbf{v}-\mathbf{v}_1$ is the relative velocity of the two colliding atoms. The cross section σ is obtained from the absolute square of the in-medium T matrix (see Fig 2). At low temperature, the in-medium cross section σ is strongly enhanced compared to the free-space one, $\sigma_{\text{free}} = 4\pi a^2/[1+(qa)^2]$, especially for head-on collisions ($\mathbf{p}_1 \sim -\mathbf{p}$) with $|\mathbf{p}| \sim |\mathbf{p}_1| \sim k_F$. This can be interpreted as a precursor effect of the pole in the T matrix at $T = T_c$ [6, 7]. (The same effect was found in nuclear physics, cf. [8]).

In Eq. (2), the collision term is written for the three-dimensional case. In two dimensions one just has to replace the integral over $d^3p_1d\Omega$ (Ω being the solid angle characterizing the direction of $\mathbf{p}'$) by an integral over $d^2p_1d\theta$ (θ being the scattering angle). The enhancement of the in-medium cross section compared to the free-space one is also found in two dimensions.

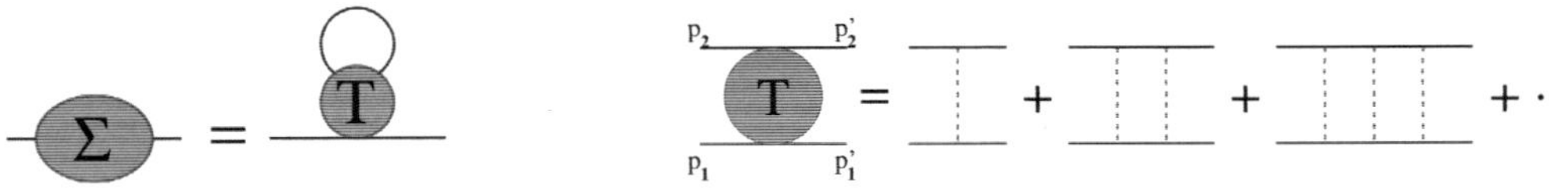

Figure 2. Feynman diagrams for the self-energy (left) and the in-medium T matrix (right) in ladder approximation.

In practice, we employ two different methods to solve Eq. (1). The first one is a numerical simulation where the continuous function $f(\mathbf{r},\mathbf{p},t)$ is replaced by a set of delta functions, the so-called test particles. So far, test-particle calculations in the context of cold atoms [9, 10, 11] have not included the mean field U and the medium modified cross section. The main difficulty is that it takes too much computation time to evaluate the in-medium cross section for each pair of possibly colliding test particles. We are presently working on a solution of this problem by combining analytical expressions for the in-medium cross section obtained in a high-temperature expansion with a parametrization fitted to the full cross-section at low temperatures. The test-particle results discussed here were obtained with the free cross section.

The second method is the method of moments. For small oscillations, the distribution function can be written in the form

$$f(\mathbf{r},\mathbf{p},t) = f_{\text{eq}}(\mathbf{r},\mathbf{p}) + \frac{df_{\text{eq}}}{d\mu}\,\Phi(\mathbf{r},\mathbf{p},t) \tag{3}$$

where f_{eq} denotes the distribution function in equilibrium, i.e., the Fermi function with chemical potential μ, and Φ is a smooth function. We make the ansatz

$$\Phi(\mathbf{r},\mathbf{p},t) = \sum_{i=1}^{n} c_i(t)\phi_i(\mathbf{r},\mathbf{p})\,, \tag{4}$$

where ϕ_i are monomials in $\mathbf{r}$ and $\mathbf{p}$ and c_i are time-dependent coefficients. We linearize the Boltzmann equation, i.e., we keep only first-order terms in Φ. By taking phase-space moments of this linearized equation, i.e., by multiplying it by each of the ϕ_i and integrating over phase space, we obtain a system of n linear equations of motion for the coefficients c_i which can easily be solved by matrix diagonalization.

3. Importance of higher-order moments

Often only the minimal set of functions ϕ_i necessary for the description of a collective mode [6, 12, 13] is used. For instance, for the breathing, quadrupole, and scissors modes, one needs second-order terms to describe the oscillation of the system shape in coordinate (e.g., $\phi_1 = xy$) and momentum space (e.g., $\phi_2 = p_x p_y$) and of the velocity field (e.g., $\phi_3 = xp_y$, $\phi_4 = yp_x$), while the simplest possible sloshing motion requires only first-order terms ($\phi_1 = x$, $\phi_2 = p_x$). However, it is important to check quantitatively how strong the corrections due to higher-order moments are.

In Fig. 3 we display a comparison between the test-particle method [10] and the method of moments at second and fourth order for the case of the quadrupole mode in a spherical trap. The peak position corresponds to the quadrupole frequency, while the width defines the

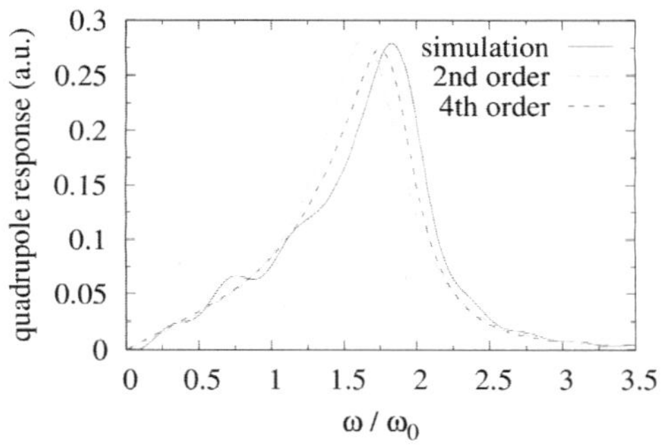

Figure 3. Quadrupole response as a function of the excitation frequency ω, obtained by a test particle simulation and by the moments method including up to second- and up to fourth-order terms [10]. The calculations were done for $N = 10^4$ atoms in a spherical trap (frequency ω_0), temperature $T/T_F = 0.4$, interaction parameter $1/k_F a = -0.5$, with the free-space cross section and without mean field.

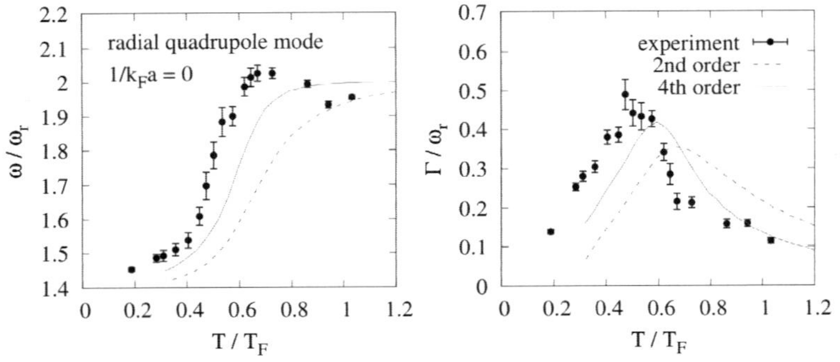

Figure 4. Frequency ω and damping rate Γ of the radial quadrupole mode as functions of temperature. The data and system parameters are taken from Ref. [13], the theoretical curves were obtained within the second- and fourth-order moments method, respectively [14].

damping rate. It can be seen that the inclusion of fourth-order moments improves significantly the agreement between the moments method and the test-particle calculation.

We also applied the fourth-order moments method to a realistic case studied in a ^{6}Li experiment at Innsbruck [13]. In this calculation [14], the medium modification of the cross section was taken into account. As an example, we display in Fig. 4 the frequency and damping rate of the quadrupole mode as functions of the temperature, obtained within the method of moments at second and fourth order, together with the experimental data. One can clearly see the smooth transition from hydrodynamic to collisionless behavior as the temperature increases. The reason is that the collision rate decreases when the gas expands and becomes more dilute at high temperature. As one can see, the second-order results overestimate the collision effects and therefore reach the collisionless regime at too high temperatures. The extension of the moments method to fourth-order terms clearly improves the agreement with the data. However, we mention that in the case of the scissors modes we did not find a clear improvement due to the inclusion of fourth-order terms [14].

4. Sloshing mode in an anharmonic trap

In a harmonic trap, the sloshing mode is an undamped center-of-mass oscillation of the atom cloud as a whole, leaving the shape of the atom cloud unchanged (Kohn theorem [15]). Since its frequency is that of the trap, independently of the interaction and other system parameters, it is very interesting for experimentalists for the precise determination of the trap frequency.

In real experiments, however, the trap potentials have more or less pronounced anharmonicities. For instance, in an optical trap, where the radial confinement is produced by a laser beam with waist w, the trap potential in radial direction is of the form $V_{\text{trap}}(r_\perp) \sim -V_0 \exp(-2r_\perp^2/w^2)$. Only if the particle energies (i.e., Fermi energy E_F and temperature T) are much smaller than the trap depth V_0, this potential can be considered as approximately harmonic. The anharmonicity of the potential changes the sloshing frequency, $\omega_{\text{slosh}} \neq \omega_r$ [if we define the radial trap frequency ω_r by $\omega_r^2 = (1/m)(d^2V_{\text{trap}}/dr_\perp^2)_{r_\perp=0}$] and leads also to a damping of the sloshing mode.

We studied the sloshing mode within the method of moments for the parameters of the experiment [13]. The main results of Ref. [16] are summarized in Fig. 5. With increasing temperature the cloud expands and extends more and more into the anharmonic region of the trap, where the trap potential is weaker than the idealized harmonic potential. Therefore, the sloshing frequency (left panel of Fig. 5) decreases, as it is observed in the experiment. The general trend is correctly reproduced by the moments method at first and at third order. Also the mean field has only a weak effect. When comparing with the experimental data, however, one should note that in the normalization of the experimental data it was assumed that the sloshing frequency extrapolated to $T = 0$ should be equal to the trap frequency, which explains why the data are shifted upwards as compared to the theoretical results.

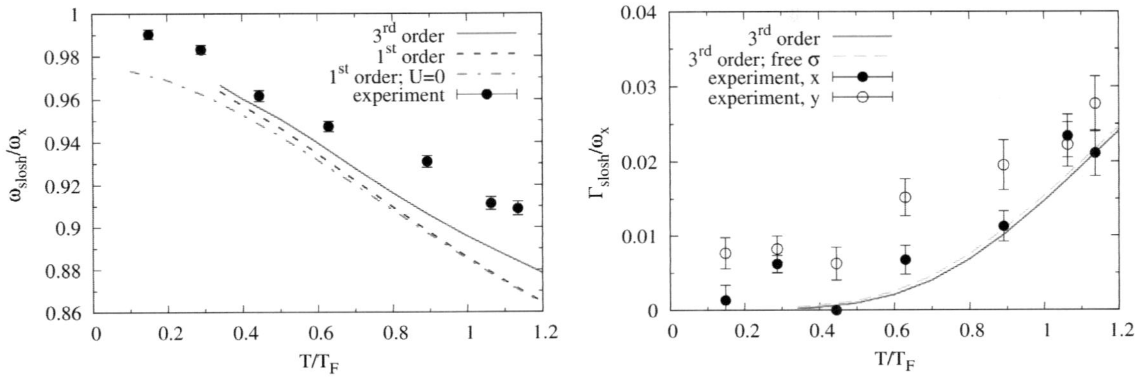

Figure 5. Left: temperature dependence of the sloshing frequency in the experiment [13] and theoretical results [16] obtained within the moments method up to third (solid line) and first order (dashes), respectively. For comparison, also the first-order result obtained without mean field ($U = 0$) is shown (dashed-dotted curve). Right: corresponding damping of the sloshing mode within the third-order moments method (solid line). For comparison, also the result obtained with the free-space cross section (free σ, dashes) is shown. The data are from [17].

While the inclusion of third-order moments does not have a dramatic effect on the sloshing frequency, it is crucial for the calculation of the damping rate of the sloshing mode. Within the first-order moments method, the sloshing mode is undamped because x and p_x are conserved during a collision. In order to describe the damping, one has to include at least third-order terms such as x^3, xy^2, xp_x^2, xp_y^2, etc. (in total, Φ contains now 18 terms [16]). As shown in the right panel of Fig. 5, the resulting damping is still very weak and it is compatible with the experiment. We note that it practically does not matter if we use the in-medium or the free-space cross section in this calculation.

In our calculation, the origin of the damping is that, as the cloud oscillates in the anharmonic potential, its form changes. This leads to deformations of the Fermi sphere in momentum space, which are damped by collisions. The change of the system shape during the sloshing motion can also be interpreted in the following way: already in a harmonic trap, there are in addition to the sloshing mode more complicated (i.e., described by moments of higher order) dipolar collective modes which are damped. The anharmonicity leads to a coupling between the sloshing mode and these higher-order modes. As examples, we display in Fig. 6 the velocity fields of the modes excited by the operators $(x^2 + y^2 - 2\langle r_\perp^2 \rangle_{\text{eq}})x$ (compressional dipole mode in the xy plane) and $(z^2 - \langle z^2 \rangle_{\text{eq}})x$ ("bending mode" in the xz plane).

5. Anharmonicity effects in the quadrupole mode in two dimensions

In a recent ^{40}K experiment done at Cambrigde [18, 19], the frequency and damping of the quadrupole mode in two dimensions were studied. Although the analysis made in [19] based on second-order moments could explain the transition of the observed frequency of the quadrupole mode from the hydrodynamic to the collisionless regime with decreasing interaction strength [increasing $\ln(k_F a_{2D})$], there was an unexplained offset in the damping rate. Since this unexplained damping persisted even in the non-interacting regime, it was reasonable to assume that it was caused by anharmonicity effects.

In fact, the experimental trap potential (see [20] for details) is not only anharmonic like an inverted Gaussian, it also has a small ellipticity and it is deformed by linear terms due to gravity and magnetic field gradients which destroy the symmetry of the potential, i.e.,

Journal of Physics: Conference Series **497** (2014) 012028 doi:10.1088/1742-6596/497/1/012028

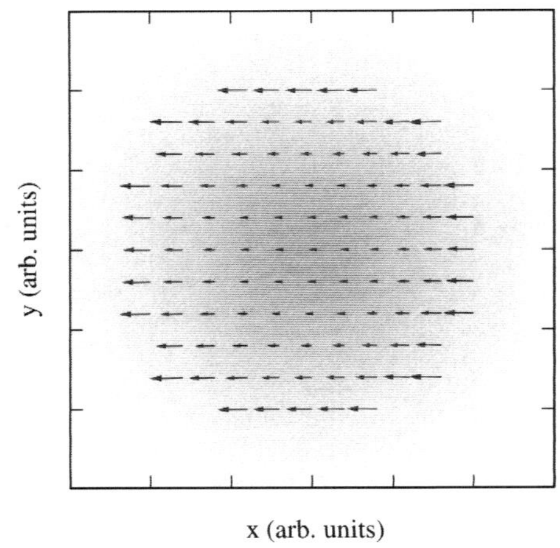

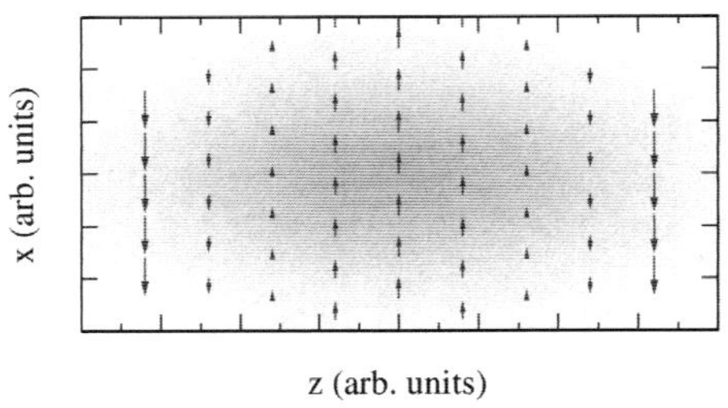

Figure 6. Velocity fields of the compressional dipole mode in the xy plane (left) and of the bending mode (right) in the xz plane, which are coupled to the sloshing mode in an anharmonic trap [16]. The background color indicates the column density of the cloud.

$V_{\text{trap}}(-\mathbf{r}) \neq V_{\text{trap}}(\mathbf{r})$. The model potential we used in our analysis [21] is shown in Fig. 7. As a consequence of the asymmetric shape of the potential, modes of different symmetries (such as sloshing, breathing, and quadrupole modes) are coupled to one another. In order to take this coupling into account, we included in our analysis all moments up to fourth order, i.e., all terms of the form $x^i y^j p_x^k p_y^l$ where i, j, k, l are non-negative integer numbers with $i+j+k+l \leq 4$ (70 terms in total) [21]. In our calculation, we include the in-medium cross section but not the mean field.

In Fig. 8 we compare our results for the frequency and damping rate of the quadrupole mode with the experimental ones. The results are shown as functions of the parameter $\ln(k_F a_{2D})$ characterizing the interaction strength, large values of $\ln(k_F a_{2D})$ corresponding to weak interaction. The frequency tends to the hydrodynamic one ($\sqrt{2}\omega_r$) in the limit of small $\ln(k_F a_{2D})$ and to the collisionless one ($2\omega_r$) in the limit of large $\ln(k_F a_{2D})$. We start with the simplest model, namely an isotropic harmonic potential (a) and include one by one the different imperfections of the realistic potential: ellipticity (b), anharmonicity (c), linear terms due to gravity and magnetic field gradients (d). It can be seen that these details have only little effect on the results: in this experiment the harmonic approximation is actually very good because the Fermi energy is small compared to the trap depth. In particular, the damping that persists in the weakly interacting regime cannot be explained by anharmonicity effects.

We note that the effect of higher-order moments is much weaker here than in the three-dimensional case discussed in Sec. 3. We thus confirm the results obtained in the analysis of

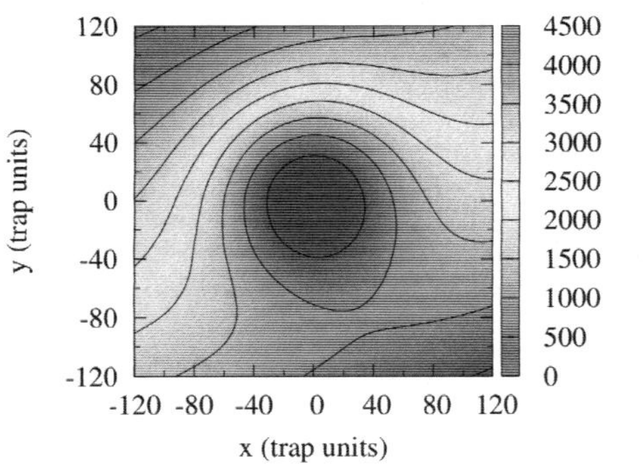

Figure 7. Contour plot of the model potential to describe the two-dimensional experiment of Refs. [18, 19]. The potential is shown in units of $\hbar\omega_r$ (ω_r being the trap frequency) as function of x and y in units of the oscillator length $l_{\text{ho}} = \sqrt{\hbar/m\omega_r}$.

Journal of Physics: Conference Series **497** (2014) 012028 doi:10.1088/1742-6596/497/1/012028

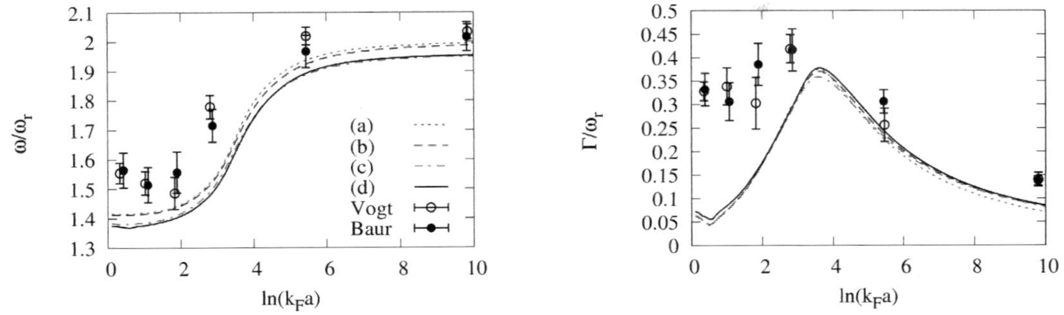

Figure 8. Frequency ω and damping rate Γ of the quadrupole mode as functions of the interaction strength (for $N = 4300$ atoms at $T = 0.47T_F$ in a two-dimensional trap). Theoretical results obtained with fourth-order moments [21] are compared with experimental data from Vogt et al. [18] and Baur et al. [19], respectively. The curves (a)–(d) correspond to different approximations of the realistic trap potential (see text).

[19] within the second-order moments method.

In contrast to what was found in an estimate based on a calculation of the shear viscosity [22], also the in-medium enhancement of the cross section has only a moderate effect on the damping. This finding, which is in agreement with the results of [19], is again very different from the three-dimensional case discussed in Sec. 3. The fact that the theory cannot explain the strong damping observed experimentally in the strongly-interacting regime [$\ln(k_F a_{2D}) \lesssim 1$] is maybe due to the breakdown of the validity of the Boltzmann equation in this regime.

6. Summary

We have studied various aspects of collective modes of trapped Fermi gases in the normal-fluid phase. The Boltzmann equation allows us to describe the transition from hydrodynamic to collisionless behavior, which is observed in both three- and two-dimensional systems. The method of moments was used to obtain approximate solutions of the linearized Boltzmann equation in the limit of small oscillations around equilibrium.

By comparing with a direct numerical solution within the test-particle method, we found that the moments method including only second-order moments overestimates the collisional effects in the case of the quadrupole mode in a spherical trap. By including also fourth-order moments, the agreement improves a lot. Also the results for the radial quadrupole mode in an elongated trap are in better agreement with experimental results if fourth-order moments are included.

Within the higher-order moments method, we were able to explain the experimentally observed damping of the sloshing mode in an anharmonic three-dimensional trap. Here the necessity to include higher-order moments (third order in this case) is obvious since at lowest order (first order) the sloshing mode is undamped.

We also studied the effects of a realistic anharmonic potential on the damping of the quadrupole mode in two dimensions. It turned out that the anharmonicity effects are by far too weak to explain the experimentally observed damping in the weakly interacting regime. It seems that the effect of higher-order moments is less important in two dimensions than in the three-dimensional cases discussed before.

It should be mentioned that the strong damping of the quadrupole mode near the superfluid-normal phase transition, observed in both two- and three-dimensional systems, still lacks a theoretical description. Probably the Boltzmann equation is not applicable in this regime.

Interesting subjects for future studies are the inclusion of the in-medium cross section and mean-field into numerical test-particle simulations, which so far have been limited to the free cross section without mean field. Work in this direction is in progress. The description of spin modes and modes in imbalanced Fermi gases is more difficult since it requires a model for the cross section and for the mean field in the imbalanced case, which is non-trivial because of the failure of the usual ladder approximation in this case [23, 24, 25].

References

[1] Schunck C H, Zwierlein M W, Schirotzek A and Ketterle W 2007 *Phys. Rev. Lett.* **98** 050404
[2] Altmeyer A, Riedl S, Wright M J, Kohstall C, Denschlag J H and Grimm R 2007 *Phys. Rev.* A **76** 033610
[3] Urban M 2008 *Phys. Rev.* A **78** 053619
[4] Cao C, Elliott E, Joseph J, Wu H, Petricka J, Schäfer T and Thomas J E 2011 *Science* **331** 58
[5] Wright M J, Riedl S, Altmeyer A, Kohstall C, Sánchez Guajardo E R, Hecker Denschlag J and Grimm R 2007 *Phys. Rev. Lett.* **99** 150403
[6] Chiacchiera S, Lepers T, Davesne D and Urban M 2009 *Phys. Rev.* A **79** 033613
[7] Bruun G M and Smith H 2005 *Phys. Rev.* A **72**, 043605
[8] Alm T, Röpke G and Schmidt M 1994 *Phys. Rev.* C **50** 31
[9] Toschi F, Vignolo P, Succi S and Tosi M P 2003 *Phys. Rev.* A **67** 041605
[10] Lepers T, Davesne D, Chiacchiera S and Urban M 2010 *Phys. Rev.* A **82** 023609
[11] Goulko O, Chevy F and Lobo C 2011 *Phys. Rev.* A **84** 051605
[12] Bruun G M and Smith H 2007 *Phys. Rev.* A **76**, 045602
[13] Riedl S, Sánchez Guajardo E R, Kohstall C, Altmeyer A, Wright M J, Hecker Denschlag J, Grimm R, Bruun G M and Smith H 2008 *Phys. Rev.* A **78** 053609
[14] Chiacchiera S, Lepers T, Davesne D, Urban M 2011 *Phys. Rev.* A **84** 043634
[15] Kohn W 1961 *Phys. Rev.* **123** 1242
[16] Pantel P-A, Davesne D, Chiacchiera S and Urban M 2012 *Phys. Rev.* A **86** 023635
[17] Sidorenkov L and Grimm R 2012 (unpublished)
[18] Vogt E, Feld M, Fröhlich B, Pertot D, Koschorreck M and Köhl M 2012 *Phys. Rev. Lett.* **108** 070404
[19] Baur S, Vogt E, Köhl M and Bruun G M 2013 *Phys. Rev.* A **87** 043612
[20] E. Vogt 2013 PhD Thesis (University of Cambridge)
[21] Chiacchiera S, Davesne D, Enss T and Urban M 2013 *Phys. Rev.* A **88** 053616
[22] Enss T, Küppersbusch C and Fritz L 2012 *Phys. Rev.* A **86** 013617
[23] Xia-Ji Liu and Hui Hu 2006 *Europhys. Lett.* **75** 364
[24] Parish M M, Marchetti F M, Lamacraft A and Simons B D 2007 *Nature Phys.* **3** 124
[25] Kashimura T, Watanabe R and Ohashi Y 2012 *Phys. Rev.* A **86** 043622

Journal of Physics: Conference Series **497** (2014) 012029 doi:10.1088/1742-6596/497/1/012029

A Phenomenological Model of the Growth of Two–Species Atomic Bose–Einstein Condensates

R W Pattinson, N G Parker, N P Proukakis

Joint Quantum Centre (JQC) Durham–Newcastle, School of Mathematics and Statistics, Newcastle University,Newcastle upon Tyne, NE1 7RU, United Kingdom

E-mail: r.w.pattinson@ncl.ac.uk

Abstract. We introduce a phenomenological mean-field model to describe the growth of immiscible two-species atomic Bose–Einstein condensates towards some equilibrium. Our model is based on the coupled Gross–Pitaevskii equations with the addition of dissipative terms to account for growth. While our model may be applied generally, we take a recent Rb–Cs experiment [McCarron et al., Phys. Rev. A 84 011603(R) (2011)] as a case study. As the condensates grow, they can pass through ranging transient density structures which can be distinct from the equilibrium states, although such a model always predicts the predominance of one condensate species over longer evolution times.

1. Introduction

First achieved experimentally in 1997 [1], binary mixtures of Bose–Einstein condensates (BECs) have been a topic of intense research both experimentally and theoretically. Experimental realisations have been achieved with two hyperfine states of the same atomic species [1–11], different isotopes of the same atomic species [12] and with different atomic species [13–15, 73]. Many aspects of these mixtures have been studied theoretically, including steady state solutions [16–28], dark–bright solitons [30–34], vortices [6, 35–47] and finite temperature models [34, 48–56, 76, 78]. One of the defining aspects of these condensates is their ability to interact with each other. Depending on the relative strength of self–interactions (interactions between two particles of the same species) and inter–species interactions, the condensates can exhibit either miscible or immiscible behaviour [17]. This introduces new physics not present in the single species condensate, such as coupled stability criteria [62–66] and modulational instabilities [67–72]. A recent experimental study of a immiscible two–species condensate [73], involving the sympathetic cooling of one condensate by the other, suggests that the growth of the condensates towards equilibrium plays an important role in determining their structure [29].

Finite temperature models have been a topic of intense theoretical research in BECs [79–82] and here we give a brief summary of their application to condensate mixtures. The Hartree–Fock model [48–51] and Popov approximation [52] have been extended from one to two–species condensates in order to study the impact of thermal clouds on the density distributions in harmonic traps, while the Hartree–Fock–Bogoliubov–Popov approximation has been applied to a homogeneous two-species system [53]. C–field methods have also been used to model the finite-temperature behaviour of miscible condensates in periodic boxes [55, 56]. Simpler finite temperature models have been employed through the addition of a dissipative term to the coupled Gross-Pitaevskii equations such that the number of particles in the system is no longer conserved (and the thermal cloud is not itself physically modelled): this led to simulations of growth and ensuing modulational instability in Ref. [54] and the dissipative motion dark–bright solitons in Ref. [34].

The aim of this work is to investigate the coupled dynamics of a two-species condensate during its growth towards equilibrium through a phenomenological mean-field model based on the dissipative coupled Gross–Pitaevskii equations. Our focus is on the recent Rb–Cs mixture experiment [73]. There, three distinct density profiles were observed depending on relative condensed particle numbers in each species. We first describe the stationary mean-field solutions of the system, showing that the observed structures can be qualitatively reproduced with the inclusion of a weak additional linear potential to the harmonic trapping potentials [29]. Next we apply our phenomenological growth model to study the coupled dynamics during evolution of the system towards equilibrium, and the transient and final states obtained.

2. Theoretical Framework

2.1. Coupled Gross-Pitaevskii Equations

In the limit of zero temperature and weak interactions, a two-species BEC can be described through the mean-field model of the coupled Gross-Pitaevskii equations (CGPEs),

$$\mathrm{i}\hbar\frac{\partial\psi_1}{\partial t} = \left(-\frac{\hbar^2}{2m_1}\nabla^2 + V_1 + g_{11}\left|\psi_1\right|^2 + g_{12}\left|\psi_2\right|^2 - \mu_1\right)\psi_1 \quad (1)$$

$$\mathrm{i}\hbar\frac{\partial\psi_2}{\partial t} = \left(-\frac{\hbar^2}{2m_2}\nabla^2 + V_2 + g_{22}\left|\psi_2\right|^2 + g_{12}\left|\psi_1\right|^2 - \mu_2\right)\psi_2, \quad (2)$$

where ψ_j are the condensate wavefunctions, m_j the atomic masses, and g_{jk} the strength of interactions. The latter are related to the s–wave scattering lengths a_{jk} through $g_{jj} = 4\pi\hbar^2 a_{jj}/m$ and $g_{jk} = 2\pi\hbar^2(m_1 + m_2)a_{12}/m_1 m_2$. We only consider repulsive interactions throughout this study, i.e. $g_{jj} > 0$ and $g_{jk} > 0$.

The trapping potentials V_j are typically harmonic and cylindrically symmetric, with the form,

$$V_j = \frac{1}{2}m_j\left[\omega^2_{j,\perp(j)}\left(x^2+y^2\right) + \omega^2_{j,z}z^2\right], \quad (3)$$

where $\omega_{j,\perp}$ denotes the harmonic trap frequency in the perpendicular (xy) direction and $\omega_{j,z}$ is the harmonic trap frequency in the axial (z) direction.

2.2. Phenomenological Growth Model

We describe the growth of the two-species BEC through the following *dissipative* coupled Gross–Pitaevskii equations,

$$\mathrm{i}\hbar\frac{\partial\psi_1}{\partial t} = (1-\mathrm{i}\gamma_1)\left(-\frac{\hbar^2}{2m_1}\nabla^2 + V_1 + g_{11}\left|\psi_1\right|^2 + g_{12}\left|\psi_2\right|^2 - \mu_1\right)\psi_1 \quad (4)$$

$$\mathrm{i}\hbar\frac{\partial\psi_2}{\partial t} = (1-\mathrm{i}\gamma_2)\left(-\frac{\hbar^2}{2m_2}\nabla^2 + V_2 + g_{22}\left|\psi_2\right|^2 + g_{12}\left|\psi_1\right|^2 - \mu_2\right)\psi_2. \quad (5)$$

The damping term γ_j is introduced here phenomenologically to *qualitatively* account for the dissipative role of finite temperature. It allows for the loss or gain of particles in the system, and the amplitude of γ_j determines the growth/decay rate of each species. For simplicity, we consider $\gamma_1 = \gamma_2$. The inclusion of such a term was originally proposed, using general arguments, by Pitaevskii [57, 58]. First implemented to trapped Bose gases in [59], the phenomenological damped GPE has been to study, e.g. vortex lattice formation [60,61] and dark soliton decay [75] in single species condensates, and dissipative dark–bright soliton dynamics in two–species condensates [34]. A constant value of γ_i is typically chosen phenomenologically such that the dissipative processes match those observed experimentally.

Because of the way that the dissipative terms appear in Eqs. 4 and 5, i.e. as a global factor to the right-hand side of the CGPEs, the dynamics evolve towards the equilibrium state of the

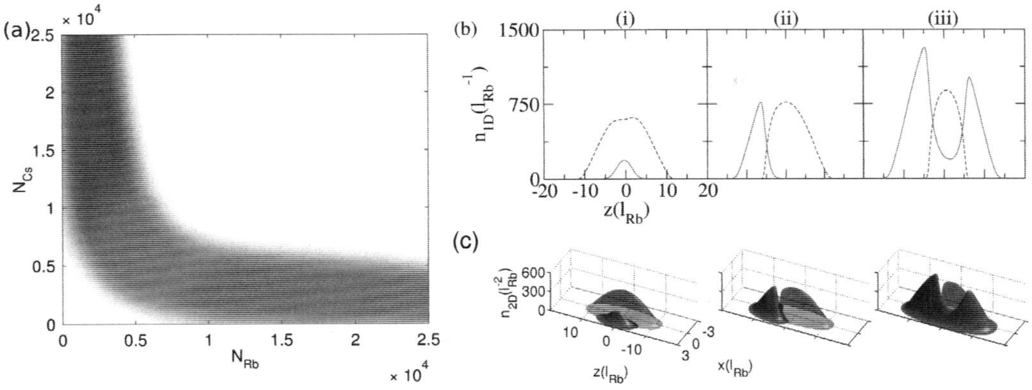

Figure 1. (a) Schematic showing the band in the $N_{\rm Rb} - N_{\rm Cs}$ plane where the experimentally observed profiles [73] fell. The three coloured regions represent the regimes associated with different observed density structures (as described in the main text). (b) and (c) show representative steady state solutions of the CGPEs for each of the three structural regimes. These three cases correspond to (i) $N_{\rm Rb} = 840$ and $N_{\rm Cs} = 8570$, (ii) $N_{\rm Rb} = 3680$ and $N_{\rm Cs} = 8510$, and (iii) $N_{\rm Rb} = 15100$ and $N_{\rm Cs} = 6470$. (b) shows these solutions through integrated 1D density profiles and (c) through integrated 2D density profiles. Rb ((Solid) red curve) and Cs ((Dashed) blue curve)

system (set by the chemical potentials). Depending on the initial state, this may be either through decay or growth towards the equilibrium state (we are interested in the latter). A closely related model was introduced in [54], although there the dissipative terms $i\hbar\gamma_j$ were introduced into the Hamiltonian. There, in contrast, they acted as gain terms, causing the (unlimited) exponential growth of both condensates.

To create growth in our simulations, we take as an initial condition for ψ_j the steady state solutions of the coupled Gross-Pitaevskii equations (1) and (2) with low particle numbers N_1 and N_2. The corresponding chemical potentials are $\mu_1^{(0)}$ and $\mu_2^{(0)}$. For $t > 0$, the chemical potentials are suddenly increased to higher values, i.e. $\mu_i \geq \mu_i^{(0)}$. Since we are considering only repulsive interactions, a larger chemical potential corresponds to a larger atom number, and so the dissipative CGPEs evolve towards this new chemical potential through the growth of the system.

3. Case Study: the Durham Rb–Cs Experiment [73]

We direct our model towards the recent ^{87}Rb-^{133}Cs two-species condensate experiment reported in Ref. [73]. The intra– and inter–species s-wave scattering lengths for Rb (labelled 1) and Cs (labelled 2) are $a_{11} = 100a_0$, $a_{22} = 280a_0$ and $a_{12} = 650a_0$ where a_0 is the Bohr radius. Two-species condensates are immiscible for $g_{12} > g_{11}g_{22}$ [16]; these scattering lengths well satisfy this criterion such that the Rb-Cs condensate is highly immiscible.

The condensates experience harmonic confinement through a magnetic trap. Due to the difference in dipole moments of the species, each species experiences different trap frequencies, $\omega_{1(2),z} = 2\pi \times 3.89$ $(2\pi \times 4.55)$Hz in the axial direction and $\omega_{1(2),\perp} = 2\pi \times 32.2$ $(2\pi \times 40.2)$Hz in the transverse directions.

In this experiment, following a period of sympathetic cooling between the species, the density profiles were imaged and the atom numbers measured. These final atom numbers varied from experimental run to run but fell consistently along a band region in the $N_{\rm Rb} - N_{\rm Cs}$ plane, as depicted in Fig. 1 (a). Moreover, the integrated axial density profiles largely revealed structures, depending on where they fell on the band: for large $N_{\rm Cs}$ and small $N_{\rm Rb}$ (blue region), the Rb cloud sat in the middle of the Cs cloud; for intermediate $N_{\rm Cs}$ and $N_{\rm Rb}$ (purple region), the Rb and Cs clouds sat side-by-side; for small $N_{\rm Cs}$ and large $N_{\rm Rb}$ (red region), the Cs cloud sat in the middle of the Rb cloud.

In previous work [29], we examined the stationary solutions of the system according to the CGPEs (1) and (2) and their comparison to the experimentally observed density profiles. Using bare harmonic trapping potentials, the three qualitative density structures outlined above could not be recovered by this approach. However, the experimental trapping potential also included small asymmetrical perturbations due to an applied magnetic tilt to enhance evaporative cooling and other effects such as gravitational sag and minute beam misalignments. To first order, these effects can be modelled by introducing a linear potential to the trapping potential in each direction for just one species (which we take to be Rb),

$$V_1 = \frac{1}{2}m_1\left[\omega_{1,\perp(1)}^2\left(x^2+y^2\right)+\omega_{1,z}^2 z^2\right]+\alpha_x x+\alpha_z z. \tag{6}$$

Even small linear potentials were found to have a significant modification to the stationary solutions of the system and promote a greater range of structural forms. In particular, for linear potentials $\alpha_z = 1.5(\hbar\bar{\omega}_1/l_1)$ and $\alpha_x = 0.02(\hbar\bar{\omega}_1/l_1)$, where $\bar{\omega}_1 = (\omega_{1,z}\omega_{1,\perp}^2)^{1/3}$ and $l_1 = \sqrt{\hbar/m\bar{\omega}_1}$, a good qualitative agreement with the experimentally observed density structures was recovered in all three regimes [29]. These values are consistent with the experimentally anticipated shifts in trapping potential. Example stationary solutions from each structural regime are presented in Fig. 1 through (b) integrated 1D density profiles and (c) integrated 2D density profiles. For (i) high N_{Cs} and low N_{Rb}, the integrated axial density profile shows the Rb cloud to exist within the Cs cloud, as observed experimentally. The integrated 2D profile reveals that this structure arises from the condensates lying side-by-side in the x-direction. For (ii) intermediate N_{Rb} and N_{Cs} the condensates lie side by side in the axial direction, as observed experimentally. For (iii) low N_{Cs} and high N_{Rb}, the integrated axial density reveals the Cs cloud to lie within the Rb cloud, again consistent with the experimentally observed structure. In 2D this appears as the Rb cloud enclosing the Cs cloud asymmetrically.

The above results highlight the importance of including the small linear potentials in modelling this experiment. As such, we include these additional linear potentials ($\alpha_z = 1.5(\hbar\bar{\omega}_1/l_1)$ and $\alpha_x = 0.02(\hbar\bar{\omega}_1/l_1)$) in all subsequent results.

4. Numerical Results

4.1. Typical Growth Simulation

In Fig. 2 (a) and (b), we show a typical evolution of the condensate particle numbers in each species and the combined total particles against time based on our dissipative model of Eqs. (4) and (4). Here, we start with a large number of Cs atoms in comparison to Rb. Cs decreases until it vanishes from the system while the growth of Rb continues until a maximum is reached. The total number decreases drastically at the start but then grows again slowly over time. This suggests that a low critical number of Cs atoms is required before Rb can start to grow at a quicker pace. We also present four corresponding density profiles at $t(\bar{\omega}) = 0, 16, 40$ and 80. The initial density distribution is a symmetric density distribution where Rb and Cs sit side by side in a transverse direction. As the condensate particles numbers evolve over time, the condensate clouds become asymmetric until a side by side formation in the axial direction emerges. Finally, a one species condensate forms for long times as no Cs particles remain in the condensate.

4.2. Growth Trajectories Through N_{Cs}–N_{Rb} Plane

Figure 3 shows the trajectories of multiple condensate particle growth curves in the N_{Rb}–N_{Cs} plane. The regions indicated by the dotted black lines have been defined in accordance to the experimental results in [73] where the density profiles observed depend on condensate particle numbers. These experimental results shall be described in more detail in the following section.

The trajectory taken in each of these simulations is determined by the final chemical potentials rather than γ_i. For constant values of $\gamma_1 = \gamma_2$, the order of magnitude of γ_i determines solely the timescale over which growth/decay occurs although as this is already sensitive competing dynamical growth process, unequal γ_i's for the two species could largely

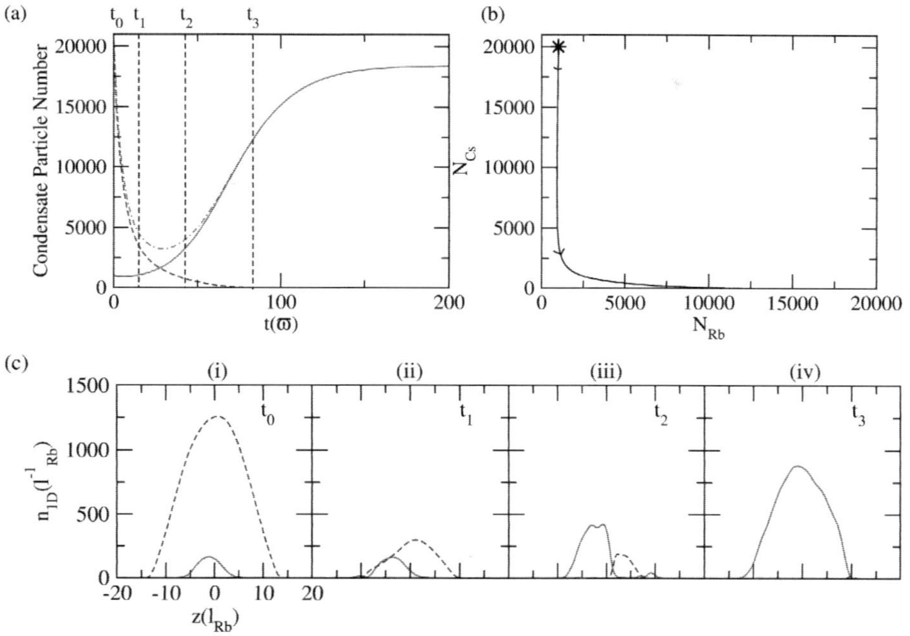

Figure 2. (a) Condensate particle numbers against time; Rb ((Solid) red curve), Cs ((Dashed) blue curve) and Total particle number ((Dotted) black curve). (b) Growth curve in the $N_{\rm Rb}$—$N_{\rm Cs}$ plane. (c) Bottom : Density profiles at times $t(\bar{\omega})$ = (i) 0, (ii) 16, (iii) 40, (iv) 80 for Rb ((Solid) red curve) and Cs ((Dashed) blue curve).

modify the dynamics [78]. In our simulations, we first note that all of the growth curves finish either on the horizontal or vertical axes i.e. with a one component condensate where the other species has vanished from the system. One component will always vanish from the system as the growth rates do not change over time, this feature is inherent in the model. Physically, this represents all of the particles being in the thermal cloud for that component (although this is only implicit within our simplified model). Nevertheless, this is still broadly consistent with the experimental findings as there were numerous unpublished results showing images where only one of the species was condensed [84]. The influence of the final chemical potentials on the path taken is clear when comparing the green and orange growth curves in Fig. 3 respectively labelled by D and E which start from the same steady state initial condition and evolve to different final single species condensates. The nonlinear dynamics and competing processes do not give us a direct handle on the precise trajectories in the $N_{\rm Rb}$–$N_{\rm Cs}$ plane although extensive simulations have enabled us to probe the most common types of trajectories, as discussed in detail below and shown in Fig. 3.

A number of our simulations (Figure 3 A, C and D) do not lead to any changes in the integrated density profiles while growth/decay occurs and each species occupies the same overall position until one vanishes (to the thermal cloud) leaving a single species condensate. In A, the initial density profile is asymmetric with Rb and Cs sat side by side in the axial direction. Once growth/decay begins, Rb vanishes rapidly from the system leaving a condensate with only Cs present. The growth curve for B have similar initial particle numbers and similar density profile changes during phenomenological growth. In this case, Rb decays while Cs grows in the centre of the trapping potentials. This growth of Cs splits the Rb into two distinct parts either side of Cs. Over longer time evolutions, Rb vanishes leaving a Cs condensate. For D, we start with a symmetric density profile and Rb will decay over time. The asymmetric intermediate density profiles in D are due to the asymmetry present in the initial steady state solution where, as Rb slowly decreases, the right peak of Rb vanishes before the left peak giving a side–by–side density profile. We obtain a Cs only condensate over longer time scales. In E, we start with

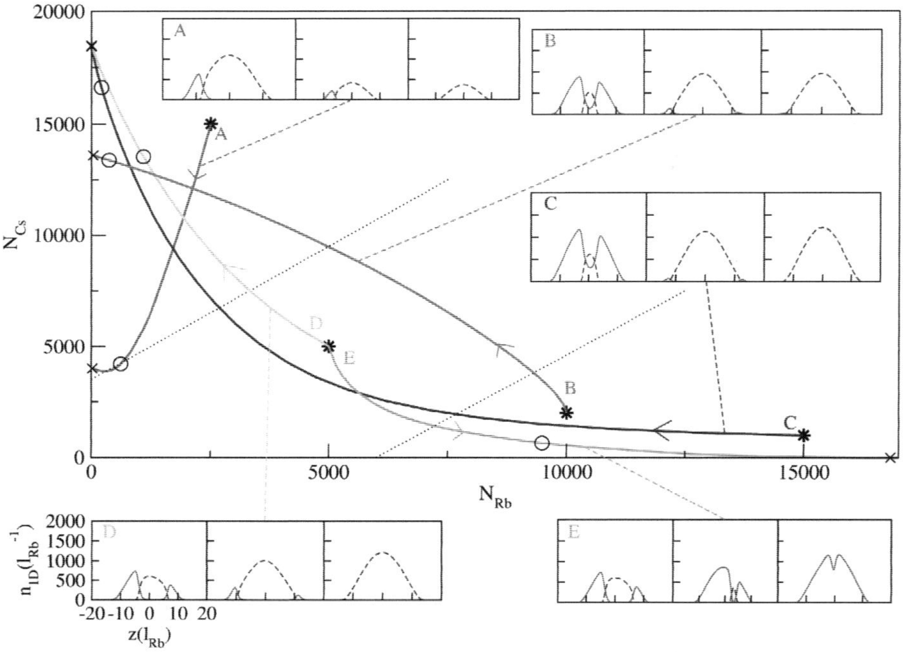

Figure 3. Condensate particle growth curves in N_{Rb} (horizontal) and N_{Cs} (vertical) plane. Initial (steady state), intermediate and final density distributions depicted by stars circles and crosses respectively. Dotted lines — boundaries of experimental regions. Each set of three density profiles corresponds to initial, intermediate and final 1D integrated density profiles for Rb ((Solid) red curve) and Cs ((Dashed) blue curve).

the same initial condition as D but choose different final chemical potentials for each of these simulations resulting in different growth curves. In the intermediate plot for the green growth curve E, we observe a dark–bright soliton which oscillates in the trapping potential until all the Cs bright component no longer exists in the system. Finally in B, we observe similar density evolution to D but over a different growth path in the N_{Rb}–N_{Cs} plane.

5. Conclusions, Discussion

We have investigated the role of growth in a two–component immiscible system modelled via a phenomenologically damped coupled GPEs. Starting with a range of steady state density distributions, the addition of a non–zero damping term simulates growth. Perturbing the chemical potential of each condensate induces competing dynamical evolution with the number of particles in the system eventually leaving one of the condensates to vanish entirely, thus leaving only a single species condensate. In a few cases, we have seen the density profiles to change from a symmetric density profile to an asymmetric one during growth provided additional linear potentials are present.

We stress that the asymmetries caused by harmonic trap offsets between the two trap centres in the axial and transverse directions are small relative to the size of the condensate clouds which explains why asymmetric density profiles can be observed in two–species experiments with different atomic species as two magnetic traps are required which are extremely hard to align perfectly. Our work shows a large number of distinct density profiles are possible during the coupled condensates dynamics. When comparing to a recent two–species experiment [73] on which our parameters are based, we observed some qualitative agreement between condensate density profiles while considering growth. However not only is the model used here a toy model with static growth, there is no explicit consideration of the thermal cloud dynamics but moreover the coupled experimental results available did not analyse the non–equilibrium structures in detail to be able to investigate a more detailed comparison .

These results can be improved by using more accurate models for finite–temperature non–equilibrium Bose gases [80]. In 1D, this system has already been looked at with the Stochastic Projected GPE [76]. However, even our simple model is enough to reveal that modelling the two transverse directions is essential in the distribution of the condensate densities. Work is currently under way between the group of [76] and ourselves, whilst there are also other efforts for modelling multicomponent condensates with the ZNG formalism [83] to look at this system in 3D with Stochastic Projected GPE.

Acknowledgments

We gratefully acknowledge active collaboration on various aspects of this work with the groups of S. L. Cornish (experiments, Durham University, UK) and S. A. Gardiner (theory, Durham University, UK) and our new collaborators I. K. Liu And S.–C. Gou (Taiwan) in our attempts to extend beyond the simplified dissipative mean–field model. N. P. Proukakis acknowledges funding from EPSRC.

References

[1] Myatt C J, Burt E A, Ghrist R W, Cornell E A and Wieman C E 1997 *Phys. Rev. Lett.* **78**(4) 586–589
[2] Hall D S, Matthews M R, Ensher J R, Wieman C E and Cornell E A 1998 *Phys. Rev. Lett.* **81**(8) 1539–1542
[3] Matthews M R, Anderson B P, Haljan P C, Hall D S, Wieman C E and Cornell E A 1999 *Phys. Rev. Lett.* **83**(13) 2498–2501
[4] Maddaloni P, Modugno M, Fort C, Minardi F and Inguscio M 2000 *Phys. Rev. Lett.* **85**(12) 2413–2417
[5] Delannoy G, Murdoch S G, Boyer V, Josse V, Bouyer P and Aspect A 2001 *Phys. Rev. A* **63**(5) 051602
[6] Schweikhard V, Coddington I, Engels P, Tung S and Cornell E A 2004 *Phys. Rev. Lett.* **93**(21) 210403
[7] Mertes K M, Merrill J W, Carretero-González R, Frantzeskakis D J, Kevrekidis P G and Hall D S 2007 *Phys. Rev. Lett.* **99**(19) 190402
[8] Anderson R P, Ticknor C, Sidorov A I and Hall B V 2009 *Phys. Rev. A* **80**(2) 023603
[9] Tojo S, Taguchi Y, Masuyama Y, Hayashi T, Saito H and Hirano T 2010 *Phys. Rev. A* **82**(3) 033609
[10] Busch T, Cirac J I, Pérez-García V M and Zoller P 1997 *Phys. Rev. A* **56**(4) 2978–2983
[11] Miesner H J, Stamper-Kurn D M, Stenger J, Inouye S, Chikkatur A P and Ketterle W 1999 *Phys. Rev. Lett.* **82**(11) 2228–2231
[12] Papp S B, Pino J M and Wieman C E 2008 *Phys. Rev. Lett.* **101**(4) 040402
[13] Ferrari G, Inguscio M, Jastrzebski W, Modugno G, Roati G and Simoni A 2002 *Phys. Rev. Lett.* **89**(5) 053202
[14] Modugno G, Modugno M, Riboli F, Roati G and Inguscio M 2002 *Phys. Rev. Lett.* **89**(19) 190404
[15] Thalhammer G, Barontini G, De Sarlo L, Catani J, Minardi F and Inguscio M 2008 *Phys. Rev. Lett.* **100**(21) 210402
[16] Ho T L and Shenoy V B 1996 *Phys. Rev. Lett.* **77**(16) 3276–3279
[17] Pu H and Bigelow N P 1998 *Phys. Rev. Lett.* **80**(6) 1130–1133
[18] Timmermans E 1998 *Phys. Rev. Lett.* **81**(26) 5718–5721
[19] Ao P and Chui S T 1998 *Phys. Rev. A* **58**(6) 4836–4840
[20] Trippenbach M, Gral K, Rzazewski K, Malomed B and Band Y B 2000 *Journal of Physics B: Atomic, Molecular and Optical Physics* **33** 4017
[21] Barankov R A 2002 *Phys. Rev. A* **66**(1) 013612
[22] Van Schaeybroeck B 2008 *Phys. Rev. A* **78**(2) 023624
[23] Gautam S and Angom D 2010 *Journal of Physics B: Atomic, Molecular and Optical Physics* **43** 095302
[24] Jezek D M and Capuzzi P 2002 *Phys. Rev. A* **66**(1) 015602
[25] Gordon D and Savage C M 1998 *Phys. Rev. A* **58**(2) 1440–1444
[26] Tanatar B and Erkan K 2000 *Phys. Rev. A* **62**(5) 053601
[27] Kim J G and Lee E K 2002 *Phys. Rev. E* **65**(6) 066201
[28] Balaž A and Nicolin A I 2012 *Phys. Rev. A* **85**(2) 023613
[29] Pattinson R W, Billam T P, Gardiner S A, McCarron D J, Cho H W, Cornish S L, Parker N G and Proukakis N P 2013 *Phys. Rev. A* **87**(1) 013625
[30] Busch T and Anglin J R 2001 *Phys. Rev. Lett.* **87**(1) 010401
[31] Kevrekidis P G, Frantzeskakis D J, Malomed B A, Bishop A R and Kevrekidis I G 2003 *New Journal of Physics* **5** 64
[32] Becker C, Stellmer S, Soltan-Panahi P, Dörscher S, Baumert M, Richter E M, Kronjäger J, Bongs K and Sengstock K 2008 *Nature Physics* **4** 9
[33] Hamner C, Chang J J, Engels P and Hoefer M A 2011 *Phys. Rev. Lett.* **106**(6) 065302

[34] Achilleos V, Yan D, Kevrekidis P G and Frantzeskakis D J 2012 *New Journal of Physics* **14** 055006
[35] Feder D L, Clark C W and Schneider B I 1999 *Phys. Rev. Lett.* **82**(25) 4956–4959
[36] Garcia-Ripoll J J and Perez-Garcia V M 2000 *Phys. Rev. Lett.* **84**(19) 4264–4267
[37] Chui S T, Ryzhov V N and Tareyeva E E 2001 *Phys. Rev. A* **63**(2) 023605
[38] Jezek D M, Capuzzi P and Cataldo H M 2001 *Phys. Rev. A* **64**(2) 023605
[39] Öhberg P and Santos L 2002 *Phys. Rev. A* **66**(1) 013616
[40] Park Q H and Eberly J H 2004 *Phys. Rev. A* **70**(2) 021602
[41] Stuhler J, Griesmaier A, Koch T, Fattori M, Pfau T, Giovanazzi S, Pedri P and Santos L 2005 *Phys. Rev. Lett.* **95**(15) 150406
[42] Woo S J, Choi S, Baksmaty L O and Bigelow N P 2007 *Phys. Rev. A* **75**(3) 031604
[43] Kasamatsu K and Tsubota M 2009 *Phys. Rev. A* **79**(2) 023606
[44] Yakimenko A I, Zaliznyak Y A and Lashkin V M 2009 *Phys. Rev. A* **79**(4) 043629
[45] Catelani G and Yuzbashyan E A 2010 *Phys. Rev. A* **81**(3) 033629
[46] Law K J H, Kevrekidis P G and Tuckerman L S 2010 *Phys. Rev. Lett.* **105**(16) 160405
[47] Kuopanportti P, Huhtamäki J A M and Möttönen M 2012 *Phys. Rev. A* **85**(4) 043613
[48] Esry B D, Greene C H, Burke Jr J P and Bohn J L 1997 *Phys. Rev. Lett.* **78**(19) 3594–3597
[49] Öhberg P and Stenholm S 1998 *Phys. Rev. A* **57**(2) 1272–1279
[50] Esry B D and Greene C H 1999 *Phys. Rev. A* **59**(2) 1457–1460
[51] Zhang C H and Fertig H A 2007 *Phys. Rev. A* **75**(1) 013601
[52] Öhberg P 1999 *Phys. Rev. A* **61**(1) 013601
[53] Pires M O C and de Passos E J V 2008 *Phys. Rev. A* **77**(3) 033606
[54] Ronen S, Bohn J L, Halmo L E and Edwards M 2008 *Phys. Rev. A* **78**(5) 053613
[55] Berloff N and Yin C 2006 *Journal of Low Temperature Physics* **145** 187–207 ISSN 0022-2291
[56] Perez-Garcia V M, Berloff N G, Kevrekidis P G, Konotop V V and Malomed B A 2009 *Physica D: Nonlinear Phenomena* **238** 1289 – 1298 ISSN 0167-2789 Nonlinear Phenomena in Degenerate Quantum Gases¡/ce:title¿
[57] Pitaevskii L P 1958 *Zh. Eksp. Teor. Fiz.* **35** 408
[58] Pitaevskii L P Phys-JETP S 1959 *Zh. Eksp. Teor. Fiz.* **35** 282
[59] Choi S, Morgan S A and Burnett K 1998 *Phys. Rev. A* **57**(5) 4057–4060
[60] Tsubota M, Kasamatsu K and Ueda M 2002 *Phys. Rev. A* **65**(2) 023603
[61] Kasamatsu K, Tsubota M and Ueda M 2003 *Phys. Rev. A* **67**(3) 033610
[62] Law C K, Pu H, Bigelow N P and Eberly J H 1997 *Phys. Rev. Lett.* **79**(17) 3105–3108
[63] Öhberg P 1999 *Phys. Rev. A* **59**(1) 634–638
[64] Adhikari S K 2001 *Phys. Rev. A* **63**(4) 043611
[65] Cazalilla M A and Ho A F 2003 *Phys. Rev. Lett.* **91**(15) 150403
[66] Kadokura T, Aioi T, Sasaki K, Kishimoto T and Saito H 2012 *Phys. Rev. A* **85**(1) 013602
[67] Shukla P K, Sten L and Fedele R 2001 *Physica* **553** 6–7
[68] Kourakis I, Shukla P K, Marklund M and Stenflo L 2005 *European Physical Journal B* **46** 381–384
[69] Kasamatsu K and Tsubota M 2004 *Phys. Rev. Lett.* **93**(10) 100402
[70] Kasamatsu K and Tsubota M 2005 *Journal of Low Temperature Physics* **138** 669–674
[71] Kasamatsu K and Tsubota M 2006 *Phys. Rev. A* **74**(1) 013617
[72] Raju T S, Panigrahi P K and Porsezian K 2005 *Phys. Rev. A* **71**(3) 035601
[73] McCarron D J, Cho H W, Jenkin D L, Köppinger M P and Cornish S L 2011 *Phys. Rev. A* **84**(1) 011603
[74] Proukakis N P, Parker N G, Barenghi C F and Adams C S 2004 *Phys. Rev. Lett.* **93**(13) 130408
[75] Proukakis N P, Parker N G, Barenghi C F and Adams C S 2004 *Phys. Rev. Lett.* **93**(13) 130408
[76] Liu I K, Cho H W, Su S-W, McCarron D J, Cornish S L and Gou S-C *Poster presented at Internation Conference on Atomic Physics (ICAP) 2012*
[77] Liu I K, Pattinson R W *et al. In preperation*
[78] Liu I K, Pattinson R W *et al. In preperation*
[79] Proukakis N P and Jackson B 2008 *J. Phys. B: At. Mol. Opt. Phys.* **41** 203002
[80] Proukakis N P, Gardiner S A, Davis M and Szymanska M 2013 *Quantum Gases: Finite Temperature and Non-Equilibrium Dynamics*
[81] Griffin A, Nikuni T and Zaremba E. 2009 *Bose-condensed gases at finite temperatures*
[82] Blakie P. B., Bradley A. S., Davis M. J., Ballagh R. J. and Gardiner C. W. 2008 *Advances in Physics* **57** 363
[83] Edmonds M *et al. In preperation*
[84] McCarron D. J. *et al. Private Communication*

Journal of Physics: Conference Series **497** (2014) 012030 doi:10.1088/1742-6596/497/1/012030

Exact quantum dynamics of yrast states in the nite 1D Bose gas

Eriko Kaminishi[1] , **Jun Sato**[2], **Tetsuo Deguchi**[2]

[1] Department of Physics, Graduate School of Science, The University of Tokyo, 7-3-1 Hongo, Bunkyo-ku, Tokyo 113-0033, Japan

[2] Department of Physics, Graduate School of Humanities and Sciences, Ochanomizu University, 2-1-1 Ohtsuka, Bunkyo-ku, Tokyo 112-8610, Japan

E-mail: `kaminishi@cat.phys.s.u-tokyo.ac.jp`

Abstract. We demonstrate that the quantum dynamics of yrast states in the one-dimensional (1D) Bose gas gives an illustrative example to equilibration of an isolated quantum many-body system. We rst formulate the energy spectrum of yrast states in terms of the dressed energy by applying the method of nite-size corrections. We then review the exact time evolution of quantum states constructed from yrast states shown by the Bethe ansatz. In time evolution the density pro le of an initially localized quantum state constructed from yrast states collapses into a flat pro le in the case of a large particle number such as $N = 1000$, while recurrence of the localized state occurs in the case of a small particle number such as $N = 20$. We suggest that the dynamical relaxation behavior for the large N case is consistent with the viewpoint of typicality for generic quantum states: the expectation values of local operators evaluated in most of quantum states are very close to those of the micro-canonical ensemble.

1. Introduction

Ultracold quantum gases have attracted much interest due to their potential application to the testing of the quantum fluctuations in many-body systems [1, 2, 3, 4]. Furthermore, it has been shown that they are also quite useful for studying nontrivial dynamical properties of isolated quantum interacting systems not only theoretically [5] but also experimentally [6, 7] (see for reviews, Ref. [8, 9]). The nonequilibrium behavior of isolated systems is closely related to fundamental aspects of quantum statistical mechanics such as quantum ergodic theorem and the concept of typicality among quantum states [10, 11, 12, 13, 14]. In this report we illustrate how the dynamics of yrast states in the one-dimensional (1D) Bose gas gives quite an illustrative example consistent with typicality of quantum states in isolated quantum many-body systems.

et us now introduce yrast states. In quantum many-body systems such as rotating nuclei, it is often important to study such eigenstates that have the lowest energy among all the eigenstates with the same given angular momentum [15]. In fact, they may provide an insight into ensembles under rotation. We call the lowest-energy state of a given angular momentum the *yrast state*. Yrast states are studied in several quantum systems such as the quantum Hall systems [16, 17]. In two dimensions, yrast states are associated with dislocations or quantum vortices [18, 19]. In the Gross-Pitaevskii mean-field picture, if the angular momentum is given by an integral multiple of the number of particles, N, the vortex core is located on the rotating axis (the centered single-vortex state), while if not, it is not (an off-centered vortex state) [18].

In order to study the 1D bosons of cold atoms we consider the ieb- iniger model [20, 21, 22]. It has various applications in the study of many-body phenomena with a wide range from the quasicondensate regime [2, 23, 24] to the strongly interacting Tonks-Girardeau regime [6, 25]. Assuming the periodic boundary conditions (PBC), the 1D system of length L is equivalent to that of the circle of circumference L. The periodic geometry has been experimentally realized as a circular waveguide or elliptic trap [23, 24]. In the 1D system under PBC, the state with a given angular momentum is equivalent to that of the corresponding momentum [26]. Furthermore, the yrast states in the 1D Bose gas are associated with one-hole excitations [27]. It has been pointed out that the dispersion relation of yrast states is very close to that of dark solitons [28]. Dark solitons are experimentally realized in 1D trap [29]. We have a conjecture that quantum states of a dark soliton in the 1D Bose gas should be constructed from the yrast states [30].

Some fundamental aspects of quantum statistical mechanics particularly for isolated many-body systems have attracted much interest quite recently [8, 9], and have been studied from the viewpoint of typicality of quantum states [10, 11, 12, 13, 14]. Although one of the earliest exact studies of nonequilibrium dynamics of quantum many-body systems goes back to 1970s [31], the problem of quantum quench has been investigated extensively through theoretical approaches quite recently during the last decade, partially motivated by the experimental realization of quantum many-body systems in cold atomic systems [32, 33, 34, 36, 35]. Relaxation and thermalization phenomena have been observed not only theoretically in the study of quantum quench but also experimentally in cold atomic systems [7]. Thus, it is an interesting problem to study many-body phenomena in the 1D Bose gas theoretically and to compare the theoretical results with experimental ones.

It is one of the greatest achievements in exactly solvable models that finite-temperature specific heat and magnetic susceptibility are explicitly evaluated by the Bethe-ansatz method [37, 38, 39]. On the other hand, until quite recently it was not easy to evaluate correlation functions of quantum integrable systems even numerically for large systems. However, by combining Slavnov's formula of scalar products [40, 41] with the formulas of the Quantum Inverse Scattering Problem [42], or by improving some formulas of form factors for the 1D Bose gas [43, 44, 45], one can evaluate numerically correlation functions and form factors of some operators of the XX spin chain [46, 47, 48] and the 1D Bose gas [44, 45] by the Bethe ansatz method (for numerical values of finite-temperature XXX correlation functions, see Ref. [49]).

The main purpose of the present report is to review interesting dynamical properties in the quantum dynamics associated with yrast states in the 1D Bose gas [50]. The contents of the present report consist of the following: In section 2 we express the energy spectrum of yrast states in terms of the dressed energy. We assume that the system size is finite for yrast states, and generalize the method of finite-size corrections in the the ground-state energy of the 1D Bose gas. Here some details given in section 2 have not been presented in Ref. [27]. In section 3 we review that some linear combination of yrast states leads to a quantum state which is localized in terms of the density operator, i.e., it has a density notch in the density profile, and also that the density profile collapses into a flat profile in time evolution [50].

2. Yrast energy spectrum via finite-size corrections in the 1D Bose gas

2.1. Lieb-Liniger model and the Bethe-ansatz method

The Hamiltonian of the ieb- iniger ( ) model [20] is given by

$$\mathcal{H}_{\rm LL} = -\frac{\hbar^2}{2m} \sum_{j=1}^{N} \frac{\partial^2}{\partial {x_j}^2} + g_{\rm 1D} \sum_{j<k}^{N} \delta(x_j - x_k) \, . \tag{1}$$

Here $g_{\rm 1D}$ gives the 1D coupling constant. We assume the periodic boundary conditions: We identify the position of coordinate x with that of $x + L$.

Journal of Physics: Conference Series **497** (2014) 012030 doi:10.1088/1742-6596/497/1/012030

An exact eigenvector of the Hamiltonian (1) is derived by the Bethe-ansatz method, if a set of quasi-momenta k_j's for $j \in \{1, 2, \ldots, N\}$ satisfies the Bethe-ansatz equations (BAE):

$$Lk_j = 2\pi I_j - \sum_{\alpha \neq j}^{N} 2\tan^{-1}\left(\frac{k_j - k_\alpha}{c}\right) \quad \text{for} \quad j = 1, 2, \ldots, N. \tag{2}$$

Here $c = mg_{1D}/\hbar^2$ is the coupling constant with the dimension of wavenumber, and quantum numbers I_j are integers for odd N and half-integers for even N.

For the repulsive interaction: $c > 0$, it has been shown that the Bethe-ansatz eigenvectors are complete [51] (for some generalization, see Ref. [52]). Therefore, each eigenvector of the Hamiltonian (1) is specified by a set of quantum numbers I_j. The ground-state solution, for instance, corresponds to the set of integers: $I_j^{(g)} = j - (N+1)/2$ for $j \in \{1, 2, \ldots, N\}$. Hereafter we consider only the case of repulsive interaction.

Thus, by solving the BAE for a given set of I_j, we obtain a set of quasi-momenta k_j. The eigenvalue of the Hamiltonian (1) is given by the sum of k_j^2 over all j [20, 21]. The total angular momentum is given by $Y = R\hbar \sum_{j=1}^{N} k_j$, or, according to the BAE, alternatively by $Y = \hbar \sum_{j=1}^{N} I_j$.

2.2. *Particle-hole excitations*

We derive an excitation with one hole and one particle from the ground state $I_j^{(g)}$. We make a hole at I_h and a particle at I_p. We denote by $k_j^{(1)}$ the quasi-momenta of the particle-hole excitation satisfying the Bethe ansatz equations (2) with integers $I_j^{(1)}$ for $j = 1, 2, \ldots N-1$ and $I_p^{(1)} = I_p$, which are given by

$$I_j^{(1)} = I_j^{(g)} + H(j - j_h) \quad \text{for} \quad j = 1, 2, \ldots, N-1, \tag{3}$$

and $I_p^{(1)} = I_p > I_N^{(g)}$. Here integer j_h denotes the hole position: $j_h = I_h + (N+1)/2$ and the symbol $H(x)$ denotes the Heaviside step function: $H(x) = 1$ for $x \geq 0$, and $H(x) = 0$ otherwise.

The angular momentum $Y^{p,h}$ of the particle-hole excited state with a particle at I_p and a hole at I_h is given by

$$Y^{p,h} = \hbar(I_p - I_h)\,. \tag{4}$$

If $Y^{p,h}$ is nonzero, the dispersion relation is not symmetric. It seems that it is not straightforward to apply the method of finite-size corrections to the spectrum of yrast states.

et us define the counting function $Z^{(1)}(k)$ for the particle-hole excitation by

$$Z^{(1)}(k) = \frac{k}{2\pi} + \frac{1}{L}\sum_{\alpha=1}^{N-1} \frac{1}{2\pi} 2\tan^{-1}\left(\frac{k - k_\alpha^{(1)}}{c}\right) + \frac{1}{L}\frac{1}{2\pi} 2\tan^{-1}\left(\frac{k - k_p^{(1)}}{c}\right). \tag{5}$$

In terms of $Z^{(1)}(k)$ the Bethe-ansatz equations are written as follows: $Z^{(1)}(k_j^{(1)}) = I_j^{(1)}/L$ for $j = 1, 2, \ldots, N-1$; $Z^{(1)}(k_p^{(1)}) = I_p^{(1)}/L$. We define a continuous counting function $\hat{Z}(k)$ by

$$\hat{Z}(k) = Z^{(1)}(k) - \frac{1}{L}H(k - \tilde{k}_h). \tag{6}$$

Here we determine $\tilde{k}_h$ by $\hat{Z}(\tilde{k}_h) = I_h/L$ [27]. We introduce the Fermi parameters $Q^{\pm}$ by

$$\hat{Z}(Q^+) = \frac{I_{N-1}^{(1)} + 1/2}{L} \quad , \quad \hat{Z}(Q^-) = \frac{I_1^{(1)} - 1/2}{L}\,. \tag{7}$$

Journal of Physics: Conference Series **497** (2014) 012030 doi:10.1088/1742-6596/497/1/012030

We define the root density $\hat{\rho}(k)$ by the derivative of $\hat{Z}(k)$ with respect to k: $\hat{\rho}(k) = d\hat{Z}/dk$. Shifting $\hat{\rho}(k)$ as $\rho_{p,h}(k) = \hat{\rho}(k) + \delta(k - k_p^{(1)})/L$, through formula (A.2) we have

$$\rho_{p,h}(k|Q^{\pm}) = \frac{1}{2\pi} + \frac{1}{L}\left(\delta(k - k_p^{(1)}) - \delta(k - \tilde{k}_h)\right) + \frac{1}{2\pi L}K(k - k_p^{(1)}) + \int_{Q^-}^{Q^+} K(k-q)\rho_{p,h}(q|Q^{\pm})\frac{dq}{2\pi} \tag{8}$$

Here, the integral kernel $K(k)$ is given by $K(k) = 2c/(k^2 + c^2)$.

2.3. Dressed energy and formal solutions to the integral equations

We define the dressed energy [37] by

$$\epsilon(k) = \epsilon_0(k) + \int_{Q^-}^{Q^+} \frac{1}{2\pi}K(k-q)\epsilon(q)dq\,. \tag{9}$$

Here $\epsilon_0(k)$ denotes the bare energy. et us define the integral operator $\hat{K}$ by

$$\left(\hat{K}f\right)(k) = \int_{Q^-}^{Q^+} K(k-q)f(q)dq\,. \tag{10}$$

We express it also by $\langle k|\hat{K}f\rangle$. Writing the dressed energy and the bare energy as $\epsilon(k) = \langle k|\epsilon\rangle$ and $\epsilon_0(k) = \langle k|\epsilon_0\rangle$, respectively, we formulate the integral equation (9) as follows: $|\epsilon\rangle = |\epsilon_0\rangle + \hat{K}|\epsilon\rangle/2\pi$. It has the solution of an infinite series:

$$|\epsilon\rangle = |\epsilon_0\rangle + \left(\frac{1}{2\pi}\hat{K}\right)|\epsilon_0\rangle + \left(\frac{1}{2\pi}\hat{K}\right)^2|\epsilon_0\rangle + \cdots = \frac{I}{I - \frac{1}{2\pi}\hat{K}}|\epsilon_0\rangle\,. \tag{11}$$

Here I denotes the identity operator. We define the bulk part of the root density of the particle-hole excitation, $\rho(k|Q^{\pm})$, by the solution of the integral equation:

$$\rho(k|Q^{\pm}) = \frac{1}{2\pi} + \int_{Q^-}^{Q^+} \frac{1}{2\pi}K(k-q)\rho(q|Q^{\pm})dq\,. \tag{12}$$

et us express the constant function of variable k as $\langle k|I\rangle = 1$. Then, the root density $\rho(k|Q^{\pm})$ is written formally as

$$\rho(k|Q^{\pm}) = \frac{1}{2\pi}\langle k|\frac{I}{I - \frac{1}{2\pi}\hat{K}}|I\rangle\,. \tag{13}$$

Thus, the bulk part of the particle-hole excitation energy is expressed in terms of the dressed energy as

$$\int_{Q^-}^{Q^+} \epsilon_0(k)\rho(k|Q^{\pm})dk = \int_{Q^-}^{Q^+} \epsilon(k)\frac{dk}{2\pi}\,. \tag{14}$$

et us now solve integral equation (8). We define function $\eta(k;q|Q^{\pm})$ by the solution of the integral equation:

$$\eta(k;q|Q^{\pm}) = \delta(k-q) + \int_{Q^-}^{Q^+} \frac{1}{2\pi}K(k-k')\eta(k';q|Q^{\pm})dk'\,. \tag{15}$$

Formulating the solution of eq. (8) as the sum of the bulk density $\rho(k|Q^{\pm})$ and corrections $\rho_{p,h}(k|Q^{\pm})$: $\rho_{p,h}(k|Q^{\pm}) = \rho(k|Q^{\pm}) + \rho_{p,h}(k|Q^{\pm})$, we have

$$\rho_{p,h}(k|Q^{\pm}) = \frac{1}{L}\left(\delta(k - k_p^{(1)}) + \int_{Q^-}^{Q^+} \eta(k;k'|Q^{\pm})\frac{1}{2\pi}K(k' - k_p^{(1)})dk' - \eta(k;\tilde{k}_h|Q^{\pm})\right)\,. \tag{16}$$

2.4. Energy spectrum of the particle-hole excitation

We introduce two agrange multipliers, chemical potential μ and angular velocity Ω, respectively. We shift the Hamiltonian as $\mathcal{H}_{Y,N} = \mathcal{H}_{\rm LL} - \mu N - \Omega Y$. We now define the dressed energy $\epsilon(k)$ by eq. (9) with the new bare energy $\epsilon_0(k) = (\hbar k)^2/(2m) - \mu - \Omega R \hbar k$, where the Fermi parameters $Q^{\pm}$ are given by eqs. (7). We determine parameters μ and Ω by the following conditions: $\epsilon(Q^+) = \epsilon(Q^-) = 0$.

The energy of the particle-hole excitation with a particle at I_p and a hole at I_h is given by

$$E_{Y,N}^{p,h} = \sum_{\alpha=1}^{N-1} \epsilon_0(k_\alpha^{(1)}) + \epsilon_0(k_p^{(1)}) \, . \tag{17}$$

For large system size L, the particle-hole excitation energy $E_{Y,N}^{p,h}$ is expressed in terms of the root density through formula (A.2).

$$\begin{aligned} E_{Y,N}^{p,h} &= L \int_{Q^-}^{Q^+} \epsilon_0(k) \rho_{p,h}(k|Q^{\pm}) dk + \epsilon_0(k_p^{(1)}) \\ &= L \int_{Q^-}^{Q^+} \epsilon_0(k) \rho(k|Q^{\pm}) dk + \left(\epsilon_0(k_p^{(1)}) + L \int_{Q^-}^{Q^+} \epsilon_0(k) \quad \rho_{p,h}(k|Q^{\pm}) dk \right) . \end{aligned} \tag{18}$$

Here we recall $\rho_{p,h}(k|Q^{\pm}) = \rho(k|Q^{\pm}) + \quad \rho_{p,h}(k|Q^{\pm})$. With the solution (16) we have

$$\epsilon_0(k_p^{(1)}) + L \int_{Q^-}^{Q^+} \epsilon_0(k) \quad \rho_{p,h}(k|Q^{\pm}) dk = \epsilon(k_p^{(1)}) - \epsilon(\tilde{k}_h) \, . \tag{19}$$

Thus, through relation (14) we express the particle-hole excitation energy $E_{Y,N}^{p,h}$ in terms of the dressed energy as follows.

$$E_{Y,N}^{p,h} = L \int_{Q^-}^{Q^+} \epsilon(k) \frac{dk}{2\pi} + \epsilon(k_p^{(1)}) - \epsilon(\tilde{k}_h) - \frac{\pi v_F}{6L} + O(L^{-2}). \tag{20}$$

The exact expression (20) of the asymptotic expansion with respect to the inverse of system size L contains both the ieb's type I and type II excitations [21] together with the finite-size corrections to the yrast energy of a given angular momentum $Y^{p,h}$ [27]. The spectrum of yrast excited states are shown by making use of expression (20) in Ref. [27].

3. Non-equilibrium dynamics of localized quantum states

3.1. Second quantized Hamiltonian of the Lieb-Liniger model

et us employ the unit system with $2m = \hbar = 1$. In terms of the canonical Bose field $\hat{\psi}(x,t)$ the second quantized Hamiltonian of the ieb- iniger model is given by

$$\mathcal{H}_{\rm NLS} = \int_0^L dx [\partial_x \hat{\psi}^\dagger \partial_x \hat{\psi} + c \hat{\psi}^\dagger \hat{\psi}^\dagger \hat{\psi} \hat{\psi}]. \tag{21}$$

The second quantized Hamiltonian $\mathcal{H}_{\rm NLS}$ leads to the quantum nonlinear Schrödinger equation: $i\partial_t \hat{\psi} = -\partial_x^2 \hat{\psi} + 2c \hat{\psi}^\dagger \hat{\psi} \hat{\psi}$.

The bulk properties of the model are characterized by a single parameter $\gamma = c/n$, where $n = N/L$ is the number density of particles. We fix the particle density as $n = 1$ and vary the coupling constant c. The unit time in our simulation is proportional to L^{-2}.

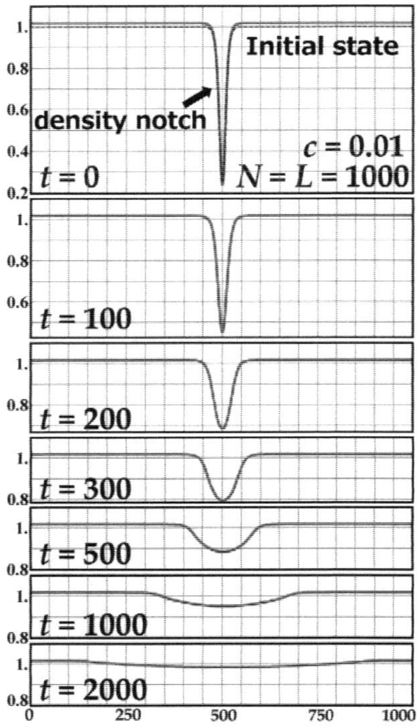

Figure 1. Snapshots of the exact time evolution of density profile $\rho(x, t)$ with $c = 0.01$ for $N = 1000$, $L = 1000$ (red solid line).

3.2. Exact time evolution of density notch: collapse into a flat profile

et us construct an initially localized quantum state [50]. We denote by $|P\rangle$ the one-hole excitation with total momentum $P = 2\pi p/L$. Here, integer p takes an integer value from the set $\{0, 1, \ldots, N-1\}$. It corresponds to the particle-hole excitation with $I_h = (N-1)/2 - p$ and $I_p = (N+1)/2$. Here, integer q satisfies $0 \leq q \leq N-1$. We define quantum states with a density notch, $|X\rangle$, by the discrete Fourier transformation of $|P\rangle$ as follows.

$$|X\rangle = \frac{1}{\sqrt{N}} \sum_{p=0}^{N-1} \exp(-2\pi i pq/N)|P\rangle \,. \tag{22}$$

Here, $|X\rangle$ has the coordinate $X = qL/N$, and its density profile has a density notch at position x with $x = X + L/2$.

In order to study the dynamics of the 1D Bose gas, we evaluate form factors of the density operator of the 1D Bose gas. Hereafter we consider the Schrödinger picture. The time evolution of the quantum state is given by $|X(t)\rangle = \exp(-i\mathcal{H}t)|X\rangle$. We evaluate the expectation value of the density operator $\hat{\rho}(x) = \hat{\psi}^\dagger(x)\hat{\psi}(x)$ with respect to state $|X(t)\rangle$ at time t as

$$\langle X(t)|\rho(x)|X(t)\rangle = \sum_{p,p'=0}^{N-1} e^{-2\pi i(p-p')q/N} e^{i(P-P')x - i(E_p - E_{p'})t} \langle P'|\rho(0)|P\rangle, \tag{23}$$

Here we recall that $P = 2\pi p/L$ and $P' = 2\pi p'/L$ denote the total momenta. The energy eigenvalues E_p of type-II excitations are readily evaluated by solving the BAE numerically.

Journal of Physics: Conference Series **497** (2014) 012030 doi:10.1088/1742-6596/497/1/012030

We calculate the form factors $\langle P'|\rho(0)|P\rangle$ in eq. (23) by making use of Slavnov's formula [40, 41] and the Gaudin-Korepin norm formula [37]. Slavnov's formula is given by

$$\langle P'|\rho(0)|P\rangle = i^N (P-P')(\prod_{j,\ell}^{N} \frac{k_j - k_\ell + ic}{k'_j - k_\ell}) \det{}_{N-1} U(k,k'), \tag{24}$$

where pseudomomenta $\{k_1, \cdots, k_N\}$ and $\{k'_1, \cdots, k'_N\}$ correspond to $|P\rangle$ and $|P'\rangle$, respectively, and the matrix elements of an $(N-1)\times(N-1)$ matrix $U(k,k')$ are given for $j,l = 1,2,\ldots,N-1$, as follows.

$$U(k,k')_{j,\ell} = 2\delta_{j\ell}\mathrm{Im}\left[\prod_{a=1}^{N} \frac{k'_a - k_j + ic}{k_a - k_j + ic}\right] + \frac{\prod_{a=1}^{N}(k'_a - k_j)}{\prod_{a\neq j}^{N}(k_a - k_j)} \left(K(k_j - k_\ell) - K(k_N - k_\ell)\right). \tag{25}$$

It is shown in Fig. 1 that the density notch in the initial state collapses into a flat profile, which gives the equilibrium state in the density profile. The collapse of the density notch shows relaxation, or more precisely, equilibration of the expectation value of the density operator, which occurs in most of isolated quantum many-body systems due to typicality of quantum states. Here, from the typicality viewpoint, we assume that most of quantum states are close to be in equilibrium: the expectation values of local operators in most of quantum states are very close to those of the micro-canonical ensemble.

More intuitively, we can explain it as follows. Suppose that a given quantum state corresponds to a representative point in the phase space and it moves in the phase space during time evolution. Then, the trajectory starting from a nonequilibrium quantum state passes through many equilibrium quantum states, since the majority of quantum states are close to be in equilibrium. It thus follows that the expectation value of a local physical quantity approaches that of the (micro-)canonical ensemble.

We now give some comments on relevant researches. In the weak coupling region the correlation length (healing length) of the 1D Bose gas increases as the coupling constant γ approaches zero. Here, quite a novel finite-size scaling behavior of condensate fraction holds [53]. For a large system with L being infinite, the collapse of an initially localized density profile can also be considered as the collapse of an initial wave packet due to the nonlinear dispersion relation of matter waves. Here we remark that coherent states can be constructed in a finite system size L [54]. Typicality of quantum states should be associated with the eigenstate thermalization hypothesis, which is studied also in the 1D Bose gas [55]. Quite recently, relaxation time is rigorously evaluated for typical isolated quantum systems [56].

3.3. Recurrence of the density notch

For the small number of particles such as $N = 20$, we observed recurrence phenomenon [50]. In particular, in the free-fermionic and the free-bosonic regimes, there are many quantum states which show recurrence [57].

Finally, we remark that the quantum system we have investigated has a finite number of particles, not an infinite one. Thus, the relaxation behavior observed in Fig. 1 is not necessarily complete. It might recur after a very long period of time such as the age of the universe.

Acknowledgment

The authors would like to thank H. Fujishima, R. Hatakeyama, R. Kanamoto and M. Ueda for useful comments. This work was partially supported by by Grant-in-Aid for Scientific Research No. 24540396. E. K. is supported by SPS.

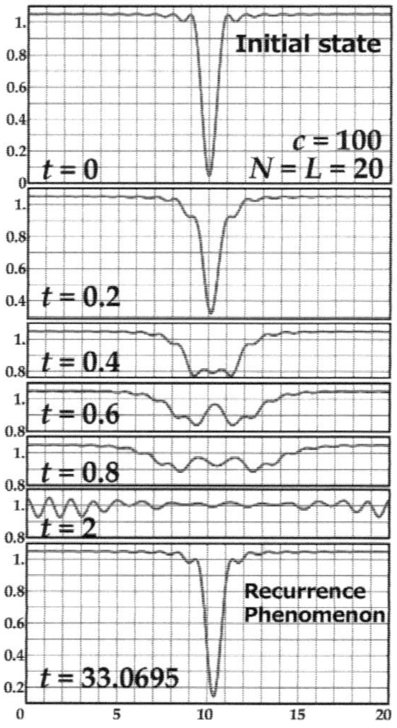

Figure 2. Snapshots of density profiles in time evolution after the initial state constructed from yrast states ($c = 100$, $N = 20$, $L = 20$). The density notch appears again at $t = 33.0695$.

Appendix A. The Euler-Maclaurin formula

Making use of the Euler-Maclaurin formula for any given analytic function $f(x)$

$$\frac{1}{L}\sum_{j=n_1}^{n_2} f(j/L) = \int_{(n_1-1/2)/L}^{(n_2+1/2)/L} f(x)dx - \frac{1}{24L^2}\left(\frac{df}{dx}(\frac{n_2+1/2}{L}) - \frac{df}{dx}(\frac{n_1-1/2}{L})\right) + o(\frac{1}{L^2}), \quad (A.1)$$

we can approximate the sum of $f(k_j)$ over an interval of integer j as an integral of k:

$$\frac{1}{L}\sum_{\alpha=1}^{N-1} f(k_\alpha^{(1)}) = \int_{Q^-}^{Q^+} f(k)\hat{\rho}(k)dk + O(1/L). \quad (A.2)$$

[1] S. Inouye, M.R. Andrews, J. Stenger, H.-J. Miesner, D.M. Stamper-Kurn, and W. Ketterle, Nature **392**, 151 (1998).
[2] A. Gorlitz, J.M. Vogels, A.E. Leanhardt, C. Raman, T.L. Gustavson, J.R. Abo-Shaeer, A.P. Chikkatur, S. Gupta, S. Inouye, T. Rosenband and W. Ketterle, Phys. Rev. Lett. **87**, 130402 (2001).
[3] M. Greiner, I. Bloch, O. Mandel, T.W. Hansch, and T. Esslinger, Phys. Rev. Lett. **87**, 160405 (2001).
[4] I. Bloch, J. Dalibard, and W. werger, Rev. Mod. Phys. **80**, 885 (2008).
[5] M. Rigol, V. Dunjko, V. Yurovsky and M. Olshanii, Phys. Rev. Lett **98**, 050405 (2007); M. Rigol, V. Dunjko and M. Olshanii, Nature **452**, 854 (2008).
[6] T. Kinoshita, T. Wenger and D.S. Weiss, Science **305**, 1125 (2004); Phys. Rev. Lett. **95**, 190406 (2005); Nature **440**, 900 (2006).
[7] M. Gring, M. Kuhnert, T. Langen, T. Kitagawa, B. Rauer, M. Schreitl, I. Mazets, D. Adu Smith, E. Demler, J. Schmiedmayer, Science **337**, 1318 (2012).
[8] V.I. Yukalov, Laser Phys. Lett. **8**, 485 (2011).
[9] A. Polkovnikov, K. Senguputa, A. Silva, and M. Vengalattore, Rev. Mod. Phys. **83**, 863 (2011)
[10] H. Tasaki, Phys. Rev. Lett. **80**, 1373 (1998).

[11] S. Goldstein, J.L. Lebowitz, R. Tumulka, and N. anghi, Phys. Rev. Lett. **96**, 050403 (2006)
[12] R. Tumulka, Eur. Phys. J. H **35**, 201 (2010) [J. von Neumann, . Phys. 57, 30 (1929)]
[13] P. Reimann, Phys. Rev. Lett. **101**, 190403 (2008).
[14] A. Sugita, Nonlinear Phenom. Complex Syst. **10**, 192 (2007).
[15] G.F. Bertsch and T. Papenbrock, Phys. Rev. Lett. **83**, 5412 (1999).
[16] N. Gemelke, E. Sarajlic, and S. Chu, ar iv:1007.2677.
[17] E.H. Rezayi, F.D.M. Haldane, and K. Yang, Phys. Rev. Lett. **83**, 1219 (1999).
[18] D.A. Butts and D.S. Rokhsar, Nature **397**, 327 (1999).
[19] S. Viefers and M. Taillefumier, ar iv:1005.2520.
[20] E. H. Lieb and W. Liniger, Phys. Rev. **130**, 1605 (1963).
[21] E.H. Lieb, Phys. Rev. **130**, 1616 (1963).
[22] M. Olshanii, Phys. Rev. Lett. **81**, 938 (1998).
[23] S. Gupta, K.W. Murch, K.L. Moore, T.P. Purdy, and D.M. Stamper-Kurn, Phys. Rev. Lett. **95**, 143201 (2005); A.S. Arnold, C.S. Garvie, and E. Riis, Phys. Rev. A **73**, 041606(R) (2006); S.R. Muniz, S.D. Jenkins, T.A.B. Kennedy, D.S. Naik, and C. Raman, Opt. Express **14**, 8947 (2006).
[24] C. Ryu, M.F. Andersen, P. Clade, V. Natarajan, K. Helmerson and W.D. Phillip, Phys. Rev. Lett. **99**, 260401 (2007).
[25] M. Girardeau, J. Math. Phys. (NY) **1**, 516 (1960); B. Paredes, A. Widera, V. Murg, O. Mandel, S. Folling, I. Cirac, G.V. Shlyapnikov, T.W. Hansch, and I. Bloch, Nature **429**, 277 (2004).
[26] R. Kanamoto, L. D. Carr and M. Ueda, Phys. Rev. A **81**, 023625 (2010).
[27] E. Kaminishi, R. Kanamoto, J. Sato and T. Deguchi, Phys. Rev. A **83**, 031601 (R) (2011).
[28] M. Ishikawa and H. Takayama, J. Phys. Soc. Jpn. **49**, 1242 (1980).
[29] C. Becker, S. Stellmer, P. Soltan-Panahi, S. Dorscher, M. Baumert, E.-M. Richter, J. Kronjager, K. Bongs, and K. Sengstock, Nature Phys. **4**, 496 (2008).
[30] J. Sato, R. Kanamoto, E. Kaminishi and T. Deguchi, Quantum dark solitons in the 1D Bose Gas and the superfluid velocity, ar iv:1204.3960 [cond-mat.quantum-gas].
[31] E. Barouch, B.M. McCoy and M. Dresden, Phys. Rev. **2**, 1075 (1970).
[32] P. Calabrese and J. Cardy, Phys. Rev. Lett. **96**, 136801 (2006); J. Stat. Mech. (2007) P06008.
[33] D. Rossini, S. Suzuki, G. Mussardo, G.E. Santoro and A. Silva, Phys. Rev. B **82**, 144302 (2010).
[34] J. Mossel and J.-S. Caux, New J. Phys. **12** (2010) 055028.
[35] A.C. Cassidy, C.W. Clark, and M. Rigol, Phys. Rev. Lett. **106**, 140405 (2011).
[36] P. Calabrese, F.H.L. Essler, and M. Fagotti, Phys. Rev. Lett. **106**, 227203 (2011).
[37] V.E. Korepin, N.M. Bogoliubov and A.G. Izergin, *Quantum Inverse Scattering Method and Correlation Functions* (Cambridge University Press, Cambridge, 1993)
[38] M. Takahashi, *Thermodynamics of one-dimensional solvable models* (Cambridge University Press, Cambridge, 1999).
[39] F.H.L. Essler, H. Frahm, F. Gohmann, A. Klumper and V.E. Korepin, *The one-dimensional Hubbard model* (Cambridge University Press, Cambridge, 2005).
[40] N. A. Slavnov, Teor. Mat. Fiz. **79**, 232 (1989).
[41] N. A. Slavnov, Teor. Mat. Fiz. **82**, 389 (1990).
[42] N. Kitanine, J.M. Maillet, V. Terras, Nucl. Phys. B **554** [FS], 647 (1999).
[43] T. Kojima, V.E. Korepin, N.A. Slavnov, Commun. Math. Phys. **188**, 657 (1997).
[44] J.-S. Caux, P. Calabrese and N. A. Slavnov, J. Stat. Mech. (2007) P01008.
[45] P. Calabrese and J.-S. Caux, J. Stat. Mech. (2007) P08032.
[46] D. Biegel, M. Karbach and G. Muller, Europhys. Lett. **59** (2002) 882.
[47] J. Sato, M. Shiroishi and M. Takahashi, J. Phys. Soc. Jpn. **73** (2004) 3008.
[48] J.-S. Caux, R. Hagemans and J.M. Maillet, J. Stat. Mech. (2005) P09003.
[49] J. Sato, B. Aufgebauer, H. Boos, F. Gohmann, A. Klumper, M. Takahashi, C. Trippe, Phys. Rev. Lett. **106**, 257201 (2011).
[50] J. Sato, R. Kanamoto, E. Kaminishi, and T. Deguchi, Phys. Rev. Lett. **108**, 110401 (2012).
[51] T.C. Dorlas, Commun. Math. Phys. **154**, 347 (1993).
[52] E. Emsiz, Lett. Math. Phys. **91**, 61 (2010).
[53] J. Sato, E. Kaminishi, and T. Deguchi, Finite-size scaling behavior of Bose-Einstein condensation in the 1D Bose Gas, ar iv:1303.2775 [cond-mat.quantum-gas].
[54] I.V. Volovich and A.S. Trushechkin, Izuvestiya: Mathematics **76** (2012) 43-84.
[55] T. N. Ikeda, Y. Watanabe, and M. Ueda, Phys. Rev. E **84**, 021130 (2011).
[56] S. Goldstein, T. Hara, and H. Tasaki, Phys. Rev. Lett. **111**, 140401 (2013).
[57] E. Kaminishi, J. Sato, and T. Deguchi, Recurrence time in the quantum dynamics of the 1D Bose gas, ar iv:1305.3412 [cond-mat.quantum-gas].

Strongly correlated excitation of a quasi 1D Rydberg gas

N Malossi[a,b,1], M M Valado[a,b], S Scotto[a,2], O Morsch[a,b], E Arimondo[a,b,c] and D Ciampini[a,b,c]

[a] Dipartimento di Fisica 'E. Fermi', Università di Pisa, Largo Pontecorvo 3, 56127 Pisa, IT
[b] INO-CNR, Via G. Moruzzi 1, 56124 Pisa, IT
[c] CNISM UdR Pisa, Dipartimento di Fisica 'E. Fermi', Università di Pisa, Largo Pontecorvo 3, 56127 Pisa, IT

E-mail: `ciampini@df.unipi.it`

Abstract. Rydberg excitation dynamics of a 87-Rb cold atom cloud is investigated in an e ective one-dimensional geometry. We measure the excitation dynamics and the full counting statistics for resonant and o -resonant excitation to the $70S$ state. While for a resonant laser excitation the counting distributions have a strong sub-Poissonian character, we nd strongly bimodal counting distributions in the o -resonant regime. The n-th central moments, up to $n = 4$, of the counting distributions are derived from the measured counting distributions.

1. Introduction

Atoms in high-lying Rydberg states strongly interact with each other via the dipole-dipole or van-der-Waals potential [1, 2, 3]. The interaction-induced level shifts can suppress resonant optical excitation of more than one Rydberg atom within a certain blockade volume containing tens or hundreds of atoms [3, 4]. This long-range interaction is at the heart of many quantum information processing schemes [5, 6] including quantum gates between individual atoms [7, 8, 9, 10]. A system of strongly interacting Rydberg atoms holds also promise for studies of dynamics and disorder in many-body systems with tunable interactions. Dynamical crystallisation with long range order over the full sample can be produced by the application of properly shaped laser pulses [11, 12]. The short-range organisation of Rydberg excitation was recently observed for Rydberg atoms created in an optical lattice [13].

For resonant laser excitation an important consequence of the dipole blockade is the suppression of fluctuations in the number of Rydberg excitations [14, 15, 16, 17], leading to highly sub-Poissonian excitation statistics with a negative large Mandel Q-parameter [18]. Such fluctuation suppression evidences the creation of quantum correlations in the whole sample.

The present work reports experimental observations for the off-resonant excitation scheme, with the excitation laser having a finite detuning from the Rydberg state. In such an off-resonant excitation scheme, a pair of atoms undergoes a pair excitation if the atomic interaction matches

[1] Present address: Scuola di Scienze e Tecnologie, Sezione di Fisica, Università di Camerino, I-62032 (MC), IT.
[2] Present address: Laboratoire National des Champs Magnetiques Intenses (UPR 3228, CNRS-UPS-UJF-INSA), 31400 Toulouse, FR.

Journal of Physics: Conference Series **497** (2014) 012031 doi:10.1088/1742-6596/497/1/012031

the laser energy defect excess. After such event, neighbouring ground-state atoms at the resonant distance have an enhanced excitation probability, owing to the interaction induced energy shifts. As a consequence an already excited Rydberg atom pair can shift other atoms into resonance, in a domino effect, leading to an increasing overall Rydberg excitation probability. We demonstrate that the strong atomic correlations result in a super-Poissonian excitation statistics with positive Mandel Q-parameter. The off-resonant excitation of Rydberg pairs was examined by previous authors [19, 20] without a study of the excitation statistics. A super-Poissonian statistics was observed for a quantum dot optically coupled to a detuned cavity, where the cooling of initial photons is inhibited, but once an initial photon is coupled the coupling of subsequent photons is increased via multiphoton processes [21]. In the investigation of single-electron tunneling in semiconductor quantum dots, the ability to detect individual charges in real-time makes it possible to count electrons, measure their counting statistics and investigate correlations between charge carriers [22].

Both resonant and off-resonant Rydberg excitation produce strong spatial and temporal correlations in the sample, created by the long-range interactions [16, 17, 23, 24]. In this work we study the interaction dynamics of Rydberg atoms in an effective 1D geometry in which excitations can occur only along one spatial direction. The dynamics in the off-resonant excitation regime is governed by a complex interplay between off-resonant single-atom and pair excitations mediated by the Rydberg-Rydberg interaction. Our measured Rydberg counting distribution shows a bimodal structure, with a peak at zero excitations corresponding to the probability that no pairs of atoms have been excited and a second peak at a large number of excitations, as reproduced by our theoretical analysis based on a Dicke-type model, described in [24].

2. Rydberg blockade

The $5S$ atoms are excited to the $70S$ Rydberg state using a two-photon excitation process with detuning via the intermediate $6P$ level (see [25, 16, 26] for details). In the Rydberg excitation scheme the excitation lasers are detuned by 2 GHz from the $6P$ state, which can therefore be adiabatically eliminated and the laser interaction will be described by a two-level effective system with Rabi frequency Ω and laser detuning . Since rubidium atoms in the $70S$ state have repulsive interactions described by the van-der-Waals coefficient $C_6 > 0$, a positive interaction energy has to be added to the excitation energy. The ensemble of N_{db}^{at} atoms within the blockade radius volume, carrying only one excitation that can be located at any of the atoms, is excited collectively to a Wigner state, evolving with a collective Rabi frequency $\Omega^{coll} = \sqrt{N_{db}^{at}}\Omega$ [3, 27]

For an off-resonant excitation using a laser detuned by   from the singly-Rydberg excited state, a pair of Rydberg atoms at distance R can be excited, if the pair van-der-Waals interaction energy matches twice the energy mismatch $\hbar\ $ of the laser excitation, i.e., $C_6/R^6 = 2\hbar\ $. The timescale for the off-resonant excitation of a single atom pair is determined by the Rabi frequency $\Omega_{off} = \Omega^2/(2\ )$ of this two-atom two-photon process.

3. Experimental realization

In our experiment we create small clouds of ultracold 87-Rb atoms in a magneto-optical trap (MOT). The cold cloud has approximately a Gaussian density profile with typical widths $\sigma = 40\,\mu$m and containing $N^{at} = 4 \times 10^4$ atoms. The resulting peak density, $n = 5 \times 10^{10}\,\mathrm{cm}^{-3}$, corresponds to a mean inter-particle spacing of around $2.6\,\mu$m. The first step laser at 420 nm is sent through a single mode optical fiber and focused to a $w = 6\,\mu$m Gaussian waist, which is smaller than the dipole blockade radius for the $70S$ state. As a consequence, the probability to excite more than one atom in the radial direction is suppressed by the exponentially decreasing two-photon Rabi frequency as a function of the radial coordinate r ($\Omega(r) \propto e^{-2r^2/w^2}$) and hence the Rydberg excitations are created in an effective 1D geometry. The second step laser at

Journal of Physics: Conference Series **497** (2014) 012031 doi:10.1088/1742-6596/497/1/012031

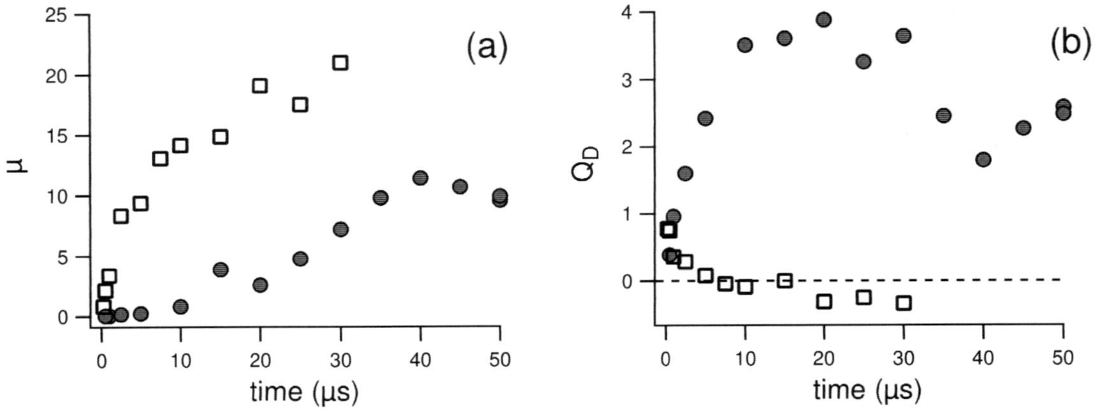

Figure 1. Experimental results for $70S$ excitation dynamics and statistics of the detected ions for resonant ($= 0$) and off-resonant excitation ($/2\pi = 7.7$ MHz). Detected ion mean number, in (a), and detected Mandel Q-factor of the detected ions, in (b), vs excitation time. Black open squares for on resonance and blue full circles for off-resonance.

1013 nm has a Gaussian waist of 80 μm, larger than the MOT size. For the atoms in the highest laser intensity region $\Omega/2\pi$ is 900 kHz. For our number $N_{db}^{at} \approx 20$ of atoms inside the blockade volume, the resonant collective Rabi frequency is $\Omega^{coll} = \sqrt{N_{db}^{at}}\Omega \approx 2\pi \times 4$ MHz.

For the $/2\pi = 7.7$ MHz detuning of the present investigation, the resonance condition for pair excitation corresponds to an interatomic separation of $R = 6.5$ μm, comparable with w, the Gaussian beam waist of the blue excitation laser. The off-resonant excitation timescale is given by $\Omega_{off} \approx 2\pi \times 50$ kHz in the highest laser intensity region. Ω_{off} is proportional to the intensity of the blue laser and quickly decreases in the radial direction, as $\Omega_{off}(r) \propto e^{-4r^2/w^2}$. Therefore, for both the resonant and off-resonant excitations our system can be considered as an effective quasi-1D system.

After the excitation pulse (of duration between 0.2 and 50 μs) an electric field is applied for 2 μs in order to field ionize the Rydberg atoms and to accelerate the resulting ions towards a channeltron. The overall detection efficiency is $\eta \approx 40\%$ [25]. During the entire excitation and detection sequence the MOT beams are switched off. The excitation-detection cycle is repeated 500 times, in order to measure the counting distribution of the detected ions. The experiments described in this paper are in the non-dissipative regime, since the longest excitation time used here (50 μs) is shorter than the Rydberg state lifetime ($\sim$ 200 μs for the $70S$ state). On the other hand, since the excitation times are typically longer than the coherence time of the laser (limited by the finite linewidth of the excitation laser of approximately 0.5 MHz), most of the system dynamics takes place in an incoherent regime.

4. Results

From the counting distributions of the detected ions we derived the probability $P(N_{ion}, t)$ for the detection of N_{ion} after an interaction time t for resonant and off-resonant excitation. From that probability we calculated the n-th order central moments $\mu_n = \langle (N_{ion} - \mu)^n \rangle$ up to $n = 4$ as well as their associated normalized quantities Mandel Q-factor $Q_D = \mu_2/\mu - 1$, Binder cumulant $B_D = 1 - \mu_4/(3\mu_2^2)$ and bimodality coefficient $b_D = (\gamma^2 + 1)/(\mu_4/\mu_2^2)$ as a function of the mean number μ. For the purposes of this paper, we focus on the 2nd and 4th order moments. All the

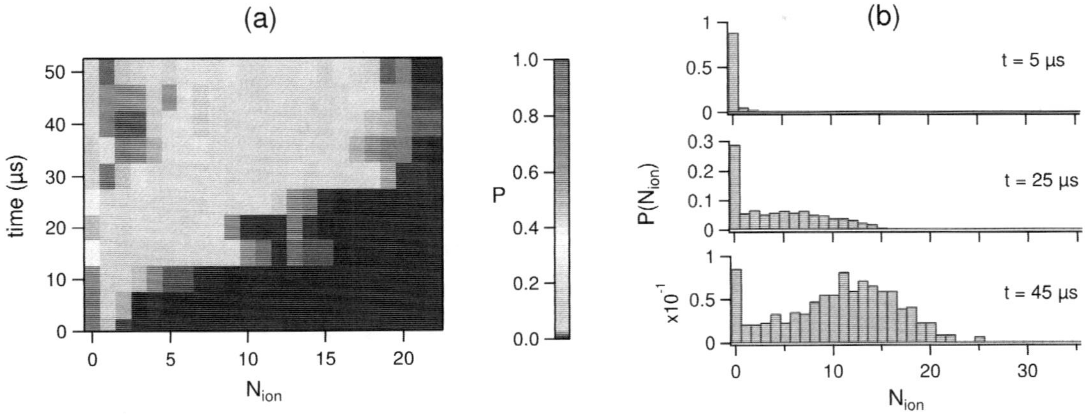

Figure 2. Histograms of $P(N_{ion}, t)$ vs interaction time t in the $/2\pi = 7.7$ MHz off-resonant regime . In (a) global evolution of $P(N_{ion}, t)$ highlighted by a false color representation. In (b) histograms for excitation times $t = 5$, 25 and 45 μs.

above quantities are calculated from the *observed* counting distribution, i.e., the distribution of the detected ions, and the finite detection efficiency has to be taken into account for the comparison with a theoretical model (see [24] for details). For the Mandel Q-factor the simple relation $Q_D = \eta Q$ between the detected quantity and the actual one holds.

The mean number of detected ions and the detected Mandel Q-parameter are shown in Fig.1. For a $= 0$ resonant excitation, a rapid increase in the average detected ion number is concomitant with a decrease in Q_D. After few microseconds the growth in the detected ion number slows down while Q_D fluctuates around -0.3, illustrating the sub-Poissonian character of the distribution on resonance. For the $/2\pi = 7.7$ MHz off-resonant excitation, the mean number of detected ions grows much more slowly than in the resonant case, since $\Omega_{off} \ll \Omega$. The maximum value is approximately half the number of detected ions in the resonant saturated regime. In that regime the Mandel Q-parameter is large and positive, showing the super-Poissonian nature of the counting distribution.

In Fig. 2 histograms of $P(N_{ion}, t)$ in the off-resonant excitation regime are presented. The false-color representation of Fig. 2(a) provides a global view, while histograms of $P(N_{ion}, t)$ vs N_{ion} for different excitation t times are shown in Fig. 2(b). The bimodality of the ion distribution is highlighted in Fig. 2(a) and appears clearly in the plot of Fig. 2(b) at $t = 45\,\mu$s.

In Fig. 3 several statistical quantities derived from the off-resonant and resonant counting distributions are shown as a function of the mean number of detected ions. These plots allow a direct comparison between the on- and off-resonant excitation regimes for equal excitation numbers, independently of their very different timescales. In Fig. 3(a) and (b) the variance and the Mandel Q-parameter are shown. For resonant excitation, the variance grows in the full explored range of parameters and the Mandel Q-factor becomes negative when the saturation is approached. For off-resonant excitation, the variance reaches its maximum for the largest number of detected ions reported in Fig. 1(a). The Mandel parameter Q_D reaches its maximum at smaller values of detected ions because it is obtained from the variance applying a normalization by the mean number. Figure 3(c) shows the bimodality parameter as a function of the mean number of detected ions. In the explored range, it is always larger for off-resonant excitation, emphasizing the qualitatively different nature of the on- and off-resonant counting distributions. Figure 3(d) reports the values of the Binder cumulant for resonant and off-resonant

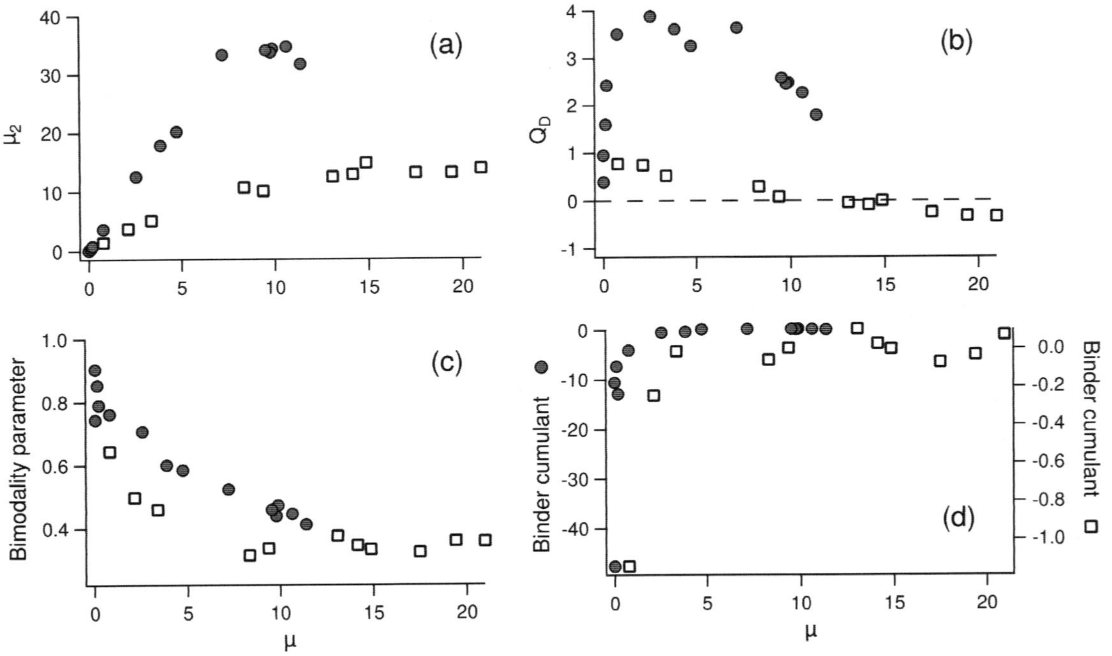

Figure 3. Statistical quantities calculated from the off-resonant (blue full circles) and resonant (black open squares) counting distributions are shown as a function of the mean number of detected ions. In (a) second-order moment μ_2, in (b) Mandel Q-factor, in (c) bimodality parameter, in (d) the Binder cumulant. The off-resonant excitation detuning is $/2\pi = 7.7$ MHz.

excitation. As originally pointed out by Binder [28], this quantity allows one to extract the critical parameters and the critical exponents for a system experiencing a phase transition, for instance the Ising quantum phase transition predicted for Rydberg experimental configurations in [29, 30, 31].

5. Conclusions

We have studied strongly correlated Rydberg excitations in an effective 1D geometry. Our system is quasi-1D for both resonant excitation and also the off-resonant excitation at the explored detuning from the $70S$ state. The explored excitation times are longer than the coherence time, limited by the finite linewidth of the excitation lasers, but shorter than the Rydberg lifetime. Therefore the presented experimental observations take place in an incoherent regime, but not in the dissipative regime explored in [24].

We have measured the the full counting statistics of the Rydberg excitations and derived from the statistical distributions the central moments up to $n = 4$. The counting distribution at resonance is characterized by a strong sub-Poissonian character, while in the off-resonant regime the histograms of the counting distribution are clearly bimodal. The statistical moments calculated from the counting distribution reveal characteristic features that are not evident in the mean and the variance (as the bimodality for off-resonant excitation), and will allow a more precise test of theoretical predictions.

The steady state of our system for off-resonant excitations and for excitation times longer than the Rydberg lifetime, has peculiar characteristics, theoretically analyzed for ultracold atoms

Journal of Physics: Conference Series **497** (2014) 012031 doi:10.1088/1742-6596/497/1/012031

confined in a crystalline structure with a reduced dimensionality [30, 31, 32, 33]. A system of regularly spaced atoms excited to Rydberg states has a formal analogy with a dissipative Ising model with a transverse field and is predicted to have a dynamical phase transition between a paramagnetic and a ferromagnetic phase [29, 30, 31], with a bimodal distribution at the interfaces between dynamical phases. Our experimental results show a bimodal character for the counting distribution well before entering in the dissipative regime and in the absence of an externally imposed crystalline structure (as for instance, by an optical lattice potential). However, since the off-resonance excitation of couples of Rydberg atoms is spatially selective, over time the Rydberg excitations are expected to arrange themselves in regular arrays. Our investigation demonstrates the possibility of using ultra-cold Rydberg atoms as quantum simulators of strongly interacting many-body systems presenting locally ordered structures.

Acknowledgments

This work was supported by the E.U. through Grant No. 265031-ITN-COHERENCE and by the MIUR PRIN-2009 QUANTUM GASES BEYOND EQUI IBRIUM.

References

[1] Gallagher T F 1994 *Rydberg Atoms* (Cambridge: Cambridge University Press)
[2] Gallagher T F and Pillet P 2008 *Adv. At. Mol. Opt. Phys.* **56** 161
[3] Comparat D and Pillet P 2010 *J. Opt. Soc. Am.* B **27** A208
[4] Sa man M, Walker T G, Mølmer K 2010 *Rev. Mod. Phys.* **82** 2313
[5] Jaksch D, Cirac J I, oller P, Rolston S L, Côte R and Lukin M D 2000 *Phys. Rev. Lett.* **85** 2208
[6] Lukin M D, Fleischhauer M, Côte R, Duan L M, Jaksch D, Cirac J I and oller P 2001 *Phys. Rev. Lett.* **87** 037901
[7] Urban E, Johnson T A, Henage T, Isenhower L, Yavuz D D, Walker T G and Sa man M 2009 *Nature Phys.* **5** 110
[8] Isenhower L, Urban E, hang L, Gill A T, Henage T, Johnson T A, Walker T G and Sa man M 2010 *Phys. Rev. Lett.* **104** 010503
[9] Gaetan A, Miroshnychenko Y, Wilk T, Chotia A, Viteau M, Comparat D, Pillet P, Browaeys A and Grangier P 2009 *Nature Phys.* **5** 115
[10] Wilk T, Gaetan A, Evellin C, Wolters J, Miroshnychenko Y, Grangier P and Browaeys A 2010 *Phys. Rev. Lett.* **104** 010502
[11] Pohl T, Demler E and Lukin MD 2010 *Phys. Rev. Lett.* **104** 043002
[12] Schachenmayer J, Lesanovsky I, Micheli A and Daley A J 2010 *NJP* **12** 103044
[13] Shauß P, Cheneau M, Endres M, Fukuhara T, Hild T, Omran A, Pohl T, Gross C, Kuhr S and Bloch I 2012 *Nature* **491** 87
[14] Cubel Liebisch T, Reinhard A, Berman P R and Raithel G 2005 *Phys. Rev. Lett.* **95** 253002 and Erratum 2007*ibid.* **98** 109903
[15] Reinhard A, Younge K C and Raithel G 2008 *Phys. Rev.* A **78** 060702(R)
[16] Viteau M, Huillery P, Bason M G, Malossi N, Ciampini D, Morsch O, Arimondo E, Comparat and Pillet P 2012 *Phys. Rev. Lett.* **109** 053002
[17] Hofmann C S, Gunter G, Schempp H, Robert-de-Saint-Vincent M, Garttner M, Evers J, Whitlock S and Weidemuller M 2013 *Phys. Rev. Lett.* **110** 203601
[18] Mandel L 1982 *Phys. Rev. Lett.* **49** 136
[19] Ates C, Pohl T, Pattard T and Rost J M 2007 *Phys. Rev. Lett.* **98** 023002
[20] Amthor T, Giese C, Hofmann C S and Weidemuller M 2010 *Phys. Rev. Lett.* **104** 013001
[21] Majumdar A, Bajcsy M, Rundquist A and Vuckovic J 2012 *Phys. Rev. Lett.* **108** 183601
[22] Gustavsson S, Leturcqa R, Studera M, Shorubalkoa I, Ihna T, Ensslina K, Driscollb D C and Gossardb A C 2009 *Surf. Sci. Rep.* **64** 191
[23] Schempp H, Gunter G, Robert-de-Saint-Vincent M, Hofmann C S, Breyel D, Komnik A, Schonleber D W, Garttner M, Evers J, Whitlock S and Weidemuller M 2014 *Phys. Rev. Lett.* **112** 013002
[24] Malossi N, Valado M M, Scotto S, Huillery P, Pillet P, Ciampini D, Arimondo E and Morsch O 2013 *Preprint* ar iv:1308.1854
[25] Viteau M, Radogostowicz J, Chotia A, Malossi N, Fuso F, Ciampini D, Morsch O, Ryabtsev I and Arimondo E 2010 *J. Phys.* B **43** 155301

Journal of Physics: Conference Series **497** (2014) 012031 doi:10.1088/1742-6596/497/1/012031

[26] Viteau M, Bason M, Radogostowicz J, Malossi N, Morsch O, Ciampini D and Arimondo E 2013 *Laser Phys.* **23** 015502
[27] Dudin Y O, Li L, Bariani F and Kuzmich A 2012 *Nature Phys.* **8** 790
[28] Binder K 1981 *Z. Phys.* B **43** 119
[29] Weimer H, Low R, Pfau T and Buchler H P 2008 *Phys. Rev. Lett.* **101** 250601
[30] Ates C, Olmos B, Garrahan J P, and Lesanovsky I 2012 *Phys. Rev.* A **85** 043620
[31] Lee T E, Ha ner H, and Cross M C 2012 *Phys. Rev. Lett.***108** 023602
[32] Hu A, Lee T E, and Clark C 2013 *Phys. Rev.*A **88** 053627
[33] Carr C, Ritter R, Weatherill K J,and Adams C S 2013 *Phys. Rev. Lett.***111** 113901

Dynamical entanglement in coupled systems

K. Ziegler
Institut für Physik, Universität Augsburg, D-86135 Augsburg, Germany

Abstract. We present a comparative analysis on the evolution of two coupled bosonic many-body quantum systems. Considering photons in two coupled cavities and bosonic atoms in a double-well potential, the effect of the boson-boson interaction on the spectral properties and the dynamical behavior is studied. In particular, we analyze the evolution of a Fock state to a N00N state, which is a superposition of two complementary Fock states. Such an entangled state appears in the evolution only if tunneling and many-body interaction are balanced.

1. Introduction

The dynamics of isolated many-body quantum systems has been a subject of intense research in atomic physics during recent years, in experiment [1] as well as in theory [2]. This interest has two major aspects. One is to prepare the system in a well-defined initial state and, secondly, to study its evolution for a period of time due to quantum tunneling and particle-particle interaction. The preparation of the initial state as the groundstate of a certain Hamiltonian H_0 and the evolution with $\exp(-iHt)$ for a different Hamiltonian H involves a sudden change $H_0 \to H$, which is usually called a quench. Such a quench can be realized in atomic systems by changing the potential wells in which the atoms are trapped [3]. For instance, bosonic atoms are prepared in a Fock state, where a definite number of atoms are localized in deep optical potential wells (right picture in Fig. 1). Then the potential barrier between a pair of neighboring wells is suddenly reduced such that the atoms can tunnel between these wells [4, 5]. From the theoretical point of view the statistical properties of this problem have been studied intensively using the Hubbard model and related models [6–9]. An essential element of the dynamical analysis is the evaluation of the spectral weights with respect to the initial state [10, 11], which links directly the quantum evolution with the spectral properties of the underlying Hamiltonian.

Besides systems of ultracold atoms an alternative approach of controllable bosons is based on photons. Then the role of the potential wells is played by microwave or optical cavities (left picture in Fig. 1). The experimental preparation of Fock states in a optical cavity has been achieved recently [12, 13]. This is a crucial step towards a systematic study of correlated many-body systems with photonic states. The interaction between the photons is indirectly mediated by atoms inside the optical cavities, which interact directly with the photons [14–16]. Once a Fock state with N photons has been prepared inside an optical cavity, we can couple the latter with another optical cavity by a waveguide or an optical fiber. Then the photons can tunnel between the two optical cavities, leading to a quantum evolution of the initial Fock state $|N, 0\rangle$ within the Hilbert space that is spanned by the eigenstates of the Hamiltonian of the new system. This new Hamiltonian can be approximated, for instance, by the Hubbard Hamiltonian, as suggested recently by several groups [17–20]. This type of system, including atomic degrees of freedom, was studied within a Hartree-Fock approximation [21].

Journal of Physics: Conference Series **497** (2014) 012032 doi:10.1088/1742-6596/497/1/012032

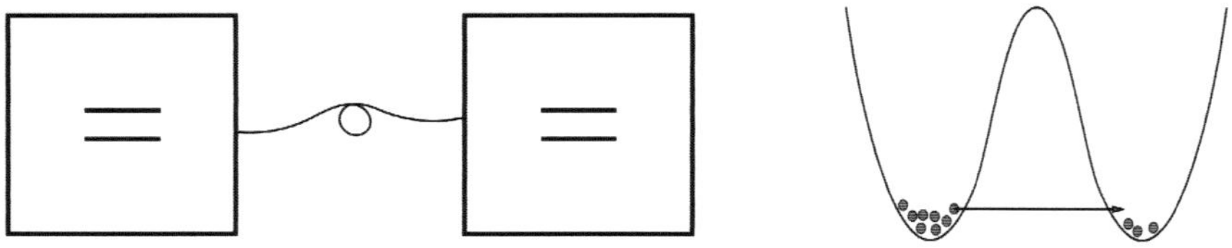

Figure 1. Two optical cavities with two-level atoms are coupled by an optical fiber (left picture). Bosonic atoms in a double-well potential present a similar system (right picture).

The evolution from the initial Fock state can, in principle, lead to an entangled state, such as the N00N state $(|N,0\rangle + \exp(i\phi)|0,N\rangle)/\sqrt{2}$. The N00N state has attracted much attention because it can be used for highly accurate interferometry and other precision measurements [22–25] and for technological application such as optical lithography [26]. Various methods for the creation of photonic N00N states have been suggested in the literature [27, 28] and indeed experimentally created up to $N = 5$ photons [25]. This clearly indicates that the creation of these entangled states is realistic. However, it still remains a problem to created N00N states for large N. The author discussed recently the dynamical creation of a N00N state from ultracold bosonic atoms in a double well [29], which shows that a balanced effect of inter-well tunneling and intra-well interaction can produce such a state with moderate probability. Since photon-photon interaction can also be mediated in a cavity by coupling the photons to atoms, we will analyze in the following the dynamical creation of entangled (N00N) states for a pair of anharmonic cavities, described by coupled Jaynes-Cummings models [30, 31], and compare the results with those of the Hubbard model.

The paper is organized as follows: In Sect. 2 we introduce the basic quantities for the dynamics of our systems and discuss the return probability, the transition probability and their relation with entangled states. As an example for non-interacting bosons we analyze two coupled harmonic cavities in Sect. 3. Then we discuss briefly the Jaynes-Cummings model for a single cavity with a two-level atom (Sect. 4.1) and derive an effective Hamiltonian for two coupled cavities (Sect. 4.2). After a brief review on bosonic atoms in a double-well potential (Sect. 5), the recursive projection method for calculating spectra and the dynamics of coupled systems is described in Sect. 6. The results of this approach for coupled anharmonic cavities and bosonic atoms in a double-well potential are discussed and compared with the results of the coupled harmonic cavities in Sect. 7. Finally, we summarize the work in Sect. 8.

2. Spectral density and the evolution of isolated systems

We consider a system which is isolated from the environment. In terms of photonic states this can be realized by an ideal optical cavity. With the initial state $|\Psi_0\rangle$ we obtain from the time evolution $|\Psi_t\rangle = e^{-iHt}|\Psi_0\rangle$ the evolution of the return probability $|\langle\Psi_0|\Psi_t\rangle|^2$ with the return amplitude $\langle\Psi_0|\Psi_t\rangle = \langle\Psi_0|e^{-iHt}|\Psi_0\rangle$. A Laplace transformation relates the return and the transition amplitude to another state $|Psi_1\rangle$ with the resolvent through the identities

$$\langle\Psi_0|\Psi_t\rangle = \int_\Gamma \langle\Psi_0|(z-H)^{-1}|\Psi_0\rangle e^{-izt}dz, \quad \langle\Psi_1|\Psi_t\rangle = \int_\Gamma \langle\Psi_1|(z-H)^{-1}|\Psi_0\rangle e^{-izt}dz \ , \qquad (1)$$

where the contour Γ encloses all the eigenvalues E_j $(j = 0, 1, ..., N)$ of H, assuming that the underlying Hilbert space is $N + 1$ dimensional. With the corresponding eigenstates $|E_j\rangle$ the spectral representation of the resolvent is a rational function of z:

$$\langle\Psi_0|(z-H)^{-1}|\Psi_0\rangle = \sum_{j=0}^{N} \frac{|\langle\Psi_0|E_j\rangle|^2}{z-E_j} = \frac{P_N(z)}{Q_{N+1}(z)}, \quad Q_{N+1}(z) = \prod_{j=0}^{N}(z-E_j) \ , \qquad (2)$$

Journal of Physics: Conference Series **497** (2014) 012032 doi:10.1088/1742-6596/497/1/012032

where $P_N(z)$, $Q_{N+1}(z)$ are polynomials in z of order N, $N+1$, respectively. These polynomials can be evaluated by the recursive projection method [32].

The expression in Eq. (2) suggests the introduction of the spectral density $\rho_{0,0}(E)$ as the imaginary part of the resolvent:

$$\rho_{0,0}(E) = \lim_{\epsilon\to 0}\frac{1}{\pi} Im\langle\Psi_0|(E-i\epsilon-H)^{-1}|\Psi_0\rangle = \lim_{\epsilon\to 0}\frac{\epsilon}{\pi}\sum_{j=0}^{N}\frac{|\langle\Psi_0|E_j\rangle|^2}{\epsilon^2+(E-E_j)^2} = \sum_{j=0}^{N}|\langle\Psi_0|E_j\rangle|^2\delta(E-E_j)\ , \tag{3}$$

where $|\Psi_0\rangle$ is a reference state. In other words, $\rho_{0,0}(E)$ is the diagonal element of the density matrix with respect to $|\Psi_0\rangle$. Then the return amplitude can be written as the Fourier transform of the spectral density

$$\langle\Psi_0|\Psi_t\rangle = \int \rho_{0,0}(E)e^{-iEt}dE = \sum_{j=0}^{N}|\langle\Psi_0|E_j\rangle|^2 e^{-iE_jt}\ . \tag{4}$$

We can also evaluate other elements of the density matrix, such as the off-diagonal element

$$\rho_{1,0}(E) = \frac{1}{\pi}\lim_{\epsilon\to 0} Im\langle\Psi_1|(E-i\epsilon-H)^{-1}|\Psi_0\rangle = \sum_j\langle\Psi_1|E_j\rangle\langle E_j|\Psi_0\rangle\delta(E-E_j)\ , \tag{5}$$

whose Fourier transforms gives the transition amplitude between the states $|\Psi_0\rangle$ and $|\Psi_1\rangle$:

$$\langle\Psi_1|\Psi_t\rangle = \int \rho_{1,0}(E)e^{-iEt}dE = \sum_{j=0}^{N}\langle\Psi_1|E_j\rangle\langle E_j|\Psi_0\rangle e^{-iE_jt}\ . \tag{6}$$

To characterize the entangled state $|\Psi_t\rangle = c_0|\Psi_0\rangle + c_1|\Psi_1\rangle$ that may appear during the evolution, we need to evaluate the amplitudes (4), (6) and count how often they realize certain values c_0, c_1 simultaneously during a long period of time. After normalization, this defines the conditional probability $P_t(c_0, c_1)$ for having $\langle\Psi_0|\Psi_t\rangle = c_0$ and $\langle\Psi_1|\Psi_t\rangle = c_1$ at a given time t.

3. Two coupled harmonic cavities

The Hamiltonian of two uncoupled harmonic cavities is $H_{hc} = \omega_0\sum_{j=1,2} a_j^\dagger a_j$, where the index $j = 1, 2$ refers to the two cavities with the photonic creation (annihilation) operators $a_j^\dagger$ (a_j). The eigenstates of this Hamiltonian are product Fock states $|N-k, k\rangle \equiv |N-k\rangle|k\rangle$ ($k = 0, .., N$). Coupling two cavities by an optical fiber is described by the tunneling Hamiltonian

$$H_{HC} = -J(a_1^\dagger a_2 + a_2^\dagger a_1) + \omega_0(a_1^\dagger a_1 + a_2^\dagger a_2) \tag{7}$$

with eigenstates $|N-k; k\rangle$. Here we assume that N is even and calculate the overlap of the eigenstates $|N-k; k\rangle$ with the initial Fock state $|N, 0\rangle$ [16]:

$$\langle N, 0|N-k; k\rangle = 2^{-N/2}\binom{N}{k}^{1/2}\ , \tag{8}$$

which is non-zero for all eigenstates. The density-matrix elements with respect to $|\Psi_0\rangle = |N, 0\rangle$ and $|\Psi_1\rangle = |0, N\rangle$ are

$$\rho_{0,0}(E) = 2^{-N}\sum_{k=0}^{N}\binom{N}{k}\delta(E+J(2k-N)),\quad \rho_{1,0}(E) = 2^{-N}\sum_{k=0}^{N}\binom{N}{k}(-1)^k\delta(E+J(2k-N))\ . \tag{9}$$

Thus, there is a binomial distribution for the spectral weight $|\langle N,0|N-k;k\rangle|^2$, with a maximal overlap for an equal number of photons in each cavity. For large N the binomial distribution becomes a Gaussian distribution, where the width of the envelope is related to the energy level spacings $\Delta E = 2J$. The Gaussian result resembles the Central Limit Theorem for independent photons. Such a behavior was also found previously for freely expanding bosons from an initial Fock state [33]. A Fourier transformation reveals a periodic behavior of the return and transition amplitudes as

$$\langle N,0|e^{-iHt}|N,0\rangle = \cos^N(Jt), \quad \langle 0,N|e^{-iHt}|N,0\rangle = (-i)^N \sin^N(Jt) \ . \tag{10}$$

Thus the evolution of the Fock state is periodic with period $2\pi/J$ but leads to a N00N state $c_0|N,0\rangle + c_N|0,N\rangle$ only with a probability that decays exponentially with N. For larger values of N the probability $P(c_0, c_N)$ indicates an anti-correlation: $P(c_0, c_N)$ vanishes as soon as both c_0 and c_N become nonzero. Therefore, the overlap of $|\Psi_t\rangle$ with a N00N state is strongly suppressed. This is a consequence of the fact that for an increasing N the photons disappear in the $(N+1)$–dimensional Hilbert space because there is no constraint due to interaction.

4. Cavities with two-level atoms

4.1. Single cavity with a two-level atom

An anharmonicity in a cavity can be created by adding an atom which interacts with the photons [14, 15, 34]. In the case of a single two-level atom we can describe the absorption and emission of photons by the atom approximately with the Jaynes-Cummings (JC) model [30, 31], whose Hamiltonian reads

$$H_{JC} = \omega_0 a^\dagger a + (\omega_0 + \Delta) c^\dagger c - g(a^\dagger c + c^\dagger a) \ . \tag{11}$$

Δ is the detuning between the atomic excitation energy and the photon energy, $c^\dagger$ (c) is the creation (annihilation) operator of the atomic excitation, and g is the coupling strength between the photons and the atom. The eigenvalues of this Hamiltonian are [30, 31]

$$E_{n,\pm} = \omega_0(n+1/2) \pm \sqrt{\Delta^2 + 4g^2(n+1)}, \quad n = 0, 1, 2, \ldots \ , \tag{12}$$

where the eigenstates $|n,\pm\rangle$ are linear combinations of the two Fock states $|n:1\rangle$ and $|n+1:0\rangle$ with n and $n+1$ photons and an atomic state with $j=0$ (atomic groundstate) and $j=1$ (excited atom). The dynamics is characterized by the energy difference $E_{n,+} - E_{n,-} = 2\sqrt{\Delta^2 + 4g^2(n+1)}$ which creates Rabi oscillations between the two Fock states $|n:0\rangle$ and $|n-1:1\rangle$ with the Rabi frequency $\Omega_R = \sqrt{\Delta^2 + 4g^2(n+1)}$.

The JC model describes an extreme case of Hilbert-space localization, where the system is constraint to a two-dimensional subspace. It is enforced by the fact that the eigenstates of the JC model are linear combinations of only two Fock states. In other words, if this system is initially prepared in a Fock state, it will never escape from the related two-dimensional subspace. The drawback of the extreme localization in Hilbert space is that we are not able to create dynamically entangled Fock states whose difference in photon numbers is larger than 1. In a more general case, however, the eigenstates of the Hamiltonian may be a superposition of many Fock states. Then the overlap of the eigenfunctions with the initial Fock state plays a crucial role.

4.2. Two coupled cavities with two-level atoms

Now we prepare the anharmonic cavity of Sect. 4.1 in the eigenstate of the JC Hamiltonian $|N,+\rangle$ and connect it with another anharmonic cavity which is in the state $|0,+\rangle$. After the connection the photons start to tunnel between the two cavities. This system is now described by the Hamiltonian

$$H_{2JC} = -J(a_1^\dagger a_2 + a_2^\dagger a_1) + \sum_{j=1,2} [\omega_0 a_j^\dagger a_j + \omega_0 c_j^\dagger c_j - g(a_j^\dagger c_j + c_j^\dagger a_j)] \ , \tag{13}$$

where the first term describes the tunneling of photons between the cavities with rate J and the second term represents the cavity levels as well as the absorption and emission of photons by the two-level atom inside each cavity. Here we have assumed no detuning: $\Delta = 0$. For the initial state we choose a product of JC eigenstates $|N, \sigma; 0, \sigma'\rangle \equiv |N, \sigma\rangle|0, \sigma'\rangle$.

The probability for a dynamically changing of the atomic levels during the photonic tunneling process is characterized by the ratio [35]

$$\frac{\langle k-1, -|a^\dagger|k, +\rangle}{\langle k-1, -|a^\dagger|k, -\rangle} = \frac{\sqrt{k+2} - \sqrt{k+1}}{\sqrt{k+2} + \sqrt{k+1}} \approx 0 \ . \tag{14}$$

Thus, for sufficiently many photons k the change of the atomic level is strongly suppressed. With these approximations we decouple the dynamics of $\pm$ states into states where the atomic level is fixed. Then we are left with states $|k, N-k\rangle$, $|N-k, k\rangle$ which only depend on the number of photons in each cavity. The dynamics is described by a Hubbard-like model, where the n_j^2 interaction is replaced by a $\sqrt{n_j}$ photon-photon interaction [35]:

$$H_{eff} = -J(a_1^\dagger a_2 + a_2^\dagger a_1) + \omega_0(a_1^\dagger a_1 + a_2^\dagger a_2) \pm 2g(\sqrt{a_1^\dagger a_1} + \sqrt{a_2^\dagger a_2}) \ . \tag{15}$$

The sign of the coupling depends on whether both atoms are in the ground state (-) or in the excited state (+). This Hamiltonian has a two-fold degeneracy for $J = 0$ due to the equivalence of the two cavities. The interaction is weaker than the n_j^2 interaction of the corresponding Hubbard model in Eq. (16). This indicates that the properties of the coupled JC models may resemble the behavior of the Hubbard model in a double well [29], with less pronounced interaction features though.

5. Bosonic atoms in a double-well potential

Another interesting realization of coupled quantum systems are bosonic atoms which are trapped in a double-well potential. This has also been considered as a Josephson junction for ultracold gases. The main advantage of such a system is based on the fact that the trapping potential, including the barrier between the wells, is easy to control by external Laser fields. The dynamics has been studied experimentally in great detail [36–38] and has revealed several regimes, either dominated by tunneling or by the many-body interaction. From the theoretical point of view the physics is described by the Hubbard-Hamiltonian

$$H_{DW} = -J(a_1^\dagger a_2 + a_2^\dagger a_1) + g[(a_1^\dagger a_1)^2 + (a_2^\dagger a_2)^2] \ , \tag{16}$$

in contrast to the effective photon-photon interaction in Eq. (15). This indicates a strong many-body interaction for atoms. Some aspects can be described by a mean-field (Gross-Pitaevskii) approximation, other aspects are of genuine quantum nature. The formation of entangled states, such as a N00N state, belongs to the latter. Of particular interest for entanglement is the intermediate regime, where tunneling and many-body interaction are strongly competing, which restricts the available Hilbert space for the dynamical formation of new states but also mixes different regions of the space [29, 38].

6. Recursive projection method

The evaluation of the resolvents as rational functions presented in Eq. (2) can be performed analytically within the recursive projection method. This method is based on the idea that we divide the $N+1$–dimensional Hilbert space (N even) into two-dimensional subspaces spanned by the states $|k, N-k\rangle$, $|N-k, k\rangle$ ($k = 0, ..., N/2$), since we are interested in matrix elements with respect to the two states $|0, N\rangle$ and $|N, 0\rangle$. Then the recursive projection method is constructed as a directed walk through all subspaces, beginning with $k = N/2$ and terminating with $k = 0$. For this purpose we define

$$a_{N/2} = \langle N,0|(z-H)^{-1}|N,0\rangle = \langle 0,N|(z-H)^{-1}|0,N\rangle \tag{17}$$

and

$$b_{N/2} = \langle 0,N|(z-H)^{-1}|N,0\rangle = \langle N,0|(z-H)^{-1}|0,N\rangle \ . \tag{18}$$

Then $a_{N/2}$ and $b_{N/2}$ are obtained from the iteration of the recurrence relation (for details cf. Ref. [29])

$$a_{k+1} = \frac{z - \tilde{f}_{k+1} - J^2 a_k (N/2+k+1)(N/2-k)}{\left[z - \tilde{f}_{k+1} - J^2 a_k (N/2+k+1)(N/2-k)\right]^2 - J^4 b_k^2 (N/2+k+1)^2 (N/2-k)^2} \tag{19}$$

$$b_{k+1} = \frac{J^2 b_k (N/2+k+1)(N/2-k)}{\left[z - \tilde{f}_{k+1} - J^2 a_k (N/2+k+1)(N/2-k)\right]^2 - J^4 b_k^2 (N/2+k+1)^2 (N/2-k)^2} \tag{20}$$

with initial values $b_0 = 0$, $a_0 = 1/(z - \tilde{f}_0)$. The term $\tilde{f}_{k+1}$ depends on the interaction of the model:

$$\tilde{f}_{k+1} = \begin{cases} g[(N/2+k+1)^2 + (N/2-k-1)^2] & \text{Hubbard model in Eq. (16)} \\ \pm 2g[\sqrt{N/2+k+1} + \sqrt{N/2-k-1}] & \text{coupled JC model of Eq. (15)} \end{cases} \ . \tag{21}$$

The recurrence relation terminates after $N/2$ steps, giving us $a_{N/2}$ and $b_{N/2}$ in the form of ratios of polynomials $P_N(z)/Q_{N+1}(z)$, as described in Eq. (2) .

7. Results

The properties of two coupled anharmonic cavities in Sect. 4.2 and the bosonic atoms in the double-well potential in Sect. 5 are characterized by tunneling between the subsystems and by a many-body interaction. The iteration of Eqs. (19), (20) gives us the following four matrix elements of the resolvent

$$\langle N,0|(z-H)^{-1}|N,0\rangle, \quad \langle 0,N|(z-H)^{-1}|0,N\rangle, \quad \langle 0,N|(z-H)^{-1}|N,0\rangle = \langle N,0|(z-H)^{-1}|0,N\rangle \ , \tag{22}$$

which allows us to evaluate the spectral density $\rho_{0,0}(E)$ in Eq. (3) and the off-diagonal element $\rho_{1,0}(E)$ of the density matrix in (5). After a Fourier transformation we obtain the return amplitude $\langle\Psi_0|\Psi_t\rangle$ and the transition amplitude $\langle\Psi_1|\Psi_t\rangle$ from Eqs. (4), (6) as functions of time.

Here it should be noticed that there exists an invariance of the recurrence relation under the following simultaneous sign changes in Eqs. (19) and (20)

$$z \to -z, \quad g \to -g, \quad a_j \to -a_j, \quad b_j \to -b_j \ . \tag{23}$$

This implies that a change from $g \to -g$ in the initial states results in a mirror image with respect to energy of $\rho_{0,0}(E,g)$ and $\rho_{N,0}(E,g)$:

$$\rho_{0,0}(E,g) = \rho_{0,0}(-E,-g), \quad \rho_{N,0}(E,g) = \rho_{N,0}(-E,-g) \ . \tag{24}$$

Journal of Physics: Conference Series **497** (2014) 012032 doi:10.1088/1742-6596/497/1/012032

This relation enables us to evaluate both atomic states of the coupled JC models with the same recurrence relation. Moreover, the density-matrix elements are invariant with respect to the harmonic frequency ω_0 of the cavities, except for a global energy shift. This reflects an important universality of the density matrix that allows us to separate the harmonic from the anharmonic properties of the cavities.

In the absence of tunneling ($J = 0$) the spectrum two harmonic cavities is completely degenerate for harmonic cavities

$$E_{N-k,k} = \omega_0(N-k) + \omega_0 k = \omega_0 N$$

but is only two-fold degenerate for anharmonic cavities

$$E_{N-k,k} = \omega_0 N \pm g(\sqrt{N-k} + \sqrt{k})$$

or for bosonic atoms in a double-well potential:

$$E_{N-k,k} = g[(N-k)^2 + k^2] \ .$$

After connecting the subsystems by a tunneling term the degeneracies are completely lifted. For harmonic cavities an equidistant spectrum appears with level spacing $\Delta E = 2J$ according to Eq. (9). Tunneling also lifts the two-fold degeneracies of the anharmonic cavities and the bosons in the double well, as depicted in Fig. 2. However, the levels are more irregularly distributed and their spacing is smaller than $2J$ for pairs of levels. Here it should be noticed that the spectrum of the bosonic atoms indicates a spectral fragmentation with an almost two-fold degenerate high energy part and a non-degenerate low energy part, which cannot be seen in the photonic spectrum.

The difference between non-interacting and interacting systems is even more pronounced for the dynamics of the return and transition amplitudes. While there is only a periodic behavior with $T = \pi/J$ in Eq. (10), anharmonic cavities have a more dynamic behavior (cf. Figs. 3, 4). In particular, on the time scale considered in Figs. 3, 4, there is no periodic behavior but oscillations on much shorter scales than $\pi/J \approx 4$. This is a consequence of the fact that the individual energy levels $E_k = J(2k - N)$ in Eq. (9) are invisible in the dynamics of the harmonic cavities due to

$$\sum_{k=0}^{N} \binom{N}{k} e^{iJ(2k-N)t} = (e^{iJt} + e^{-iJt})^N, \quad \sum_{k=0}^{N} \binom{N}{k} (-1)^k e^{iJ(2k-N)t} = (-e^{iJt} + e^{-iJt})^N \ . \qquad (25)$$

Such kind of interference effect is accidental for the non-interacting system and does not occur in the presence of interaction. Therefore, we can distinguish the individual levels in the dynamics only for the latter.

The return amplitude decays for both types of systems rapidly (cf. Fig. 3) but it recovers much earlier for interacting systems, not to the full value though. Remarkable is also the behavior of the transition amplitude. The time T_t it takes to reach the state $|N,0\rangle$ from $|0,N\rangle$ for the first time is about the same for all three types of systems, indicating that $T_t \approx \pi/J$ must be solely determined by the tunneling rate J. This provides a method to measure the tunneling rate J in the dynamics of the system, regardless of the interaction.

Our main goal, the dynamical creation of an entangled state from a pure state, is also strongly affected by T_t, since entanglement in terms of a N00N state is not possible for times shorter than T_t. For times larger than T_t only the interacting systems can reach the state $|N,0\rangle$ while maintaining a non-zero overlap with the initial state (cf. Figs. 3, 4). On the other hand, only for a small number of photons (e.g., $N = 2$) the harmonic cavities are capable to create

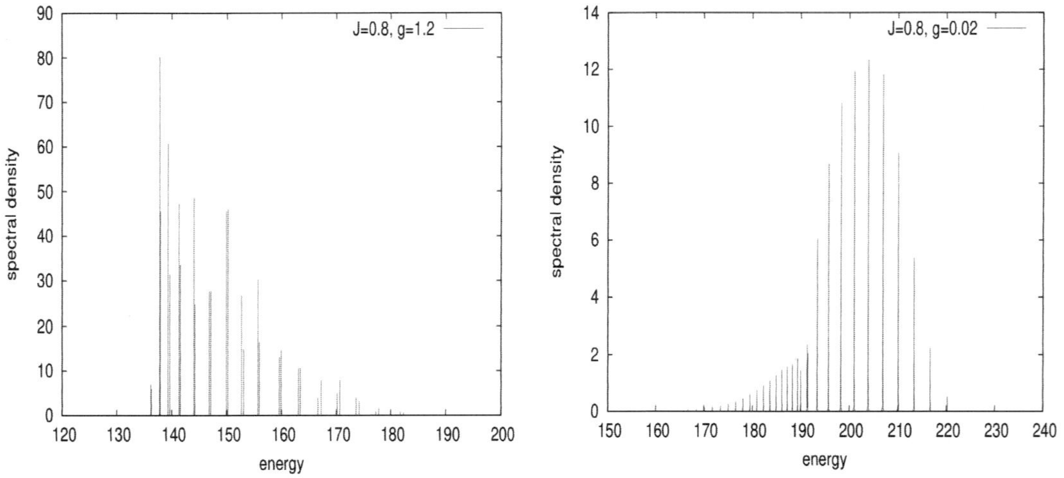

Figure 2. Spectral density $\rho_{0,0}$ for 100 bosons with tunneling rate $J = 0.8$. Left panel: Two coupled cavities where the photons are coupled with two-level atoms with coupling strength $g \approx 1.2$. Right panel: Bosonic atoms in a double-well potential with Hubbard coupling constant $g = 0.02$. All energies of the cavities are measured in units of the cavity frequency $\hbar\omega_0$, the energies of the bosonic atoms in units of the double-well potential.

a N00N state dynamically, as discussed in Sect. 3. For a discrete sequence of time steps we have counted the occurrence of certain values of the return amplitude ($\langle N, 0|e^{-iHt}|N, 0\rangle$ and the transition amplitude $\langle 0, N|e^{-iHt}|N, 0\rangle$) for interacting bosons. This yields the conditional probability $P_t(c_0, c_N)$, which was defined at the end of Sect. 2. Examples are plotted in Fig. 6. The plots demonstrate that the dynamical creation of a N00N state is feasible for anharmonic cavities even for $N = 100$. However, in comparison with bosonic atoms in a double well with Hubbard interaction this probability is quite small.

Phenomenological damping is described by an imaginary parameter $i\epsilon$ in the complex energy $z = E + i\epsilon$ of Eq. (1), which produces exponential damping on the time scale $T = \hbar/\epsilon$. In Fig. 5 the effect of damping on the transition dynamics is plotted for coupled cavities. As expected, the damping is crucial only on time scales larger that T. This implies that only strong damping may play a role in the suppression of dynamically created entanglement.

8. Discussion and Conclusions

We have considered a pair of optical cavities, where the photons in each cavity couple to a two-level atom and compared this system with bosonic atoms in a double-well potential. The cavities are described as JC models, while the atomic system is described by a Bose-Hubbard Hamiltonian. For a large number of photons the coupled JC models can be approximated by an effective Hamiltonian that resembles the Bose-Hubbard Hamiltonian, where the interaction term depends on the square root of the density.

After the initial preparation with all bosons in one subsystem the quantum system is allowed to evolve due to tunneling between the two subsystems. We have found that the boson-boson interaction plays a crucial role in terms of the spectral properties as well as for the dynamic properties: While the spectral density and the dynamics are very regular for harmonic cavities, the spectrum of the interacting systems are less regular with smaller level spacings, and the corresponding dynamics is also quite irregular. For harmonic cavities (i.e. in the absence

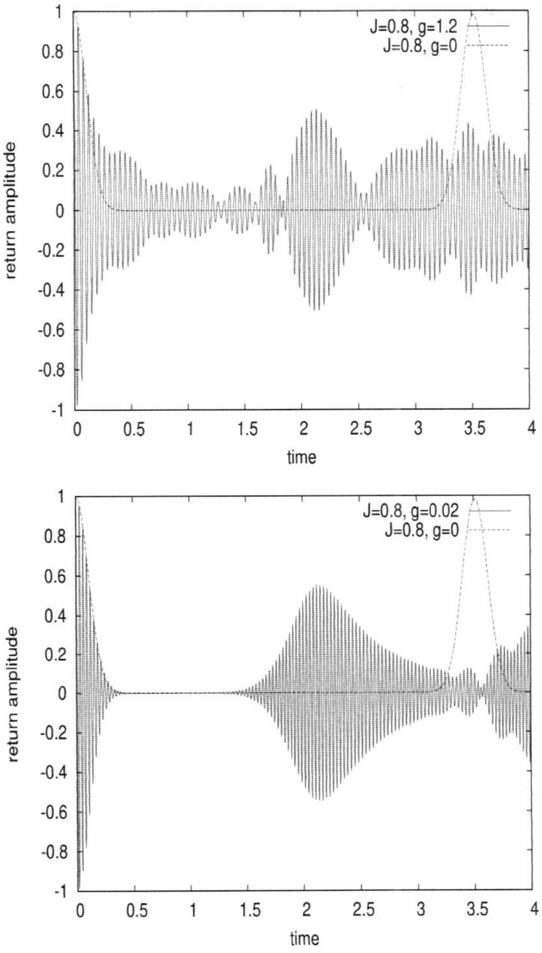

Figure 3. Return amplitude $\langle\psi_0|\psi_t\rangle$ as a function of time for two coupled harmonic cavities (dashed curve), for two coupled cavities with two-level atoms (full curve in the upper panel) and for bosonic atoms in a double-well potential (full curve in the lower panel). The parameters are the same as in the previous Figure, the time is measured in units of $1/\omega_0$.

of the two-level atoms) the distribution of the levels with spectral weights $p_j = |\langle\Psi_0|E_j\rangle|^2$ is binomial for equidistant energy levels with spacing $\Delta E = 2J$. The resulting evolution is periodic and is reminiscent of Rabi oscillations with a single frequency J. Such a behavior was observed experimentally for weakly interacting bosonic atoms [4, 5] and should also be accessible in experiments with photons in harmonic cavities.

The amplitudes for visiting the initial Fock state $|N,0\rangle$ or the complimentary Fock state $|0,N\rangle$ vary as $\cos^N(Jt)$ or $(-i)^N\sin^N(Jt)$, respectively. This implies for a large number N of bosons that (i) these states are visited only for a very short period of time and (ii) the two Fock states are visited at well separated times. Thus the dynamical creation of a N00N state from a Fock state $|N,0\rangle$ is very unlikely for harmonic cavities, unless the number of photons is small. The reason is that the photons can travel without seeing each other through the entire Hilbert space. A simultaneous overlap of $|\Psi_t\rangle$ with both Fock states $|N,0\rangle$ and $|0,N\rangle$ is very

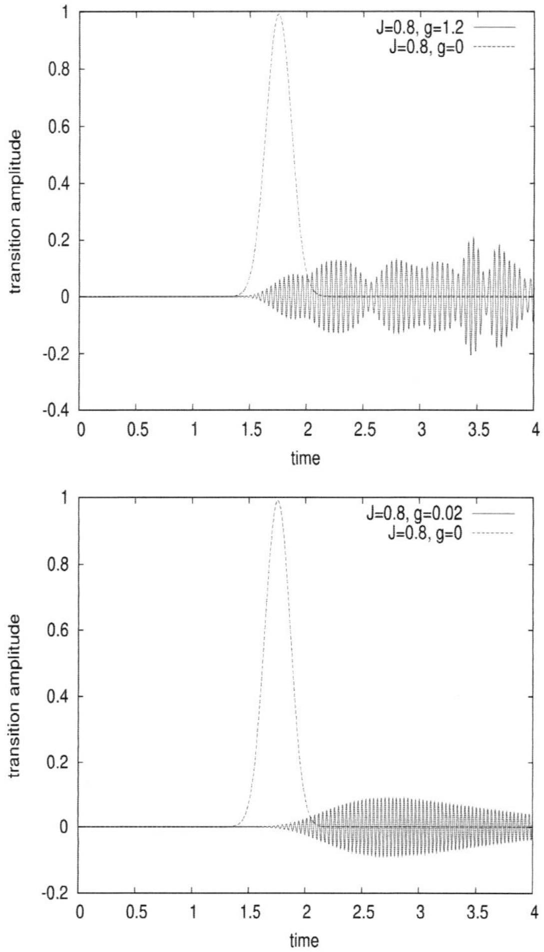

Figure 4. Transition amplitude $\langle\psi_1|\psi_t\rangle$ as a function of time for two coupled harmonic cavities (dashed curve), for two coupled cavities with two-level atoms (full curve in the upper panel) and for bosonic atoms in a double-well potential (full curve in the lower panel). The parameters are the same as in the previous Figure.

unlikely then. This is a situation in which it is very difficult to control and follow the quantum evolution. On the other hand, applications of finite quantum systems, such as in quantum information processing [39, 40], require a controllable evolution, in which only certain parts of the available Hilbert space can be visited with reasonable probability. In terms of our two-cavity system this means that the spectral weight $p_j = |\langle\Psi_0|E_j\rangle|^2$ with respect to the initial state $|\Psi_0\rangle$ is small for most eigenstates $|E_j\rangle$ and has only a few pronounced maxima that can be used for information storage. We have found that such a structured spectral density appears for anharmonic cavities, created by coupling two-level atoms to the cavity photons. Then the photons experience a mutual influence which restricts their individual random walks in Hilbert space significantly and, what is even more important here, they can have a simultaneous overlap with both states $|N, 0\rangle$ and $|0, N\rangle$. This effect enables the system to create dynamically a N00N state. The situation is even better for bosonic atoms in a double-well potential, where the

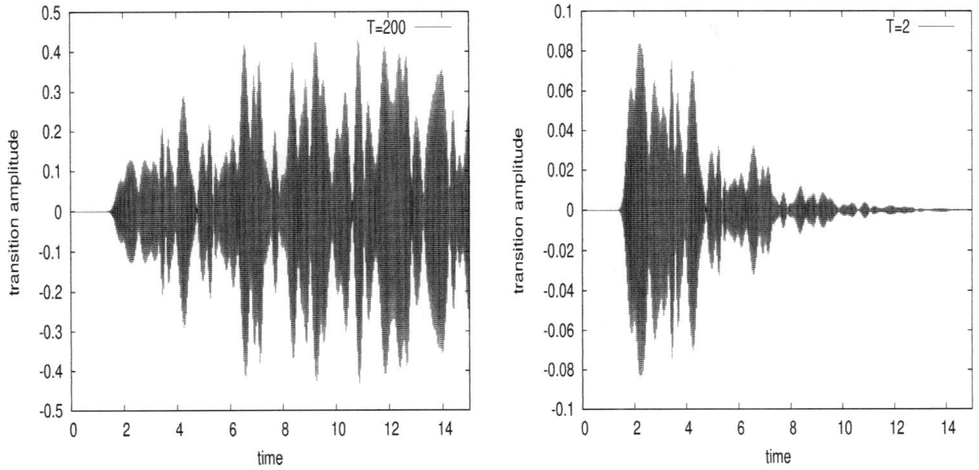

Figure 5. Effect of an increased damping on the transition amplitude $\langle\psi_1|\psi_t\rangle$ for coupled cavities with decay times $T = 200$ and $T = 2$. The other parameters are the same as in the previous Figures.

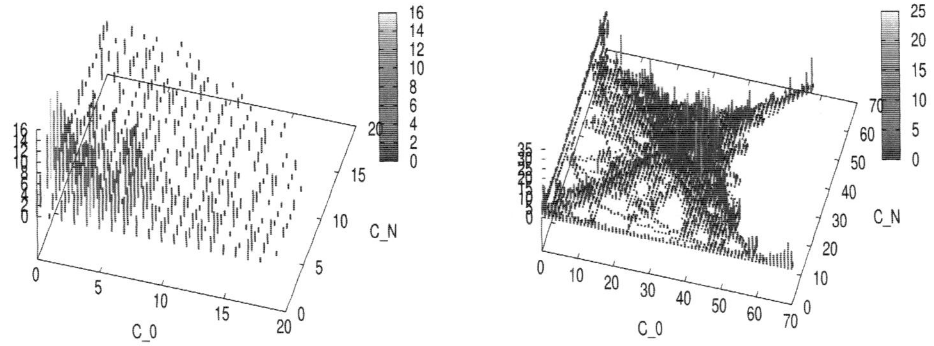

Figure 6. Probability $P_t(c_0, c_N)$ for creating a N00N state for large times in coupled anharmonic cavities with $N = 100$ photons, $J = 0.8$, $g = 1.2$ (left panel) and for $N = 20$ bosonic atoms in a double-well potential with $U/J \approx 0.026$ (right panel, from Ref [29]). The axes are scaled by a factor 100.

stronger interaction effects supports the formation of a N00N state. This allows us to conclude that the complex quantum dynamics of two coupled anharmonic optical cavities and bosonic atoms in a double-well potential offer an approach for quantum information processing as it has also been proposed for ultracold atoms [39] and cold trapped ions [40].

References

[1] Bloch I 2008 Science **319** 202
[2] Yukalov V I 2009 Laser Phys. **19** 1
[3] Trotzky S *et al.* 2008 Science **319** 295
[4] Shin Y *et al.* 2004 Phys. Rev. Lett. **92** 150401
[5] Albiez M. *et al.* 2005 Phys. Rev. Lett. **95** 010402
[6] Kollath C, Läuchli A M and Altman E 2007 Phys. Rev. Lett. **98** 180601
[7] Rigol M, Dunjko V, Yurovsky V and Olshanii M 2007 Phys. Rev. Lett. **98** 050405
[8] Yukalov V I 2011 Laser Phys. Lett. **8** 485
[9] Yukalov V I, Rakhimov A and Mardonov S 2011 Laser Phys. **21** 264
[10] Roux G 2009 Phys. Rev. A **79** 021608
[11] Rigol M 2010 Phys. Rev. A **82** 037601
[12] Brune M *et al.* 2008 Phys. Rev. Lett. **101** 240402
[13] Wang H *et al.* 2008 Phys. Rev. Lett. **101** 240401
[14] Carmichael H 2008 Nature Phys. **4** 346
[15] Schuster I *et al.* 2008 Nature Phys. **4** 382
[16] Ziegler K 2012 Laser Physics **22** 331
[17] Imamoglu A, Schmidt H, Woods G and Deutsch M 1997 Phys. Rev. Lett. **79** 1467
[18] Grangier P, Walls D F and Gheri K M 1998 Phys. Rev. Lett. **81** 2833
[19] Hartmann M J, Brandão F G S L and Plenio M B 2006 Nature Phys. **2** 849
[20] Hartmann M J, Brandão F G S L and Plenio M B 2008 Laser & Photon. Rev. **2** 527556
[21] Ji A-C, Sun Q, Xie X C, and Liu W M 2009 Phys. Rev. Lett. **102** 023602
[22] Lee H, Kok P and Dowling J P 2002 Journal of Modern Optics **49** 2325
[23] Walther P, Pan J-W, Aspelmeyer M, Ursin R, Gasparoni S and Zeilinger A 2004 Nature **429** 158
[24] Mitchell M W, Lundeen J S and Steinberg A M 2004 Nature **429** 161
[25] Afek I, Ambar O and Silberberg Y 2010 Science **328** 879
[26] Boto A and Kok P 2000 Phys. Rev. Lett. **99** 2733
[27] Kok P, Lee H and Dowling J P 2002 Phys. Rev. **65** 052104
[28] Cable H and Dowling J P 2007 Phys. Rev. Lett. **99** 163604
[29] Ziegler K 2011 J. Phys. B: At. Mol. Opt. Phys. **44** 145302
[30] Jaynes E T and Cummings F W 1963 Proc. Inst. Elect. Eng. **51** 89
[31] Cummings F W 1965 Phys. Rev. **140** A1051
[32] Ziegler K 2010 Phys. Rev. A **81** 034701
[33] Cramer M, Dawson C M, Eisert J, and Osborne T J 2008 Phys. Rev. Lett. **100** 030602
[34] Kleppner D 1981 Phys. Rev. Lett. **47** 233
[35] Ziegler K 2013 arXiv:1304.2655.
[36] Albiez M *et al.* 2005 Phys. Rev. Lett. **95** 010402
[37] Gati R *et al.* 2006 New J. Phys. **8** 189
[38] Gati R and Oberthaler M K 2007 J. Phys. B: At. Mol. Opt. Phys. **40** R61
[39] Garca-Ripoll J J, Zoller P, and Cirac J I 2005 J. Phys. B: At. Mol. Opt. Phys. **38** S567
[40] Wineland D J and Leibfried D 2011 Laser Phys. Lett. **8** 175

Entangling two distinguishable quantum bright solitons via collisions

Thomas P. Billam[1], **Caroline L. Blackley**[2], **Bettina Gertjerenken**[3], **Simon L. Cornish**[4], **Christoph Weiss**[4]

[1] Jack Dodd Center for Quantum Technology, Department of Physics, University of Otago, Dunedin 9016, New Zealand
[2] Joint Quantum Centre (JQC) Durham–Newcastle, Department of Chemistry, Durham University, Durham DH1 3LE, United Kingdom
[3] Institut für Physik, Carl von Ossietzky Universität, D-26111 Oldenburg, Germany
[4] Joint Quantum Centre (JQC) Durham–Newcastle, Department of Physics, Durham University, Durham DH1 3LE, United Kingdom

E-mail: `Christoph.Weiss@durham.ac.uk`

Abstract. The generation of mesoscopic Bell states via collisions of distinguishable bright solitons has been suggested in *Phys. Rev. Lett.* **111**, 100406 (2013). Here, we extend our former proposal to two hyperfine states of ^{85}Rb instead of two different atomic species, thus simplifying possible experimental realisations. A calculation of the s-wave scattering lengths for the hyperfine states $(f, m_f) = (2, +2)$ and $(3, +2)$ identifies parameter regimes suitable for the creation of Bell states with an advantageously broad Feshbach resonance. We show the generation of Bell states using the truncated Wigner method for the soliton's centre of mass and demonstrate the validity of this approach by a comparison to a mathematically rigorous effective potential treatment of the quantum many-particle problem.

1. Introduction

Bright solitons are a promising candidate to generate quantum entanglement for a mesoscopic number of atoms. Such bright solitons are realised experimentally in Bose-Einstein condensates [1–6]. These experiments have thus far been modelled by a mean-field description. However, going to lower particle numbers naturally requires a fuller quantum mechanical treatment. The *quantum bright solitons* described by such a treatment provide an excellent model system with which to investigate the "middle-ground" between quantum and classical physics [7, 8].

Scattering bright solitons off a single barrier was recently investigated in [9–18] and references therein; with two barriers a soliton diode was suggested in [19]. In the regime of very low kinetic energies [20–22], scattering a quantum bright soliton [23–28] off a barrier can even lead to Schrödinger cat states [20, 21] that can be detected using their interference properties [18, 20].

Schrödinger-cat states are highly non-classical superpositions[1] which are relevant for

[1] In a measurement, all particles would be on one side of the barrier; before the measurement, they were in a quantum superposition of all being on the right and all being on the left.

Journal of Physics: Conference Series **497** (2014) 012033 doi:10.1088/1742-6596/497/1/012033

quantum-enhanced interferometry [29]. The focus of our paper are mesoscopic Bell states

$$|\psi_{\rm Bell}\rangle \equiv \frac{1}{\sqrt{2}}\Big(|{\rm A},{\rm B}\rangle + e^{i\alpha}|{\rm B},{\rm A}\rangle\Big), \tag{1}$$

where $|A,B\rangle$ ($|B,A\rangle$) signifies that the BEC A is on the left (right) and the BEC B is on the right (left). While it might sound tempting to realise such mesoscopic quantum superpositions as, say, the ground states of Bose-Einstein condensate in a double well with carefully chosen signs and strengths of interactions, such an approach will not be successful in the presence of tiny asymmetries (cf. [30]) and decoherence. Suggestions of how such a state can be realised dynamically for Bose-Einstein condensates can be found in Refs. [30–35] and references therein.

Rather than using a potential to generate mesoscopic entanglement [20,21], we have suggested to scatter two *distinguishable* quantum bright solitons off each other to generate mesoscopic Bell states [36]. Two colliding distinguishable bright solitons behave very differently from two colliding but initially indistinguishable solitons [37, 38]: for indistinguishable solitons, either higher order nonlinear terms [37] (cf. [39]) or additional harmonic confinement [38] are needed to generate entanglement. Quantum bright solitons have also been discussed in the context of symmetry breaking states [26]; for more general treatment of symmetry breaking in finite quantum systems see [40] and references therein.

In this paper we discuss the generation of a mesoscopic Bell state via scattering two distinguishable bright solitons. While our original proposal [36] scattered two solitons of different species (^{85}Rb and ^{133}Cs), we now suggest to use two hyperfine states of ^{85}Rb. This allows the generation of mesoscopic Bell states closer to the case of many photons which is an area of current theoretical and experimental research [41,42]. In addition to their inherent fundamental interest, such states are of potential application as a resource in quantum information [42].

Our paper is organised as follows: We first introduce the many-particle quantum model used to describe the two colliding solitons in sec. 2 before justifying our use of a classical field approach to describe mesoscopic quantum superpositions in sec. 3. In sec. 4 we describe a new Feshbach resonance, offering excellent control over distinguishable soliton collisions, which we use for our numerics in sec. 5. In sec. 6 we present signatures that distinguish quantum superpositions from statistical mixtures. The paper ends with the conclusions in sec. 7.

2. Model

In order to model two distinguishable solitons on the many-particle quantum level, we use the same approach as [36] and set $m_{\rm A} = m_{\rm B}$ at the end, where $m_{\rm A}$ ($m_{\rm B}$) is the atomic mass of species A (B) (as we have two hyperfine states of the same species). For our quasi-1D system, we consider an experimentally motivated harmonic confinement $\omega = 2\pi f$. Mixtures of ultracold gases can be confined in a common optical trap with the same trap frequencies [43], yielding

$$\omega = \frac{2\pi}{T}\,; \quad \lambda_{\rm A} = \sqrt{\frac{\hbar}{m_{\rm A}\omega}}\,; \quad \lambda_{\rm B} = \sqrt{\frac{\hbar}{m_{\rm B}\omega}}, \tag{2}$$

where $\lambda_{\rm A}$ and $\lambda_{\rm B}$ are the harmonic oscillator lengths of the two species; the interactions $g_S = hf_\perp a_S$ are set by the scattering lengths a_S (S = A,B or AB) and the perpendicular trapping-frequency, $f_\perp$ [44].

We use the Lieb-Liniger model [45,46] for two species with additional harmonic confinement

$$\begin{aligned}\hat{H} = &-\sum_{j=1}^{N_{\rm A}}\frac{\hbar^2}{2m_{\rm A}}\partial_{x_j}^2 + \sum_{j=1}^{N_{\rm A}-1}\sum_{n=j+1}^{N_{\rm A}} g_{\rm A}\delta(x_j - x_n) - \sum_{j=1}^{N_{\rm B}}\frac{\hbar^2}{2m_{\rm B}}\partial_{y_j}^2 + \sum_{j=1}^{N_{\rm B}-1}\sum_{n=j+1}^{N_{\rm B}} g_{\rm B}\delta(y_j - y_n)\\ &+\sum_{j=1}^{N_{\rm A}}\sum_{n=1}^{N_{\rm B}} g_{\rm AB}\delta(x_j - y_n) + \sum_{j=1}^{N_{\rm A}}\frac{1}{2}m_{\rm A}\omega^2 x_j^2 + \sum_{j=1}^{N_{\rm B}}\frac{1}{2}m_{\rm B}\omega^2 y_j^2\,,\end{aligned} \tag{3}$$

where x_j (y_j) and $g_{\rm A} < 0$ ($g_{\rm B} < 0$) are the atomic coordinates and intra-species interactions of species A (B), and $g_{\rm AB} \geq 0$ is the inter-species interaction.

We suggest to prepare the two solitons independently; for weak harmonic confinement a single soliton has the ground state energy (cf. [47])

$$E_{\rm S}(N_{\rm S}) = -\frac{1}{24}\frac{m_{\rm S} g_{\rm S}^2}{\hbar^2} N_{\rm S}(N_{\rm S}^2 - 1)\,; \quad {\rm S} \in \{{\rm A}, {\rm B}\}\,. \tag{4}$$

Thus, our system has the total ground-state energy

$$E_0 = E_{\rm A}(N_{\rm A}) + E_{\rm B}(N_{\rm B})\,. \tag{5}$$

The total kinetic energy related to the centre-of-mass momenta $\hbar K_{\rm S}$ (${\rm S} \in \{{\rm A}, {\rm B}\}$) of the two solitons reads

$$E_{\rm kin} = \frac{\hbar^2 K_{\rm A}^2}{2N_{\rm A} m_{\rm A}} + \frac{\hbar^2 K_{\rm B}^2}{2N_{\rm B} m_{\rm B}}. \tag{6}$$

We extend the low-energy regime investigated for single-species solitons in Refs. [20, 22, 48] to two species:

$$E_{\rm kin} < \min\{\Delta_{\rm A}, \Delta_{\rm B}\}, \quad \Delta_{\rm S} = |E_{\rm S}(N_{\rm S} - 1) - E_{\rm S}(N_{\rm S})|\,. \tag{7}$$

In this energy regime, each of the quantum matter-wave bright solitons is energetically forbidden to break up into two or more parts. Highly entangled states are characterised by a roughly 50:50 chance of finding the soliton A (B) on the left/right combined with a left/right correlation close to one indicating that whenever soliton A is on the one side, soliton B is on the other:

$$\begin{aligned} \gamma(\delta) \equiv \;& \int_\delta^\infty dx_1 \ldots \int_\delta^\infty dx_{N_{\rm A}} \int_{-\infty}^{-\delta} dy_1 \ldots \int_{-\infty}^{-\delta} dy_{N_{\rm B}} |\Psi|^2 \\ & + \int_{-\infty}^{-\delta} dx_1 \ldots \int_{-\infty}^{-\delta} dx_{N_{\rm A}} \int_\delta^\infty dy_1 \ldots \int_\delta^\infty dy_{N_{\rm B}} |\Psi|^2\,, \end{aligned} \tag{8}$$

where $\Psi = \Psi(x_1, \ldots, x_{N_{\rm A}}, y_1, \ldots, y_{N_{\rm B}})$ is the many-particle wave function (normalised to one) and $\delta \geq 0$. The correlation $\gamma(\delta)$ will serve as an indication of entanglement: Bell states (1) are characterised by $\gamma \simeq 1$ combined with a 50:50 chance to find soliton A either on one side or on the other.

Behaviour for larger particle numbers can be described by the Gross-Pitaevskii equation (GPE) (cf. [49–52])

$$\begin{aligned} i\hbar\partial_t\varphi_{\rm A}(x,t) = \;& \left[-\tfrac{\hbar^2}{2m_{\rm A}}\partial_x^2 + \tfrac{g_{\rm A}}{2}|\varphi_{\rm A}(x,t)|^2\right]\varphi_{\rm A}(x,t) + \left[\tfrac{1}{2}m_{\rm A}\omega^2x^2 + \tfrac{g_{\rm AB}}{2}|\varphi_{\rm B}(x,t)|^2\right]\varphi_{\rm A}(x,t) \\ i\hbar\partial_t\varphi_{\rm B}(x,t) = \;& \left[-\tfrac{\hbar^2}{2m_{\rm B}}\partial_x^2 + \tfrac{g_{\rm B}}{2}|\varphi_{\rm B}(x,t)|^2\right]\varphi_{\rm B}(x,t) + \left[\tfrac{1}{2}m_{\rm B}\omega^2x^2 + \tfrac{g_{\rm AB}}{2}|\varphi_{\rm A}(x,t)|^2\right]\varphi_{\rm B}(x,t)\,, \end{aligned}$$

where the single-particle density $|\varphi_{\rm S}(x,t)|^2$ is normalised to $N_{\rm S}$ (${\rm S} \in \{{\rm A}, {\rm B}\}$).

3. Justifying Truncated Wigner for the centre of mass

When hitting a barrier, the generic behaviour of a mean-field bright soliton is to break into two parts; the fraction of the atoms transmitted decreases for increasing potential strength (cf. [12, 14]). An analogous behaviour also occurs when two distinguishable mean-field bright solitons collide with each other, as shown in the Supplemental Material of [36]. Only at very low kinetic energies [20, 22, 36] do mesoscopic quantum superpositions occur as a result of such collisions.

To describe low kinetic energy collisions of two distinguishable bright solitons, taking into account the formation of mesoscopic quantum superpositions, we combine mean-field calculations via the GPE with Truncated-Wigner Approximation (TWA) for the centre of mass degree of freedom in order to model true quantum behaviour [36]. The truncated-Wigner approximation (TWA) describes quantum systems by averaging over realisations of an appropriate classical field equation (in this case, the GPE) with initial noise appropriate to either finite [53] or zero temperatures [12]. While the GPE assumes both position and momentum are well-defined, this is not true for a single quantum particle of finite mass for which, in general, both position and momentum involve quantum noise satisfying the uncertainty relation. Our TWA calculations for the soliton centre-of-mass wave function use Gaussian probability distributions for both (satisfying minimal uncertainty) [36].

This centre-of-mass TW technique can be justified by comparison to the rigorously proved [54] effective potential approach [20, 48]: In fig. 1 we compare the single-particle effective potential treatment [fig. 1(a)] for the case of a low-mass bright soliton colliding with a heavy bright soliton with a centre-of-mass TW GPE simulation [fig. 1(b)] using the same effective single-particle potential. In the low kinetic energy regime considered, the low-mass bright soliton is either completely reflected or completely transmitted in any individual realisation. The good level of agreement up to the time where the solitons re-collide confirms that the centre-of-mass TW technique can successfully capture the dynamical formation of quantum superpositions in the centre-of-mass coordinate, as required.

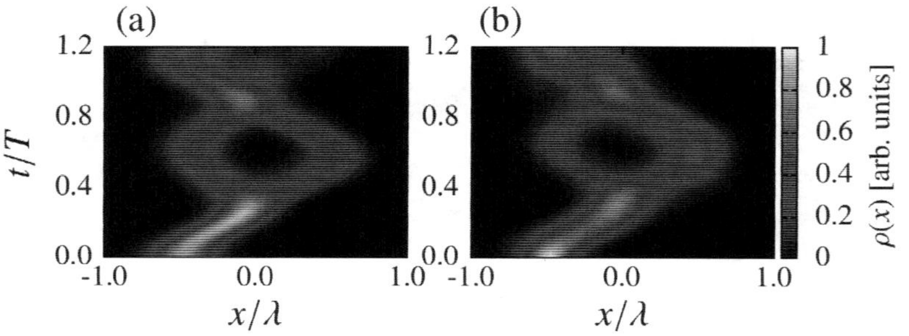

Figure 1. (a) Single-particle density for a N-particle quantum bright soliton (soliton A) hitting a narrow, heavy non-moving soliton (soliton B), computed using the effective potential approach, as in ref. [36]. (b) GPE simulation, using centre-of-mass TW technique, of a single N-particle quantum bright soliton colliding with the same single-particle potential due to soliton B as in the effective potential treatment. Taking $m_{\mathrm{A}} = m_{\mathrm{B}} = m$ and $N_{\mathrm{B}}g_{\mathrm{B}} = 10N_{\mathrm{A}}g_{\mathrm{A}}$, the system can be described in terms of the harmonic oscillator length $\lambda \equiv \lambda_{\mathrm{A}}$; we choose parameters such that the mean initial displacement of the soliton -0.48λ and the single-particle potential $V(x) = A\hbar\omega\mathrm{sech}^2(3x/2\lambda)$ with $A \approx 1.2$ [36]. $N_{\mathrm{A}} = 100$. TW results averaged over 1000 realisations.

4. Suitable Feshbach resonance

Using mixed states of the same atomic species allows for the creation of distinguishable solitons while removing the need for a dual-species laser cooling apparatus. The physical requirements for the experiment are a negative background scattering length for each of the two distinguishable soliton states, and a wide Feshbach resonance in the mixed-state scattering length.

Coupled-channels calculations were performed as detailed in Ref [55] on each of the $(f_a, f_b) = (2, 3)$ hyperfine manifold of $^{85}Rb_2$, using the MOLSCAT program [56] adapted to handle collisions in external fields [57]. A wide tunable resonance was found in the $(f_a, m_{f_a})(f_b, m_{f_b}) = (2, 2)(3, 2)$ channel. The resonance has a width of Δ=14 G determined by the difference between the zero-crossing and the pole in the scattering length. Whilst excited-state resonances are subject to decay from inelastic collisions [58] the resonance has $a_{\mathrm{res}} > 10,000\ a_0$ making it ‘pole-like’ from an experimental point of view. In the excited states the complex scattering length is given by $a(B) = \alpha(B) - i\beta(B)$, where $\alpha(B)$ is the real part of the scattering length, and $\beta(B)$ the imaginary part of the scattering length is proportional to the rate-coefficient for 2-body losses due to inelastic collisions, $K_{\rm loss} = \frac{2h}{\mu} g_n \beta(B)$, where g_n=1 (2) for a BEC of distinguishable (indistinguishable) particles. The real part of the scattering length and associated plots of $K_{\rm loss}$, of both the mixed-state and the individual states, are shown in fig. 2. Note that $K_{\rm loss} = 0$ for the absolute internal ground state $(f, m_f) = (2, +2)$.

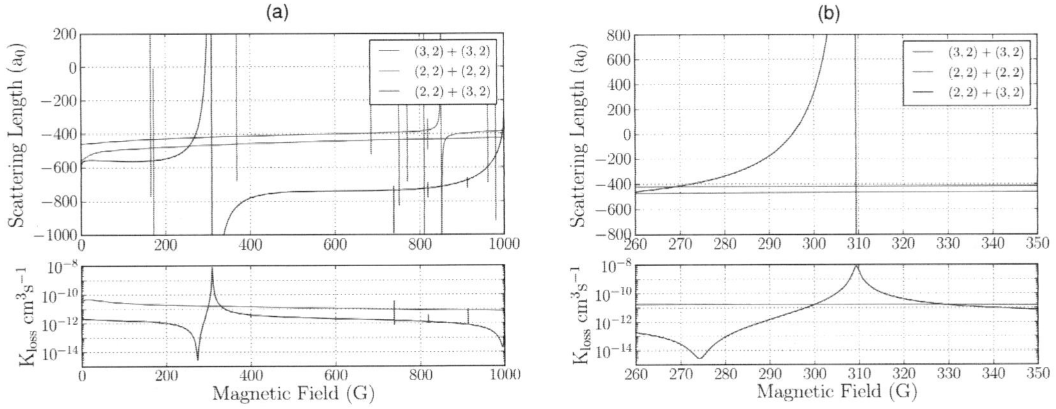

Figure 2. The s-wave scattering lengths for the $(f, m_{\rm f})$= (2,+2), (3,+2) and (2,+2)+(3,+2) states of ^{85}Rb. (a) The scattering length is split into real and imaginary components, the real part is shown in the top plot, the imaginary part is proportional to the inelastic decay rate-coefficent $K_{\rm loss}$, shown in the lower graph. (b) Zoom of (a), the wide resonance in the mixed spin state allows for tuning of the scattering length.

The three-dimensional scattering calculations can be converted into a one dimensional interaction parameter g by taking account of the trapping frequency $(f_\perp)$. With the introduction of the trapping parameters it is possible to cause a confinement induced resonance (CIR) as predicted in [59] when $a_\perp \approx Ca_{\rm 3D}$. However, given the confinement parameters for this problem ($f_\perp = 50\,{\rm Hz}$ and $f = 2\,{\rm Hz}$, see fig. 3), the CIR would occur when $a_{\rm 3D} \approx 3.5 \times 10^5 a_0$ which would not interfere with any practical implementation.

5. Truncated Wigner for the centre of mass for two distinguishable bright solitons

Using the Feshbach resonance described in the previous section we perform a centre-of-mass TW GPE simulation for the two-component GPE using parameters for a mixture of the $(f, m_f) = (2, +2)$ and $(3, +2)$ hyperfine states of ^{85}Rb. The resulting average density profiles for the two components, and the left/right correlation $\gamma(0)$ are shown in fig. 3. The high (≈ 1) value of $\gamma(0)$ subsequent to the first collision indicates the formation of a Bell state with high fidelity. Compared to the ^{85}Rb – ^{133}Cs scheme suggested in ref. [36], the present scheme is feasible at higher atom numbers, less sensitive to magnetic bias field strength, and generates higher-fidelity Bell states. These factors make the present scheme an even more experimentally attractive proposal to generate Bell states of distinguishable bright solitons.

6. Distinguishing quantum superpositions from statistical mixtures

Bell inequalities, which are both interesting because they allow to fundamentally test our understanding of quantum mechanics [60, 61] and because of their importance for quantum

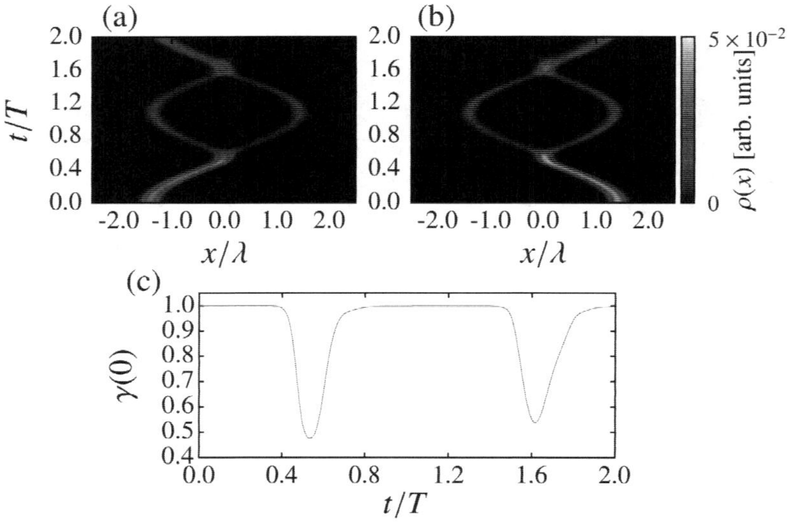

Figure 3. Centre-of-mass TW GPE simulation of a two-component collision of solitons in the $(f, m_f) = (2, +2)$ and $(3, +2)$ hyperfine states of ^{85}Rb. Parameters are $a_{(2,2)} = -410a_0$, $a_{(3,2)} = -460a_0$, $N_{(2,2)} = N_{(3,2)} \approx 90$, $f = 2\,\mathrm{Hz}$, $f_\perp = 50\,\mathrm{Hz}$, and $a_{(2,2)/(3,2)} \simeq 30.0a_0$ (conveniently reached at around 295 G, see fig. 2). The initial displacement of the solitons is $\approx \pm 10.1\mu$m. Panels (a), (b) and (c) respectively show the average single-particle densities of the $(2,2)$ and $(3,2)$ components, and the left/right correlation $\gamma(0)$. 1000 realisations were performed.

cryptography [62], are still a topic of current research [63]. For mesoscopic Bell states, related separability conditions are available [42, 64]. For a bipartite photonic system a violation of the inequality

$$\sum_{k=1}^{3} \Delta S_k^2 / \langle S_0 \rangle \geq 2 \tag{9}$$

has been shown to be a sufficient condition of non-separability and has been used to identify polarisation entanglement for squeezed vacuum pulses [42]. Here, $S_k = S_k^A + S_k^B$ denote the Stokes parameters [64] and $\langle S_0 \rangle$ is the total photon number. To convey condition (9) to our situation the properties left and right would take on the role of horizontal and vertical polarisation.

In addition to the above, in the collisions we consider here the interference properties discussed in [36] for two different species would also be available to distinguish quantum superpositions from statistical mixtures.

7. Conclusion

We have investigated numerically the generation of mesoscopic Bell states via the collision of two distinguishable quantum bright solitons. For experimentally realistic parameters, we have used Truncated Wigner for the centre of mass [36] (which we justified further) to predict entanglement generation. We have in particular extended the scheme suggested in [36] for two bright solitons of two different species to two solitons of two distinct hyperfine states of the same species, providing several advantages compared to the original suggestion [36]:

(i) We predict a much broader Feshbach resonance (fig. 2 b) then for the two-species case investigated in [36]. This will considerably simplify future experiments.

(ii) We predict a higher left/right correlation in the Bell state (fig. 3 c), potentially aiding experimental detection.

(iii) Only a Bose-Einstein condensate of one species is required; the two distinguishable bright solitons could be produced from a single initial Bose-Einstein condensate.

(iv) The current situation is closer to the mesoscopic Bell states for photons of refs. [41, 42].

Acknowledgments

We thank S. A. Gardiner, J. L. Helm, J. M. Hutson, C. R. Le Sueur, L. Khaykovich for discussions. We thank the Marsden Fund of New Zealand (Contract No. UOO162) and the Royal Society of New Zealand (Contract No. UOO004) (T. P. B.), the Faculty of Science at Durham University (C. L. B.), and the UK EPSRC (Grant No. EP/G056781/1 and EP/K03250X/1) (C. W.) for funding.

References

[1] Khaykovich L, Schreck F, Ferrari G, Bourdel T, Cubizolles J, Carr L D, Castin Y and Salomon C 2002 *Science* **296** 1290
[2] Strecker K E, Partridge G B, Truscott A G and Hulet R G 2002 *Nature (London)* **417** 150
[3] Eiermann B, Anker T, Albiez M, Taglieber M, Treutlein P, Marzlin K P and Oberthaler M K 2004 *Phys. Rev. Lett.* **92** 230401
[4] Cornish S L, Thompson S T and Wieman C E 2006 *Phys. Rev. Lett.* **96** 170401
[5] Pollack S E, Dries D, Olson E J and Hulet R G 2010 DAMOP: Conference abstract, http://meetings.aps.org/link/BAPS.2010.DAMOP.R4.1
[6] Marchant A L, Billam T P, Wiles T P, Yu M M H, Gardiner S A and Cornish S L 2013 *Nat. Commun.* **4** 1865
[7] Hackermuller L, Hornberger K, Brezger B, Zeilinger A and Arndt M 2004 *Nature (London)* **427** 711
[8] Zurek W H 2003 *Rev. Mod. Phys.* **75** 715
[9] Ernst T and Brand J 2010 *Phys. Rev. A* **81** 033614
[10] Wang C H, Hong T M, Lee R K and Wang D W 2012 *Opt. Express* **20** 22675
[11] Damgaard Hansen S, Nygaard N and Mølmer K 2012 *Preprint* arXiv:1210.1681
[12] Martin A D and Ruostekoski J 2012 *New J. Phys.* **14** 043040
[13] Helm J L, Billam T P and Gardiner S A 2012 *Phys. Rev. A* **85** 053621
[14] Cuevas J, Kevrekidis P G, Malomed B A, Dyke P and Hulet R G 2013 *New J. Phys.* **15** 063006
[15] Álvarez A, Cuevas J, Romero F R, Hamner C, Chang J J, Engels P, Kevrekidis P G and Frantzeskakis D J 2013 *J. Phys. B* **46** 065302
[16] Polo J and Ahufinger V 2013 *Phys. Rev. A* **88** 053628
[17] Fogarty T, Kiely A, Campbell S and Busch T 2013 *Phys. Rev. A* **87** 043630
[18] Gertjerenken B 2013 *Phys. Rev. A* **88** 053623
[19] Asad-uz-zaman M and Khawaja U A 2013 *EPL (Europhys. Lett.)* **101** 50008
[20] Weiss C and Castin Y 2009 *Phys. Rev. Lett.* **102** 010403
[21] Streltsov A I, Alon O E and Cederbaum L S 2009 *Phys. Rev. A* **80** 043616
[22] Gertjerenken B, Billam T P, Khaykovich L and Weiss C 2012 *Phys. Rev. A* **86** 033608
[23] Carter S J, Drummond P D, Reid M D and Shelby R M 1987 *Phys. Rev. Lett.* **58** 1841
[24] Lai Y and Haus H A 1989 *Phys. Rev. A* **40** 854
[25] Drummond P D, Shelby R M, Friberg S R and Yamamoto Y 1993 *Nature (London)* **365** 307
[26] Castin Y and Herzog C 2001 *C. R. Acad. Sci. Paris, Ser. IV* **2** 419 (*Preprint* arXiv:cond-mat/0012040)
[27] Carr L D and Brand J 2004 *Phys. Rev. Lett.* **92** 040401
[28] Delande D, Sacha K, Płodzień M, Avazbaev S K and Zakrzewski J 2013 *New J. Phys.* **15** 045021
[29] Giovannetti V, Lloyd S and Maccone L 2004 *Science* **306** 1330
[30] Weiss C and Teichmann N 2007 *Laser Phys. Lett.* **4** 895
[31] Micheli A, Jaksch D, Cirac J I and Zoller P 2003 *Phys. Rev. A* **67** 013607
[32] Mahmud K W, Perry H and Reinhardt W P 2005 *Phys. Rev. A* **71** 023615
[33] Dagnino D, Barberan N, Lewenstein M and Dalibard J 2009 *Nature Phys.* **5** 431
[34] Garcia-March M A, Dounas-Frazer D R and Carr L D 2011 *Phys. Rev. A* **83** 043612
[35] Mazzarella G, Salasnich L, Parola A and Toigo F 2011 *Phys. Rev. A* **83** 053607
[36] Gertjerenken B, Billam T P, Blackley C L, Le Sueur C R, Khaykovich L, Cornish S L and Weiss C 2013 *Phys. Rev. Lett.* **111** 100406

[37] Lewenstein M and Malomed B A 2009 *New J. Phys.* **11** 113014
[38] Holdaway D I H, Weiss C and Gardiner S A 2013 *Phys. Rev. A* **87** 043632
[39] Khaykovich L and Malomed B A 2006 *Phys. Rev. A* **74** 023607
[40] Birman J, Nazmitdinov R and Yukalov V 2013 *Phys. Rep.* **526** 1
[41] Stobińska M, Töppel F, Sekatski P and Chekhova M V 2012 *Phys. Rev. A* **86** 022323
[42] Iskhakov T S, Agafonov I N, Chekhova M V and Leuchs G 2012 *Phys. Rev. Lett.* **109** 150502
[43] Safronova M S, Arora B and Clark C W 2006 *Phys. Rev. A* **73** 022505
[44] Bergeman T, Moore M G and Olshanii M 2003 *Phys. Rev. Lett.* **91** 163201
[45] Lieb E H and Liniger W 1963 *Phys. Rev.* **130** 1605
[46] Seiringer R and Yin J 2008 *Commun. Math. Phys.* **284** 459
[47] McGuire J B 1964 *J. Math. Phys.* **5** 622
[48] Sacha K, Müller C A, Delande D and Zakrzewski J 2009 *Phys. Rev. Lett.* **103** 210402
[49] Pu H and Bigelow N P 1998 *Phys. Rev. Lett.* **80** 1134
[50] Timmermans E 1998 *Phys. Rev. Lett.* **81** 5718
[51] Öhberg P and Santos L 2001 *Phys. Rev. Lett.* **86** 2918
[52] He, ZM, Wang, DL, Ding, JW and Yan, XH 2012 *Eur. Phys. J. D* **66** 139
[53] Bienias P, Pawlowski K, Gajda M and Rzazewski K 2011 *EPL (Europhys. Lett.)* **96** 10011
[54] Weiss C and Castin Y 2012 *J. Phys. A* **45** 455306
[55] Blackley C L, Le Sueur C R, Hutson J M, McCarron D J, Köppinger M P, Cho H W, Jenkin D L and Cornish S L 2013 *Phys. Rev. A* **87** 033611
[56] Hutson J M and Green S 1994 MOLSCAT computer program, version 14 distributed by Collaborative Computational Project No. 6 of the UK Engineering and Physical Sciences Research Council
[57] González-Martínez M L and Hutson J M 2007 *Phys. Rev. A* **75** 022702
[58] Hutson J M 2007 *New J. Phys.* **9** 152
[59] Olshanii M 1998 *Phys. Rev. Lett.* **81** 938
[60] Clauser J F and Shimony A 1978 *Reports on Progress in Physics* **41** 1881
[61] Aspect A, Dalibard J and Roger G 1982 *Phys. Rev. Lett.* **49** 1804
[62] Ekert A K 1991 *Phys. Rev. Lett.* **67** 661
[63] Torlai G, McKeown G, Marek P, Filip R, Jeong H, Paternostro M and De Chiara G 2013 *Phys. Rev. A* **87** 052112
[64] Simon C and Bouwmeester D 2003 *Phys. Rev. Lett.* **91** 053601

22nd International Laser Physics Worksop (LPHYS 13) IOP Publishing
Journal of Physics: Conference Series **497** (2014) 012034 doi:10.1088/1742-6596/497/1/012034

Quantum probabilities and entanglement for multimode quantum systems

V I Yukalov[1,2], E P Yukalova[1,3] and D Sornette[1,4]
[1]DMTEC, ETH Zürich, Swiss Federal Institute of Technology, Zürich CH-8092, Switzerland
[2]Bogolubov Laboratory of Theoretical Physics, Joint Institute for Nuclear Research, Dubna 141980, Russia
[3]Laboratory of Information Technologies, Joint Institute for Nuclear Research, Dubna 141980, Russia
[4]Swiss Finance Institute, c/o University of Geneva, 40 blvd. Du Pont d'Arve, CH 1211 Geneva 4, Switzerland

E-mail: yukalov@theor.jinr.ru

Abstract. Quantum probabilities are defined for several important physical cases characterizing measurements with multimode quantum systems. These are the probabilities for operationally testable measurements, for operationally uncertain measurements, and for entangled composite events. The role of the prospect and state entanglement is emphasized. Numerical modeling is presented for a two-mode Bose-condensed system of trapped atoms. The interference factor is calculated by invoking the channel-state duality.

1. Introduction

Multimode quantum systems are of great interest because of their importance for a variety of applications, such as quantum electronics and quantum information processing [1]. There exists a number of multimode-system types of different physical nature. These, for instance, can be atomic systems with several excited electron levels, molecular systems with several roto-vibrational modes, quantum dots with several populated exciton levels, spin assemblies with several spin projections, trapped ions and trapped neutral atoms with several energy states, and so on [2].

For the purpose of quantum information processing and quantum computing, it is necessary to have well defined notions of quantum probabilities, including the probability of composite events. Quantum probabilities of separate events, describing quantum measurements, have been defined by von Neumann [3]. The development of this notion can be found in the recent reviews [4,5]. Note that, starting from von Neumann [3], the theory of quantum measurements is considered as an analogue of the quantum decision theory [6,7]. Bohr [8-10] has argued that quantum theory is an appropriate tool for describing human decision making. It has been shown [11-15] that quantum decision theory, for both quantum measurements as well as for human decision making, can be developed on the same grounds employing quantum techniques. Mathematical problems appearing when introducing quantum probabilities for composite events are discussed in Ref. [16].

In the present paper, after recalling the notion of the quantum probabilities for operationally testable measurements (Sec. 2), we introduce the notion of operationally uncertain

Journal of Physics: Conference Series **497** (2014) 012034 doi:10.1088/1742-6596/497/1/012034

measurements (Sec. 3). Then, in Sec. 4, following the theory of Ref. [16], we define the joint quantum probabilities for composite events, emphasizing the application to measurements under uncertainty. In Sec. 5, we discuss how one can estimate the average value of the interference factor. We show that this value, under a rather general assumption of its symmetric distribution, equals 1/4. The necessity of entanglement for the appearance of coherent interference is stressed in Sec. 6. An example of calculating the interference factor, employing the channel-state duality, for the case of a trapped Bose-Einstein condensate with generated coherent modes, is given in Sec. 7, also containing a brief conclusion.

2. Operationally testable measurements

First, let us recall the definition of quantum probabilities for separate measurements representing quantum events [3]. Quantum events, obeying the Birkhoff-von Neumann quantum logic [17] form a non-commutative non-distributive ring $\mathcal{R}$. The nonempty collection of all subsets of the event ring $\mathcal{R}$, including $\mathcal{R}$, which is closed with respect to countable unions and complements, is the event sigma algebra Σ. The algebra of quantum events is the pair $\{\Sigma, \mathcal{R}\}$ of the sigma algebra Σ over the event ring $\mathcal{R}$.

Observables in quantum theory are represented by self-adjoint operators $\hat{A}$ on a Hilbert space $\mathcal{H}$, pertaining to the algebra of local observables $\mathcal{A}$. From the eigenproblem

$$\hat{A}|n\rangle = A_n|n\rangle \; , \tag{1}$$

one defines the projectors

$$\hat{P}_n \equiv |n\rangle\langle n| \; , \tag{2}$$

allowing for the spectral decomposition

$$\hat{A} = \sum_n A_n \hat{P}_n \; . \tag{3}$$

The projectors are summed to the unity operator in $\mathcal{H}$:

$$\sum_n \hat{P}_n = \hat{1}_{\mathcal{H}} \; . \tag{4}$$

The operator spectrum $\{A_n\}$ can be discrete or continuous, degenerate or not. In what follows, we assume, for simplicity, a discrete non-degenerate spectrum. The generalization to an arbitrary spectrum is straightforward. Assuming a non-degenerate spectrum, we keep in mind the von Neumann suggestion [3] of avoiding degeneracy by lifting it with additional small terms breaking the operator symmetry responsible for the spectrum degeneracy, and at the end removing these terms. In contrast, a discrete spectrum is typical of finite quantum systems [2].

A physical system is characterized by a statistical operator $\hat{\varrho}$, called the *system state*, which is a trace-class positive operator normalized to one, so that the trace over $\mathcal{H}$ is $\mathrm{Tr}\hat{\varrho} = 1$. As a result of a measurement with an operator $\hat{A}$, one can observe an eigenvalue A_n, which can be termed the event A_n. The probability of this event is

$$p(A_n) \equiv \mathrm{Tr}\hat{\rho}\hat{P}_n \equiv \langle \hat{P}_n \rangle \; , \tag{5}$$

with the trace over $\mathcal{H}$. By this definition,

$$\sum_n p(A_n) = 1 \; , \qquad 0 \leq p(A_n) \leq 1 \; ,$$

hence the family $\{p(A_n)\}$ forms a probability measure. According to the Gleason theorem [18], this measure is unique for a Hilbert space of dimensionality larger than two.

Journal of Physics: Conference Series **497** (2014) 012034 doi:10.1088/1742-6596/497/1/012034

In the theory of quantum measurements, the projectors $\hat{P}_n$ play the role of observables, so that, for an event A_n, one has the correspondence

$$A_n \rightarrow \hat{P}_n \equiv |n\rangle\langle n| \; . \tag{6}$$

Because of the projector orthogonality property

$$\hat{P}_m \hat{P}_n = \delta_{mn} \hat{P}_n \; , \tag{7}$$

the events with non-coinciding indices are also orthogonal,

$$A_m \bigcap A_n = \delta_{mn} A_m \; . \tag{8}$$

For a union of mutually orthogonal events, there is the correspondence

$$\bigcup_n A_n \rightarrow \sum_n \hat{P}_n \; , \tag{9}$$

which results in the additivity of the probabilities:

$$p\left(\bigcup_n A_n\right) = \sum_n p(A_n) \; . \tag{10}$$

The procedure described above is called the standard measurement [19].

3. Operationally uncertain measurements

Situations can exist when the result of a measurement is not well defined, so that one cannot tell that a particular event has occurred, but it is only known that some of the events A_n could be realized. This is what is called an uncertain, inconclusive, or generalized measurement [4,5].

Suppose the observed event A is a set $\{A_n\}$ of possible events. Though the events A_m and A_n are orthogonal for $m \neq n$, they form not a standard union but an *uncertain union* that we shall denote as

$$A \equiv \{A_n\} \equiv \biguplus_n A_n \tag{11}$$

in order to distinguish it from the standard union $\bigcup_n A_n$. To the uncertain event A, there corresponds the wave function

$$|A\rangle = \sum_n a_n |n\rangle \; , \tag{12}$$

where $|a_n|^2$ play the role of weights for the events A_n. Instead of the correspondence (9) for the standard union, we now have the correspondence

$$\biguplus_n A_n \rightarrow \hat{P}_A \equiv |A\rangle\langle A| \; . \tag{13}$$

Generally, the proposition operator $\hat{P}_A$ is not a projector. Since

$$\hat{P}_A \hat{P}_B = \langle A|B\rangle \; |A\rangle\langle B| \; ,$$

it becomes an orthogonal projector only when functions (12) for different A and B are normalized and orthogonal, which is not necessarily required. This proposition operator is connected with the projectors $\hat{P}_n$ by the relation

$$\hat{P}_A = \sum_n |a_n|^2 \hat{P}_n + \sum_{m \neq n} a_m^* a_n |n\rangle\langle m| \; .$$

Journal of Physics: Conference Series **497** (2014) 012034 doi:10.1088/1742-6596/497/1/012034

The probability of the uncertain event A reads as

$$p(A) = p\left(\biguplus_n A_n\right) = \sum_n |a_n|^2 p(A_n) + q(A) \;, \tag{14}$$

where the second term

$$q(A) \equiv \sum_{m \neq n} a_m^* a_n \langle m|\hat{\rho}|n\rangle \tag{15}$$

is caused by the interference of the uncertain subevents A_n that are called *modes*. As we see the probability of an uncertain union is not additive, contrary to the probability of the standard union (10).

4. Quantum joint probabilities

To define the joint probability of different events, we follow the theory of Ref. [16]. Let us be interested in two observables $\hat{A}$ and $\hat{B}$ that can be commuting or not. Spanning the eigenfunctions of $\hat{A}$, we construct a Hilbert space $\mathcal{H}_A$. Similarly, the observable $\hat{B}$ possesses the eigenvalues and eigenfunctions given by the eigenproblem

$$\hat{B}|\alpha\rangle = B_\alpha|\alpha\rangle \;. \tag{16}$$

Spanning the eigenfunctions of $\hat{B}$, we have a Hilbert space $\mathcal{H}_B$. These two observables are treated as a tensor product $\hat{A} \bigotimes \hat{B}$ acting on the tensor-product Hilbert space

$$\mathcal{H}_{AB} \equiv \mathcal{H}_A \bigotimes \mathcal{H}_B \;. \tag{17}$$

The composite event of observing A_n and B_α induces the correspondence

$$A_n \bigotimes B_\alpha \;\rightarrow\; \hat{P}_n \bigotimes \hat{P}_\alpha \;, \tag{18}$$

where

$$\hat{P}_\alpha \equiv |\alpha\rangle\langle\alpha| \tag{19}$$

is a projector in $\mathcal{H}_B$.

The system state is now a statistical operator $\hat{\varrho}$ on the tensor-product space (17). The joint probability of the composite event (18) reads as

$$p(A_n \bigotimes B_\alpha) = \mathrm{Tr}\hat{\rho}\hat{P}_n \bigotimes \hat{P}_\alpha \equiv \langle \hat{P}_n \bigotimes \hat{P}_\alpha\rangle \;, \tag{20}$$

with the trace over space (17). The composite event (18) is the simplest composite event, which enjoys the factor form, being composed of two elementary events. More complicated structures arise when at least one of the events is a union. It is important to emphasize the difference between the standard union and the uncertain union introduced in Sec. 3.

Considering the composite event, being a product of an elementary event A_n and the standard union of mutually orthogonal events B_α, we use the known property [20] of composite events:

$$A_n \bigotimes \bigcup_\alpha B_\alpha = \bigcup_\alpha A_n \bigotimes B_\alpha \;. \tag{21}$$

In the right-hand side of Eq. (21), we have the union of mutually orthogonal composite events, since B_α are mutually orthogonal [20]. Hence

$$p\left(A_n \bigotimes \bigcup_\alpha B_\alpha\right) = \sum_\alpha p(A_n \bigotimes B_\alpha) \;. \tag{22}$$

Journal of Physics: Conference Series **497** (2014) 012034 doi:10.1088/1742-6596/497/1/012034

Therefore the probability of a composite event, with one of the factors being the standard union, is additive.

However, the situation is essentially different when dealing with an uncertain union introduced in Sec. 3. Suppose we have such an uncertain union

$$B \equiv \{B_\alpha\} \equiv \biguplus_\alpha B_\alpha \tag{23}$$

corresponding to a function

$$|B\rangle = \sum_\alpha b_\alpha |\alpha\rangle \; . \tag{24}$$

We may construct a composite event, or prospect

$$\pi_n = A_n \bigotimes B = A_n \bigotimes \biguplus_\alpha B_\alpha \; , \tag{25}$$

which corresponds to the prospect state

$$|\pi_n\rangle = |n\rangle \bigotimes |B\rangle = \sum_{n\alpha} b_\alpha |n\alpha\rangle \; . \tag{26}$$

Then prospect (25) induces the correspondence

$$\pi_n \quad \rightarrow \quad \hat{P}(\pi_n) \equiv |\pi_n\rangle\langle\pi_n| \tag{27}$$

defining the prospect operator $\hat{P}(\pi_n)$.

Note that the prospect states (26), in general, are not orthogonal and normalized to one. Because of this, the prospect operators, generally, are not projectors. However, the resolution of unity is required:

$$\sum_n \hat{P}(\pi_n) = \hat{1}_{AB} \; , \tag{28}$$

where $\hat{1}_{AB}$ is the unity operator in space (17). The family $\{\hat{P}(\pi_n)\}$ composes a positive operator-valued measure.

The prospect probability

$$p(\pi_n) \equiv \mathrm{Tr}\hat{\rho}\hat{P}(\pi_n) \equiv \langle\hat{P}(\pi_n)\rangle \; ,$$

with the trace over space (17), becomes the sum of two terms,

$$p(\pi_n) = f(\pi_n) + q(\pi_n) \; , \tag{29}$$

the first term

$$f(\pi_n) \equiv \sum_\alpha |b_\alpha|^2 p(A_n \bigotimes B_\alpha) \; , \tag{30}$$

containing diagonal elements with respect to α, and the second term

$$q(\pi_n) \equiv \sum_{\alpha \neq \beta} b_\alpha^* b_\beta \langle n\alpha|\hat{\rho}|n\beta\rangle \; , \tag{31}$$

formed by the nondiagonal elements. By constructions and due to the resolution of unity (28), the prospect probability (29) satisfies the properties

$$\sum_n p(\pi_n) = 1 \; , \qquad 0 \leq p(\pi_n) \leq 1 \; , \tag{32}$$

Journal of Physics: Conference Series **497** (2014) 012034 doi:10.1088/1742-6596/497/1/012034

making the family $\{p(\pi_n)\}$ a probability measure.

Expression (31) is caused by the quantum nature of the considered events producing interference of the modes composing the uncertain union (23). Therefore quantity (31) can be called the *interference factor* or *coherence factor*.

According to the quantum-classical correspondence principle, going back to Bohr [21], classical theory is to be the limiting case of quantum theory, when quantum effects vanish. In the present case, this implies that when the interference (coherence) factor tends to zero, the quantum probability has to tend to a classical probability. Keeping in mind this decoherence process, we require the validity of the quantum-classical correspondence principle in the form

$$p(\pi_n) \rightarrow f(\pi_n) , \qquad q(\pi_n) \rightarrow 0 , \tag{33}$$

assuming that the decoherence process leads to the classical probability $f(\pi_n)$. Being a probability, this classical factor needs to be normalized, so that to satisfy the conditions

$$\sum_n f(\pi_n) = 1 , \qquad 0 \leq f(\pi_n) \leq 1 . \tag{34}$$

As a consequence of Eqs. (32) and (34), the interference factor enjoys the properties

$$\sum_n q(\pi_n) = 0 , \qquad -1 \leq q(\pi_n) \leq 1 . \tag{35}$$

As is seen, the appearance of the interference factor is due to the structure of the composite event (25) containing an uncertain union. The occurrence of such a structure can be due to two reasons. One possibility can occur when one accomplishes measurements with the observable $\hat{B}$, which provide an uncertain result, and then realizes measurements with the observable $\hat{A}$ leading to an operationally testable result. The second possibility can correspond to an operationally testable measurement with $\hat{A}$, when it is known that the measured system has been subject to some not well controlled perturbations that could be of external or internal origin. External perturbations can be caused by the influence of surrounding, and internal perturbations can be produced by the measurer and measuring devices. In the case when uncertainty is induced by uncontrolled perturbations, one may either trust the results of the final measurement for $\hat{A}$, characterizing the trust by B_1, or one may not trust these measurements, denoting this by B_2. Then the uncertain union is $B_1 \uplus B_2$. Since this uncertainty is not caused by another measurement for another observable, but is due to uncontrolled perturbations and the measurer trust, it can be termed *hidden uncertainty*. Such a hidden uncertainty can also arise in quantum decision theory [11-15], when a decision maker chooses between several lotteries whose setup is not trustable, i.e. dependent on uncertain factors.

5. Average interference factor

The interference factor is, of course, a contextual random quantity, depending on the kind of measurements, measured observables, and experimental setup. But we may try to evaluate its expected value.

Being a random quantity, the interference factor should be characterized by a probability distribution $\varphi(q)$. The latter has to be normalized on the interval $[-1, 1]$, so that

$$\int_{-1}^{1} \varphi(q)\, dq = 1 . \tag{36}$$

And the alternation condition (35) can be written as

$$\int_{-1}^{1} q\varphi(q)\, dq = 0 . \tag{37}$$

Journal of Physics: Conference Series **497** (2014) 012034 doi:10.1088/1742-6596/497/1/012034

In the following, using the notation

$$q_+ \equiv \int_0^1 q\varphi(q)\, dq \,, \qquad q_- \equiv \int_{-1}^0 q\varphi(q)\, dq \,, \tag{38}$$

one has

$$q_+ + q_- = 0 \,. \tag{39}$$

It is easy to see [12,15] that, for the case of non-informative prior, when the distribution $\varphi(q)$ is uniform, we have the *quarter law*, when $q_+ = 1/4$, $q_- = -1/4$.

This result can be generalized showing that the quarter law is valid for a wide class of distributions possessing some symmetry properties. As an example, let us take the often used beta-distribution that we define here on the interval $[-1, 1]$ in the following way:

$$\varphi(q) = \frac{\lambda_+}{B(\alpha,\beta)}\, q^{\alpha-1}(1-q)^{\beta-1} \qquad (0 \leq q \leq 1) \,,$$

$$\varphi(q) = \frac{\lambda_-}{B(\mu,\nu)}\, |q|^{\mu-1}(1-|q|)^{\nu-1} \qquad (-1 \leq q \leq 0) \,, \tag{40}$$

where the parameters $\alpha, \beta, \mu, \nu, \lambda_+, \lambda_-$ are positive and

$$B(\alpha,\beta) = \frac{\Gamma(\alpha)\Gamma(\beta)}{\Gamma(\alpha+\beta)} \,, \qquad B(\mu,\nu) = \frac{\Gamma(\mu)\Gamma(\nu)}{\Gamma(\mu+\nu)} \,.$$

The normalization condition (36) gives

$$\lambda_+ + \lambda_- = 1 \,.$$

The quantities (38) are given by

$$q_+ = \frac{\alpha\lambda_+}{\alpha+\beta} \,, \qquad q_- = -\,\frac{\mu\lambda_-}{\mu+\nu} \,.$$

And equation (37) results in the equality

$$\frac{\alpha\lambda_+}{\alpha+\beta} = \frac{\mu\lambda_-}{\mu+\nu} \,.$$

Meanwhile, the values of q_+ and q_- are not uniquely defined. But, if we assume a symmetric distribution, such that $\lambda_+ = \lambda_- = 1/2$ and $\alpha = \beta$, then it follows that $\mu = \nu$ and the quarter law is valid for arbitrary positive α and μ:

$$q_+ = \frac{1}{4} \,, \qquad q_- = -\,\frac{1}{4} \,.$$

The non-informative prior is a particular case of expression (40), when $\alpha = \beta = 1, \mu = \nu = 1$, which yields $\lambda_+ = \lambda_- = 1/2$ and gives the uniform distribution $\varphi(q) = 1/2$.

6. Prospect and state entanglement

Entanglement plays an important role in quantum information processing [1] and has thus been widely studied for many physical systems of different nature (see, e.g., Refs. [22-27]).

As is demonstrated in Sec. 3, the interference factor arises if the considered prospect is composite and contains an uncertain union, as in prospect (25). This prospect can be termed entangled. It is principally different from the composite prospect (21) that yields the additive probability (22) containing no interference. Thus the existence of an entangled prospect is a necessary condition for the appearance of an interference factor.

Another necessary condition is that the system state $\hat{\varrho}$ should also be entangled. To show that a disentangled state does not produce interference, let us take the system state in the disentangled product form

$$\hat{\rho} = \hat{\rho}_A \bigotimes \hat{\rho}_B \; . \tag{41}$$

Then the interference factor (31) becomes

$$q(\pi_n) = \sum_{\alpha \neq \beta} b_\alpha^* b_\beta \langle n|\hat{\rho}_A|n\rangle\langle\alpha|\hat{\rho}_B|\beta\rangle \; . \tag{42}$$

In view of property (35) and taking into account the normalization

$$\mathrm{Tr}_A\hat{\rho}_A = \sum_n \langle n|\hat{\rho}_A|n\rangle = 1 \; , \tag{43}$$

we get

$$\sum_n q(\pi_n) = \sum_{\alpha \neq \beta} b_\alpha^* b_\beta \langle\alpha|\hat{\rho}_B|\beta\rangle = 0 \; . \tag{44}$$

Using Eqs. (42) and (44), we find

$$q(\pi_n) = \langle n|\hat{\rho}_A|n\rangle \sum_n q(\pi_n) = 0 \; . \tag{45}$$

So, the disentangled state (41) does not allow for a nontrivial interference factor.

Though the state entanglement is a necessary condition for the occurrence of interference, it is not sufficient. As a counterexample, we can consider a maximally entangled state such as the multimode state

$$\hat{\rho} \equiv |\psi\rangle\langle\psi| = \frac{1}{M}\sum_{mn} |mm\rangle\langle nn| \tag{46}$$

composed of the multimode function

$$|\psi\rangle = \frac{1}{\sqrt{M}} \sum_m |mm\rangle \; , \tag{47}$$

with the number of modes M,

$$M = \sum_m 1 = \mathrm{dim}\mathcal{H}_A = \mathrm{dim}\mathcal{H}_B \; . \tag{48}$$

The entanglement-production measure [28,29] for the statistical operator (46) is

$$\varepsilon(\hat{\rho}) = \log M \; . \tag{49}$$

But the interference factor (31) is zero:

$$q(\pi_n) = \frac{1}{M}\sum_{\alpha \neq \beta} b_\alpha^* b_\beta \delta_{\alpha n}\delta_{\beta n} = 0 \; . \tag{50}$$

In this way, the entanglement of the prospect and of the system state is a necessary, but not sufficient, condition for the occurrence of the mode interference.

Journal of Physics: Conference Series **497** (2014) 012034 doi:10.1088/1742-6596/497/1/012034

7. Two-mode Bose system

As an illustration of the approach, let us consider a gas of trapped Bose atoms at ultracold temperatures, when practically all atoms pile down to a Bose-Einstein condensed state. At the temperatures close to zero and in the presence of very weak interactions, the trapped Bose system can be described by the nonlinear Schrödinger (NLS) equation with a discrete spectrum due to the atomic confinement [30,31]. The functions that are the solutions to the stationary NLS equation are termed the coherent modes. In equilibrium, all atoms settle down to the collective ground-state energy level. By applying an alternating external field, which either modulates the trapping potential or induces the oscillation of the scattering length by means of Feshbach resonance techniques [32], the atomic gas goes into a nonequilibrium state. Tuning the frequency of the modulating field to the resonance with a transition frequency between the atomic energy levels, it is feasible to generate one or several excited modes. Thus it is possible to create a two-mode Bose system [30,31]. Other ways of generating two-mode (or multimode) Bose systems could be by splitting condensates with an external beam [33] or loading condensates into a double-well (or multi-well) potential, where the ground-state energy level splits into two (or several) energy levels, with the splitting being regulated by adjusting the parameters of the potential [34].

Let us denote the modes, that are the normalized solutions to the stationary NLS equation, as $|n\rangle$. An event A_n symbolizes the observation of a $|n\rangle$-th mode. The probabilities of separate events A_n are calculated in the standard way, as in Sec. 2. But our aim is to show how one could define the joint probabilities in the case of entangled prospects of type (25). The prospects of this type can arise owing to two reasons.

One possibility corresponds to the situation when, at a given time, one accomplishes a measurement for observing events A_n while, at the previous times, realizing uncertain measurements for the events denoted as B_α. The measurements can be treated as nondestructive, yielding only phase shifts, but not directly influencing the level populations [35].

The other interpretation could be as follows. At a given time, one accomplishes measurements for observing events A_n, while one is aware that, at the previous times, the system has been subject to uncontrolled perturbations. It is clear that these two cases are analogous to each other, since uncertain measurements can be mathematically described as uncontrolled perturbations.

According to Sec. 4, the interference factor, caused by the interference of modes in the process of uncertain measurements, is

$$q(\pi_n) = p(\pi_n) - f(\pi_n) \; . \tag{51}$$

Resorting to the channel-state duality, the picture can be translated into a temporal representation [16,35]. Then, for the prospect probability, we have the correspondence

$$p(\pi_n) \;\rightarrow\; p_n(t) \; , \tag{52}$$

respectively, for the interference factor,

$$q(\pi_n) \;\rightarrow\; q_n(t) \; , \tag{53}$$

where the quantities $p_n(t)$ and $q_n(t)$ are defined through a channel picture. The influence of measurements can be described as the action of a random noise of strength σ. Without the noise, entangling modes, the probability is given by

$$f_n(t) \equiv \lim_{\sigma\rightarrow 0} p_n(t) \; . \tag{54}$$

Then, the interference factor can be defined as

$$q_n(t) = p_n(t) - f_n(t) \; . \tag{55}$$

Considering a Bose-condensed system with two modes, generated by means of the trap modulation [30,31], and keeping in mind nondestructive measurements [35], we have the mode populations

$$p_1(t) = \frac{1 - s(t)}{2} , \qquad p_2(t) = \frac{1 + s(t)}{2} , \tag{56}$$

in which $s(t)$ is the population imbalance, satisfying the equation of motion

$$\frac{ds}{dt} = -b \sqrt{1 - s^2} \sin x , \tag{57}$$

and $x(t)$ is the phase difference described by the equation

$$dx = s \left(1 + \frac{b}{\sqrt{1 - s^2}} \cos x \right) dt + \sigma dW_t . \tag{58}$$

Here b is a parameter quantifying the amplitude of the pumping field modulating the trap, W_t is the standard Wiener process with the standard deviation σ, and time t is dimensionless.

In the absence of random perturbations ($\sigma = 0$), there exist two regimes in the dynamics of Eqs. (57) and (58), depending on the value of the pumping amplitude b with respect to the critical value

$$b_c = \frac{s_0^2}{2(1 + \sqrt{1 - s_0^2} \cos x_0)} , \tag{59}$$

where $s_0 \equiv s(0), x_0 \equiv x(0)$ are the initial conditions. The subcritical regime, when $b < b_c$, corresponds to the Rabi oscillations of $s(t)$, while the supercritical regime, when $b > b_c$, is the regime of Josephson oscillations [30,31]. The value of b_c varies in the diapason

$$0 \leq b_c \leq \frac{1}{2} \qquad (0 \leq s_0^2 \leq 1) . \tag{60}$$

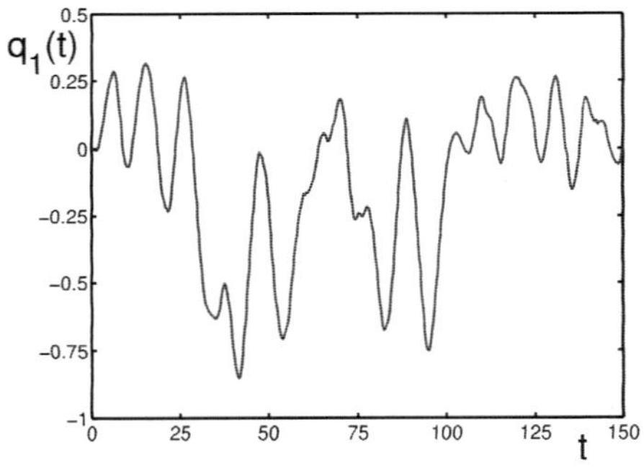

Figure 1. Interference factor $q_1(t)$ as a function of dimensionless time for the subcritical regime, with $b = 0.25 < b_c = 0.282$.

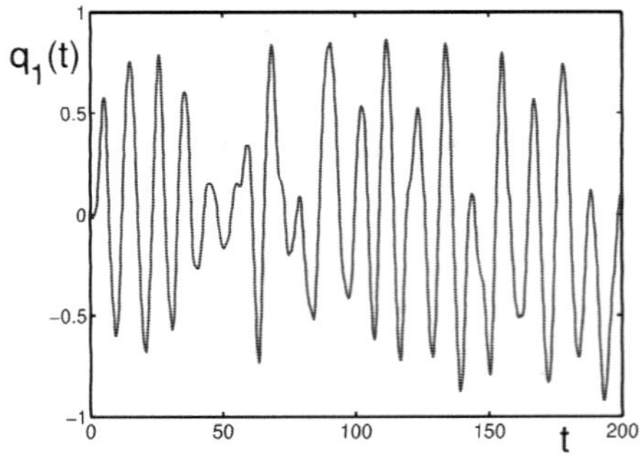

Figure 2. Interference factor $q_1(t)$ as a function of dimensionless time for the supercritical regime, with $b = 0.5 > b_c = 0.282$.

Solving Eqs. (57) and (58) in the presence of the random perturbations, we take the initial conditions $s_0 = -0.9, x_0 = 0$, which define the critical pumping parameter $b_c = 0.282$. The results of the numerical solution for the interference factor (55), with $n = 1$, are presented in

Fig. 1 for the subcritical regime and in Fig. 2 for supercritical regime. It is sufficient to present only one interference factor, since the second is given by $q_2(t) = -q_1(t)$.

One can observe that the fluctuations of the interference factor are larger in the supercritical regime.

In conclusion, we have shown how the quantum probabilities for composite events, related to quantum measurements, can be defined. A special attention has been payed to the case of operationally uncertain measurements, when there appears an interference factor. We described how the average value of the interference factor can be estimated under rather general conditions. The necessity of entanglement in the considered prospect, as well as in the system state, for the appearance of interference has been stressed. An example of calculating the interference factor, employing the channel-state duality, for a trapped Bose-Einstein condensate with generated coherent modes, was demonstrated.

Acknowledgments
Financial supports from the Swiss National Foundation and from the Russian Foundation for Basic Research are acknowledged.

[1] Keyl M 2002 *Phys. Rep.* **369** 431
[2] Birman J L, Nazmitdinov R G and Yukalov V I 2013 *Phys. Rep.* **526** 1
[3] von Neuman J 1955 *Mathematical Foundations of Quantum Mechanics* (Princeton: Princeton University)
[4] Holevo A S 2011 *Probabilistic and Statistical Aspects of Quantum Theory* (Berlin: Springer)
[5] Holevo A S and Giovanetti V 2012 *Rep. Prog. Phys.* **75** 046001
[6] Benioff P A 1972 *J. Math. Phys.* **13** 908
[7] Holevo A S 1973 *J. Multivar. Anal.* **3** 337
[8] Bohr N 1933 *Nature* **131** 421
[9] Bohr N 1933 *Nature* **131** 457
[10] Bohr N 1958 *Atomic Physics and Human Knowledge* New York: Wiley
[11] Yukalov V I and Sornette D 2008 *Phys. Lett.* A **372** 6867
[12] Yukalov V I and Sornette D 2009 *Entropy* **11** 1073
[13] Yukalov V I and Sornette D 2009 *Eur. Phys. J.* B **71** 533
[14] Yukalov V I and Sornette D 2010 *Adv. Complex Syst.* **13** 659
[15] Yukalov V I and Sornette D 2011 *Theor. Decis.* **70** 283
[16] Yukalov V I and Sornette D 2013 *Laser Phys.* **23** 105502
[17] Birkhoff G and von Neumann J 1936 *Ann. Math.* **37** 823
[18] Gleason A M 1957 *J. Math. Mech.* **6** 885
[19] Huttner B, Muller A, Gautier J D, Zbinden H and Gisin N 1996 *Phys. Rev.* A **54** 3783
[20] Niestegge G 2004 *J. Math. Phys.* **45** 4714
[21] Bohr N 1913 *Phil. Mag.* **26** 1
[22] O'Connor K M and Wooters W K 2001 *Phys. Rev.* A **63** 052302
[23] Wang X 2001 *Phys. Rev.* A **64** 012313
[24] Arnesen M C, Bose S and Vedral V 2001 *Phys. Rev. Lett.* **87** 017901
[25] Zanardi P 2002 *Phys. Rev.* A **65** 042101
[26] Gittings J R and Fisher A J 2002 *Phys. Rev.* A **66** 032305
[27] Zanardi P and Wang X 2002 *J. Phys.* A **35** 7947
[28] Yukalov V I 2003 *Phys. Rev.* A **68** 022109
[29] Yukalov V I and Sornette D 2010 *Phys. At. Nucl.* **73** 559
[30] Yukalov V I, Yukalova E P and Bagnato V S 1997 *Phys. Rev.* A **56** 4845
[31] Yukalov V I, Yukalova E P and Bagnato V S 2000 *Laser Phys.* **10** 26
[32] Chin C, Grimm R, Julienne P and Tiesinga E 2010 *Rev. Mod. Phys.* **82** 1225
[33] Stimming H P, Mauser N J, Schmiedmayer J and Mazets I E 2011 *Phys. Rev.* A **83** 023618
[34] Yukalov V I, Yukalova E P 1996 *J. Phys.* A **29** 6429
[35] Yukalov V I, Yukalova E P and Sornette D 2013 *Laser Phys. Lett.* **10** 115502

Journal of Physics: Conference Series **497** (2014) 012035 doi:10.1088/1742-6596/497/1/012035

Propagation of collective modes in non-overlapping dipolar Bose-Einstein Condensates

A. Gallemí, M. Guilleumas, R. Mayol and M. Pi

Departament d'Estructura i Constituents de la Matèria, Facultat de Física, Universitat de Barcelona, 08028-Barcelona, Spain

Institut de Nanociència i Nanotecnologia (IN^2UB), Universitat de Barcelona, 08028-Barcelona, Spain

E-mail: gallemi@ecm.ub.edu

Abstract. We investigate long-range effects of the dipolar interaction in Bose-Einstein condensates by solving the time-dependent 3D Gross-Pitaevskii equation. We study the propagation of excitations between non-overlapping condensates when a collective mode is excited in one of the condensates. We obtain the frequency shifts due to the long-range character of the dipolar coupling for the bilayer and also the trilayer system when the dipolar mode is excited in one condensate. The propagation of the monopolar and quadrupolar modes are also investigated. The coupled-pendulum model is proposed to qualitatively explain the long range effects of the dipolar coupling.

1. Introduction

Recent experiments on dipolar Bose-Einstein Condensates (dBECs) of chromium [1], dysprosium [2], and erbium atoms [3], have stimulated a growing interest in ultracold dipolar quantum gases [4] due to the long-range and anisotropic nature of such interaction. The research on this kind of interaction is expecting to give rise to important new features not only at microscopic level, but at macroscopic level as well. The development of new experimental techniques focused on the realization of molecular gases with large electric polar moment, where the dipolar effects are particularly strong, has led to the hope that challenging new frontiers in this field can be opened.

The theoretical research in dipolar gases has tended to focus mostly on the study of the anisotropic effects of the dipolar interaction [4, 5], rather than on the long-range ones, where there are only few proposals: cat states in triple-well potentials [6, 7], collisions between solitons in dual-core couplers [8] and propagation of center-of-mass modes studied by means of a variational method [9, 10].

The effect of the dipole-dipole interaction on the frequency of a collective mode has been experimentally measured in a single dBEC [11]. However, a central issue that remains opened is to understand the mechanism of the expansion and propagation of collective modes between non-overlapping condensates.

The aim of the present work is to study the propagation of dipolar, monopolar and quadrupolar modes induced by the long-range nature of the dipolar interaction by solving the full

3D time-dependent Gross-Pitaevskii equation (TDGPE). We consider two dipolar Bose-Einstein condensates harmonically trapped in a double-well configuration such that the overlap between the two clouds is negligible as well as the corresponding tunneling effect. The only effective force acting between the two dBECs is the one produced by the long-range behaviour of the dipolar interaction.

First, we study the propagation of center-of-mass (dipolar) modes between two non-overlapping pancake-shaped condensates. Our time-dependent numerical results are in agreement with previous ones obtained within the variational method [9, 10]. Then, we investigate the propagation of other excitations (monopolar and quadrupolar modes), in the bilayer system. That is, the appearance and shift of new characteristic frequencies. Finally, as an extension, we test our model in a triple-well configuration, comparing the results with the case of the double-well configuration, and finding out that the classical model gives a qualitative description of the dipolar coupling. The coupled-pendulum model is used to understand qualitatively the results obtained before.

2. The system: dipolar interaction in BECs and double-well configuration

Bose-Einstein condensates at low temperatures can be described in the mean-field framework provided a large number of particles and a weakly interacting system. The TDGPE in presence of the dipolar interaction reads:

$$i\hbar\frac{\partial\Psi(\vec{r},t)}{\partial t} = \left[-\frac{\hbar^2}{2m}\nabla^2 + V_{ext}(\vec{r}) + g|\Psi(\vec{r},t)|^2 + V_D(\vec{r})\right]\Psi(\vec{r},t)\,, \tag{1}$$

where $\Psi(\vec{r},t)$ is the wavefunction of the condensate, which is normalized to the number of particles $N = \int|\Psi(\vec{r},t)|^2 d\vec{r}$, m is the mass of the bosons, and $V_{ext}(\vec{r})$ is the trapping potential. The contact interaction potential is characterized by the coupling constant $g = 4\pi\hbar^2 a/m$, where a is the s-wave scattering length. The mean-field dipolar interaction is given by

$$V_D(|\vec{r}-\vec{r}\,'|,\theta) = d^2\,\frac{1-3\cos^2\theta}{|\vec{r}-\vec{r}\,'|^3}\,. \tag{2}$$

Here $\vec{d} = \sqrt{\mu_0/4\pi}\,\vec{\mu}_m$, where μ_0 is the magnetic permeability of vacuum and $\vec{\mu}_m$ is the magnetic dipole moment of the atoms. All the atoms are assumed to have the same dipole moment, both in magnitude and orientation given a polarization direction. The angle θ is the angle between the magnetization axis and the relative position between two dipols. We can see from Eq. (2) that in contrast to contact interaction, the dipolar interaction is anisotropic, due to the term $\cos^2\theta$, and long-range, due to the $1/|\vec{r}-\vec{r}\,'|^3$ dependence.

In this work, we consider a bilayer system confined in the following double well potential:

$$V_{ext}(x,y,z) = \frac{1}{2}m\left(\omega_\perp^2\,(y^2+z^2) + \omega_x^2\,min[(x+x_c)^2,(x-x_c)^2]\right)\,. \tag{3}$$

The shape of each single dBEC depends on the trapping frequencies $\omega_\perp$ and ω_x. The distance between the two dBECs is given by $L = 2x_c$. In the numerical calculation we have used the following parameters: $2x_c = 3\mu$m, $m = 52$ uma (^{52}Cr atoms) , total number of particles $N = 5000$ (2500 bosons in each dBEC), and trapping frequencies $\omega_x = 1500\times 2\pi$ Hz $\gg \omega_\perp = 50\times 2\pi$ Hz. Thus, each dBEC has a pancake-shaped geometry (Fig. 1). The oscillator length in the x direction that gives a longitudinal length scale is: $l_x = \sqrt{\hbar/m\omega_x^2} = 23.21$nm. The scattering length choosed for the particles is $a = 0.001\,a_B$, where a_B is the Bohr radius.

The two condensates are non-overlapping, that is, the barrier height is large enough to avoid the tunneling of particles. But they must be close enough in order to analyze the long-range properties of the interaction since it decays as $1/|\vec{r}-\vec{r}'|^3$. Finally, the dipole moments are aligned in the x-direction, parallel to the symmetry axis of the pancakes. This corresponds to the attractive configuration (the repulsive configuration would imply the dipoles to be oriented in the plane of the pancake shaped condensate). The purpose of this choice is to ensure, first, the stability of the condensate [4] (cigar-shaped geometry brings earlier to the collapse than the pancake-shaped), and second, other dipole moment orientations would decrease dipolar effects.

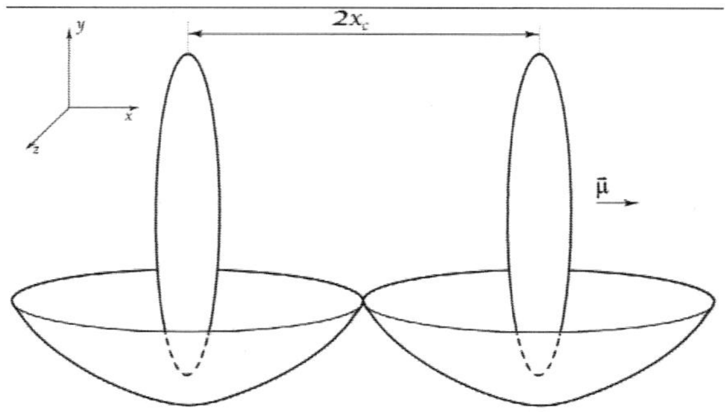

Figure 1. Illustration of the double-well configuration. The dipole moments are oriented in the x-direction, perpendicular to the pancake-shaped dBECs.

3. The coupled-pendulum model

Before presenting the numerical simulations, a qualitative description of the propagation of a collective mode between two non-overlapping dBECs can be found in the frame of classical mechanics [9]. We consider the classical analogy of two-coupled pendulums, where the natural oscillation frequency of each pendulum (ω_0) corresponds to the frequency of the excited mode, and the coupling string α corresponds to the dipolar coupling. The two natural modes of oscillation are:

$$\omega_{in} = \omega_0 \tag{4}$$

$$\omega_{out} = \omega_0\sqrt{1+2\alpha/\omega_0^2}\,, \tag{5}$$

which are the *in-phase mode* and the *out-of-phase mode*, respectively. In the dipolar drag effect when the center-of-mass mode is excited, ω_{in} is the frequency of the trap in the direction of the initial displacement (Kohn's theorem).

The parameter α, depends on the dipole moment $\vec{\mu}_m$, the distance between the two condensates $2\,x_c$ and their density gradient. Whereas the first two parameters are easily tunable, the latter is more difficult to control (in spite of the fact that it can be changed by varying all the parameters). Therefore, these two parameters and the frequency of the excited mode will detune the shift of the *out-of-phase mode* frequency ω_{out} [9].

The previous result can be generalized to n coupled pendulums. The eigenvalues of the problem are:

$$\omega_{ev} = \omega_0\sqrt{1+(2+\gamma)\alpha/\omega_0^2}\,. \tag{6}$$

Journal of Physics: Conference Series **497** (2014) 012035 doi:10.1088/1742-6596/497/1/012035

And the parameter γ can be found by solving the following equation [12]:

$$(1+\gamma)^2\xi(\gamma,n-2)-2(1+\gamma)\xi(\gamma,n-3)+\xi(\gamma,n-4)=0\,, \tag{7}$$

where

$$\xi(\gamma,n)=(-1)^n\,\frac{\sinh((n+1)\Omega(\gamma))}{\sinh\Omega(\gamma)}\,,\ \text{with}\ \ \Omega(\gamma)=\operatorname{arccosh}(-\frac{\gamma}{2})\,. \tag{8}$$

4. Numerical results

A nice long range effect that can be realized in a bilayer configuration is the dipolar drag. It is well known, that when the center of mass mode (dipole mode) is excited in a single BEC confined in a harmonic trap (by displacing the condensate out of its equilibrium position and releasing it, or giving some velocity to the center of mass of the condensate), the condensate oscillates harmonically with the same frequency as the characteristic trap frequency in the direction of the initial perturbation. This is still true in presence of long-range interactions that is for a dipolar Bose-Einstein condensate in a harmonic potential. However, when the dipole mode is excited in one dBEC of a bilayer system, although there is no tunneling between both condensates, the center of mass of the other dBEC starts to oscillate, as a result of the long-range character of the dipolar interaction, and both the *in-phase* and the *out-of-phase* dipole modes of the bilayer system are excited. Whereas the *in-phase mode* (center of mass) is independent of the dipolar interaction, the frequency of the *out-of-phase* mode depends on it. Dipolar drag is one of the so called propagation of collective modes, produced by the long-range nature of the dipolar interaction. It has been already studied in electron gases in uniform bilayer systems, under the name of Coulomb drag [13, 14].

4.1. Dipolar drag effect

In order to observe the dipolar drag effect one has to take into account the following items:

- The perturbation that produces the dipolar excitation must be in a transversal direction of the pancake-shaped dBEC, because the confinement in the longitudinal direction (ω_x) is so tight that leads to a small frequency shift, see Eq. (5). Here we excite the center of mass mode of the left condensate along y-direction.
- The dipole orientation shall be in the x-direction (attractive configuration between the two layers), otherwise, the dipolar interaction favors the interaction between the atoms of the same condensate rather than the coupling with the other condensate.

Taking into account the above considerations we have studied the dipolar drag effect by solving the TDGPE. The numerical procedure has three steps:

(i) By using the imaginary time step method, we find the stationary solution to reach the ground state of the system, which has the form $\Psi(\vec{r},t)=\Psi_{GS}(\vec{r})\,exp(-i\epsilon t/\hbar)$, where ϵ is the chemical potential.

(ii) Then, we add a momentum to the wavefunction of the left condensate displacing it out of its ground state, with the transformation (in the region $x<0$):

$$\Psi(\vec{r},0)\rightarrow\Psi_{GS}(\vec{r})\,e^{i\lambda y}\,, \tag{9}$$

where λ quantifies the magnitude of the perturbation. We remind that the kick must be in a transversal direction of the pancake, where the BEC is less confined.

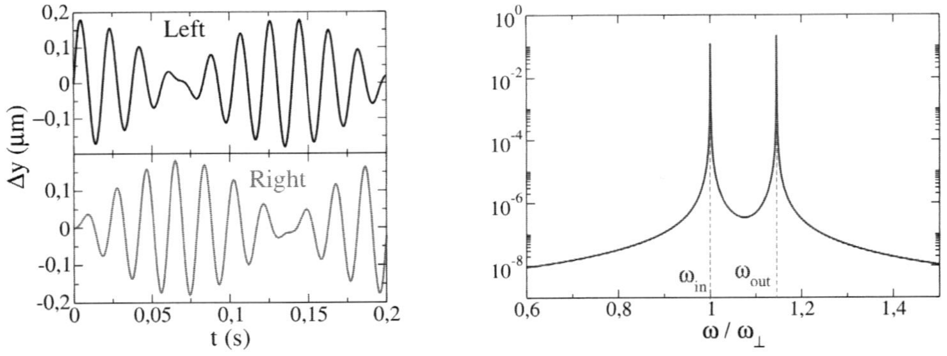

Figure 2. Left: Diplacement of the center of mass of the left (top) and right (bottom) condensate with the value of $d/\mu_B = 10$, where μ_B is the Bohr magneton. Right: Fourier transform (logarithmic scale) of the displacement of the center of mass of both condensates. The first peak corresponds to $\omega_{in} = \omega_\perp = 50\,\mathrm{Hz} \times 2\pi = 314.16\,\mathrm{Hz}$, whereas the second one is $\omega_{out} = 1.14\,\omega_\perp = 358.14\,\mathrm{Hz}$. The strength is plotted in arbitrary units.

(iii) The excited new wavefunction is evolved in time by solving Eq.(1) using Runge-Kutta and Hamming's method (predictor-modifier-corrector) [15].

In Fig. 2 we show the displacements of the centers of mass of both condensates and the corresponding Fourier analysis [16]. One can see that the right dBEC starts to oscillate due to the dipolar coupling. The beating of the oscillations comes from the presence of the two normal modes of the coupled system. The frequencies are obtained by Fourier analysis, the lower frequency corresponds to the *in-phase mode* (4) and the second peak is the *out-of-phase mode* which is slightly shifted from the excitation frequency $\omega_\perp$ (5).

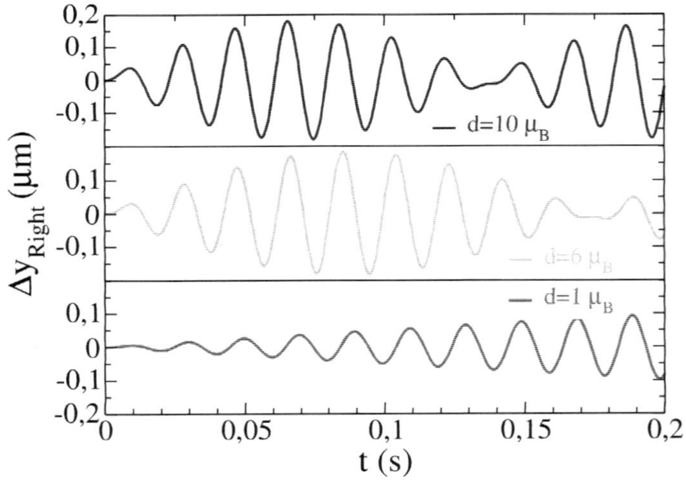

Figure 3. Displacement of the center of mass of the condensate of the right for different dipole moments. From top to bottom, $d = 10\,\mu_B$, $d = 6\,\mu_B$ and $d = 1\,\mu_B$.

We can analyse now the dependence of the frequency shift on the dipole moment and the distance between the condensates. On the one hand, in Fig. 3, we show the displacement of the center of mass of the right dBEC for different values of d: the frequency shift is larger and the time response is smaller when the dipole moment increases. We have used the values: $d/\mu_B = 1$; 6 and 10. On the other hand, in Fig. 4, we plot the frequency shift for several values of L, showing the decreasing of the shift when the dBECs get away. When they approach, a change of the potential behaviour occurs because they start to see the structure of dipols of the pancake-shaped condensates.

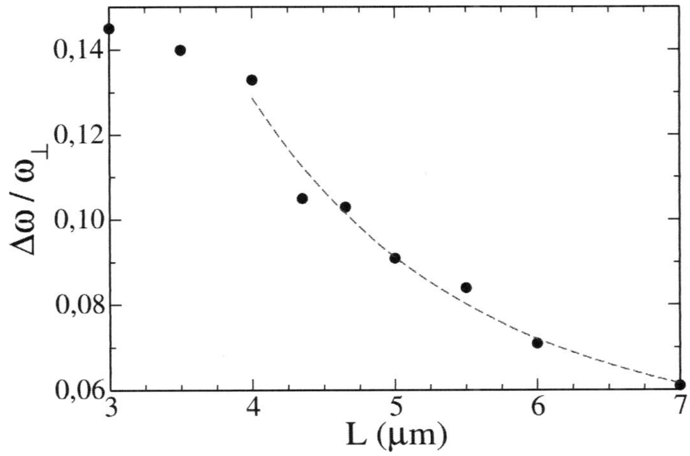

Figure 4. Frequency shift of the center of mass mode as a function of the relative distance of the bilayer system. The line shows the potential decay at large distances.

4.2. Other excitations

As an extension, we have studied the propagation of other collective excitations in the same system (for $d = 10\mu_B$), in particular, the monopolar excitation (breathing mode) and quadrupolar ones ($2x^2 - y^2 - z^2$, yz, xy).

In order to get a good approximation of the frequency of a collective mode for a single dBEC, one can use the following equations obtained for contact interaction [17, 18]:

$$\omega^2(m=0) = \omega_\perp^2 \left(2 + \frac{3}{2}\beta^2 \mp \frac{1}{2}\sqrt{9\beta^4 - 16\beta^2 + 16}\right) \tag{10}$$

$$\omega^2(m=\pm l) = l\,\omega_\perp^2 \tag{11}$$

$$\omega^2(m=\pm(l-1)) = (l-1)\,\omega_\perp^2 + \omega_x^2\,, \tag{12}$$

where $\beta = \omega_x/\omega_\perp$ is the anisotropy of the BEC (in our case $\beta = 30$), $l = 1$ for center-of-mass modes, and $l = 2$ for quadrupolar ones. The case $m = 0$ refers the monopolar mode, which has to be dealed separatedly. Since we consider a pancake-shaped condensate with large β, we will see that not all these frequencies will be excited in our simulations neither even all these modes are totally decoupled.

4.2.1. Monopolar excitations. As regards the monopolar excitation, when $\beta \gg 1$, from (10) there are two frequencies with $m = 0$: $\omega_1 = \sqrt{10/3}\,\omega_\perp \approx 1.83\,\omega_\perp$ and $\omega_2 = \sqrt{3}\,\omega_x$.

Although these results have been obtained for contact interaction, in presence of dipolar interaction, the values are slightly shifted [4, 11]. To quantify this shift, we have studied the monopolar response of a single dBEC with $\beta = 30$, by solving the TDGPE and we have found $\omega_1 = 1.9\,\omega_0$.

Now we study the propagation of the monopolar excitation between two non-overlapping dBECs. The perturbation corresponding to the monopolar excitation (breathing mode) is the following one (only applied on the left condensate):

$$\Psi(\vec{r}, 0) \rightarrow \Psi_{GS}(\vec{r})\; e^{i\lambda(x^2+y^2+z^2)} \,. \tag{13}$$

Evolving this wavefunction in time by solving the TDGPE, we find similar results to the dipolar drag case, but with different frequencies (Fig. 5). In other words, only the low-lying energy frequency (ω_1) and its corresponding shifted frequency are excited. In comparison with the dipolar drag effect, the magnitude of the shift in both cases is similar. Therefore, we see one frequency at $\omega = 1.9\,\omega_\perp$ and its corresponding shifted frequency, showing that the coupled-pendulum model can be extended also for monopolar modes.

4.2.2. Quadrupolar excitations. The frequencies corresponding to quadrupolar excitations in a pure contact interacting BEC are (11,12): $\omega_1 = \sqrt{2}\,\omega_\perp$ and $\omega_2 = \omega_x$. Three kind of quadrupolar excitations are implemented in the perturbation of the wavefunction, $2\,x^2 - y^2 - z^2$ (also known as quadrupolar), xy and yz respectively:

$$\Psi(\vec{r}, 0) \rightarrow \Psi_{GS}(\vec{r})\; e^{i\lambda(2\,x^2-y^2-z^2)} \tag{14}$$
$$\Psi(\vec{r}, 0) \rightarrow \Psi_{GS}(\vec{r})\; e^{i\lambda\,x\,y} \tag{15}$$
$$\Psi(\vec{r}, 0) \rightarrow \Psi_{GS}(\vec{r})\; e^{i\lambda\,y\,z} \,. \tag{16}$$

Looking at the results of the time evolution (Fig. 5), those of the $2\,x^2 - y^2 - z^2$ excitation leads to a surprise, owing to the fact that it is shockingly similar to the results obtained for monopolar excitations. The reason for this similarity is that in a pancake geometry, with large β, the quadrupolar excitation degenerates to the monopolar one because the x-component of the perturbation gets almost inhibited due to the strong confinement. The frequencies obtained are the same for both cases.

Monopolar	$x^2 + y^2 + z^2$	$\omega_{in} = 1.89\,\omega_\perp$	$\omega_{out} = 2.11\,\omega_\perp$
Quadrupolar	$-2x^2 + y^2 + z^2$	$\omega_{in} = 1.89\,\omega_\perp$	$\omega_{out} = 2.11\,\omega_\perp$
	xy	$\omega_{in} \approx \omega_x$	$\omega_{out} = \omega_{in}$
	yz	$\omega_{in} = 1.45\,\omega_\perp$	$\omega_{out} = 1.58\,\omega_\perp$

Table 1. Table that presents for the monopolar and each quadrupolar excitation the shape of the excitation and the value of the *in-phase* and the *out-of-phase* frequency found out in the Fourier transform.

In contrast, it is not the case for the xy and yz excitation, as can be seen in Fig. 5. The former excites only the $\omega_2 = \omega_x$ frequency, whereas the latter excites $\omega_1 = 1.45\,\omega_\perp$, and a second shifted frequency. The responsible of these differences is the kind of excitation. The xy excitation only has one term, which depends on x. It means that the relevant frequency in the trap will be ω_x. However, since the dBEC is tightly confined in the x-direction, the corresponding shift due to the dipolar coupling cannot be resolved.

Journal of Physics: Conference Series **497** (2014) 012035 doi:10.1088/1742-6596/497/1/012035

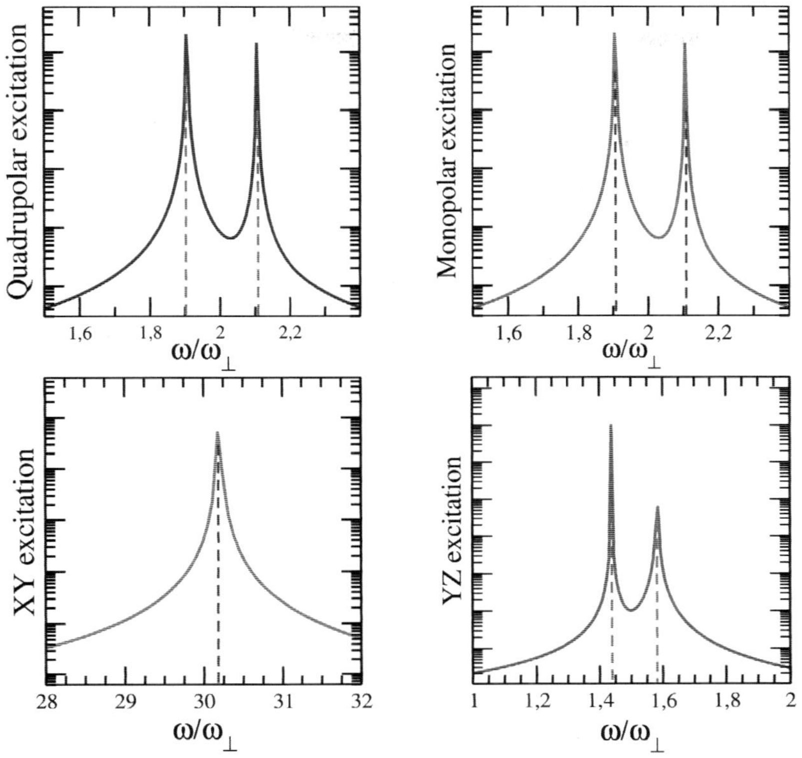

Figure 5. Fourier transform of the monopolar and quadrupolar responses ($2\,x^2 - y^2 - z^2$, xy and yz). The strength is in logarithmic scale and arbitrary units.

As regards the yz excitation, since the perturbation does not affect the x-direction, the frequency excited will be that of lower energy. In the numerical results we obtain one peak in the spectrum at $\omega_1 = 1.45\,\omega_\perp$, and, in addition, a second frequency due to the dipolar interaction will appear in this case because the frequency of the trap in the transversal direction is small.

By symmetry and geometry, xz excitation is analogous to the xy excitation, and $y^2 - z^2$ will be also analogous to the yz excitation. In a pancake geometry, one can recover one or the other simply rotating the coordinate system.

5. Dipolar drag effect in a triple-well configuration

Let us extend the dipolar drag effect to the case, of a triple-well trap. Some studies have been carried out in triple-well configurations related with dipolar effects [19], phase diagrams [6] or symmetry breaking [7]. In the classical analogy of three coupled-pendulums (6-8), the eigenvalues of the problem are: $\omega_{in} = \omega_\perp$, $\omega_{out,1} = \omega_\perp\sqrt{1+\alpha/\omega_\perp^2}$ and $\omega_{out,2} = \omega_\perp\sqrt{1+3\alpha/\omega_\perp^2}$.

In Fig. 6, we show the displacements of the centers of mass when only the left condensate is perturbed. All three frequencies appear in the movement, and, as a consequence, the displacements of the centers of mass have a more complex behaviour.

Although there are three eigenmodes, it is possible to excite only two frequencies by choosing the appropiate initial conditions. If we perturb only the condensate of the center, looking to the eigenvectors of the problem, only the two frequencies that preserve the symmetry of the

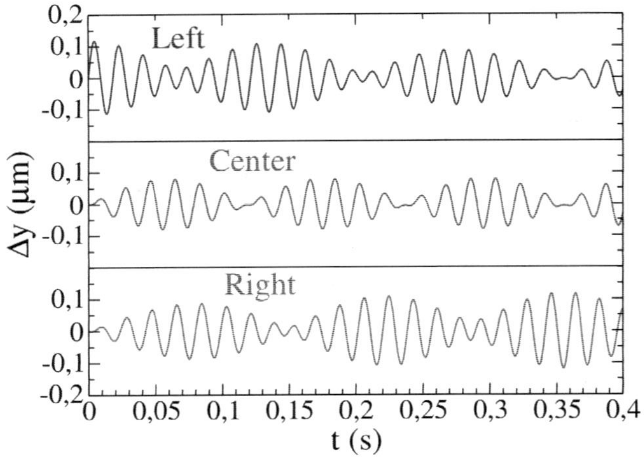

Figure 6. Displacements of the centers of mass of the three condensates in a triple-well configuration when we excite only the left condensate. From top to bottom, condensate of the left, center and right.

system, ω_{in} and $\omega_{out,2}$, will appear in the movement of the center of mass (the other frequency corresponds to a non-symmetric mode).

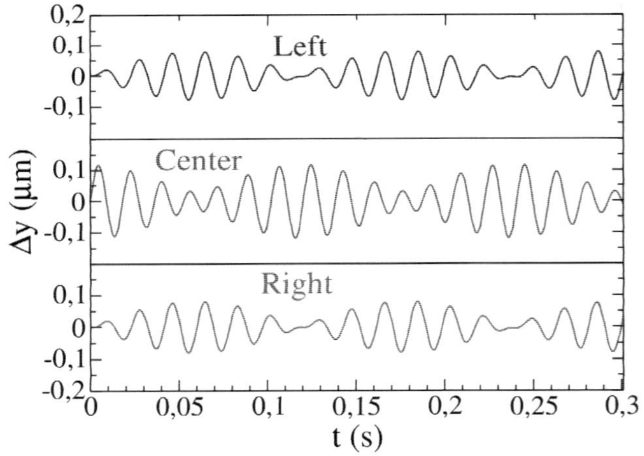

Figure 7. Displacements of the centers of mass of the three different condensates in the triple-well configuration, when we perturb only the center condensate. From top to bottom, condensate of the left, center and right.

In Fig. 7, we present the numerical results obtained by solving the TDGPE exciting only the center of mass mode of the centered dBEC. As expected, only two frequencies are involved in the movement of the three non-overlapping condensates. We have obtained the corresponding frequencies by performing a Fourier analysis and we have compared them with the case of only two dBECs:

$$\frac{(\omega_{out,2}^2 - \omega_\perp^2)_{3well}}{(\omega_{out}^2 - \omega_\perp^2)_{2well}} \approx 1.6\,. \qquad (17)$$

It is very close to the value of the ratio 1.5 obtained in the classical analogy. This means that the classical coupled-pendulum model yields a qualitative description of the dipolar coupling between few non-overlapping dBECs. Another remarkable fact is that, indeed, the magnitude of the shift is increased for three condensates compared with the case of two.

6. Summary and conclusions

In the present work, the full TDGPE has been numerically solved to study the long-range character of the dipolar interaction. We have investigated the propagation of collective modes, such as dipolar and quadrupolar excitations, between non-overlapping pancake-shaped dBECs. A classical coupled-pendulum model has been proposed.

We have analyzed the effect of different parameters in the propagation of the excitations. The dipole moment enhances the dipolar effects, whereas the enlargement of the scattering length and the distance between condensates leads the system to be dominated by the contact interaction which hides the desired long-range effects.

7. Acknowledgments

We acknowledge financial support from DGI (Spain) Grant No. FIS2011-28617-C02-01 and from Generalitat de Catalunya Grant No. SGR2009-1289.

References

[1] Griesmaier A, Werner J, Hensler S, Stuhler J and Pfau T 2005 *Phys. Rev. Lett.* **94** 160401
[2] Lu M, Burdick N Q , Youn S H and Lev B L 2011 *Phys. Rev. Lett.* **107** 190401
[3] Aikawa K, Frisch A, Mark M, Baier S, Rietzler A, Grimm R and Ferlaino F 2012 *Phys. Rev. Lett.* **108** 210401
[4] Lahaye T, Menotti C, Santos L, Lewenstein M and Pfau T 2009 *Rep. Prog. Phys.* **72** 126401
[5] Abad M, Guilleumas M, Mayol R, Pi M and Jezek D M 2010 *Phys. Rev. A.* **81** 043619
[6] Lahaye T, Pfau T and Santos L 2010 Phys. Rev. Lett. **104** 170404
[7] Yan P, Wang Y, Ji S and Liu X 2012 *Phys. Lett. A.* **376** 3141
[8] Li Y, Liu J, Pang W and Malomed B A 2013 *Phys. Rev. A.* **87** 013604
[9] Matveeva N, Recati A and Stringari S 2011 *Eur. Phys. J. D.* **65** 219
[10] Huang C and Wu W 2010 *Phys. Rev. A.* **82** 053612
[11] Bismut G, Pasquiou B, Maréchal E, Pedri P, Vernac L, Gorceix O and Laburthe-Tolra B 2010 *Phys. Rev. Lett.* **105** 040404
[12] Hu G Y and O'Connell R F 1996 *J. Phys. A: Math.* **29** 1511
[13] Jauho A P and Smith H 1993 Phys. Rev. B. **47** 4420
[14] Rojo A G 1999 *J. Phys.: Condens. Matter.* **11** R31
[15] Abad M, Guilleumas M, Mayol R, Pi M and Jezek D M 2011 *Phys. Rev. A.* **84** 035601
[16] Calvayrac F, Reinhard P G, Suraud E 1997 *Ann. Phys.* **255** 125
[17] Dalfovo F, Giorgini S, Pitaevskii L and Stringari S 1999 *Rev. Mod. Phys.* **71** 463
[18] Stringari S 1996 *Phys. Rev. Lett.* **77** 2360
[19] Peter D, Pawlowski K, Pfau T and Rzazewski K 2012 *J. Phys. B: At. Mol. Opt. Phys.* **45** 225302

Journal of Physics: Conference Series **497** (2014) 012036 doi:10.1088/1742-6596/497/1/012036

Numerical studies on vortices in rotating dipolar Bose-Einstein condensates

R. Kishor Kumar and P. Muruganandam

School of Physics, Bharathidasan University, Palkalaiperur Campus, Tiruchirappalli 620024, Tamilnadu, India

E-mail: kishor@cnld.bdu.ac.in, anand@cnld.bdu.ac.in

Abstract.
We study the formation and dynamics of vortices in quasi two-dimensional dipolar Bose-Einstein condensates (BECs). We solve the time dependent two-dimensional Gross-Pitaevskii equation by using a combined split-step Crank-Nicolson (SSCN) and Fast Fourier-Transform (FFT) based numerical scheme and investigate the vortex dynamics in dipolar BECs trapped with harmonic and optical lattice potentials. The consequence of dipole-dipole interaction on vortex nucleation and number of vortices has been analysed. We observe that the breaking of symmetry due to anisotropic dipolar interaction enormously increase the speed of creation of vortices.

1. Introduction

Bose-Einstein condensates provide an adaptable testing environment for superfluidity. An important property of a superfluid is the creation of quantized vortices. The concept of quantized vortices in superfluid was first developed theoretically by Onsager and Feynman for liquid helium [1, 2]. Their view on vortices received a strong support from the experiments of Hall and Vinen in superfluid helium II [3]. The first experiment directly aimed to detect the quantization of circulation is performed by Vinen in 1961 [4]. One of the most significant phenomena associated with BECs is the creation of vortices under rotation. The successful experimental realization of Bose-Einstein condensates of alkali atoms during last century has provided a pathway to observe the quantized vortices in a controllable quantum system [5, 6, 7, 8, 9, 10]. In BEC experiments vortices are nucleated with help of either by laser stirring or by rotating magnetic traps, and they are observed above certain critical rotational velocity. Experimentally a single vortex has been created in two-component BECs using phase engineering technique [5]. Multiply charged vortices has been studied and critical velocity also was calculated in stirred BECs of ^{87}Rb atoms confined in an axis symmetric magnetic trap [6]. Later, formation of highly ordered triangular vortex lattices contained over more than 100 number of vortices has been observed in Na atoms [7]. Also, BECs in optical lattice under rotation have been shown to exhibit various interesting properties such as vortex nucleation, vortex structures, vortex structural phase transition, pinning effect of vortices with the peak of shallow OL [11, 12, 13].

Several numerical schemes have been proposed to study the properties of BECs including vortices [14]. For instance, finite element method has been used to study stationary state of vortices and to solve Hartree-Fock-Bogoliubov treatment of a vortex lattice in rapidly rotating BECs [15, 16]. Vortices in anisotropic traps has investigated by discrete variable approximation [17] and BEC vortices in OL has studied using two-parameter continuation algorithm with spectral collocation methods [18]. The numerical schemes used in earlier studies are mainly focussed on the alkali BECs with with local and isotropic interaction. The alkali metal atoms used in early BEC experiments have negligible dipole moment. However, most bosonic atoms and molecules have large dipole moments and BECs of ^{52}Cr [19], ^{168}Er [20] and ^{164}Dy [21], with strong long-range dipolar interaction superposed, have been realized. The recent progress in the study of BECs of dipolar atoms with non-local and anisotropic interaction has exposed various fascinating physics due to the peculiar competition between short-range contact interaction and long-range dipolar interaction [22]. Due to the presence of anisotropic interaction in dipolar BECs one can expect the distinct feature during quantized circulation. Studies based on mean-field models revealed several interesting properties of dipolar BECs. For example, it has been shown that the anisotropic nature of dipolar BEC strongly influence the number, structure and stability of the vortices. The critical rotation frequency for vortex nucleation found to decrease as the strength of dipolar interaction increases [23, 24]. It has also been noted that the dipolar interaction increases the number of vortices while the contact interaction enhances the vortex stability [24].

In this paper, we present a combined split-step Crank-Nicolson and Fast Fourier-Transform (FFT) based numerical scheme to study the vortex dynamics in dipolar BECs trapped with harmonic and optical lattice potentials. In section 2 we introduce the theoretical model for dipolar Gross-Pitaevskii equation in rotating frame with different trapping potentials. We provide the numerical method for solving two-dimensional time dependent dipolar Gross-Pitaevskii equation with angular momentum in section 3. Numerical results and conclusions are given in sections 4 and 5, respectively.

2. Theoretical model

The dynamics of rotating BECs can be studied using mean field Gross-Pitaevskii (GP) equation [10]. A dipolar BEC with N atoms, each of mass m at absolute zero temperature in a rotating frame can be described by the Gross-Pitaevskii equation as [25]

$$i\frac{\partial\phi(\mathbf{r},t)}{\partial t} = \left[-\frac{1}{2}\nabla^2 + V(\mathbf{r}) + 4\pi a N|\phi(\mathbf{r},t)|^2 - \Omega L_z + N\int U_{dd}(\mathbf{r}-\mathbf{r}')|\phi(\mathbf{r}',t)|^2 d^3r'\right]\phi(\mathbf{r},t), \tag{1}$$

where $V(\mathbf{r}) = V_{ho}(\mathbf{r}) + V_{OL}(\rho)$ is the confining axially symmetric harmonic potential and optical lattice potential, $\phi(\mathbf{r},t)$ the wave function at time t with normalization $\int|\phi(\mathbf{r},t)|^2 d\mathbf{r} = 1$, a the atomic scattering length. The axial and radial trap frequencies of the harmonic potential, $V_{ho}(\mathbf{r})$, are $\Omega_z\omega$ and $\Omega_\rho\omega$, respectively, and are related with the trap aspect ratio, $\lambda = \Omega_z/\Omega_\rho$. In equation (1) length is measured in units of harmonic oscillator length $l \equiv \sqrt{\hbar/m\omega}$, frequency in units of ω, time t in units of ω^{-1}. $L_z = -i(x\partial_y - y\partial_x)$ corresponds to the z-component of the angular momentum due to the rotation of the dipolar BEC about z axis with angular velocity Ω. Here Ω is expressed in units of the radial trap frequency $\Omega_\rho\omega$. The integral term in equation (1) accounts for the dipole-dipole interaction with

$$U_{dd}(\mathbf{x}) = a_{dd}\frac{1-3\cos^2\theta}{|\mathbf{x}|^3}, \tag{2}$$

where $\mathbf{x} = \mathbf{r} - \mathbf{r}'$ determines the relative position of dipoles and θ is the angle between $\mathbf{x}$ and the direction of polarization, z. The constant $a_{dd} = \mu_0\bar{\mu}^2 m/(12\pi\hbar^2)$ is a length characterizing the strength of dipolar interaction and, its experimental value for ^{52}Cr, ^{168}Er and ^{164}Dy are $16a_0$,

$66a_0$ and $131a_0$, respectively, where a_0 is the Bohr radius [26]. $\bar{\mu}$ corresponds to the magnetic dipole moment of a single atom and μ_0 the permeability of free space.

A two-dimensional optical lattice superimposed on a three-dimensional harmonic trap is given, in dimensionless form, as

$$V(\mathbf{r}) = \frac{1}{2}\Omega_\rho^2\rho^2 + \frac{1}{2}\Omega_z^2 z^2 + V_0\left[\sin^2(kx) + \sin^2(ky)\right], \tag{3}$$

where $\mathbf{r} \equiv (\vec{\rho}, z)$, with $\vec{\rho}$ the radial coordinate and z the axial coordinate, V_0 is the depth of the OL and k is the wave number. We consider the highly oblate dipolar BEC with trap aspect ratio $\lambda = 50$ for our present study. In strong axial traps the dynamics of the rotating dipolar BEC can effectively be studied in two-dimension by simply integrating out the z dependence in equation (1), as [27, 28, 29],

$$\begin{aligned} i\frac{\partial\phi_{2D}(\vec{\rho},t)}{\partial t} &= \left[-\frac{\nabla_\rho^2}{2} + V_{2D} - \Omega L_z + \frac{4\pi a N}{\sqrt{2\pi}d_z}|\phi_{2D}(\vec{\rho},t)|^2\right. \\ &\left. + \frac{4\pi a_{dd}N}{\sqrt{2\pi}d_z}\int\frac{d^2k_\rho}{(2\pi)^2}e^{i\mathbf{k}_\rho\cdot\vec{\rho}}\,\tilde{n}(\mathbf{k}_\rho)\times h_{2D}\left(\frac{k_\rho d_z}{\sqrt{2}}\right)\right]\phi_{2D}(\vec{\rho},t). \end{aligned} \tag{4}$$

where $V_{2D} = \left(x^2+y^2\right)/2 + V_0\left[\sin^2(kx)+\sin^2(ky)\right]$ is a combined two-dimensional harmonic trap and optical lattice potential. In equation (4), $\tilde{n}(\mathbf{k}_\rho) = \int \exp(i\mathbf{k}_\rho.\vec{\rho})|\phi_{2D}(\vec{\rho})|^2 d\vec{\rho}$, $k_\rho \equiv (k_x, k_y)$, $h_{2D}(\xi) = 2 - 3\sqrt{\pi}\xi\exp(\xi^2)\,\mathrm{erfc}(\xi)$, and the dipolar term is written in Fourier space. The lattice spacing (d_{lat}) and amplitude (V_0) of optical lattice potential can be varied by tuning the frequency and intensity of lasers. In the present study, the OL spacing is chosen as $d_{lat} = \lambda_L/2 \approx 534$nm, where $\lambda_L = 1064nm$ is the wavelength of the laser used in experiments [30]. The corresponding dimensionless parameters for OL spacing, $d_{lat} = \pi/k = 0.534$ and $k = 1.87\pi$.

3. Numerical Method

The Crank-Nicolson method is a finite difference method used for numerically solving the heat equation and similar partial differential equations. The method was developed by J. Crank and P. Nicolson in the 20th century [31]. This method is a reliable and stable routine for numerically solving time-dependent Schrödinger-like equations.

We consider a split-step Crank-Nicolson based numerical scheme [32, 33] for studying rotational dynamics of dipolar BECs using quasi-2D GP equation with angular momentum.

$$\begin{aligned} i\frac{\partial}{\partial t}\varphi(x,y;t) =& \left[-\frac{1}{2}\frac{\partial^2}{\partial\rho^2} + V_{2D} + \frac{4\pi a N}{\sqrt{2\pi}d_z}\varphi(x,y;t)|^2 - \Omega L_z\right. \\ &\left. + \frac{4\pi a_{dd}N}{\sqrt{2\pi}d_z}\int\frac{d^2k_\rho}{(2\pi)^2}\exp(i\mathbf{k}_\rho\cdot\vec{\rho})\tilde{n}(\mathbf{k}_\rho)h_{2D}(\frac{k_\rho d_z}{\sqrt{2}})\right]\varphi(x,y;t), \\ \equiv& H\varphi(x,y;t). \end{aligned} \tag{5}$$

The Hamiltonian H can be conveniently broken into three pieces $H = H_1 + H_2 + H_3$, where

$$H_1 = V_{2D} + \frac{4\pi a N}{\sqrt{2\pi}d_z}|\varphi(x,y;t)|^2 + \frac{4\pi a_{dd}N}{\sqrt{2\pi}d_z}\int\frac{d^2k_\rho}{(2\pi)^2}\exp(i\mathbf{k}_\rho\cdot\vec{\rho})\tilde{n}(\mathbf{k}_\rho)h_{2D}(\frac{k_\rho d_z}{\sqrt{2}}), \tag{6}$$

$$H_2 = -\frac{\partial^2}{\partial x^2} - i\Omega y\frac{\partial}{\partial x}, \tag{7}$$

$$H_3 = -\frac{\partial^2}{\partial y^2} + i\Omega x\frac{\partial}{\partial y}. \tag{8}$$

Essentially we split (5) into

$$i\frac{\partial}{\partial t}\varphi(x,y;t) = \Bigg[V_{2D} + \frac{4\pi aN}{\sqrt{2\pi}d_z}|\varphi(x,y;t)|^2 + \frac{4\pi a_{dd}N}{\sqrt{2\pi}d_z} \times \int \frac{d^2k_\rho}{(2\pi)^2}\exp(i\mathbf{k}_\rho\cdot\vec{\rho})\tilde{n}(\mathbf{k}_\rho)h_{2D}(\frac{k_\rho d_z}{\sqrt{2}})\Bigg]\varphi(x,y;t) \equiv H_1\varphi(x,y;t) \tag{9}$$

$$i\frac{\partial}{\partial t}\varphi(x,y;t) = -\frac{1}{2}\frac{\partial^2}{\partial x^2}\varphi(x,y;t) - i\Omega y\frac{\partial}{\partial x}\varphi(x,y;t) \equiv H_2\varphi(x,y;t) \tag{10}$$

$$i\frac{\partial}{\partial t}\varphi(x,y;t) = -\frac{\partial^2}{\partial y^2}\varphi(x,y;t) + i\Omega x\frac{\partial}{\partial y}\varphi(x,y;t) \equiv H_3\varphi(x,y;t) \tag{11}$$

In equation (6) the dipolar term is directly written in momentum space and the FFT algorithm is carried out in Cartesian coordinates.

We first solve equation (9) with a initial value $\varphi(x,y;t_0)$ at $t = t_0$ to obtain an intermediate solution at $t = t_0 + \Delta$, where Δ is the time step. Then this intermediate solution is used as initial value to solve (10) and (11) yielding the final solution at $t = t_0 + \Delta$ as $\varphi(x,y;t_0+\Delta)$. This procedure is repeated n times to get the final solution at a given time $t_{\text{final}} = t_0 + n\Delta$. The time variable is discretized as $t_n = n\Delta$.

The solution, say φ^n, is advanced first over the time step Δ at time t_n by solving the GP equation (9) with $H = H_1$ to produce an intermediate solution $\varphi^{n+1/2}$. Here φ^n is the discretized wave function at time t_n. As there is no derivative in H_1, this propagation is performed essentially exactly for small Δ through the operation

$$\varphi^{n+1/2} = \bigcirc_{\text{nd}}(H_1)\varphi^n \equiv e^{-i\Delta H_1}\varphi^n, \tag{12}$$

where $\bigcirc_{\text{nd}}(H_1)$ denotes time-evolution operation with H_1 and the suffix 'nd' denotes non-derivative.

Next we perform the time propagation corresponding to the operators H_2 and H_3 numerically by the semi-implicit Crank-Nicolson scheme [34]: To employ the Crank-Nicolson scheme the GP equation (5) is mapped onto two-dimensional spatial grid points in x and y. The discretization along the spatial grid point x is carried out as follows. Equation (5) is discretized with $H = H_2$ of (7) by the following Crank-Nicolson scheme as [34, 35, 36]:

$$\frac{i(\varphi_{i,j}^{n+1} - \varphi_{i,j}^{n+1/2})}{\Delta} = -\frac{1}{2h_x^2}\Big[(\varphi_{i+1,j}^{n+1} - 2\varphi_{i,j}^{n+1} + \varphi_{i-1,j}^{n+1}) + (\varphi_{i+1,j}^{n+1/2} - 2\varphi_{i,j}^{n+1/2} + \varphi_{i-1,j}^{n+1/2})\Big] - \frac{i\Omega y}{4h_x}\Big[(\varphi_{i+1,j}^{n+1} - \varphi_{i-1,j}^{n+1}) + (\varphi_{i+1,j}^{n+1/2} - \varphi_{i-1,j}^{n+1/2})\Big], \tag{13}$$

where for the spherically-symmetric $\varphi_i^n = \varphi(x_i;t_n)$ refers to $x \equiv x_i = ih_x$, $i = 0,1,2,...,N_x$ and h_x is the space step. For the discretization along x, we choose $x \equiv x_i = -N_x h_x/2 + ih_x$, $i = 0,1,2,...,N_x$.

The above procedure results in a series of tridiagonal sets of equations (13) in $\varphi_{i+1,j}^{n+1}$, $\varphi_{i,j}^{n+1}$, and $\varphi_{i-1,j}^{n+1}$ at time t_{n+1}, which are solved using the proper boundary conditions. The tridiagonal equations emerging from (13) are written explicitly as [34]

$$A_i^-\varphi_{i-1,j}^{n+1} + A_i^0\varphi_{i,j}^{n+1} + A_i^+\varphi_{i+1,j}^{n+1} = b_i, \tag{14}$$

where

$$b_i = \frac{i\Delta}{2h_x^2}(\varphi_{i+1,j}^{n+1/2} - 2\varphi_{i,j}^{n+1/2} + \varphi_{i-1,j}^{n+1/2}) - \frac{\Delta\Omega y_i}{4h_x}(\varphi_{i+1,j}^{n+1/2} - \varphi_{i-1,j}^{n+1/2}) + \varphi_{i,j}^{n+1/2}, \tag{15}$$

Journal of Physics: Conference Series **497** (2014) 012036 doi:10.1088/1742-6596/497/1/012036

and $A_i^0 = 1 + i\Delta/h_x^2$, $A_i^- = -\Delta/(2h_x)(i/h_x + \Omega y_i/2)$, $A_i^+ = \Delta/(2h_x)(-i/h_x + \Omega y_i/2)$ All quantities in b_i refer to time step $t_{n+1/2}$ and are considered known. The only unknowns in (14) are the wave forms $\varphi_{i\pm1,j}^{n+1}$ and $\varphi_{i,j}^{n+1}$ at time step t_{n+1}. To solve (14), we assume the one-term forward recursion relation

$$\varphi_{i+1,j}^{n+1} = \alpha_i \varphi_{i,j}^{n+1} + \beta_i, \tag{16}$$

where α_i and β_i are coefficients to be determined. Substituting (16) in (14) we obtain

$$A_i^- \varphi_{i-1,j}^{n+1} + A_i^0 \varphi_{i,j}^{n+1} + A_i^+(\alpha_i \varphi_{i,j}^{n+1} + \beta_i) = b_i, \tag{17}$$

which leads to the solution

$$\varphi_{i,j}^{n+1} = \gamma_i (A_i^- \varphi_{i-1,j}^{n+1} + A_i^+ \beta_i - b_i), \tag{18}$$

with

$$\gamma_i = -1/(A_i^0 + A_i^+ \alpha_i). \tag{19}$$

From (16) and (18) we obtain the following backward recursion relations for the coefficients α_i and β_i

$$\alpha_{i-1} = \gamma_i A_i^-, \quad \beta_{i-1} = \gamma_i (A_i^+ \beta_i - b_i). \tag{20}$$

We shall use the recursion relations (18), (19) and (20) in a backward sweep of the lattice to determine α_i and β_i for i running from $N_x - 2$ down to 0. The initial values chosen are $\alpha_{N_x-1} = 0, \beta_{N_x-1} = \varphi_{N_x}^{n+1}$. This ensures the correct value of φ at the last lattice point. After determining the coefficients α_i, β_i and γ_i, we can use the recursion relation (16) from $i = 0$ to $N_x - 1$ to determine the solution for the entire space range using the starting value $\varphi_0^{n+1}(= 0)$ known from the boundary conditions. The value at the last lattice point is also taken to be known ($= 0$). Thus we have determined the solution by using two sets of recursion relations across the lattice each involving about N_x operations. A similar discretization procedure is followed for H_3 also.

All the numerical simulations in this manuscript are carried out with $dx = dy = 0.2$ (space step) and $dt = 0.004$ (time step).

4. Vortices in dipolar BECs

In this section, we show the vortex nucleation in the dipolar BECs by numerical simulations of the time-dependent GP equation (4) using the numerical scheme discussed above. For this purpose, we fix the parameters at $N = 10\,000$ atoms and $\Omega = 0.5$.

4.1. Dynamics of a rotating dipolar BEC in a harmonic trap ($V_0 = 0$)

First, we consider the harmonically trapped dipolar BECs with different dipolar strengths. Initially we prepare the ground state wavefunction by solving (4) numerically using imaginary time propagation in the presence of harmonic trap but without angular momentum ($\Omega = 0$). The ground state is then evolved in real time propagation by including angular momentum ($\Omega \neq 0$). In addition, a phenomenological dissipation is included to facilitate the smooth vortex formation [37, 38, 39]. The dissipation is introduced by replacing 'i' with '$(i - \gamma)$' in the time dependent equation (4), where γ ($\sim 10^{-5}$) accounts for the strength of dissipation.

The condensate begins to rotate after applying the rotation. During the rotation the condensate elongates and quadrupole shape deformation takes place. The boundary surfaces

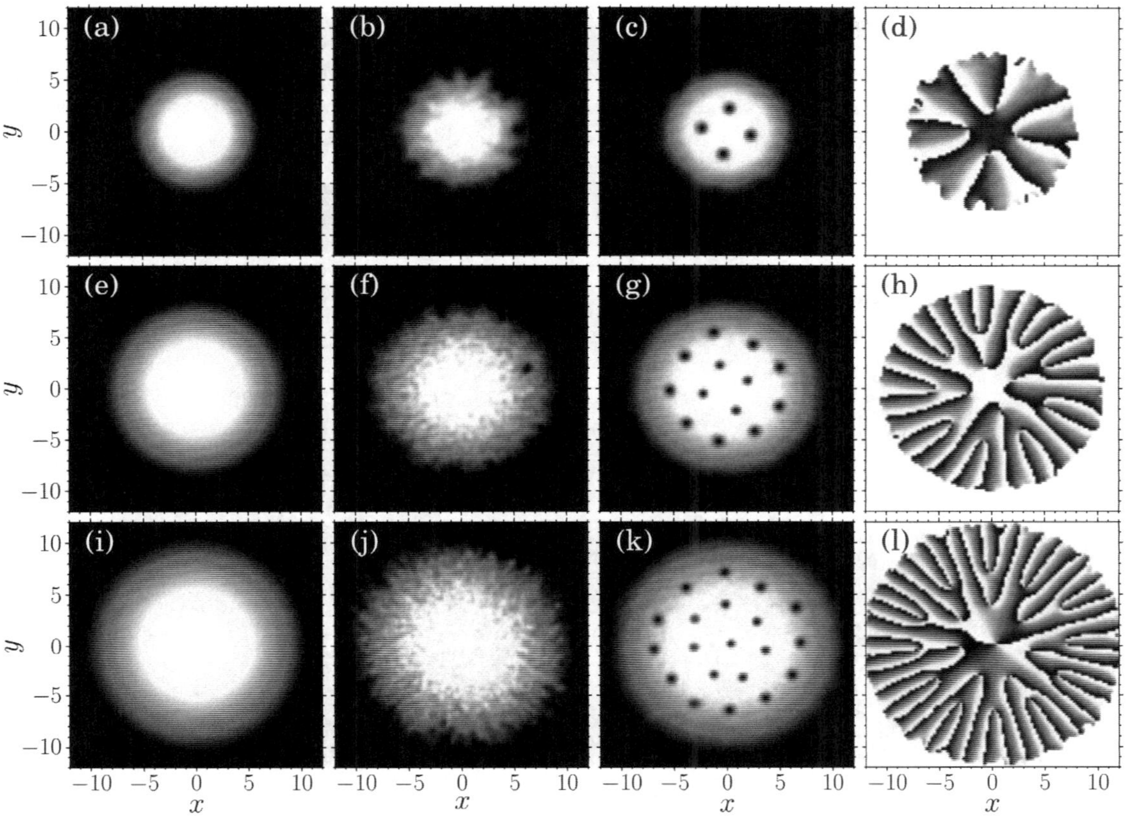

Figure 1. Contour plots of the densities, $|\phi_{2D}(\rho, t)|^2$, showing the development of vortices in a dipolar BEC $a = 0$, $N = 10\,000$ and $\lambda = 50$ after the trapping potential begins to rotate with $\Omega = 0.5$ for $a_{dd} = 16\,a_0$: (a) $t = 0$ (b) $t = 62\,000\,\omega^{-1}$, (c) $t = 160\,000\,\omega^{-1}$, for $a_{dd} = 66\,a_0$ (e) $t = 0$, (f) $t = 15\,000\,\omega^{-1}$ (g) $t = 70\,000\,\omega^{-1}$, and for $a_{dd} = 131\,a_0$ (i) $t = 0$, (j) $t = 14\,000\,\omega^{-1}$ (k) $t = 58\,000\,\omega^{-1}$. The phase distributions of condensate wave function $\phi_{2D}(\rho, t)$ are shown in (d), (h) and (l).

becomes unstable and ripples are formed and the surface ripples initiate the vortex nucleation. The nucleated vortices enter into the condensate and reach a stable configuration as time progress. In the following we compare the time taken for the creation of vortex or vortices and number of vortices corresponding for different dipolar strengths.

Figures 1(a)-(c) show the development of vortices for $a_{dd} = 16\,a_0$ (^{52}Cr BEC). In this case the surface ripples are formed at $t \sim 62\,000\omega^{-1}$. Then it attains an equilibrium state of 4 vortices at $t \sim 160\,000\omega^{-1}$ as shown in figure 1(c). The corresponding phase pattern of final wavefunction is shown in figure 1(d).

Figures 1(e)-(g) show the development of vortices in the case of $a_{dd} = 66\,a_0$ (^{168}Er BEC). Here the surface ripples are formed at time $t \sim 15\,000\omega^{-1}$, which is much faster when compared to that of ^{52}Cr BEC. Further, about 13 vortices form a stable configuration as shown in figure 1(g) and figure 1(h) shows the corresponding phase profile.

We have also shown the snapshots of vortices for the case of ^{164}Dy BEC ($a_{dd} = 131\,a_0$) in figure 1(i)-(k). The surface ripples are formed more rapidly, at $t \sim 14\,000\omega^{-1}$ as shown in figure 1(j), and a very stable pattern with 20 vortices, as seen in figure 1(k), are created for $t = 58\,000\,\omega^{-1}$. Figure 1(l) depicts the phase profile of the corresponding wave function.

Thus the above observation clearly indicates that the time taken for the creation of steady vortex pattern in dipolar BECs decreases considerably with the increase of dipolar strength. Also, number of vortices is significant with respect to dipolar strengths. Next we study the effect of optical lattice on the vortices in dipolar BEC.

4.2. Dynamics of a rotating dipolar BEC in an optical lattice

In order to see the effect of optical lattice, a square optical lattice is introduced ($V_0 \neq 0$) in addition to harmonic trapping potential. In experiments the alkali BEC is loaded into a static square optical lattice and then studied by applying rotation [11]. The presence of OL found to suppress the shape deformation and quadrupole oscillations in conventional BECs [40]. Optical lattice also helps to produce the vortices rapidly in BECs. In figure 2 we show the time development of the vortices from the contours of the density profiles $|\phi_{2D}|^2$ by fixing the OL depth at $V_0 = 10$. The dipole-dipole interaction normally ease the vortex nucleation due

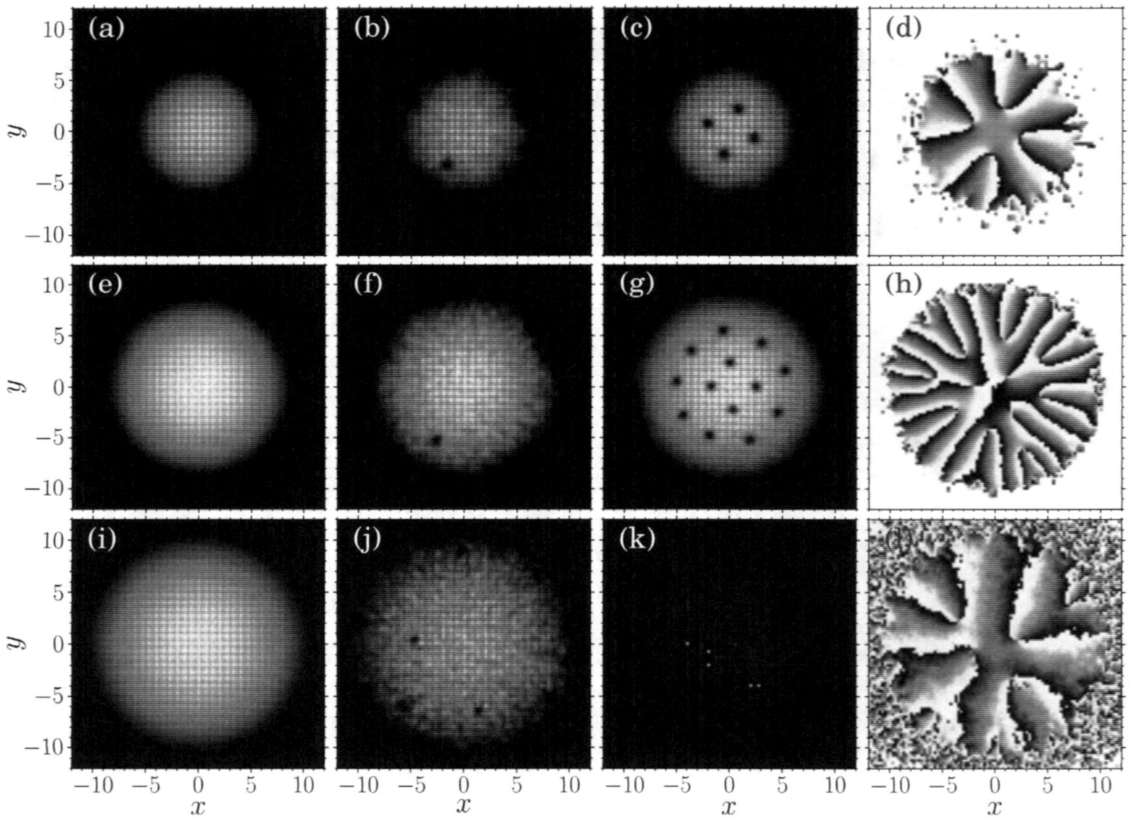

Figure 2. Contour plots of the densities, $|\phi_{2D}(\rho, t)|^2$, showing the development of vortices in a dipolar BEC loaded in OL with amplitude $V_0 = 10$, $a = 0$, $N = 10\,000$ and $\lambda = 50$ after the trapping potential begins to rotate with $\Omega = 0.5$ for $a_{dd} = 16\,a_0$: (a) $t = 0$ (b) $t = 40\,000\,\omega^{-1}$, (c) $t = 60\,000\,\omega^{-1}$, for $a_{dd} = 66\,a_0$ (e) $t = 0$, (f) $t = 8000\,\omega^{-1}$ (g) $t = 24\,000\,\omega^{-1}$ and for $a_{dd} = 131\,a_0$ (i) $t = 0$, (j) $t = 900\,\omega^{-1}$ (k) $t = 1\,000\,\omega^{-1}$. The phase distributions of condensate wave function $\phi_{2D}(\rho, t)$ are shown in (d) and (h).

Table 1. The approximate time (t_{vor}) for the creation of equilibrium vortices and the number of vortices (N_v) in the presence and absence of OL for different dipolar BECs.

BEC	t_{vor} (ω^{-1})		N_v	
	$V_0 = 0$	$V_0 = 10$	$V_0 = 0$	$V_0 = 10$
^{52}Cr	160 000	60 000	4	4
^{168}Er	70 000	24 000	13	13
^{164}Dy	58 000	—	20	—

to symmetry breaking. In addition, the presence of OL provides a supplementary symmetry breaking due to which the speed of creation of vortices has enhanced considerably. Figures 2(a)-(c) show the development of vortices for $a_{dd} = 16\,a_0$ (^{52}Cr BEC). In this case, the surface ripples are formed at $t \sim 40\,000\,\omega^{-1}$ and attain an equilibrium state of vortices pinned within the peaks of OL. About 4 vortices form a stable pattern as shown in figure 2(c). The corresponding phase pattern of final wavefunction is shown in figure 2(d). Figures 2(e)-(g) show the development of vortices in the case of $a_{dd} = 66\,a_0$ (^{168}Er BEC). Here the surface ripples are formed at time $t \sim 8000\,\omega^{-1}$, which is much faster when compared to that of ^{52}Cr BEC discussed above. Further, about 13 vortices form a stable pinned configuration as shown in figure 2(g) and figure 2(h) shows the corresponding phase profile. We have also shown the development of vortices for the case of ^{164}Dy atoms in figures 2(i)-(k), where the vortices are created very rapidly at $t \sim 900\,\omega^{-1}$. However, the strong dipolar interaction destabilized the condensate. Actually a pure dipolar BEC of with larger dipolar strength, when the contact interaction is made zero, collapses in OL during rotation.

The above observation clearly indicates that the time taken for the creation of steady vortex pattern in dipolar BECs in OL decreases considerably with the increase of dipolar strength. The numerical results are summarized and given in table 1.

5. Conclusion

We have studied numerically the influence of dipole-dipole interaction on the formation of vortices in a rotating dipolar Bose-Einstein condensate of ^{52}Cr, ^{168}Er and ^{164}Dy atoms in harmonic and optical lattice potential. We have presented a combined split-step Crank-Nicolson and FFT based numerical scheme to solve the time-dependent quasi-two dimensional Gross-Pitaevskii equation. Using the above scheme we have analysed the dynamics of the creation of vortices in dipolar BECs. The time taken for the creation of vortices in dipolar BECs has been calculated. The presence of dipolar interaction and optical lattice potential greatly enhance the fast creation of ordered vortices in a dipolar BEC. The number of vortices increases significantly with respect to dipolar strength.

Acknowledgments

The work of PM forms a part of Council of Scientific and Industrial Research (Ref. No. 03(1186)/10/EMR-II) Government of India funded research projects.

References

[1] Onsager L 1949 *Nuovo Cimento* **6** 279-287
[2] Feynman R P 1955 *Progress in Low Temperature Physics Vol. I* (North-Holland, Amsterdam)
[3] Hall H E and Vinen W F 1956 *Proc. R. Soc. Lond. A* **238** 215
[4] Vinen W F 1961 *Proc. R. Soc. Lond. A* **260** 218
[5] Matthews M R, Anderson B P, Haljan P C, Hall D S, Wieman C E and Cornell E A 1999 *Phys. Rev. Lett.* **83** 2498

[6] Madison K W, Chevy F, Wohlleben W and Dalibard J 2000 *Phys. Rev. Lett.* **84** 806
[7] Abo-Shaeer J R, Raman C, Vogels J M and Ketterle W 2001 *Science* **292** 476
[8] Pethick C and Smith H 2002 *Bose-Einstein condensation in dilute gases* (Cambridge: Cambridge University Press)
[9] Griffin A, Nikuni T and Zaremba E 2009 *Bose condensed gases at finite temperatures* (Cambridge: Cambridge University Press)
[10] Fetter A L 2009 *Rev. Mod. Phys.* **81** 647
[11] Tung s, Schweikhard V and Cornell E A 2006 *Phys. Rev. Lett.* **97** 240402
[12] Williams R A, Al-Assam S and Foot C J 2010 *Phys. Rev. Lett.* **104** 050404
[13] Reijnders J W and Duine R A 2004 *Phys. Rev. Lett.* **93** 060401
[14] Minguzzia A, Succib S, Toschib F, Tosia M P and Vignolo P 2004 *Phys. Rep.* **395** 223
[15] Aftalion A and Du Q 2001 *Phys. Rev. A* **64** 063603(R)
[16] Baksmaty L O, Liu Y, Landman U, Bigelow N P and Pu H 2009 *Math. Comput. Simulat.* **80** 131
[17] Feder D L, Clark C W and Schneider 1999 *Phys. Rev. A* **61** 011601(R)
[18] Jeng B W, Wang Y S and Chien C S 2013 *Comp. Phys. Commun.* **184** 493
[19] Griesmaier A, Stuhler J, Koch T, Fattori M, Pfau T and Giovanazzi S 2006 *Phys. Rev. Lett.* **97** 250402
Lahaye T, Menotti C, Santos L, Lewenstein M and Pfau T 2009 *Rep. Prog. Phys.* **72** 126401
[20] Aikawa K, Frisch A, Mark M, Baier S, Rietzler A, Grimm R and Ferlaino F 2012 *Phys. Rev. Lett.* **108** 210401
[21] Lu M, Burdick N Q, Seo Ho Youn and Lev B L 2011 *Phys. Rev. Lett.* **107** 190401
[22] Baranov M A 2008 *Phys. Rep.* **464** 71 ; Baranov M A, Dalmonte M, Pupillo G and Zoller P 2012 *Chem. Rev.* **112** 5012
[23] Malet F, Kristensen T, Reimann S M and Kavoulakis G M 2011 *Phys. Rev. A* **83** 033628
[24] Kumar R K and Muruganandam P 2012 *J. Phys. B: At. Mol. Phys.* **45** 215301
[25] Yi S and Pu H 2006 *Phys. Rev. A* **73** 061602(R)
[26] Youn S H, Lu M, Ray U and Lev B L 2010 *Phys. Rev. A* **82** 043425; Lu M, Burdick N Q and Lev B L 2012 *Phys. Rev. Lett.* **108** 215301
[27] Pedri P and Santos L 2005 *Phys. Rev. Lett.* **95** 200404
[28] Fischer U R 2006 *Phys. Rev. A* **73** 031602(R)
[29] Muruganandam P and Adhikari S K 2012 *Laser Phys.* **22** 813
[30] Müller S, Billy J, Henn E A L, Kadau H, Griesmaier A, Jona-Lasinio M, Santos L and Pfau T 2011 *Phys. Rev. A* **84** 053601
[31] Crank J and Nicolson P 1947 *Proc. Camb. Phil. Soc.* **43** 50
[32] Adhikari S K and Muruganandam P 2002 *J. Phys. B: At. Mol. Opt. Phys.* **35** 2831
[33] Muruganandam P and Adhikari S K 2009 *Comput. Phys. Commun.* **180** 1888
Vudragović D, Vidanović I, Balaz A, Muruganandam P and Adhikari S K 2012 *Comput. Phys. Commun.* **183** 2021
[34] Koonin S E and Meredith D C 1990 *Computational Physics: Fortran Version* (Addison-Wesley, Reading)
[35] Ames W F 1992 *Numerical Methods for Partial Differential Equations, 3rd Ed.* (Academic Press, New York)
[36] Dautray R and Lions J L 1993 *Mathematical Analysisand Numerical Methods for Science and Technology, Vol. 6, Chap. XX, Sec. 2* (SpringerVerlag, Berlin)
[37] Tsubota M, Kasamatsu K and Ueda M 2002 *Phys. Rev. A* **65** 023603
[38] Kasamatsu K, Machida M, Sasa N and Tsubota M 2005 *Phys. Rev. A* **71** 063616
[39] García-Ripoll J J and Pérez-García V M 2001 *Phys. Rev. A* **64** 053611
[40] Yasunaga M and Tsubota M 2007 *J. Low Temp. Phys.* **148** 363

LDPC Coding for QKD at Higher Photon Flux Levels Based on Spatial Entanglement of Twin beams in PDC

Fred Daneshgaran[a],**Marina Mondin**[b], **Inam Bari**[b]

[a] Elec. & Comp. Engr. Dept., California State Univ., 5151 State University Dr., 90032 Los Angeles, USA.
[b] Dip. di Elettronica e Telecom., Politecnico di Torino, C.so Duca degli Abruzzi 24, 10129 Torino, Italy.

E-mail: fdanesh@calstatela.edu, mondin.marina@polito.it[1]

Abstract.

Twin beams generated by Parametric Down Conversion (PDC) exhibit quantum correlations that has been effectively used as a tool for many applications including calibration of single photon detectors. By now, detection of multi-mode spatial correlations is a mature field and in principle, only depends on the transmission and detection efficiency of the devices and the channel. In [2, 4, 5], the authors utilized their know-how on almost perfect selection of modes of pairwise correlated entangled beams and the optimization of the noise reduction to below the shot-noise level, for absolute calibration of Charge Coupled Device (CCD) cameras. The same basic principle is currently being considered by the same authors for possible use in Quantum Key Distribution (QKD) [3, 1]. The main advantage in such an approach would be the ability to work with much higher photon fluxes than that of a single photon regime that is theoretically required for discrete variable QKD applications (in practice, very weak laser pulses with mean photon count below one are used).The natural setup of quantization of CCD detection area and subsequent measurement of the correlation statistic needed to detect the presence of the eavesdropper Eve, leads to a QKD channel model that is a Discrete Memoryless Channel (DMC) with a number of inputs and outputs that can be more than two (i.e., the channel is a multi-level DMC).

This paper investigates the use of Low Density Parity Check (LDPC) codes for information reconciliation on the effective parallel channels associated with the multi-level DMC. The performance of such codes are shown to be close to the theoretical limits.

Keywords: Quantum Key Distribution, Quantum Correlation, Polar Codes, Multi-level DMC

1. Introduction

Parametric Down Conversion (PDC) is an effective means of producing entangled photons. The state produced by spontaneous PDC exhibits perfect momentum phase matching for plane wave pump field. The state of a single bipartite transverse mode near degeneracy can be written as:

$$|\psi(\vec{q})\rangle = \sum_n C_{\vec{q}}(n)\,|n\rangle_{i,\vec{q}}\,|n\rangle_{s,-\vec{q}}, \tag{1}$$

where, i stands for the idler and s stands for the signal beams. The two modes in equation (1) are entangled in the number of photons in each pair of modes $\pm\vec{q}$. Imaging the beams in the far field of a thin lens of focal length f in a f-f arrangement as depicted in Figure 1, leads to an association of each

[1] This work has been supported by NATO SfP Project 984397 entitled "SfP - Secure Quantum Communications".

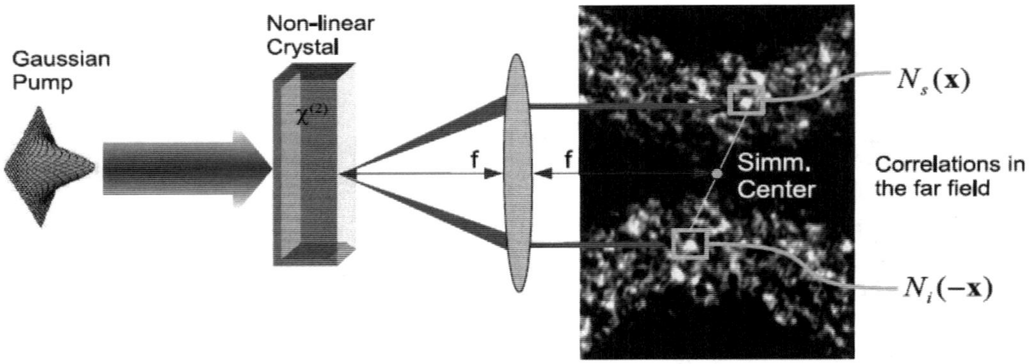

Figure 1. General block diagram of the experimental setup from [4].

mode to a unique position in the focal plane via the mapping:

$$\frac{2cf}{\omega_p}\vec{q} \to \vec{x}$$

Hence, a perfect correlation should be detected in the photon numbers $N_{i,\vec{x}}$ and $N_{s,-\vec{x}}$ where the center of symmetry relative to which the symmetric positions are identified is the pump detection plane interception point. In reality, the pump field is not a perfect plane wave, rather a Gaussian beam with spatial waist w_p inducing an uncertainty in the relative propagation direction of the twin photons on the order of the angular beam-width of the pump. This uncertainty is the coherence area of the process roughly corresponding to the transverse size of the mode in the far field and is given by:

$$A_{coh} \sim \left[\frac{2\pi cf}{\omega_p w_p}\right]^2$$

The number of spatial modes observed over a detection area that is usually taken to be much larger than the coherence area is given by $M_{spatial} = A_{det,j}/A_{coh}$, where $j = i$ or s. Similarly, it is assumed that the detection time is much larger than the coherence time hence the number of temporal modes $M_t = T_{det}/T_{coh}$ is much larger than one. The total number of modes in the detection region is $M_{tot} = M_{spatial}M_t$. The different modes in a single region are independent and thus the statistics of the detected photons is multi-thermal with mean value $\langle N_j\rangle = M_{tot}\eta_j\mu$, where, $j = i$ or s, μ is the number of photons per mode, and η_j is the overall efficiency. The variance of the number of detected photons is given by:

$$\langle \delta^2 N_j\rangle = \langle N_j\rangle\left(1+\frac{\langle N_j\rangle}{M_{tot}}\right) = \langle N_j\rangle\left(1+\mathcal{E}\right) = M_{tot}\eta_j\mu(1+\eta_j\mu) \qquad (2)$$

where $\mathcal{E}$ is the excess noise defined as fluctuations that exceed the Shot Noise Limit (SNL). The covariance between the signal and idler photon numbers is given by:

$$\langle \delta N_i \delta N_s\rangle = M_{tot}\eta_i\eta_s\mu(1+\mu). \qquad (3)$$

The correlation statistic between signal and idler is measured in terms of the fluctuations of the difference $N_- = N_s - N_i$ normalized to the corresponding level of shot noise:

$$\sigma = \frac{\langle \delta^2 N_-\rangle}{\langle N_i + N_s\rangle} = 1 - \eta_+ + \frac{\eta_-^2}{4\eta_+^2}\left(\eta_+ + \frac{\langle N_s + N_i\rangle}{M_{tot}}\right), \qquad (4)$$

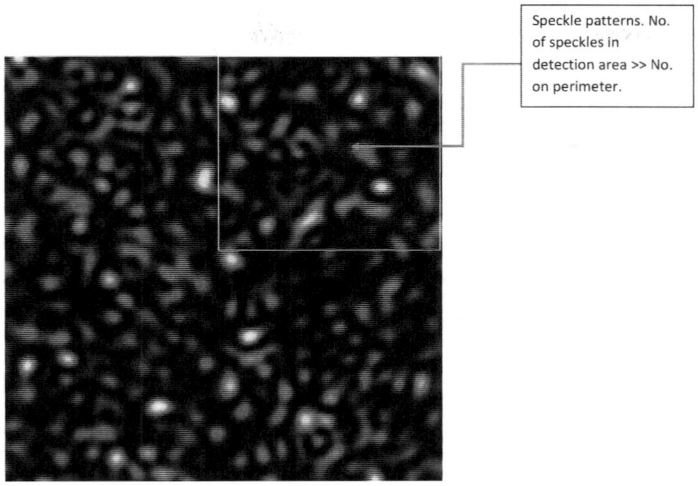

Figure 2. A detection region identified as a square on the upper right corner whereby the condition that the number of speckles in the region is much larger than that on the perimeter is satisfied.

where, $\eta_+ = (\eta_s + \eta_i)/2$ and $\eta_- = \eta_s - \eta_i$. If the losses are perfectly balanced, $\eta_s = \eta_i = \eta$ and we get $\sigma = 1 - \eta$ depending only on the quantum efficiency. In the ideal case of perfect efficiency, indeed, $\sigma \to 0$ while for classical states of light the degree of correlation is bounded by $\sigma \geq 1$ with the lower limit achieved by coherent beams leading to $\sigma = 1$.

The basic requirement on the size of the detection area is that the number of speckles in the region far exceeds the number on the perimeter. A pictorial representation of this is depicted in Figure 2. The core idea of using the spatial entanglement of twin beams for Quantum key Distribution (QKD) stems from the fact that the correlation statistic generated on two symmetric regions in the CCD focal plane must exhibit a $\sigma < 1$. Eavesdropping by Eve will lead to increase in σ above one and is therefore detectable. Indeed, the extent of interference by Eve may be measured by how much σ has increased beyond its expected value. Depending on the extent of this interference, the parties can decide to either continue generating a random key or halt the process all together. Note that under ideal circumstances $\sigma \to 0$ while practically, it is always above zero but hopefully sufficiently below one to allow for secure communication to take place. This is pictorially depicted in Figure 3. In reality, one cannot assert that eavesdropping will always lead to $\sigma > 1$. Even in the simplest situation, Eve can always reduce the level of eavesdropping for having $\sigma < 1$, but then again the information she acquires will be very small as well.

Eventually, disposing of ideal technologies, Eve can measure a certain number N of photons and reproduce it by a squeezed source in the photon number even post selected (e.g., a PDC sources), sending to Bob the signal only when she observes N Photons in the heralding channel. However, if one could detect quantum correlations in the near field and in the far field, this would correspond to having two conjugated bases. In this case, only entanglement should provide strict correlation at the same time in the two measurements, a situation that cannot be reproduced by Eve.

The process of generating a random key based on this core concept is as follows:

- Alice generates the twin beams via PDC and images one of the two beams on its CCD array and launches the second beam towards Bob;
- both Alice and Bob are assumed to have achieved timing synchronization so they know the start and end time of each firing of the laser pulses;

Journal of Physics: Conference Series **497** (2014) 012037 doi:10.1088/1742-6596/497/1/012037

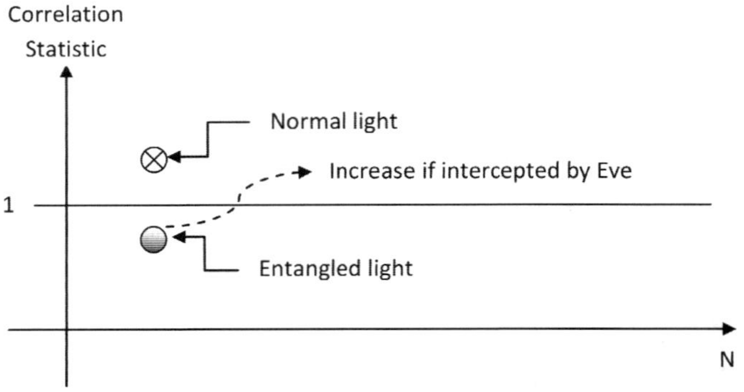

Figure 3. Pictorial representation of the impact of eavesdropping by Eve on the measured correlation statistic.

- the CCD detection area is partitioned into a number of smaller detection areas we call super-pixels, say four equal sized regions, satisfying our basic requirement that the detection area be much larger than the coherence area of the spatial modes;
- Alice and Bob make measurements on one reference quadrant (Alice and Bob's reference quadrants are at symmetric positions). Alice sends Bob her measurement results that Bob uses to estimate σ based on his a-priori knowledge of system level parameters and experimental setup uncertainties. Bob repeats the same process on a different set of measurement so that Alice can obtain a similar statistic. If the measured σ is acceptable by both parties, they continue with the protocol;
- assuming the measurements made on the reference quadrant in the previous step indicated that impact of eavesdropping by Eve is tolerable, Alice and Bob make measurements of the number of the detected photons in the other quadrants. A pair of quadrants, one at Alice and one at Bob, that are in symmetric regions of their CCDs focal plane represent a quantum channel between the two parties. In our example, three of the four quadrants would be used and constitute three parallel quantum channels that can be used to generate random keys;
- for each of the quantum channels described in the previous step, Alice and Bob make a measurement of the number of their detected photons and use binning and associate a unique label to their measurement. Ideally, for each quantum channel, Alice and Bob obtain an identical sequence of labels in their measurement that constitute a secret key. In practice, fluctuations in the number of measured photons lead to discrepancies and the label sequence at Alice and Bob don't match perfectly, hence requiring Information Reconciliation (IR) and privacy amplification to lead to distilled keys that can be used for cryptography.

2. Development of the Channel Model

The aim of this section is to outline in detail the process for generating the binary or multilevel DMC channel model for each of the parallel quantum channels described in the previous section. To this end, we need a detailed knowledge of the photon statistics in each detection area. As stated previously, the number of modes in a given detection area is very large. The number of temporal modes is dependent on the duration of the laser pulse width and the coherence time and is reported to be 5000 in [4]. The number of spatial modes is dependent on the size of the detection area that carries a trade-off in terms of the degree of the determination of the Center of Symmetry (CS), requiring a small detection area, and visibility of the quantum correlation, requiring a large detection area. The determination of the center of symmetry can be done with high resolution using small pixels and then performing the experiment with larger pixels.

In the experimental setup presented in [4], the number of spatial modes is about 150. Given the large number of detected modes, the statistic of the number of detected photons in a given region is multi-thermal.

For a single mode, the thermal distribution is geometric with Probability Mass Function (PMF) $P_1(n) = (1-\xi)\xi^n$, where, the parameter ξ is related to the mean of the geometric PMF λ via $\xi = \frac{\lambda}{1+\lambda}$. The photons in different modes are independent and therefore the PMF of the total number of detected photons in N modes is the N-fold convolution of the geometric PMF and given by:

$$P_N(k) = \frac{(k+N-1)!}{k!(N-1)!}(1-\xi)^N \xi^k.$$

The mean and variance of the photon number in this case is given by $E[X] = N\lambda$ and $\sigma_X^2 = N\lambda(1+\lambda)$.

In each detection region, at either Alice or Bob, the number of photons are counted and using a binning approach (i.e., multi-level quantization), the bin to which the number of detected photons belong is identified and the associated symbol is assigned as a component of a potential secret key. The Alice and Bob's measurements on the symmetric detection regions illuminated by the entangled twin beams may either result in identical bins, leading to identical symbols, or different bins leading to two different symbols. The derivation of the channel model requires computation of the probability that a certain number of photons is detected at Alice, and another number is detected at Bob. Once this joint probability is known, the computation of the transition probabilities associated with the resulting DMC modeling the behavior of a fictitious channel we may imagine exists between Alice and Bob and causes discrepancies between their detected symbols becomes straightforward.

The problem of determination of the number of detected photons at Alice and Bob is essentially a bi-partite detection problem. Let $P(n)$ denote the probability of generation of n photon-pairs at the source and let $Bin(k|n,\xi)$ denote the binomial PMF:

$$Bin(k|n,\xi) = \frac{n!}{k!(n-k)!}\xi^k(1-\xi)^{n-k},$$

then the probability of detecting k photons at Alice and m photons at Bob is given by:

$$P(K,m) = \sum_{n=\max(k,m)}^{\infty} P(n)Bin(k|n,\xi_A)Bin(m|n,\xi_B), \tag{5}$$

where, ξ_A and ξ_B are related to the mean of the geometric PMF for a single mode. Substituting $P_N(n)$ for $P(n)$ in above expression we get the desired PMF of detecting k photons at Alice and simultaneously m photons at Bob over our detection area.

3. LDPC Coding for Information Reconciliation

Alice and Bob need to be in possession of identical sequences of symbols before they can proceed with privacy amplification and generate secure keys for encryption. Information Reconciliation (IR) is the process of eliminating discrepancies that may exist in the sequences at Alice and Bob as much as possible. This is achieved using essentially error correction coding. The process is not perfect in that there is always some residual symbol errors left leading to very low symbol and frame error rates. By selecting a very low frame error rate threshold, one can almost guarantee that the symbol sequences at Alice and Bob are identical with very high probability.

Multitude of error correction techniques are available for information reconciliation. Broadly speaking, we can classify the available techniques into

- Forward Error Correction (FEC) techniques, and
- interactive two-way coding schemes with feedback such as the CASCADE algorithm.

Fundamentally from an information theoretic point of view , there is no advantage to two-way interactive techniques (this is reinforced in light of the given symmetry of the problem formulation presented in the introduction section). Either Alice or Bob could initiate an information reconciliation protocol since their roles are perfectly interchangeable. There is a vast body of literature on FEC techniques from the digital communications field and we shall resort to the available techniques for IR.

Forward error correction is a very mature field with performance of modern coding techniques approaching the capacity of the underlying channels at extremely long block lengths. Hence, one has a vast menu of FEC techniques to employ for the problem at hand. There are however some simple criteria one could apply in making the selection of a suitable technique:

- the code must be systematic since the data block is what Alice and Bob have direct access to, albeit, with possible discrepancies. As noted in introduction, there is no channel per-se between Alice and Bob. We may view the discrepancy between their symbol sequences as having been caused by a fictitious Q-ary DMC;
- any systematic FEC generates coded symbols that need to be communicated across a classic public channel. We assume the encoding operation is performed at Alice while decoding is performed at Bob;
- most modern decoding techniques rely on the use of soft-information processing and this is what we shall assume as well. We have however, two separate channels that lead to two different soft metrics that would need to be combined in decoding. The fictitious channel between Alice and Bob is modeled as a Q-ary DMC, the real channel over which the coded symbols propagate is a public channel we assume to be ideal.

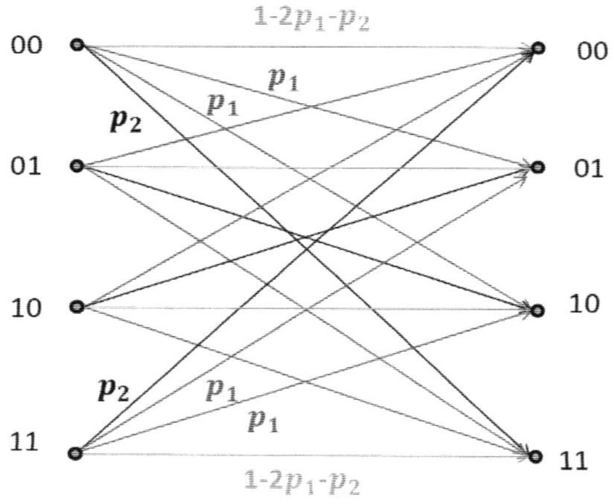

Figure 4. The considered Q-ary DMC channel model ($Q = 4$).

As an initial simplified model we will consider a Q-ary DMC channel model (with $Q = 4$) as shown in Figure 4, with input alphabet $X = \{x_k\}_{k=1}^{Q} = \{00, 01, 10, 11\}$ and output alphabet $Y = \{y_k\}_{k=1}^{Q} = \{00, 01, 10, 11\}$. We associate a binary labeling to the Q transmitted symbols, and denote as p_i the probability of having i bit errors in one transmitted Q-ary symbol (i=1,2), with the added hypothesis (justified by preliminary tests) that the probability of having one bit error per symbol is independent on the position of the error within the symbol. With reference to Figure 4 notice that, in the hypothesis of independent equally likely transmitted bits, the equivalent bit error probability (i.e. the equivalent Quantum Bit Error Rate - $QBER$) is

$$QBER = P(bit\ error) = p_1 + p_2. \tag{6}$$

The Q-ary channel model in Figure 4 is used to transmit the k bits that compose the cryptographic key. After the transmission of the key, and once the presence of Eve has been excluded as described in Section 1, additional r redundancy bits are transmitted on a parallel ideal Binary Symmetric Channel (BSC), so that the k information bits together with the r redundancy bits represent a $n = k + r$ bits codeword of a rate $R_c = k/n$ LDPC code. Note that the above channel model is obtained from proper binning of the photon numbers at the transmitter and receiver (i.e., the intervals used to define the channel symbols at input and output can be adjusted to get the transition probabilities that adhere to the model above). At the receiver a soft metric LDPC decoder is employed, operating according to a belief propagation strategy, with input soft metrics from the Q-ary channel evaluated as

$$LLR(y_k^i) = log\left(\frac{\sum_{x\in x(0)^i} P(y_k|x)}{\sum_{x\in x(1)^i} P(y_k|x)}\right), \tag{7}$$

where $LLR(y_k^i)$ is the Log Likelihood Ratio (LLR) of the i-th bit of the received symbol y_k and $x(w)^i$ are the symbols of X whose i-th bit has weight w. Applying equation (7) to our channel model we obtain

$$LLR(y_k^i) = \begin{cases} +log\left(\frac{1-p_1-p_2}{p_1+p_2}\right) & \text{if } y_k^i = 0 \\ -log\left(\frac{1-p_1-p_2}{p_1+p_2}\right) & \text{if } y_k^i = 1 \end{cases} \tag{8}$$

A rate 0.8 (1000,800) LDPC code has been applied to perform IR on the scheme described so far, selecting, for the same overall $QBER$, different values of p_1 and p_2. The decoded Bit Error Rate (BER) performances shown in Figure 5 have been obtained by Montecarlo simulation, showing that the decoded BER depends on the overall $QBER$ value and not on how the error probability is distributed on the first and second bit of the Q-ary transmitted symbols. This proves, in practice, that with the considered binary FEC LDPC codes the use of a Q-ary channel is equivalent to $log_2(Q)$ successive uses of a BSC with the same value of $QBER$.

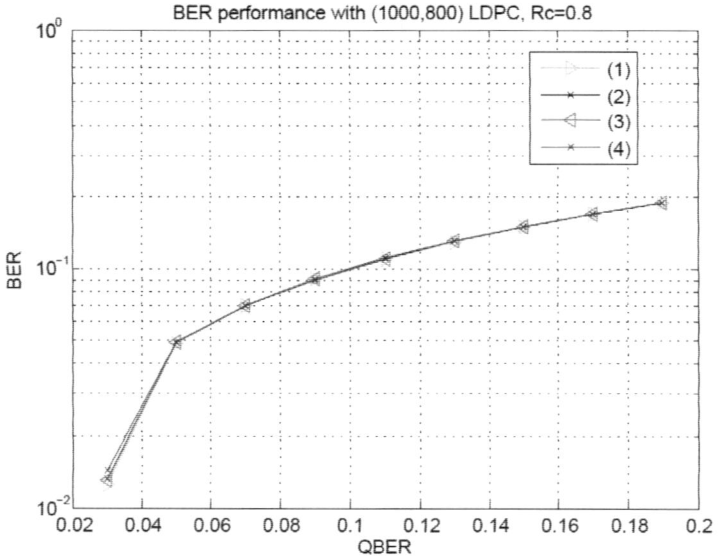

Figure 5. BER performances of a (1000,800) binary LDPC over the Q-ary channel in Figure 3 for (2) $p_1 = QBER^2$, (3) $p_1 = QBER^2/2$ and (4) $p_1 = QBER/2$ as a function of $QBER$, compared with (1) a BSC with transition probability $QBER$. In (2), (3) and (4) $p_2 = QBER - p_1$.

LDPC codes with different rates have then been simulated, obtaining the efficiency of the considered information reconciliation scheme. The efficiency, defined as [6]

$$\epsilon_{IR} = \frac{1 - R_c}{H(X|Y)} = \frac{1 - R_c}{R_c H_2(QBER)} \tag{9}$$

is shown in Figure 6, where the considered codeword lengths n are always smaller or equal to 1000. It can be noticed that the proposed scheme allows to achieve efficiency values better (i.e. closer to 1) than the CASCADE algorithm, with reasonable complexities, proving the effectiveness of the approach.

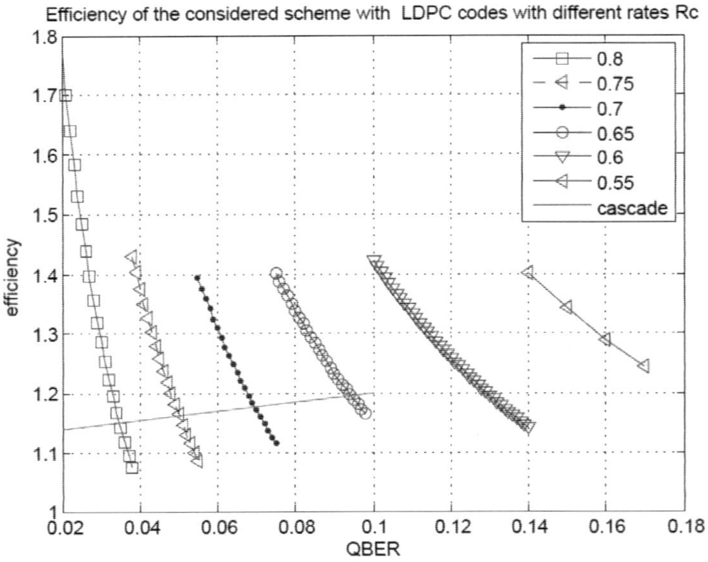

Figure 6. Reconciliation efficiency with LDPC codes for different coding rates as a function of the channel quantum bit error rate $QBER$. The red continuous line is a graphical lower bound of the efficiency of the CASCADE IR algorithm [6] .

4. Conclusions

This paper offers a preliminary investigation on the use of FEC LDPC codes for information reconciliation when the underlying channel is a Q-ary DMC, for QKD applications based on higher photon flux levels with spatial entanglement of twin beams in PDC. The simulation results show that acceptable error reconciliation efficiency values can be obtained with reasonable decoding complexity.

5. Bibliography

[1] I. P. Degiovanni M. Genovese M.Gramegna A. Avella, G. Brida and P. Traina. *Phys. Rev. A*, 82:062309, 2010.
[2] M. Genovese Brida and I. Ruo Berchera. *Nature Photonics*, 4:227, 2010.
[3] I. P. Degiovanni M. Genovese P. Traina G. Brida, A. Cavanna. *Laser Physics Letters*, 9:247, 2012.
[4] M. Genovese M. L. Rastello I. Ruo-Berchera G. Brida, Ivo P. Degiovanni. *Opt. Exp.*, 18:20572, 2010.
[5] A. Gatti M.Genovese A.Meda I.Ruo-Berchera G.Brida, L. Caspani. *Phys. Rev. Lett.*, 102:213602, 2009.
[6] V. Martin J. Martinez-Mateo, D. Elkouss. *Quantum Information and Computation*, 12:07910812, 2012.

Journal of Physics: Conference Series **497** (2014) 012038 doi:10.1088/1742-6596/497/1/012038

Properties of thermoelectric $Ce_{0.09}Fe_{0.67}Co_{3.33}Sb_{12}/FeSb_2Te$ multi-layered structures prepared by laser ablation

R Zeipl[1], M Jelínek[1,3], M Vlček[2], T Kocourek[1], J Walachová[1]

[1] Institute of Physics, Academy of Sciences of the Czech Republic, v.v.i., Na Slovance 2, 18221 Prague, Czech Republic

[2] Institute of Macromolecular Chemistry, Academy of Sciences of the Czech Republic, v.v.i., Heyrovského nám. 2, 16206 Prague, Czech Republic

[3] Czech Technical University, Faculty of Biomedical Engineering, Nám. Sítná 3105, 27201 Kladno, Czech Republic

E-mail: zeipl.radek@gmail.com

Abstract. Multi-layered $Ce_{0.09}Fe_{0.67}Co_{3.33}Sb_{12}/FeSb_{2.1}Te$ structures composed of thin equidistant layers were prepared by Pulsed Laser Deposition on fused silica quartz glass substrates. The structures were prepared at different substrate temperatures (230 °C or 250 °C) applying the laser beam energy density of 3 Jcm^{-2}. In the contribution we present some thermoelectric properties such as the in-plane electrical conductivity, the Seebeck coefficient and the power factor for the multi-layered structures in the temperature range from 300 K to 500 K. Comparison of multi-layered structure's thermoelectric properties with single thin $Ce_{0.09}Fe_{0.67}Co_{3.33}Sb_{12}$ and $FeSb_{2.1}Te$ layers is given. A cross sectional picture of the multi-layered structure made by Scanning Electron Microscope is presented for the thicker multi-layered structure.

1. Introduction

Skutterudites have been of high interest as a promising candidate for thermoelectric applications. The key advantage of skutterudites is their possible high figure of merit ZT (ZT=0.8 at T=750 K for bulk $Ce_{0.12}Fe_{0.71}Co_{3.29}Sb_{12}$ [1], ZT=1.1 at T=750 K for bulk $Ce_{0.28}Fe_{1.5}Co_{2.5}Sb_{12}$ [2] and $Ce_{0.9}Fe_3CoSb_{12}$ [3]. Theoretical maximum ZT=1.4 at 1000 K was predicted for $Ce_{0.9}Fe_3CoSb_{12}$ [3]). ZT is the essential material property for thermoelectric energy conversion on the Earth and in the cosmos. The dimensionless figure of merit is expressed by a formula:

$$ZT = \frac{S^2\sigma}{\lambda}T \tag{1}$$

where S, σ, T and λ are the Seebeck coefficient, the electrical conductivity, the ambient temperature and the thermal conductivity, respectively. The term $S^2\sigma$ is called power factor and can be used for evaluation of thermoelectric material if thermal conductivity λ is unknown.

The simplest skutterudites crystallize in a body-centered-cubic structure of the spare group Im3. The low thermal conductivity of skutterudites is obtained by filling the voids in the structure with small diameter, large-mass interstitials such as for example Yb, Ce or Ba. The lattice thermal conductivity can be also reduced by substituting Co with Fe as in our case.

A great improvement of thermoelectric properties was mathematically and also experimentally proved by preparing materials in the form of a low dimensional system [4-13] such as thin layer, superlattice or multi-layered structure, which can be prepared by simultaneous deposition of very thin layers of two or more different materials. Such improvement in comparison with bulk materials was published for the thin skutterudite layers of $CoSb_3$ [14,15] or for the skutterudite superlattices [16].

There have been just a few methods for a thin skutterudite layer or a superlattice deposition published in literature up to date. These are magnetron rf-sputtering [8,15,17], DC magnetron sputtering [18-20], ion-beam sputtering [21], electrochemical deposition [22], simultaneous evaporation of elements which are components of skutterudite followed by temperature treatment in ambient or vacuum [23-26] and Pulsed Laser Deposition (PLD) from a skutterudite target [14,16,23,27-44].

Earlier we published results on thin skutterudite layers prepared by PLD in Ar atmosphere from the $Ce_{0.1}Fe_{0.7}Co_{3.3}Sb_{12}$ hot pressed target [29]. The layers were well crystalline when deposited with the laser beam density Ds=3 Jcm^{-2} at substrate temperature Ts in the range from 200 °C to 250 °C. The best thermoelectric properties were obtained on the layers prepared at Ts=250 °C, although intensities of the skutterudite lines in X-ray diffraction analysis (XRD) diffractogram were more intensive for layers prepared at Ts=200 °C. All layers were of P-type electrical conductivity [29]. The Wavelength Dispersive analysis (WDX) proved $Ce_{0.09}Fe_{0.67}Co_{3.33}Sb_{12}$ layers stoichiometry.

Recently, we prepared thin skutterudite layers by PLD in Ar atmosphere from the $FeSb_2Te$ hot pressed target. The best thermoelectric properties were obtained on the layers prepared at Ts=250 °C and Ts=230 °C with Ds=3 Jcm^{-2}. Such layers were well crystalline and were also of P-type electrical conductivity. The WDX shoved $FeSb_{2.1}Te$ layers stoichiometry. Bulk ternary skutterudite $FeSb_2Te$ had been examined and published in details before [45,46] and was proved to be a good thermoelectric material.

In this contribution, we examine thin thermoelectric multi-layered $Ce_{0.09}Fe_{0.67}Co_{3.33}Sb_{12}/FeSb_{2.1}Te$ system composed of thin equidistant layers $Ce_{0.09}Fe_{0.67}Co_{3.33}Sb_{12}$ and $FeSb_{2.1}Te$ each 5 nm in thickness (5 nm period) prepared by PLD on a fused silica quartz glass substrate. The structures were prepared at Ts=230 °C and Ts=250 °C with Ds=3 Jcm^{-2}. A comparison of multi-layered structure's thermoelectric properties with properties of single thin $Ce_{0.09}Fe_{0.67}Co_{3.33}Sb_{12}$ and $FeSb_{2.1}Te$ layers are also given. It is expected that the preparation of $Ce_{0.09}Fe_{0.67}Co_{3.33}Sb_{12}/FeSb_{2.1}Te$ multi-layered structure can be successful because of the similar lattice constant of both materials [45-47].

2. Experimental details

PLD targets of $FeSb_2Te$ and $Ce_{0.1}Fe_{0.7}Co_{3.3}Sb_{12}$ composition where synthesized from individual elements by high-temperature solid-state reactions. Stoichiometric amounts of Fe (99.9 %), Sb (99.999 %), Te (99.999 %) and Ce (99.9 %), Fe (99.9 %), Co (99.9 %) and Sb (99.999 %) were sealed into evacuated carbon-coated silica glass tubes and heated up to 1050 °C for 48 hrs in a furnace. After quenching into a water bath, the same ampoule was placed into the furnace and annealed at 550 °C for 120 hrs. The resultant material was then ground under acetone, pelletized and heated again at 550 °C for 120 hrs. The completion of the solid-state reaction of obtained powder samples was verified by powder XRD.

The final targets for PLD deposition 20 mm in diameter and 2 mm in height were prepared by the hot pressing method (temperature 500 °C, pressure ~60 MPa for 1 hr). The measured density of pressed targets was found about 96-98 % of theoretical density.

The powder X-ray diffraction patterns of $FeSb_2Te$ ternary compounds used for the target were collected in the Bragg-Brentano geometry on a Bruker D8 Advance diffractometer. CuKα radiation was used. The crystal structure of $FeSb_2Te$ was refined by the Rietveld method for X-ray powder diffraction data using the FullProf program [23]. XRD spectrogram of target compound is depicted in figure 5 (top view).

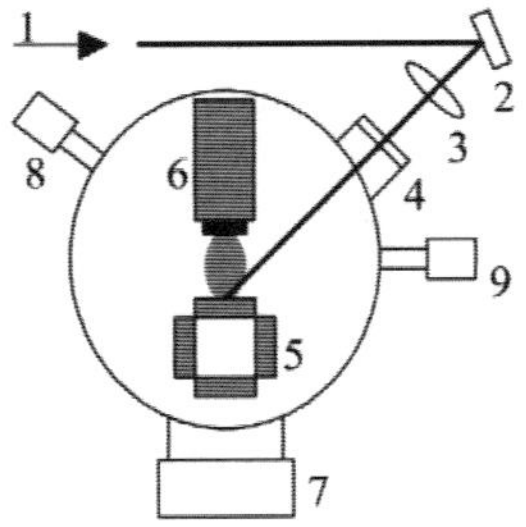

Figure 1. The basic scheme of the experimental apparatus for PLD: (1) laser beam, (2) mirrors, (3) focusing lens, (4) quartz window, (5) target holder, (6) substrate holder, (7) vacuum pump, (8, 9) Pirani and Penning vacuum gauges, respectively.

The basic schema of the experimental apparatus for PLD is depicted in figure 1. Conceptually and experimentally, PLD is an extremely simple method, probably the simplest of all thin film growth techniques. A high power pulsed excimer KrF laser (COMPexProTM 205 F) radiation (1) is used as an external energy source to vaporize materials of target (5) and to deposit a thin film. A set of optical components is used to focus the laser beam to the target surface (2, 3). After the laser pulse irradiation the temperature rises very rapidly (1011 Ks^{-1}) and the evaporation becomes non-equilibristic. In our experiment substrates were cleaned from the mechanical dirt in an ultrasonic cleaner. After that the substrates were subsequently cleaned in acetone, toluene and in ethanol. Cleaning in the vapours of boiling ethanol then completed this process. Fused silica substrates were finally annealed in an oven at a temperature around 250 °C. The layers and multi-layered structures were deposited on fused silica quartz glass substrate 10x10 mm. The deposition took place at Ar atmosphere (13 Pa). The distance of the substrate from the target was set to 40 mm.

Series of multi-layered $Ce_{0.09}Fe_{0.67}Co_{3.33}Sb_{12}/FeSb_{2.1}Te$ structures composed of thin equidistant $Ce_{0.09}Fe_{0.67}Co_{3.33}Sb_{12}$ and $FeSb_{2.1}Te$ layers of 5 nm in thickness and total thickness of about 60 nm were prepared by PLD at Ts=230 °C and Ts=250 °C with Ds=3 Jcm^{-2}. The deposition conditions were chosen based on previous results taken on single $Ce_{0.09}Fe_{0.67}Co_{3.33}Sb_{12}$ and $FeSb_{2.1}Te$ layers as the conditions giving the best thermoelectric properties.

Transport properties, such as the in-plain electrical resistivity and the Seebeck coefficient, were measured on each multi-layered structure and on single layers in the temperature range from 300 K up to 500 K. The power factor was then calculated. Four square shaped contacts for the measurements were prepared by evaporating Ti. Pressed Pt/PtRh thermocouples with diameter of 0.07 mm were used as leads. A conventional DC van der Pauw's method was used for the electrical conductivity measurement. The experimental error of this method is about 10 % for the conductivity measurement.

The Seebeck coefficient was determined from the variation of the electromotive force for different temperature gradients across the layer. The both sides of the sample were in the thermal contact with an independent wire resistant sub-heater that supplies the heat and induces the sample temperature gradient. The thermocouple junctions were bonded to each corner of the square shaped sample. The experimental error of the Seebeck coefficient measurement is about 20 %.

The thickness of the layers was measured using a mechanical stylus profile-meter (α-Step 500, Tencor Instruments, U.S.A). The uncertainty in the thickness estimation is about 10 % in the examined range of thicknesses partly because of the fact that the layer thickness in the centre of the sample is higher than on its edge.

Scanning Electron Microscope (SEM) images were measured using Leo 1550 device (magnification up to 500 k). To see the edge of the layer, the sample was scanned under 45°.

3. Results and discussion

3.1. Single thin layers

The best single layer prepared from $Ce_{0.1}Fe_{0.7}Co_{3.3}Sb_{12}$ target was prepared at Ts=250 °C with Ds=3 Jcm^{-2}. It showed semi-conducting P-type temperature behaviour of the electrical resistivity. The measured value was roughly 8.5 times higher at room temperature and roughly 5 times higher at 500K in comparison with published bulk $Ce_{0.12}Fe_{0.71}Co_{3.29}Sb_{12}$ data [1]. The room temperature value of measured Seebeck coefficient of the layer was roughly 1.4 times higher in comparison with bulk $Ce_{0.12}Fe_{0.71}Co_{3.29}Sb_{12}$ [1] but we observed lower temperature slope, so the measured value is comparable with the published bulk material value at 500K [1]. Temperature dependencies of electrical resistivity and Seebeck coefficient are depicted in figure 2.

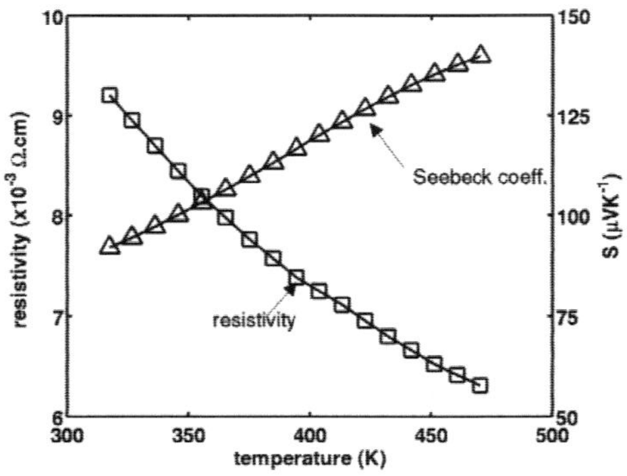

Figure 2. Temperature dependencies of the electrical resistivity (squares) and the Seebeck coefficient (triangles) for the 64nm thick P-type layer prepared from $Ce_{0.1}Fe_{0.7}Co_{3.3}Sb_{12}$ target at Ts=250 °C with Ds=3 Jcm^{-2}.

The WDX composition study done on 439 nm thick single layer proved an average stoichiometry of $Ce_{0.09}Fe_{0.67}Co_{3.33}Sb_{12}$. The maximal deviation was in both cases less than +/-3 % for all elements.

The XRD measurement done on 439 nm thick single layer proved that the layer was well crystalline with $CoSb_3$ structure.

The temperature dependencies of the electrical resistivity and the Seebeck coefficient for single thin layers prepared from $FeSb_2Te$ target at Ts=230 °C and Ts=250 °C with Ds=3 Jcm^{-2} are depicted in figure 3.

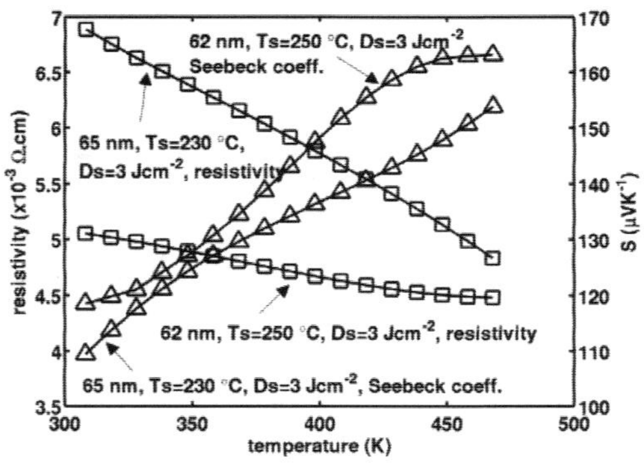

Figure 3. Temperature dependencies of the electrical resistivity (squares) and the Seebeck coefficient (triangles) for the 65 nm thick P-type layer prepared from $FeSb_2Te$ target at Ts=230 °C and 62 nm Ts=250 °C with Ds=3 Jcm^{-2}.

Both layers showed semi-conducting P-type behaviour of the electrical resistivity (the electrical resistivity decreases with rising temperature). The room temperature values of electrical resistivity of the layers were higher than the value published for bulk $FeSb_2Te$ material ($1.59x10^{-3}$ Ohmcm at 25 °C) [20] and also higher than value measured on PLD target ($2.50x10^{-3}$ Ohmcm at 25 °C), see table 1.

Table 1 The comparison of the room temperature values of the electrical resistivity, the Seebeck coefficient and the power factor of PLD layers prepared at Ts=230°C with Ds=3 Jcm^{-2} and Ts=250°C with Ds=3 Jcm^{-2} from $FeSb_2Te$ target with bulk $FeSb_2Te$ material [45] and used PLD target.

Temp 25°C	Bulk $FeSb_2Te$ [20]	PLD $FeSb_2Te$ target	$FeSb_2Te$ layer, Ts=250°C	$FeSb_2Te$ layer, Ts=230°C
R (mOhmcm)	1.59	2.50	5.08	7.02
S (μV/K)	54.70	112.96	119.00	104.57
S^2/R (10^{-9}x$WK^{-2}cm^{-1}$)	1881.82	5140.63	2786.87	1557.45

The room temperature values of Seebeck coefficient of $FeSb_2Te$ single layers prepared at Ts=230 °C with Ds=3 Jcm^{-2} and Ts=250 °C with Ds=3 Jcm^{-2} are about twice higher than published values for bulk $FeSb_2Te$ material [45]. Prepared layers shoved higher Seebeck coefficients in the whole studied temperature range from room temperature up to 500 K in comparison with values for bulk $FeSb_2Te$ material [46]. The layer prepared at Ts=230 °C with Ds=3 Jcm^{-2} had lower room temperature value of power factor 1557.45x10^{-9} $WK^{-2}cm^{-1}$ than the value published for bulk material [45, 46] and also than layers prepared at Ts=250 °C with Ds=3 Jcm^{-2}. But the power factor's slope for the layer prepared at Ts=230 °C with Ds=3 Jcm^{-2} suggests that it may be comparable with layers prepared at Ts=250 °C for higher temperatures (above 500 K).

WDX experiments that were done on the thicker layers proved average $FeSb_{2.1}Te$ stoichiometry that corresponds very well to the target composition.

XRD experiments done on the thicker layers with thickness 240 nm and 235 nm prepared at Ts=230 °C and Ts=250 °C with Ds=3 Jcm^{-2}, respectively, proved that both layers were well crystalline with peaks corresponding to the used PLD target.

The resulting dependency of power factor versus temperature for the best single PLD layers is together with the published bulk $Ce_{0.12}Fe_{0.71}Co_{3.29}Sb_{12}$ data [1] depicted in figure 4. The power factors of our best prepared single layers are much smaller (roughly twice smaller at room temperature and roughly 5 times smaller at 500K) in comparison with the bulk $Ce_{0.12}Fe_{0.71}Co_{3.29}Sb_{12}$ [1]. Temperature dependency of power factor of bulk $FeSb_2Te$ has never been published. For the bulk material there was only room temperature value of power factor about 1881.8x$10^{-9}$$WK^{-2}cm^{-1}$ published [45,46].

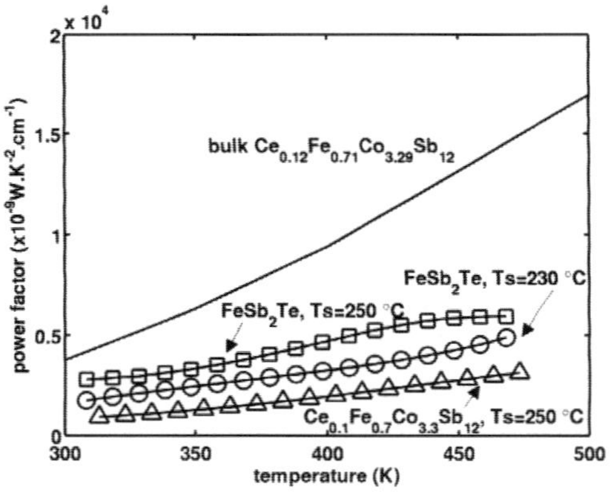

Figure 4. Temperature dependencies of the power factor for the single layers prepared from $Ce_{0.1}Fe_{0.7}Co_{3.3}Sb_{12}$ target at Ts=250 °C with Ds=3 Jcm^{-2}, thickness 64 nm (triangles) and from $FeSb_2Te$ target at Ts=230 °C with Ds=3 Jcm^{-2}, 65 nm thickness (circles) and Ts=250 °C with Ds=3 Jcm^{-2}, 62 nm thickness (squares) together with the published data for bulk $Ce_{0.12}Fe_{0.71}Co_{3.29}Sb_{12}$ [1].

3.2. Multi-layered structures

The multi-layered $Ce_{0.09}Fe_{0.67}Co_{3.33}Sb_{12}/FeSb_{2.1}Te$ structures (about 60 nm in total thickness) composed of thin equidistant $Ce_{0.09}Fe_{0.67}Co_{3.33}Sb_{12}$ and $FeSb_{2.1}Te$ layers (5 nm in thickness) were prepared by PLD at Ts=250 °C with Ds=3 Jcm^{-2} and at Ts=230 °C with Ds=3 Jcm^{-2} with aim to improve thermoelectric properties of single thin $Ce_{0.09}Fe_{0.67}Co_{3.33}Sb_{12}$ and $FeSb_{2.1}Te$ layers prepared before.

The in-plain temperature dependencies of the electrical resistivity of $Ce_{0.09}Fe_{0.67}Co_{3.33}Sb_{12}/FeSb_{2.1}Te$ multi-layered structures (60 nm in total thickness) are given in figure 5. Both structures showed semi-conducting P-type behaviour of the electrical resistivity (the decrease of electrical resistivity with the increase of temperature). The multi-layered structures had lower electrical resistivity than the best above mentioned single layers $Ce_{0.09}Fe_{0.67}Co_{3.33}Sb_{12}$ and $FeSb_{2.1}Te$ in the whole studied temperature range.

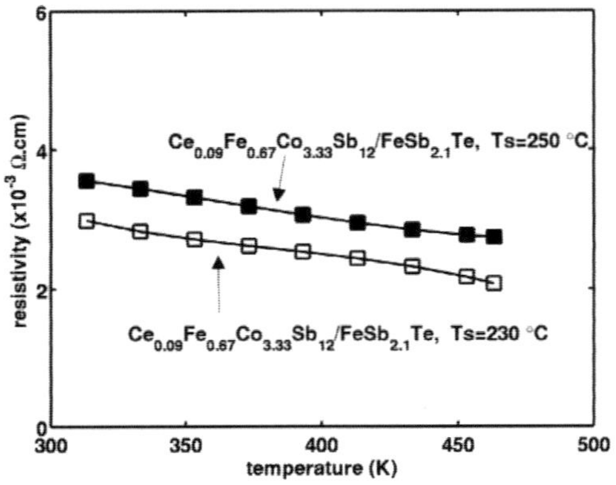

Figure 5. Temperature dependency of electrical resistivity for 60 nm thick $Ce_{0.09}Fe_{0.67}Co_{3.33}Sb_{12}/FeSb_{2.1}Te$ multi-layered structures prepared with Ds=3 Jcm^{-2} at Ts=230 °C (empty squares) and at Ts=250 °C (black filled squares).

The multi-layered structures showed also much lower in-plain Seebeck coefficient in the whole measured temperature range than the single layers. Due to the low Seebeck coefficient, the resultant power factor of all prepared multi-layered structures is lower than the power factor of the best prepared single layers.

When we compared the measured values of power factor of the $Ce_{0.09}Fe_{0.67}Co_{3.33}Sb_{12}/FeSb_{2.1}Te$ multi-layered structures with published bulk $Ce_{0.12}Fe_{0.71}Co_{3.29}Sb_{12}$ material [1], we found that multi-layered system values are roughly four times lower at room temperature and roughly ten times lower at about 500 K. If the decrease of thermal conductivity on multi-layered structure in comparison to bulk material is taken into account, we may speculate that even better values of ZT for prepared multi-layered systems may be achieved in comparison with the bulk $Ce_{0.12}Fe_{0.71}Co_{3.29}Sb_{12}$ material. It is assumed that cross-sectional electrical conductivity and Seebeck coefficient are not much influenced by interfaces in the multi-layered structures. Temperature dependency of power factor of bulk $FeSb_2Te$ has never been published, so the power factor of multiple-structures can not be compared. Only the room temperature value of power factor of about $1881.8x10^{-9}WK^{-2}cm^{-1}$ was published [45,46].

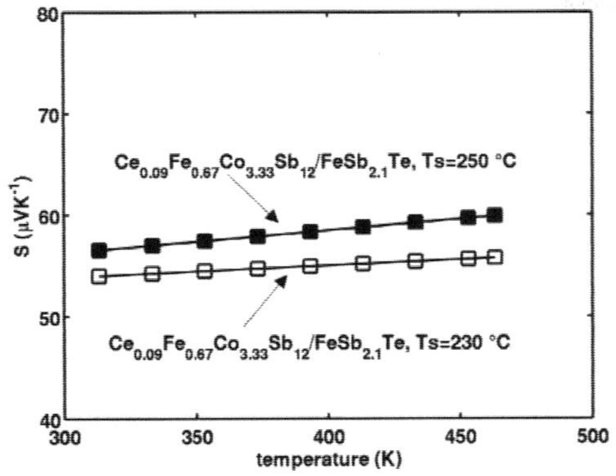

Figure 6. The temperature dependency of Seebeck coefficient for 60 nm thick $Ce_{0.09}Fe_{0.67}Co_{3.33}Sb_{12}/FeSb_{2.1}Te$ multi-layered structures prepared with Ds=3 Jcm^{-2} at Ts=230 °C (empty squares) and at Ts=250 °C (black filled squares).

The temperature dependencies of Seebeck coefficient and resultant power factor for 60 nm thick $Ce_{0.09}Fe_{0.67}Co_{3.33}Sb_{12}/FeSb_{2.1}Te$ multi-layered structures composed of thin equidistant layers (5 nm in thickness) prepared at Ts=230 °C and 250 °C with Ds=3 Jcm^{-2} are given in figure 6 and Figure 7, respectively.

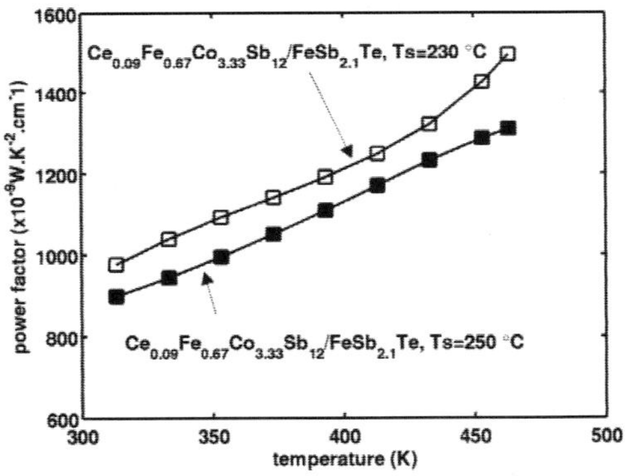

Figure 7. The temperature dependency of the power factor for 60 nm thick $Ce_{0.09}Fe_{0.67}Co_{3.33}Sb_{12}/FeSb_{2.1}Te$ multi-layered structures prepared with Ds=3 Jcm^{-2} at Ts=230 °C (empty squares) and at Ts=250 °C (black filled squares).

The SEM study was done on 286 nm thick multi-layered structure composed of altering equidistant layers of $Ce_{0.09}Fe_{0.67}Co_{3.33}Sb_{12}$ (thickness 17.5 nm) and $FeSb_{2.1}Te$ (thickness 18.3 nm). The SEM cross sectional picture of multi-layered structure is presented in figure 8, where individual layers were visible.

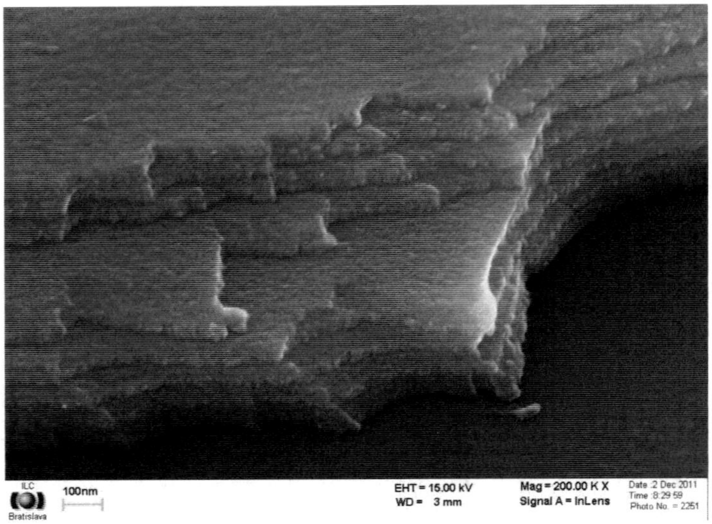

Figure 8. The SEM cross sectional picture of $Ce_{0.09}Fe_{0.67}Co_{3.33}Sb_{12}/FeSb_{2.1}Te$ multi-layered structure (total structure thickness 286 nm) composed of altering equidistant layers of $Ce_{0.09}Fe_{0.67}Co_{3.33}Sb_{12}$ (17.5 nm in thickness) and $FeSb_{2.1}Te$ (18.3 nm in thickness).

4. Conclusions

Multi-layered $Ce_{0.09}Fe_{0.67}Co_{3.33}Sb_{12}/FeSb_{2.1}Te$ structures composed of thin equidistant $Ce_{0.09}Fe_{0.67}Co_{3.33}Sb_{12}$ and $FeSb_{2.1}Te$ layers (5 nm in thickness) were successfully prepared by PLD at Ts=250 °C and Ts=230 °C with Ds=3 Jcm^{-2}. The resultant in-plain power factors of multi-layered structures were lower than the power factors of both used materials in a form of single layer. To make an overall evaluation of thermoelectric properties the Harman measurement of thermoelectric figure merit and a measurement of the thermal conductivity, which is expected to decrease doe to number of interfaces in the structure, is necessary.

5. Acknowledgments

The project has been supported by Czech Grant Agency under GAČR P108/10/1315 and P108/13-33056S.

References

[1] Tang X F, Chen L, Goto T, Hirai T and Yuan R Z 2001 *J. Mater. Sci.* **36** 5435
[2] Tang X F, Zhang Q, Chen L, Goto T and Hirai T 2005 *J. Appl. Phys.* **97** 093712
[3] Sales B C, Mandrus D, Chakoumakos B C, Keppens V and Thompson J R 1997 *Phys. Rev.* B **56** 15081
[4] Pichanusakorn P and Bandaru P R 2010 *Mater. Sci. Eng. R. Rep.* **67** 19
[5] Hicks L D and Dresselhaus M S 1993 *Phys. Rev.* B. **47** 12727
[6] Hicks L D and Dresselhaus M S 1993 *Phys. Rev.* B. **47** 16631
[7] Hicks L D, Harman T C, Sun X and Dresselhaus M S 1996 *Phys. Rev.* B. **53** 10493
[8] Hicks L D, Harman T C, Sun X and Dresselhaus M S 1996 *Proc. of 15th Int. Conf. on Thermoelectrics*, March 26-29, Pasadena, CA USA, p 450
[9] Ohta H, Kim S, Mune Y, Mizoguchi T, Nomura K, Ohta S, Nomura T, Nakanishi Y, Ikuhara Y, Hirano M, Hosono H and Koumoto K 2007 *Nature Materials* **6** 129
[10] Dragoman M and Dragoman D 2007 *Appl. Phys. Lett.* **91** 203116
[11] Venkatasubramanian R, Siivola E, Colpitts T and Quinn B O 2001 *Nature* **413** 597
[12] Kim W, Zide J, Gossard A, Klenov D, Stemmer S, Shakouri A and Majumdar A 2006 *Phys. Rev. Lett.* **96** 045901

Journal of Physics: Conference Series **497** (2014) 012038 doi:10.1088/1742-6596/497/1/012038

[13] Yamasaki I, Yamanaka R, Mikami M, Sonobe H, Mori Y and Sasaki T 1998 *Proc. of the 17th Int. Conf. on Thermoelectrics*, May 24-28, Nagoya, Japan, p 210
[14] Durand H A, Nishimoto K, Ito K and Kataoka I 2000 *Appl. Surf. Sci.* **154** 387
[15] Anno H, Sakakibara T, Notora Y, Tashiro H, Koyanagi T, Kaneko H and Matsubara K 1997 *Proc. of 16th Int. Conf. on Thermoelectrics*, August 26-29, Dresden, Germany, p 338
[16] Song D W, Liu W L, Zeng T, Borca-Tasciuc T, Chen G, Caylor J C and Sands T D 2000 *Appl. Phys. Lett.* **77** 3854
[17] Arnache O, Girata D, Perez F, Castro L F, Prieto P and Lopera W 2005 *Solid State Commun.* **133** 343
[18] Savchuk V, Bolouz A, Chakraborty S, Schumann J and Vinzelberg H 2002 *J. Appl. Phys.* **92** 5319
[19] Schupp B, Bacher I, Hecker M, Mattern N, Savchuk V and Schumann J 2003 *Thin Solid Films* **434** 75
[20] Savchuk V, Schumann J, Schupp B, Behr G and Mattern N 2003 *J. Alloy. Compd.* **351** 248
[21] Yamashita Y, Matsui T and Morii K 2000 *J. Jpn. I. Met.* **64** 351
[22] Vidu, Li S, Quach D V and Stroeve P 2012 *J. Appl. Electrochem.* **42** 333
[23] Sellinschegg H, Stuckmeyer S L, Hornbostel J D and Johnson D C 1998 *Chem. Mater.* **10** 1096
[24] Hornbostel M D, Hyer E J, Thiel J, Evaldson J H and Johnson D C 1997 *Inorg. Chem.* **36** 4270
[25] Williams J R, Smalley A L E, Sellinschegg H, Daniels-Hafer C, Harris J, Johnson M B and Johnson D C 2003 *J. Am. Chem. Soc.* **125** 10335
[26] Smalley A L E, Kim S and Johnson D C 2003 *Chem. Mater.* **15** 3847
[27] Caylor J C, Dander M S, Stacey A M, Harper J S, Gronsky R and Sands T 2001 *J. Mater. Res.* **16** 2467
[28] Dauscher A, Puyet M, Lonoir B, Colceag D and Dinescu M 2004 *Appl. Phys.* A **79** 1465
[29] Zeipl R, Walachova J, Lorincik J, Leshkov S, Josiekova M, Jelinek M, Kocourek T, Jurek K, Navratil J, Benes L and Plechacek T 2010 *J. Vac. Sci. Technol.* A. **28** 523
[30] Caylor J C, Stacy A M and Bandaru P 1998 Symp. on Advances in Laser Ablation of Materials at the 1998 MRS Spring Meeting, April 13-16, San Francisco, CA, p 399
[31] Durand H A, Ito K and Kataoka I 1998 Proc. of Conf. on Laser Processing of Materials and Industrial Applications II, September 16-19, Beijing, China, p 120
[32] Caylor J C, Stacy A M and Sands T 1999 *Symp. on Thermoelectrical Materials*, November 30-December 3, Boston, Massachusetts, USA, p 327
[33] Durand H A, Suzuki A and Nishimoto K 2000 *Symp. on New Methods, Mechanisms and Models of Vapor Deposition*, April 24-26, San Francisco, CA, p 177
[34] Caylor J C, Stacy A M, Gronsky R and Sands T 2000 *Abstract of Papers of the American Chemical Society* **219** U883
[35] Suzuki A 2001 *Proc. of 20th Int. Conf. of Thermoelectrics*, June 08-11, Beijing, China, p 318
[36] Caylor J C, Stacey A M, Gronsky R and Sands T 2001 *J. Appl. Phys.* **89** 3508
[37] Caylor J C, Dander M S, Stacey A M, Harper J S, Gronsky R and Sands T 2001 *J. Mater. Res.* **16** 2467
[38] Suzuki A 2003 *Jpn. J. Appl. Phys.* **42** 2843
[39] Baumbach R E, Yuhasz W M and Maple M B 2006 *Appl. Phys.* A **84** 227
[40] Zeipl R, Navratil J, Benes L, Kocourek T, Jelinek M, Lorincik J, Vanis J, Zelinka J and Walachova J 2006 *Proc. of 25th Int. Conf. on Thermoelectrics*, August 06-10, Vienna, Austria, p 451
[41] Colceag D, Dauscher A, Lenoir B, Da Ros V, Birjega R, Moldovan A and Dinescu M 2007 *Appl. Surf. Sci.* **253** 8097
[42] Baumbach R E, Yuhasz W M and Maple M B 2008 *Phys.* B **403** 800
[43] Zeipl R, Jelinek M, Navratil J, Kocourek T, Yatskiv R, Josiekova M, Leshkov S, Jurek K and

Walachova J 2010 *J. Optoelectron. Adv. M.* **12** 572
[44] Kumar S R S, Alyamani A, Graff J W, Tritt T M and Alshareef H N 2011 *J. Mater. Res.* **26** 1836
[45] Fleurial J P US Patent 5.747.728. May 5. 1998
[46] Fleurial J P 1997 *Proc. of the 16th Int. Conf. on Thermoelectrics*, August 26-29, Dresden, Germany, p 91
[47] Morelli D T, Meisner G P, Chen B, Hu S and Uher C 1997 *Phys. Rev.* B. **56** 7376

Schwinger pair creation in constant and time-dependent fields

Adolfo Huet[1], **Sang Pyo Kim**[2], **Christian Schubert**[3]

[1]Centro de Física Aplicada y Tecnología Avanzada, (CFATA) UNAM Juriquilla, Boulevard Juriquilla No. 3001, C.P. 76230, Querétaro, México.

[2]Department of Physics, Kunsan National University, Kunsan 573-701, Korea.

[3]Instituto de Física y Matemáticas, Universidad Michoacana de San Nicolás de Hidalgo, Apdo. Postal 2-82, C.P. 58040, Morelia, Michoacán, Mexico.

E-mail: ahuet@fata.unam.mx,sangpyokim79@gmail.com,schubert@ifm.umich.mx

Abstract. The effect of pair creation from the vacuum in a strong electric field, predicted by Schwinger in 1951, may conceivably be confirmed experimentally with some of the ultrastrong lasers presently under construction. However, it is not easy to calculate the effect for realistic laser fields. We will shortly review here the intermediate case of a purely time-dependent but otherwise generic electric field, which is still relatively amenable to calculation.

1. Introduction

As early as 1931 Sauter [1] understood that Dirac's theory of the electron predicts that an electric field of sufficient strength and extent can induce spontaneous creation of electron – positron pairs from the vacuum. In the vacuum tunneling picture, one imagines that, by a statistical fluctuation, a virtual pair separates out far enough to draw its rest mass energy from the field (fig. 1).

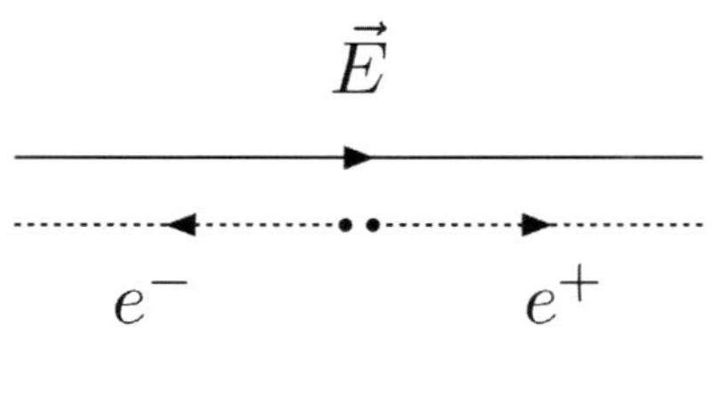

Figure 1. The tunneling picture of pair creation by an external field.

Twenty years later Schwinger [2] developed the modern approach to pair creation, which relates the total pair production probability P (probability of the decay of the vacuum) to the imaginary part of the effective action $\Gamma[E]$:

$$P = 1 - e^{-2\mathrm{Im}\Gamma(E)} \approx 2\mathrm{Im}\Gamma(E)\,. \tag{1.1}$$

Journal of Physics: Conference Series **497** (2014) 012039 doi:10.1088/1742-6596/497/1/012039

For the constant field case, Schwinger found his well-known representation of this imaginary part in terms of an infinite sum of exponentials [2],

$$\mathrm{Im}\mathcal{L}(E) = \frac{m^4}{8\pi^3}\Bigl(\frac{eE}{m^2}\Bigr)^2 \sum_{n=1}^{\infty} \frac{1}{n^2} \exp\Bigl[-n\pi\frac{m^2}{eE}\Bigr] . \tag{1.2}$$

Here the nth term relates to coherent creation of n pairs in one Compton volume. $\mathrm{Im}\mathcal{L}(E)$ depends on E nonperturbatively, which confirms the tunneling picture.
Schwinger also derived the corresponding formula for scalar QED [2]:

$$\mathrm{Im}\mathcal{L}_{\mathrm{scal}}(E) = \frac{m^4}{16\pi^3}\Bigl(\frac{eE}{m^2}\Bigr)^2 \sum_{n=1}^{\infty} \frac{(-1)^{n+1}}{n^2} \exp\Bigl[-n\pi\frac{m^2}{eE}\Bigr] . \tag{1.3}$$

Starting with the work of Ritus in the seventies, also radiative corrections to the one-loop formula (1.2) have been extensively studied [3, 4, 5, 6, 7, 8]. It turns out that the two-loop correction to Schwinger's formula implies an enhancement of pair creation probability. More interestingly, Lebedev and Ritus [5] showed in 1984 that for a weak (subcritical) field, this enhancement can be effectively described by a field - dependent mass,

$$m(E) \approx m - \frac{\alpha}{2}\frac{eE}{m} . \tag{1.4}$$

This again fits into the tunneling picture: a virtual pair turns real at a finite distance and thus gets born with a negative Coulomb energy that subtracts from the energy to be drawn from the field. The second term on the rhs of (1.4) is easily shown to correspond precisely to this energy. For a constant field the pair creation probability is exponentially small for

$$E \ll E_{\mathrm{crit}} \approx 10^{18}\mathrm{V/m} . \tag{1.5}$$

Although present laser technology fails this threshold only by a few orders of magnitude, it is generally believed that, to have any chance at seeing pair creation soon, complicated laser configurations must be used to lower the pair creation threshold. The calculation of pair creation rates for such fields will usually require approximative methods. Presently, there exist essentially four methods that can be attempted for the calculation of the pair creation rate in a generic electric field:

- Until recent years, practically the only approximation method used in this context was WKB [9, 10, 11].
- A more sophisticated version of WKB is the *worldline instanton formalism*, which was invented in 1982 by Affleck et al. [4] but applied to non-constant fields only much later [12, 13, 14].
- The quantum kinetic approach, based on a *quantum Vlasov equation* ("QVE") [15, 16, 17, 18, 19].
- The *Wigner formalism* [20, 21].

In the following we will discuss the worldline instanton and QVE approaches.

Journal of Physics: Conference Series **497** (2014) 012039 doi:10.1088/1742-6596/497/1/012039

2. The worldline instanton approach

The worldline instanton method is based on Feynman's worldline representation of the one-loop effective action in Scalar [22] and Spinor QED [23]. Since spin will not be important in the following, let us write down this path integral just for the scalar case:

$$\begin{aligned}\Gamma_{\rm scal}[A] &= \int_0^\infty \frac{dT}{T}\,\mathrm{e}^{-m^2T}\int \mathcal{D}x(\tau)\,\mathrm{e}^{-S[x(\tau)]}\,,\\ S[x(\tau)] &= \int_0^T d\tau\Bigl(\frac{\dot{x}^2}{4}+ieA\cdot\dot{x}\Bigr)\,.\end{aligned} \tag{2.1}$$

Here m and T are the mass and proper time of the loop scalar, and the path integral $\int \mathcal{D}x(\tau)$ runs over closed trajectories in (Euclidean) spacetime. In the instanton approach of [4] this path integral is replaced by a single stationary trajectory, which is simply a solution of the Lorentz force equation with periodic boundary conditions. For a constant electric field field pointing in the x_3 direction the nth worldline instanton is a circle in the $x_3 - x_4$ plane, of radius m/eE and winding number n:

$$x^{\rm inst}(u) = \frac{m}{eE}\Bigl(x_1, x_2, \cos(2n\pi u), \sin(2n\pi u)\Bigr)\,. \tag{2.2}$$

The worldline action evaluated on the instanton with winding number n is

$$S[x^{\rm inst}] = n\pi\,\frac{m^2}{eE} \tag{2.3}$$

and thus reproduces the nth exponent in Schwinger's representations (1.2),(1.3). Affleck et al. [4] moreover showed how to calculate the global prefactor in the scalar formula (1.3) in terms of the fluctuation determinant around the stationary trajectory [4]. In [12] it was further shown how to adapt all this to the spinor loop case. Thus for the constant field the instanton approximation yields, in both Scalar and Spinor QED, the *exact* imaginary part of the effective action. This exactness is due to the fact that the worldline path integral for a constant field is gaussian [24]. For generic fields there will be corrections, and the instanton approximation can a priori be expected to be good only in the limit where the mass is much larger than all mass scales contained in the external field.

Proceeding to non-constant fields, a well-studied case is the one of the "spatial bump", a field defined by

$$E(x_3) = \frac{E}{\cosh^2(kx_3)} \tag{2.4}$$

and thus localized in the x_3 direction. Nikishov [25] in 1970 was able to derive for this case an exact formula for the pair production rate as an integral over momenta of a known function. The worldline instantons for this case can still be given in terms of trigonometric functions [12],

$$\begin{aligned}x_3^{\rm inst}(u) &= \frac{m}{eE}\frac{1}{\tilde{\gamma}}\,\mathrm{arcsinh}\Bigl(\frac{\tilde{\gamma}}{\sqrt{1-\tilde{\gamma}^2}}\,\sin(2n\pi u)\Bigr)\,,\\ x_4^{\rm inst}(u) &= \frac{m}{eE}\frac{1}{\tilde{\gamma}\sqrt{1-\tilde{\gamma}^2}}\,\arcsin(\tilde{\gamma}\cos(2n\pi u))\,,\end{aligned} \tag{2.5}$$

Journal of Physics: Conference Series **497** (2014) 012039 doi:10.1088/1742-6596/497/1/012039

where $\tilde{\gamma}$ is the "inhomogeneity parameter"

$$\tilde{\gamma} \equiv \frac{mk}{eE} . \tag{2.6}$$

The corresponding stationary action is

$$S[x^{\text{inst}}] = n \frac{m^2\pi}{eE} \left(\frac{2}{1 + \sqrt{1 - \tilde{\gamma}^2}} \right) . \tag{2.7}$$

It *increases* with decreasing inhomogeneity, so that the pair creation rate *decreases*. Its form suggests that something special should happen for $\tilde{\gamma} = 1$, and indeed it turns out that real instanton solutions exist only for $\tilde{\gamma} < 1$. The form of these solutions is shown in fig. 2.

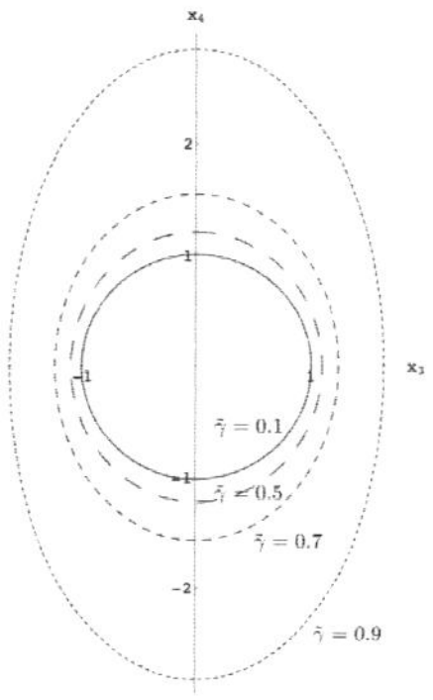

Figure 2. Plot of instanton paths for the spacelike bump.

Since for this case our the worldline instanton approximation has no reason to be exact, from the absence of instantons we cannot yet conclude that there is no pair creation at all for $\tilde{\gamma} > 1$. However, the total absence of pair creation for $\tilde{\gamma} > 1$ is supported also by a numerical evaluation of Nikishov's formula mentioned above, as well as by a direct Monte Carlo evaluation [26] of Feynman's path integral (2.1). This fits again into the vacuum tunneling picture - it is easy to check that for $\tilde{\gamma} > 1$ the field has insufficient extent to provide a virtual particle with its rest mass energy.

Coming to our main topic, the purely time-dependent field $\mathbf{E}(t) = (0, 0, E(t))$, here the Lorentz force equation reduces to the following first-order ordinary DGL [12], where

$$\begin{aligned} \dot{x}_3 &= -\frac{iea}{m} A_3(x_4) , \\ \dot{x}_4 &= a\sqrt{1 + \left(\frac{e\,A_3(x_4)}{m}\right)^2} , \end{aligned} \tag{2.8}$$

with

$$a = \frac{m}{eE} 2n\pi . \tag{2.9}$$

At least numerically, the DGL (2.8) can always be solved. Then

$$\mathrm{Im}\Gamma(E) \overset{m\,\mathrm{large}}{\approx} N\,\mathrm{e}^{-S[x^{\mathrm{inst}}]} \tag{2.10}$$

provides a large-mass approximation for the imaginary part, where a general method for the computation of the prefactor determinant N was found in [13]. Again the simplest non-constant field to consider is the time-like analogue of the bump (2.4),

$$E(t) = \frac{E}{\cosh^2(\omega\, t)}\,. \tag{2.11}$$

The corresponding worldline instantons are related to the ones for the spatial bump (2.5) by the analytical continuation $\tilde{\gamma} \to -i\gamma$, where $\gamma \equiv \frac{m\omega}{eE}$, with the consequence that, unlike the spatial case, for the time-like bump real instanton solutions exist for all values of the inhomogeneity parameter γ. And the stationary action,

$$S[x^{\mathrm{inst}}] \;=\; n\,\frac{m^2\pi}{eE}\left(\frac{2}{1+\sqrt{1+\gamma^2}}\right) \tag{2.12}$$

decreases with increasing inhomogeneity, thus the pair creation rate *increases*.

Thus, the worldline instanton formalism suggests the following general rule:

- Inhomogeneity in space tends to reduce the pair creation rate.
- Inhomogeneity in time tends to enhance the pair creation rate.

This also leads one to expect that a purely time-dependent field should *always* give a non-zero pair creation rate.

3. The quantum Vlasov equation

A very different approach to the purely time-dependent electric field case is based on the observation that for it the spatial momentum $\mathbf{k}$ is a good quantum number. Thus one can fix it, and use a Bogoliubov transformation between the vacua at initial time and time t to derive an evolution equation for the density of pairs $\mathcal{N}(t)$ with fixed momenta. This is the QVE. Restricting ourselves to the scalar QED case for definiteness, here it takes the form [15, 16, 18, 17, 19]

$$\dot{\mathcal{N}}(t) \;=\; \frac{\dot\omega(t)}{2\omega(t)}\int_{-\infty}^{t} dt'\frac{\dot\omega(t')}{\omega(t')}(1+2\mathcal{N}(t'))\cos\Big[2\int_{t'}^{t}dt''\omega(t'')\Big] \tag{3.1}$$

where

$$\omega^2(t) \;=\; (k_{\|} - qA_{\|}(t))^2 + \mathbf{k}_\perp^2 + m^2\,. \tag{3.2}$$

Here the gauge $\dot{\mathbf{A}}(t) = -\mathbf{E}(t)$ has been used, and $k_{\|}$ denotes the momentum component along the field. $\mathcal{N}(t)$ has to be set to zero at $t = -\infty$, and for $t \to \infty$ turns into the density of created pairs with fixed momentum $\mathbf{k}$; no direct physical meaning should be ascribed to this quantity at finite time, as has been emphasized in [18, 27]. For the constant field case, the QVE (3.1) and its spinor QED analogue can be solved in closed form [27], leading to the behaviour of $\mathcal{N}(t)$ shown in fig. 3 (this figure is taken from [27]).

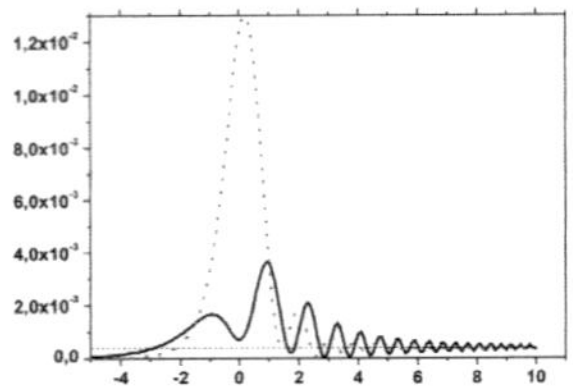

Figure 3. $\mathcal{N}(t)$ for the constant field case (solid line = fermion, dotted line = scalar).

Note the strongly oscillatory behaviour which is typical for solutions of the QVE.

In [28] two of the authors derived an alternative form of (3.1) which is equivalent to it (the precise relation between the two forms of the QVE will be discussed in a forthcoming publication [29]), but more amenable to an analytic treatment. This led to the discovery of an infinite set of purely time-dependent electric field configurations for which the QVE can be solved in terms of the well-known soliton solutions of the Korteweg-de Vries equation [28, 30]. The simplest one of these field configurations is given by

$$qA(t) = k_{\parallel} - \sqrt{k_{\parallel}^2 + \frac{2\omega_0^2}{\cosh^2(\omega_0 t)}} \tag{3.3}$$

($\omega_0 = \sqrt{\mathbf{k}^2 + m^2}$). The solitonic character of the associated solutions shows itself in the property that $\lim_{t\to\infty}\mathcal{N}(t) = 0$ for the $\mathbf{k}$ used in the definition of the field (3.3). This does not mean that such an electric field does not pair-create at all - as was already mentioned pair-creation seems inevitable for purely time-dependent fields - but it can be tuned not to pair-create at a given momentum. In fig. 4 we show a plot of $\mathcal{N}(t)$ for the field (3.3).

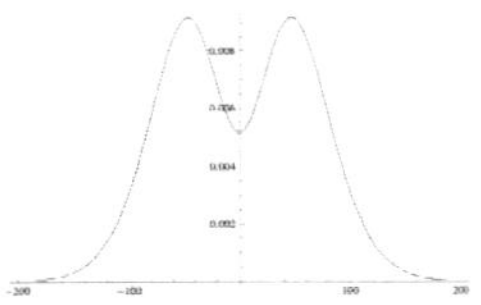

Figure 4. $\mathcal{N}(t)$ in Scalar QED for the simplest solitonic gauge field.

Note the symmetry with respect to $t = 0$ and the absence of oscillations. More extensive applications of the alternative Vlasov equation of [28] will be given elsewhere.

References

[1] F. Sauter, Z. Phys. **69** (1931) 742.
[2] J. Schwinger, Phys. Rev. **82** (1951) 664.
[3] V. I. Ritus, Zh. Eksp. Teor. Fiz **69** (1975) 1517 [Sov. Phys. JETP **42** (1975) 774].
[4] I.K. Affleck, O. Alvarez and N.S. Manton, Nucl. Phys. **B 197** (1982) 509.
[5] §.L. Lebedev and V.I. Ritus, Zh. Eksp. Teor. Fiz. **86** (1984) 408 [JETP **59** (1984) 237].
[6] W. Dittrich and M. Reuter, *Effective Lagrangians in Quantum Electrodynamics*, Springer 1985.
[7] M. Reuter, M. G. Schmidt and C. Schubert, Ann. Phys. (N.Y.) **259** (1997) 313, hep-th/9610191.
[8] G. V. Dunne and C. Schubert, JHEP 0206 (2002) 042, arXiv:hep-th/0205005.
[9] E. Brézin and C. Itzykson, Phys. Rev. D **2**, 1191 (1970).
[10] N. B. Narozhny, A. I. Nikishov, Yad. Fis. **11** (1970) 1072 [Sov. Journ. Phys. **11** (1970) 596].
[11] V. S. Popov, Zh. Eksp. Teor. Fiz. **62** (1972) 1248 [Sov. Phys. JETP **35** (1972) 569].
[12] G. V. Dunne and C. Schubert, Phys. Rev. D **72** (2005) 105004, arXiv:hep-th/0507174.

[13] G. V. Dunne, Q.-h. Wang, H. Gies and C. Schubert, Phys. Rev. D **73** (2006) 065028, hep-th/0602176.
[14] G. V. Dunne and Q.-h. Wang, Phys. Rev. D **74** (2006) 065015, hep-th/0608020.
[15] Y. Kluger, J. M. Eisenberg, B. Svetitsky, F. Cooper and E. Mottola, Phys. Rev. Lett. **67** (1991) 2427, Phys. Rev. D **45** (1992) 4659.
[16] Y. Kluger, J. M. Eisenberg and E. Mottola, Phys. Rev. D **58** (1998) 125015.
[17] S. Schmidt, D. Blaschke, P. Röpke, A. V. Prozorkevich, S. A. Smolyansky and V. D. Toneev, Phys. Rev. D **59** (1999) 094005, hep-ph/9810452.
[18] F. Hebenstreit, R. Alkofer and H. Gies, Phys. Rev. D **78**, 061701 (2008), arXiv:0807.2785.
[19] C. K. Dumlu, Phys. Rev. D **79** (2009) 065027, arXiv:0901.2972 [hep-th].
[20] F. Hebenstreit, "Schwinger effect in inhomogeneous electric fields", PhD thesis, Graz University, 2011, arXiv:1106.5965 [hep-ph].
[21] F. Hebenstreit, A.Ilderton, M. Marklund and J. Zamanian, Phys. Rev. D **83** (2011) 065007, arXiv:1011.1923 [hep-ph].
[22] R.P. Feynman, Phys. Rev. **80** (1950) 440.
[23] R.P. Feynman, Phys. Rev. **84** (1951) 108.
[24] M.G. Schmidt and C. Schubert, Phys. Lett. B **318** (1993) 438, hep-th/9309055.
[25] A. I. Nikishov, Nucl. Phys. B **21** (1970) 346.
[26] H. Gies and K. Klingmüller, Phys. Rev. D **72** (2005) 065001, hep-ph/0505099.
[27] A. M. Fedotov, E. G. Gelfer, K.Yu. Korolev and S. A. Smolyansky, Phys. Rev. D **83** (2011) 025011, arXiv:1008.2098 [hep-ph].
[28] S. P. Kim and C. Schubert, Phys .Rev. D **84** (2011) 125028, arXiv:1110.0900 [hep-th].
[29] A. Huet, S. P. Kim and C. Schubert, in preparation.
[30] S. P. Kim, "Schwinger Pair Production in Solitonic Gauge Fields", arXiv:1110.4684 [hep-th].

AUTHOR INDEX

AUTHOR INDEX